Graduate Texts in Mathematics

54

Jack E. Graver
Mark E. Watkins

Combinatorics with Emphasis on the Theory of Graphs

Springer-Verlag
New York Heidelberg Berlin

Jack E. Graver
Department of Mathematics
Syracuse University
Syracuse, NY 13210
USA

Mark E. Watkins
Department of Mathematics
Syracuse University
Syracuse, NY 13210
USA

AMS Subject Classification: 05-xx

Library of Congress Cataloging in Publication Data

Graver, Jack E 1935–
Combinatorics with emphasis on the theory of graphs.
(Graduate texts in mathematics ; 54)
Bibliography: p.
Includes index.
1. Combinatorial analysis. 2. Graph theory.
I. Watkins, Mark E., 1937– joint author.
II. Title. III. Series.
QA164.G7 511′.6 77–1200

Printed in the United States of America.

9 8 7 6 5 4 3 2 1

ISBN 0-387-90245-7 Springer-Verlag New York

ISBN 3-540-90245-7 Springer-Verlag Berlin Heidelberg

To our Fathers,

Harold John Graver

Misha Mark Watkins (in memory)

Preface

Combinatorics and graph theory have mushroomed in recent years. Many overlapping or equivalent results have been produced. Some of these are special cases of unformulated or unrecognized general theorems. The body of knowledge has now reached a stage where approaches toward unification are overdue. To paraphrase Professor Gian-Carlo Rota (Toronto, 1967), "Combinatorics needs fewer theorems and more theory."

In this book we are doing two things at the same time:

A. We are presenting a unified treatment of much of combinatorics and graph theory. We have constructed a concise algebraically-based, but otherwise self-contained theory, which at one time embraces the basic theorems that one normally wishes to prove while giving a common terminology and framework for the development of further more specialized results.

B. We are writing a textbook whereby a student of mathematics or a mathematician with another specialty can learn combinatorics and graph theory. We want this learning to be done in a much more unified way than has generally been possible from the existing literature.

Our most difficult problem in the course of writing this book has been to keep A and B in balance. On the one hand, this book would be useless as a textbook if certain intuitively appealing, classical combinatorial results were either overlooked or were treated only at a level of abstraction rendering them beyond all recognition. On the other hand, we maintain our position that such results can all find a home as part of a larger, more general structure.

To convey more explicitly what this text is accomplishing, let us compare combinatorics with another mathematical area which, like combinatorics, has been realized as a field in the present century, namely topology. The basic unification of topology occurred with the acceptance of what we now call a "topology" as the underlying object. This concept was general enough to encompass most of the objects which people wished to study, strong enough to include many of the basic theorems, and simple enough so that additional conditions could be added without undue complications or repetition.

We believe that in this sense the concept of a "system" is the right unifying choice for combinatorics and graph theory. A system consists of a finite set of objects called "vertices," another finite set of objects called "blocks," and an "incidence" function assigning to each block a subset of the set of vertices. Thus graphs are systems with blocksize two; designs are systems with constant blocksize satisfying certain conditions; matroids are also systems; and a system is the natural setting for matchings and inclusion-exclusion. Some important notions are studied in this most general setting, such as connectivity and orthogonality as well as the parameters and vector spaces of a system. Connectivity is important in both graph theory and matroid theory, and parallel theorems are now avoided. The vector spaces of a system have important applications in all of these topics, and again much duplication is avoided.

One other unifying technique employed is a single notation consistent throughout the book. In attempting to construct such a notation, one must face many different levels in the hierarchy of sets (elements, sets of elements, collections of sets, families of collections, etc.) as well as other objects (systems, functions, sets of functions, lists, etc.). We decided insofar as possible to use different types of letters for different types of objects. Since each topic covered usually involves only a few types of objects, there is a strong temptation to adopt a simpler notation for that section regardless of how it fits in with the rest of the book. We have resisted this temptation. Consequently, once the notational system is mastered, the reader will be able to flip from chapter to chapter, understanding at glance the diverse roles played in the middle and later chapters by the concepts introduced in the earlier chapters.

An undergraduate course in linear algebra is prerequisite to the comprehension of most of this book. Basic group theory is needed for sections *II*E and *XI*C. A deeper appreciation of sections *III*E, *III*G, *VII*C, and *VII*D will be gained by the reader who has had a year of topology. All of these sections may be omitted, however, without destroying the continuity of the rest of the text.

The level of exposition is set for the beginning graduate student in the mathematical sciences. It is also appropriate for the specialist in another mathematical field who wishes to learn combinatorics from scratch but from a sophisticated point of view.

It has been our experience while teaching from the notes that have evolved into this text, that it would take approximately three semesters of three hours classroom contact per week to cover all of the material that we have presented. A perusal of the Table of Contents and of the "Flow Chart of the

Sections" following this Preface will suggest the numerous ways in which a subset of the sections can be covered in a subset of three semesters. A List of Symbols and an Index of Terms are provided to assist the reader who may have skipped over the section in which a symbol or term was defined.

As indicated in the figure below, a one-semester course can be formed from Chapters I, II, IX, and XI. However, the instructor must provide some elementary graph theory in a few instances. The dashed lines in the figure below as well as in the Flow Chart of the Sections indicate a rather weak dependency.

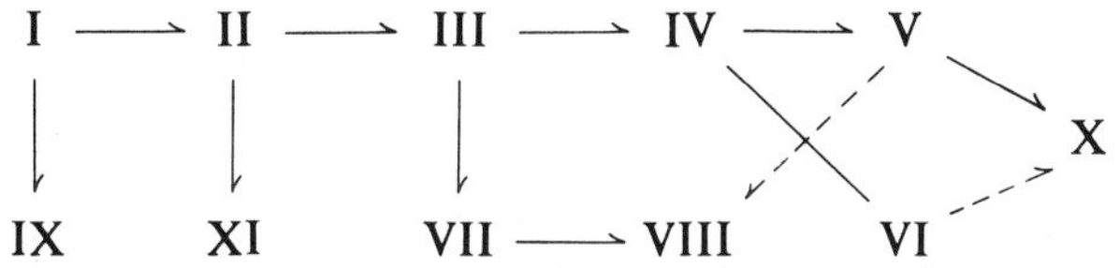

If a two-semester sequence is desired, we urge that Chapters I, II, and III be treated in sequence in the first semester, since they comprise the theoretical core of the book. The reader should not be discouraged by the apparent dryness of Chapter II. There is a dividend which is compounded and paid back chapter by chapter. We recommend also that Chapters IV, V, and VI be studied in sequence; they are variations on a theme, a kind of minimax or maximin principle, which is an important combinatorial notion. Since Chapter X brings together notions from the first six chapters with allusions to Chapters VII and IX, it would be a suitable finale.

There has been no attempt on our part to be encyclopedic. We have even slighted topics dear to our respective hearts, such as integer programming and automorphism groups of graphs. We apologize to our colleagues whose favorite topics have been similarly slighted.

There has been a concerted effort to keep the technical vocabulary lean. Formal definitions are not allotted to terms which are used for only a little while and then never again. Such terms are often written between quotation marks. Quotation marks are also used in intuitive discussions for terms which have yet to be defined precisely.

The terms which do form part of our technical vocabulary appear in **bold-face** type when they are formally defined, and they are listed in the Index.

There are two kinds of exercises. When the term "**Exercise**" appears in bold-face type, then those assertions in italics following it will be invoked in subsequent arguments in the text. They almost always consist of straightforward proofs with which we prefer not to get bogged down and thereby lose too much momentum. The word "*Exercise*" (in italics) generally indicates a specific application of a principle, or it may represent a digression which the limitations of time and space have forced us not to pursue. In principle, all of the exercises are important for a deeper understanding of and insight into the theory.

Chapters are numbered with Roman numerals; the sections within each chapter are denoted by capital letters; and items (theorems, exercises, figures, etc.) are numbered consecutively regardless of type within each section. If an item has more than one part, then the parts are denoted by lower case Latin letters. For references within a chapter, the chapter number will be suppressed, while in references to items in other chapters, the chapter number will be italicized. For example, within Chapter III, Euler's Formula is referred to as F2b, but when it is invoked in Chapter VII, it is denoted by *III*F2b.

Relatively few of the results in this text are entirely new, although many represent new formulations or syntheses of published results. We have also given many new proofs of old results and some new exercises without any special indication to this effect. We have done our best to give credit where it is due, except in the case of what are generally considered to be results "from the folklore".

A special acknowledgement is due our typist, Mrs. Louise Capra, and to three of our former graduate students who have given generously of their time and personal care for the well-being of this book: John Kevin Doyle, Clare Heidema, and Charles J. Leska. Thanks are also due to the students we have had in class, who have learned from and taught us from our notes. Finally, we express our gratitude to our families, who may be glad to see us again.

Syracuse, N.Y.
April, 1977

Jack E. Graver
Mark E. Watkins

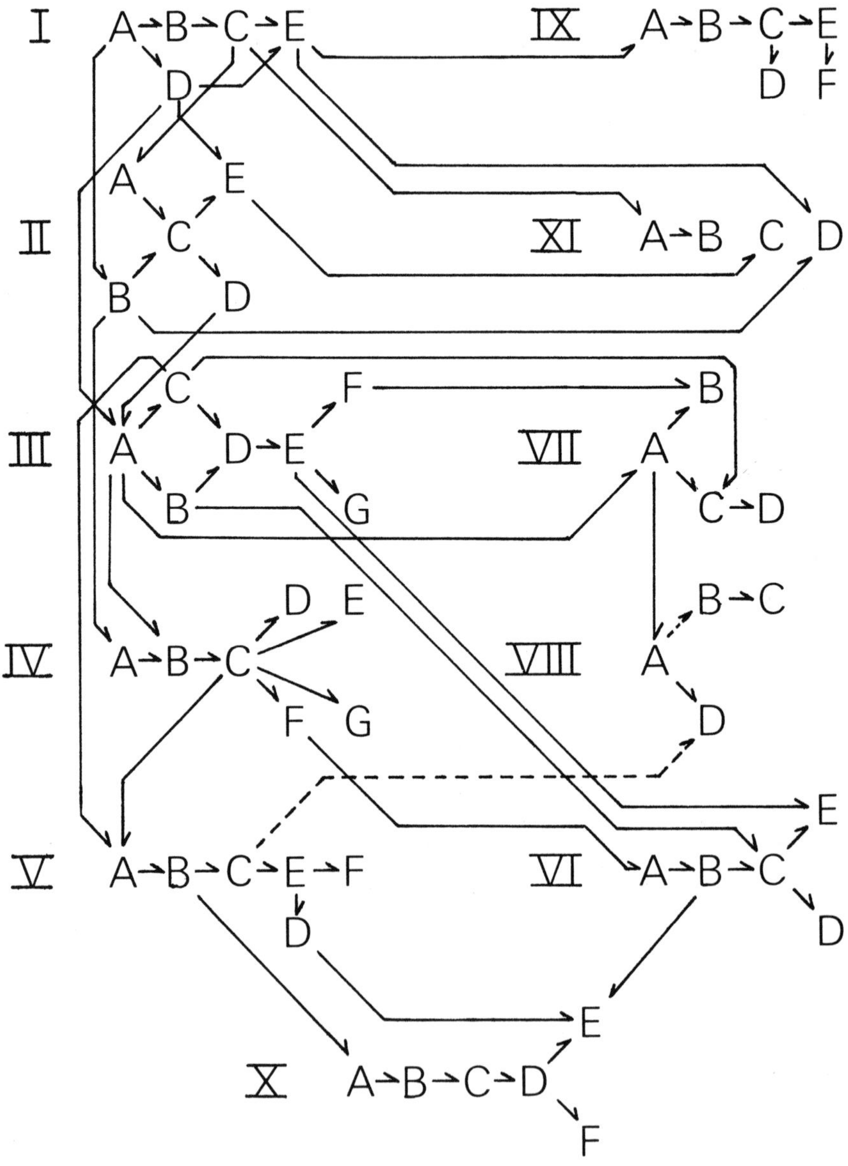

Flow Chart of the Sections

Contents

CHAPTER I

Finite Sets

IA Conventions and Basic Notation

The symbols $\mathbb{N}$, $\mathbb{Z}$, $\mathbb{Q}$, $\mathbb{R}$, $\mathbb{K}$ will always denote, respectively, the natural numbers (including 0), the integers, the rational numbers, the real numbers, and the field of order 2. In each of these systems, 0 and 1 denote, respectively, the additive and multiplicative identities.

If U is a set, $\mathscr{P}(U)$ will denote the collection of all subsets of U. It is called the **power set** of U. In general, the more common, conventional terminology and notation of set theory will be used throughout except occasionally as noted. One such instance is the following usage: while "$U \subseteq W$" will continue to mean that U is a subset of W, we shall write "$U \subset W$" when $U \subseteq W$ and $U \neq W$. (Thus U can be empty if W is not empty.) The cardinality of the set U will be denoted by $|U|$, and $\mathscr{P}_m(U)$ will denote the collection of all subsets of U with cardinality m. A set of cardinality m is called an **m-set**.

The binary operation of **sum** (Boolean sum) of sets S and T in $\mathscr{P}(U)$ is denoted by $S + T$, where

$$S + T = \{x: x \in S \cup T; x \notin S \cap T\}.$$

In particular, $S + U$ is the complement of S in U, and no other notation for complementation will be required. Since the sum is the most frequently used set-operation in this text, we include a list of properties which can be easily verified.

For $R, S, T \in \mathscr{P}(U)$,

A1 $\quad S + T = T + S$

A2 $\quad (R + S) + T = R + (S + T)$

A3 $S + T = S \Leftrightarrow T = \varnothing$

A4 $S + T = \varnothing \Leftrightarrow S = T$

A5 $S + T = (S \cup T) + (S \cap T)$

A6 $R \cup (S + T) \supseteq (R \cup S) + (R \cup T)$

A7 $R \cap (S + T) = (R \cap S) + (R \cap T)$

A8 $R + (S \cap T) \supseteq (R + S) \cap (R + T)$

A9 $(R + S) \cap (R + T) \subseteq R + (S \cup T) \subseteq (R + S) \cup (R + T)$

A10 *Exercise.* Show that the inclusions in A6, A8, and A9 cannot, in general, be reversed.

Because of A1 and A2, the sum $\sum_{S \in \mathscr{S}} S$ where $\mathscr{S} \subseteq \mathscr{P}(U)$ is well-defined if $\mathscr{S} \neq \varnothing$. If $\mathscr{S} = \varnothing$, we understand this sum to be $\varnothing$.

As usual, the **cartesian product** of sets $X_1, \ldots, X_m$ will be denoted by $X_1 \times \ldots \times X_m$. Thus

$$X_1 \times \ldots \times X_m = \{(x_1, \ldots, x_m): x_i \in X_i \text{ for } i = 1, \ldots, m\}.$$

A **function** f from X into Y is a subset of $X \times Y$ such that $|f \cap (\{x\} \times Y)| = 1$ for all $x \in X$. Following established convention, $f: X \to Y$ will mean that f is a function from X into Y. For each $x \in X$, $f(x)$ is the second component of the unique element of $f \cap (\{x\} \times Y)$. We shall adhere to the terms **injection** if $|f \cap (X \times \{y\})| \leq 1$ for all $y \in Y$; **surjection** if $|f \cap (X \times \{y\})| \geq 1$ for all $y \in Y$; and **bijection** if $|f \cap (X \times \{y\})| = 1$ for all $y \in Y$.

We say sets X and Y are **isomorphic** if there exists a bijection $b: X \to Y$, and we note that X and Y are isomorphic if and only if $|X| = |Y|$.

A (**binary**) **relation** on U is a subset of $U \times U$. Let R_i be a relation on U_i for $i = 1, 2$. We say that (U_1, R_1) is **isomorphic** to (U_2, R_2) if there exists a bijection $b: U_1 \to U_2$ such that $(x, y) \in R_1$ if and only if $(b(x), b(y)) \in R_2$. A binary relation R on U is **reflexive** if $(u, u) \in R$ for all $u \in U$; R is **symmetric** if $(u, v) \in R$ implies $(v, u) \in R$ for all $u, v \in U$; R is **transitive** if $(u, v) \in R$ and $(v, w) \in R$ together imply $(u, w) \in R$ for all $u, v, w \in U$. R is an **equivalence relation** if it is reflexive, symmetric, and transitive.

Problems involving categories being outside the scope of this book, we find it best to ignore them, and we shall freely use such terms as "equivalent" and "equivalence relation" in regard to objects from various categories and not only to elements of some given set. Such disregard for categorical problems will be particularly flagrant as we treat in turn various notions of "isomorphism." For example, the "relation" of "is isomorphic to" is clearly an "equivalence relation" on the category of sets.

We denote the set of all functions from X into Y by Y^X. Since $\varnothing \times Y = \varnothing$, $Y^{\varnothing}$ consists of a single function $\varnothing$ which is an injection; in case $Y = \varnothing$, it is a bijection, of course. If $S \subseteq X$, then the **restriction of** f **to** S, denoted by $f_{|S}$, belongs to Y^S and satisfies $f_{|S}(x) = f(x)$ for all $x \in S$.

A bijection $b: U \to U$ is called a **permutation** of U. The set of all permutations of U is denoted by $\Pi(U)$. The **identity** on U is the function $1_U \in \Pi(U)$ given by $1_U(x) = x$ for all $x \in U$.

The function $f: X \to Y$ induces two corresponding functions between $\mathscr{P}(X)$ and $\mathscr{P}(Y)$. One of these is also denoted by f, and $f: \mathscr{P}(X) \to \mathscr{P}(Y)$ is given by

$$f[S] = \{f(x): x \in S\}, \quad \text{for all } S \in \mathscr{P}(X).$$

(Note that the choice of parentheses or brackets to surround the argument determines which of the two functions denoted by the symbol f is intended.) The set $f[S]$ is the **image of** S **under** f. In particular, $f[X]$ is the **image of** f. The other function induced by f is the function $f^{-1}: \mathscr{P}(Y) \to \mathscr{P}(X)$ given by

$$f^{-1}[T] = \{x: f(x) \in T\}, \quad \text{for all } T \in \mathscr{P}(Y).$$

If f is a bijection, its inverse, also denoted by f^{-1}, is a function $f^{-1}: Y \to X$. By our convention, if $y \in Y$, $f^{-1}[y]$ $(= f^{-1}[\{y\}])$ denotes a subset of X, but if f is a bijection, $f^{-1}(y)$ denotes an element of X. f **maps** S **into** T if $f[S] \subseteq T$ and **onto** T if $f[S] = T$. We say f is a **constant function** if $|f[X]| \leq 1$.

Let $f: X \to Y$; $S, T \in \mathscr{P}(X)$; $U, W \in \mathscr{P}(Y)$. The following basic properties of functions and sets are readily verified:

A11 $$f[S \cup T] = f[S] \cup f[T]$$

A12 $$f[S \cap T] \subseteq f[S] \cap f[T]$$

A13 $$f^{-1}[U \cup W] = f^{-1}[U] \cup f^{-1}[W]$$

A14 $$f^{-1}[U \cap W] = f^{-1}[U] \cap f^{-1}[W]$$

A15 $$f[S + T] \supseteq f[S] + f[T]$$

A16 $$f^{-1}[U + W] = f^{-1}[U] + f^{-1}[W]$$

A17 *Exercise*. Show that the inclusions in A12 and A15 cannot, in general, be reversed.

Let X, Y, and Z be sets. Let $f \in Y^X$ and $g \in Z^Y$. The composite of f by g will be denoted by gf. Clearly $gf \in Z^X$. We conclude the present section with a rapid review of some elementary properties of functions and some terminology.

A18 If both f and g are injections (respectively, surjections, bijections), then so is gf.

A19 $$(gf)^{-1} = f^{-1}g^{-1} \in \mathscr{P}(X)^{\mathscr{P}(Z)}.$$

A20 g is an injection if and only if there exists $h \in Y^Z$ such that $hg = 1_Y$.

A21 Let g be an injection. If $gf_1 = gf_2$ for $f_1, f_2 \in Y^X$, then $f_1 = f_2$. The converse holds if $|X| \geq 2$.

A22 f is a surjection if and only if there exists $j \in X^Y$ such that $fj = 1_Y$.

A23 Let f be a surjection. If $g_1 f = g_2 f$ for $g_1, g_2 \in Z^Y$, then $g_1 = g_2$. The converse holds if $|Z| \geq 2$.

A24 f is a bijection if and only if there exists $b \in X^Y$ such that $bf = 1_X$ and $fb = 1_Y$. In this case $b = f^{-1}$, and so b is unique.

A25 If X is finite and $h \in X^X$, then h is a surjection if and only if h is an injection.

If $S \subseteq X$ and $h \in X^X$, we say h **fixes** S if $h[S] \subseteq S$. If $h_{|S} = 1_S$, we say h **fixes** S **pointwise.**

If $*$ is a binary operation on Y, then $*$ induces a binary operation on Y^X which is also denoted by $*$. Thus

$$(f_1 * f_2)(x) = f_1(x) * f_2(x), \quad \text{for all } f_1, f_2 \in Y^X, x \in X.$$

Note that if $*$ on Y enjoys any of the properties of associativity, commutativity, or existence of an identity, then that property is also enjoyed by $*$ on Y^X.

One final important convention: henceforth, **all arbitrarily chosen sets will be finite unless explicitly stated otherwise.**

A26 *Exercise.* Let $f: X \to Y$. Show that if f is an injection (respectively, surjection, bijection), then so is the induced function $f: \mathscr{P}(X) \to \mathscr{P}(Y)$, and conversely.

A27 *Exercise.* Let $f: X \to Y$. Show that if f is an injection (respectively, surjection, bijection), then $f^{-1}: \mathscr{P}(Y) \to \mathscr{P}(X)$ is a surjection (respectively, injection, bijection), and conversely.

IB Selections and Partitions

Let U be a set and let $S \in \mathscr{P}(U)$. The **characteristic function** of S is the function

$$c_S: U \to \mathbb{K}$$

given by

B1
$$c_S(x) = \begin{cases} 1 & \text{if } x \in S; \\ 0 & \text{if } x \in U + S. \end{cases}$$

B2 Proposition. *The function* $\sigma: \mathbb{K}^U \to \mathscr{P}(U)$ *given by*

$$\sigma(c) = \{x \in U: c(x) \neq 0\} \quad \textit{for all } c \in \mathbb{K}^U$$

is a bijection. Moreover, $\sigma^{-1}(S) = c_S$ *for all* $S \in \mathscr{P}(U)$.

PROOF. Clearly σ is an injection. If $S \in \mathscr{P}(U)$, then $\sigma(c_S) = S$. Hence σ is a surjection. □

B3 Exercise. Let $S, T \in \mathscr{P}(U)$. Prove that

$$c_S + c_T = c_{S+T} \quad \textit{and} \quad c_S c_T = c_{S \cap T},$$

and express $c_{S \cup T}$ in terms of c_S and c_T.

For a set U, a function $s \in \mathbb{N}^U$ is called a **selection** from U. If $x \in U$, the number $s(x)$ is the "number of times x is selected by s". The number

$$|s| = \sum_{x \in U} s(x)$$

is the **cardinality (weight) of the selection** s. If $|s| = m$, we say that s is an m-**selection.** The set of all m-selections from U is denoted by $\mathbb{S}_m(U)$, and we let

$$\mathbb{S}(U) = \bigcup_{m=0}^{\infty} \mathbb{S}_m(U) = \mathbb{N}^U.$$

If $S \in \mathscr{P}(U)$, we define the **characteristic selection** of S by

B4 $$s_S(x) = \begin{cases} 1 & \text{if } x \in S; \\ 0 & \text{if } x \in U + S. \end{cases}$$

The difference between B1 and B4 is subtle but important. In B4, the symbols 0 and 1 denote elements of $\mathbb{N}$ rather than $\mathbb{K}$. Of course, c_S and s_S are closely related, but since $1 + 1$ gives a different "answer" in $\mathbb{N}$ than in $\mathbb{K}$, the characteristic function and characteristic selection are not the same thing. In particular, the correspondence $S \to s_S$ gives a natural injection of $\mathscr{P}(U)$ into $\mathbb{S}(U)$ under which $S + T$ is not necessarily mapped onto $s_S + s_T$, even though $S \cap T$ is always mapped onto $s_S s_T$ for all $S, T \in \mathscr{P}(U)$. (Cf. B3.)

A subcollection $\mathscr{Q} \subseteq \mathscr{P}(U)$ of nonempty subsets of U is called a **partition** of U if

$$\sum_{Q \in \mathscr{Q}} Q = U$$

and

$$Q \cap R = \varnothing, \quad \text{for all } Q, R \in \mathscr{Q};\ Q \neq R.$$

The elements of $\mathscr{Q}$ are called the **cells** of $\mathscr{Q}$. If $|\mathscr{Q}| = m$, we call $\mathscr{Q}$ an m-**partition** of U. The collection of all m-partitions of U is denoted by $\mathbb{P}_m(U)$; $\mathbb{P}(U)$ denotes the collection of all partitions of U. A fundamental identity satisfied by any partition $\mathscr{Q} \in \mathbb{P}(U)$ is

B5 $$|U| = \sum_{Q \in \mathscr{Q}} |Q|.$$

There is a natural multiplication on $\mathbb{P}(U)$. Let $\mathcal{Q}, \mathcal{R} \in P(U)$ and let $\mathcal{Q}\mathcal{R}$ be the collection of nonempty subsets of the form $Q \cap R$ where $Q \in \mathcal{Q}$ and $R \in \mathcal{R}$.

B6 Exercise. Prove that *if* $\mathcal{Q} \in \mathbb{P}_m(U)$ *and* $\mathcal{R} \in \mathbb{P}_n(U)$, *then* $\mathcal{Q}\mathcal{R} \in \mathbb{P}_p(U)$ *for some* $p \leq mn$. Show, moreover, that this multiplication is commutative and associative and admits an identity in $\mathbb{P}(U)$.

The next result delineates the fundamental relationship between partitions and equivalence relations.

B7 Proposition. *A necessary and sufficient condition that a relation* R *on a set* U *be an equivalence relation is that there exist a partition* $\mathcal{Q} \in \mathbb{P}(U)$ *such that* $(x, y) \in R$ *if and only if* x *and* y *are elements of the same cell of* $\mathcal{Q}$.

PROOF. Let R be an equivalence relation on U. For each $x \in U$ let $S_x = \{w \in U: (x, w) \in R\}$. Since R is reflexive, $x \in S_x$ and so $S_x \neq \varnothing$ for each $x \in U$. Let $x, y \in U$ and suppose $w \in S_x \cap S_y$. Thus (x, w) and $(y, w) \in R$. Since R is symmetric, $(w, y) \in R$, and since R is transitive, $(x, y) \in R$. Now let $z \in S_y$; hence $(y, z) \in R$. Again by transitivity, $(x, z) \in R$ and $z \in S_x$. This proves that $S_y \subseteq S_x$. By a symmetrical argument we see that $S_x \subseteq S_y$. Thus exactly one of the following holds for any $x, y \in U$: $S_x = S_y$ or $S_x \cap S_y = \varnothing$. If $\mathcal{Q} = \{S: S = S_x \text{ for some } x \in U\}$, then $\mathcal{Q} \in \mathbb{P}(U)$.

Conversely, let $\mathcal{Q} \in \mathbb{P}(U)$. Define the relation R on U by: $(x, y) \in R$ if $x, y \in Q$ for some $Q \in \mathcal{Q}$. One readily verifies that R is an equivalence relation. □

B8 Proposition. *Let* $f: B \to U$. *Then* $\{f^{-1}[x]: x \in f[B]\}$ *is a* $|f[B]|$*-partition of* B.

PROOF. For each $b \in B$, $b \in f^{-1}[x]$ if and only if $x = f(b)$. Hence $\sum_{x \in f[B]} f^{-1}[x] = B$ and $f^{-1}[x] \cap f^{-1}[y] = \varnothing$ for $x \neq y$. Finally, $f^{-1}[x] \neq \varnothing$ if and only if $x \in f[B]$. □

B9 Proposition. *Let* $f: B \to U$. *Let* $s: U \to \mathbb{N}$ *be defined by* $s(x) = |f^{-1}[x]|$. *Then* s *is a* $|B|$*-selection from* U.

PROOF. Clearly $s \in \mathbb{S}(U)$. We have that

$$|s| = \sum_{x \in U} |f^{-1}[x]| = \sum_{x \in f[B]} |f^{-1}[x]| = |B|.$$

The first equality here is the definition of $|s|$; the second follows from the fact that $|\varnothing| = 0$ and $f^{-1}[x] = \varnothing$ for $x \notin f[B]$; the third equality follows from B5 and B8. □

If $f: B \to U$, then the **partition of** f is $\{f^{-1}[x]: x \in f[B]\}$, and the **selection of** f is the function $s: U \to \mathbb{N}$ given by $s(x) = |f^{-1}[x]|$.

B10 *Exercise.* Prove that the functions $f: B \to U$ and $g: C \to U$ have the same selection if and only if there is a bijection $b: B \to C$ such that $f = gb$.

B11 Exercise. Prove that *the functions $f: B \to U$ and $h: B \to W$ have the same partition if and only if there is a bijection $b: f[B] \to h[B]$ such that $bf = h$.*

B12 *Exercise.* Let $f: X \to Y$. Define $f_1: \mathbb{S}(Y) \to \mathbb{S}(X)$ by $f_1(s) = sf$ for all $s \in \mathbb{S}(Y)$. Show that f is an injection (respectively, surjection, bijection) if and only if f_1 is a surjection (respectively, injection, bijection).

B13 *Exercise.* Let $f: X \to Y$. Define $f_2: \mathbb{P}(Y) \to \mathbb{P}(X)$ as follows: if $\mathscr{Q} \in \mathbb{P}(Y)$, then $f_2(\mathscr{Q})$ consists of the nonempty members of the collection $\{f^{-1}[Q]: Q \in \mathscr{Q}\}$. First verify that $f_2(\mathscr{Q}) \in \mathbb{P}(X)$; then show that f is an injection (respectively, surjection, bijection) if and only if f_2 is a surjection (respectively, injection, bijection).

The remainder of this section is concerned with the notion of "isomorphism" between objects of the kinds we have been considering.

Functions $f: B \to U$ and $g: C \to W$ are **isomorphic** if there exist bijections $p: B \to C$ and $q: U \to W$ such that $f = q^{-1}gp$. The pair (p, q) is called a **function-isomorphism.** The selections $s \in \mathbb{S}(U)$ and $t \in \mathbb{S}(W)$ are **isomorphic** if there exists a bijection $q: U \to W$ such that $s = tq$. Such a bijection is called a **selection-isomorphism.** (These two definitions are illustrated by the commutative diagrams B14. In this and other such diagrams bijections are indicated by the symbol $\cong$.) Partitions $\mathscr{Q} \in \mathbb{P}(B)$ and $\mathscr{R} \in \mathbb{P}(C)$ are **isomorphic** if there exists a bijection $p: B \to C$ such that $p[Q] \in \mathscr{R}$ for all $Q \in \mathscr{Q}$. The bijection p is a **partition-isomorphism.**

B14

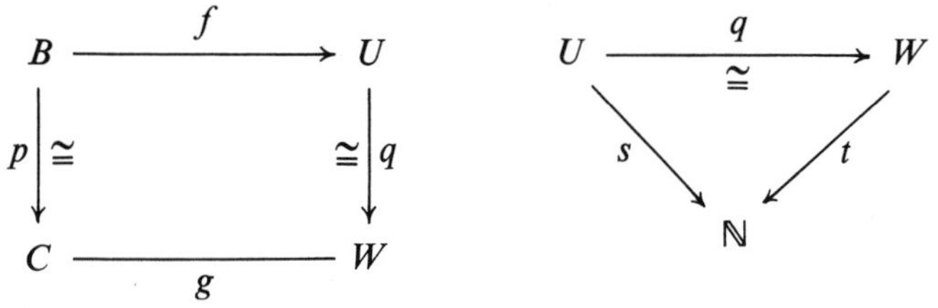

B15 Exercise. Prove that *in each of the above definitions, "isomorphism" is an equivalence relation.*

B16 Proposition. *Let $f: B \to U$ and $g: C \to W$. Let $p: B \to C$ and $q: U \to W$ be bijections.*

(a) *If (p, q) is a function-isomorphism from f to g, then p is a partition-isomorphism from the partition of f to the partition of g and q is a selection-isomorphism from the selection of f to the selection of g.*

(**b**) *If q is a selection-isomorphism from the selection of f to the selection of g, then there exists a bijection $p': B \to C$ such that (p', q) is a function-isomorphism from f to g.*

(**c**) *If p is a partition-isomorphism from the partition of f to the partition of g and if $|U| = |W|$, then there exists a bijection $q': U \to W$ such that (p, q') is a function-isomorphism from f to g.*

PROOF. (a) Let S be a cell of the partition of f, i.e., $S = f^{-1}[x]$ for some $x \in U$. By A19, $p[S] = p[f^{-1}[x]] = g^{-1}[q(x)]$, which is a cell of the partition of g. Let s be the selection of f and t the selection of g. Let $x \in U$. By definition and A19,

$$t(q(x)) = |g^{-1}[q(x)]| = |p^{-1}[g^{-1}[q(x)]]| = |f^{-1}[x]| = s(x).$$

Thus $tq = s$.

(b) With s and t as in the proof of (a), we assume $tq = s$. For any $x \in U$,

$$|f^{-1}[x]| = s(x) = tq(x) = |g^{-1}[q(x)]|.$$

Hence there exists a bijection $p_x: f^{-1}[x] \to g^{-1}[q(x)]$. These bijections for all $x \in U$ determine a bijection $p': B \to C$ by $p'(w) = p_x(w)$ where $w \in f^{-1}[x]$. Clearly $f = q^{-1}gp'$.

(c) Since p is a partition-isomorphism from the partition of f to the partition of g, we have

$$\{g^{-1}[x]: x \in W\} = \{p[f^{-1}[x]]: x \in U\}.$$

We may define $q'': f[B] \to g[C]$ by choosing $q''(x)$ to be the unique $y \in W$ such that $g^{-1}[y] = p[f^{-1}[x]]$. Clearly q'' as defined is a bijection, and since $|U| = |W|$, it may be extended to a bijection $q': U \to W$. One may easily verify that $q'f = gp$. □

A more succinct but somewhat weaker formulation of the above proposition is the following.

B17 Corollary. *Let $f: B \to U$ and $g: C \to W$. Then the following statements are equivalent:*

(**a**) *f and g are isomorphic;*

(**b**) *the selections of f and g are isomorphic;*

(**c**) *$|U| = |W|$ and the partitions of f and g are isomorphic.*

We return briefly to cartesian products presented in the first section and list some readily verifiable properties. Let W, X, and Y be sets. Then

B18 $X \times Y$ and $Y \times X$ are set-isomorphic.

B19 $W \times (X \times Y)$ and $(W \times X) \times Y$ are set-isomorphic to $W \times X \times Y$.

B20 $\mathcal{Q} \in \mathbb{P}(Y)$ if and only if $\{X \times Q: Q \in \mathcal{Q}\} \in \mathbb{P}(X \times Y)$.

B21 If $\{X_1, \ldots, X_m\} \in \mathbb{P}(X)$, then the function $f \mapsto (f_{|X_1}, \ldots, f_{|X_m})$ is a set-isomorphism between Y^X and $Y^{X_1} \times \ldots \times Y^{X_m}$.

Given the cartesian product $X_1 \times \ldots \times X_m$, the ith-**coordinate projection** is the function from $X_1 \times \ldots \times X_m$ into X_i given by $(x_1, \ldots, x_m) \mapsto x_i$.

B22 *Exercise.* Describe the selections and partitions of the coordinate projections of the cartesian product $X \times Y$.

IC Fundamentals of Enumeration

We begin this section with a list of some of the more basic properties of finite cardinals. Some of these were mentioned in the preceding sections.

C1 If $S \in \mathscr{P}(U)$, then $|S| \leq |U|$.

C2 If $\mathscr{Q} \in \mathbb{P}(U)$, then $|U| = \sum_{Q \in \mathscr{Q}} |Q|$.

For sets X and Y,

C3 $$|X \cup Y| + |X \cap Y| = |X| + |Y|$$

C4 $$|X \cup Y| - |X \cap Y| = |X + Y|$$

C5 $$|X + Y| + 2|X \cap Y| = |X| + |Y|$$

C6 $$|X \times Y| = |X||Y|.$$

C7 Proposition. *For any sets X and Y,*

$$|Y^X| = |Y|^{|X|}.$$

Proof. Let X be an m-set. We first dispense with the case where $m = 0$. If also $Y = \varnothing$, then the Proposition holds if we adopt the convention that $0^0 = 1$. If $Y \neq \varnothing$, then $|Y|^{|\varnothing|} = 1$, as required.

Now suppose $m > 0$, and consider the m-partition $\{\{x_1\}, \ldots, \{x_m\}\}$ of X. By B21 and C6,

$$|Y^X| = |Y^{\{x_1\}} \times \ldots \times Y^{\{x_m\}}| = |Y^{\{x_1\}}| \cdot \ldots \cdot |Y^{\{x_m\}}|.$$

Clearly $|Y^{\{x_i\}}| = |Y|$ for all i, and so $|Y^X| = |Y|^m = |Y|^{|X|}$. □

C8 Corollary. $|\mathscr{P}(U)| = 2^{|U|}$ *for any set U.*

Proof. Use C7 and B2. □

Because of C8, one often finds in the literature the symbol 2^U in use in place of the symbol $\mathscr{P}(U)$.

C9 *Exercise.* Let $S \in \mathscr{P}(U)$. How many functions in U^U fix S? How many fix S pointwise?

C10 *Exercise.* Let $S \in \mathscr{P}(X)$ and $T \in \mathscr{P}(Y)$. How many functions in Y^X map S into T?

C11 *Exercise.* Let $\{S_1, \ldots, S_m\} \in \mathbb{P}(X)$ and $\{T_1, \ldots, T_m\} \in \mathbb{P}(Y)$. How many functions in Y^X map S_i into T_i for all $i = 1, \ldots, m$?

C12 *Exercise.* Let $S, T \in \mathscr{P}(U)$. How many subsets of U contain S? How many avoid S (R avoids S if $R \cap S = \varnothing$)? How many meet S (R meets S if $R \cap S \neq \varnothing$)? How many meet both S and T?

Three important cardinality questions about the set Y^X are how many elements are injections, how many are surjections, and how many are bijections. For convenience we denote

$$\operatorname{inj}(Y^X) = \{f \in Y^X : f \text{ is an injection}\}$$

$$\operatorname{sur}(Y^X) = \{f \in Y^X : f \text{ is a surjection}\}$$

$$\operatorname{bij}(Y^X) = \{f \in Y^X : f \text{ is a bijection}\}.$$

We now proceed to resolve the first and third of these questions. The second question is deceptively more complicated and will not be resolved until §E. By convention, $0! = 1$ and $n! = n(n-1)!$ for $n \in \mathbb{N} + \{0\}$.

C13 Proposition. *For sets X and Y,*

$$|\operatorname{inj}(Y^X)| = \begin{cases} 0 & \text{if } |X| > |Y|; \\ \dfrac{|Y|!}{(|Y| - |X|)!} & \text{if } |X| \leq |Y|. \end{cases}$$

PROOF. Obviously $\operatorname{inj}(Y^X) = \varnothing$ if $|X| > |Y|$. Suppose $|X| \leq |Y|$. If $X = \varnothing$, then both $|\operatorname{inj}(Y^X)|$ and $|Y|!/(|Y| - |X|)!$ equal 1. If $|X| = 1$, then $\operatorname{inj}(Y^X) = Y^X$, and by C7, $|\operatorname{inj}(Y^X)| = |Y|^{|X|} = |Y| = |Y|!/(|Y| - 1)!$.

We continue by induction on $|X|$, assuming the proposition to hold whenever $|X| \leq m$ for some integer $m \geq 1$. Suppose $|X| = m + 1$. Fix $x \in X$ and let $X' = X + \{x\}$. Let $Y = \{y_1, \ldots, y_n\}$ and let

$$Y_j = Y + \{y_j\}, \quad j = 1, \ldots, n.$$

Since $m = |X'| = |X| - 1 \leq |Y| - 1 = |Y_j|$, the induction hypothesis implies that

$$\textbf{C14} \quad |\operatorname{inj}(Y_j^{X'})| = \frac{|Y_j|!}{(|Y_j| - |X'|)!} = \frac{(n-1)!}{(n-1-m)!} \quad (j = 1, \ldots, n).$$

If we define

$$I_j = \{f \in \operatorname{inj}(Y^X) : f(x) = y_j\}, \quad (j = 1, \ldots, n),$$

it is clear that $\{I_1, \ldots, I_n\} \in \mathbb{P}(\operatorname{inj}(Y^X))$. Moreover, the correspondence $f \mapsto f_{|X'}$ is clearly a bijection from I_j onto $\operatorname{inj}(Y_j^{X'})$ for each $j = 1, \ldots, n$.

Combining this fact with C2 and C14, we obtain

$$\begin{aligned}|\text{inj}(Y^X)| &= \sum_{j=1}^{n} |I_j| \\ &= \sum_{j=1}^{n} |\text{inj}(Y_j^{X'})| \\ &= n\cdot\frac{(n-1)!}{(n-1-m)!} = \frac{n!}{(n-(m+1))!} \\ &= \frac{|Y|!}{(|Y|-|X|)!}.\end{aligned}$$ □

From the above formula one immediately obtains

C15 Corollary. *For sets X and Y,*

$$|\text{bij}(Y^X)| = \begin{cases} 0 & \text{if } |X| \neq |Y|; \\ |Y|! & \text{if } |X| = |Y|. \end{cases}$$

Since $\text{bij}(X^X) = \Pi(X)$ we have

C16 Corollary. *If X is an n-set, $|\Pi(X)| = n!$*

C17 *Exercise.* Let X be an n-set and let $S \in \mathscr{P}_k(X)$. How many permutations of X fix S pointwise? How many fix S (set-wise)? How many map some given point $x \in X$ onto some point of S?

For $m, n \in \mathbb{N}$ it is conventional to write

$$\binom{n}{m} = \begin{cases} \dfrac{n!}{m!\,(n-m)!} & \text{if } m \leq n; \\ 0 & \text{if } m > n. \end{cases}$$

Observe that $\binom{n}{m} = \binom{n}{n-m}$ if $m \leq n$.

C18 Corollary. *For any set X, $|\mathscr{P}_m(X)| = \binom{|X|}{m}$.*

Proof. Let M be some fixed m-set. For each $S \in \mathscr{P}_m(X)$, let $B_S = \text{bij}(S^M)$. Then clearly $\{B_S : S \in \mathscr{P}_m(X)\} \in \mathbb{P}(\text{inj}(X^M))$. By C13, C2, and then C15,

$$\frac{|X|!}{(|X|-m)!} = |\text{inj}(X^M)| = \sum_{S \in \mathscr{P}_m(X)} |B_S| = |\mathscr{P}_m(X)|m!.$$ □

Numbers of the form $\binom{n}{m}$ are called **binomial coefficients** because they arise also from the binomial theorem of elementary algebra, as will presently be demonstrated. A vast amount of literature has been devoted to proving "binomial identities." The following corollary and some of the ensuing exercises in this section provide examples of some of the easier and more useful such identities.

C19 Corollary.

$$\sum_{i=0}^{n} \binom{n}{i} = 2^n.$$

PROOF. Let X be an n-set. Then $\{\mathscr{P}_i(X) : i = 0, 1, \ldots, n\} \in \mathbb{P}(\mathscr{P}(X))$. The result follows from C2 and C8. □

C20 Corollary.

$$\binom{n-1}{m-1} + \binom{n-1}{m} = \binom{n}{m}.$$

PROOF. Let U be an n-set and choose $x \in U$. The collection of m-subsets of U which do not contain x is precisely $\mathscr{P}_m(U + \{x\})$, while the collection of those that do is set-isomorphic to $\mathscr{P}_{m-1}(U + \{x\})$. Hence $|\mathscr{P}_{m-1}(U + \{x\})| + |\mathscr{P}_m(U + \{x\})| = |\mathscr{P}_m(U)|$. □

Of course one could also have obtained this corollary from the definition by simple computation. It is, however, of interest to see a combinatorial argument as well.

C21 Binomial Theorem. *Let a and b be elements of a commutative ring with identity. Then*

$$(a + b)^n = \sum_{i=0}^{n} \binom{n}{i} a^i b^{n-i}.$$

PROOF. To each function $f: \{1, 2, \ldots, n\} \to \{a, b\}$ there corresponds a unique term of the product $(a + b)^n$, namely $a^{|f^{-1}[a]|} b^{|f^{-1}[b]|}$. Thus

$$(a + b)^n = \sum_f a^{|f^{-1}[a]|} b^{|f^{-1}[b]|}, \quad \text{where } f \in \{a, b\}^{\{1,2,\ldots,n\}}.$$

Hence

$$\begin{aligned}(a + b)^n &= \sum_{i=0}^{n} |\{f : |f^{-1}[a]| = i\}| a^i b^{n-i} \\ &= \sum_{i=0}^{n} |\mathscr{P}_i(\{1, 2, \ldots, n\})| a^i b^{n-i} \\ &= \sum_{i=0}^{n} \binom{n}{i} a^i b^{n-i}.\end{aligned}$$

□

By choosing the ring to be $\mathbb{Z}$ and letting $a = -1$, and $b = 1$ above, we obtain the following identity:

C22 Corollary.

$$\sum_{i=0}^{n} (-1)^i \binom{n}{i} = 0^n,$$

for all $n \in \mathbb{N}$; *equivalently,*

$$\sum_{R \in \mathscr{P}(U)} (-1)^{|R|} = 0^{|U|},$$

for any set U.

C23 *Exercise.* How many subsets in $\mathscr{P}(U)$ have even (respectively, odd) cardinality?

As we have indicated, we will evaluate $|\text{sur}(Y^X)|$ in §E after having developed more powerful techniques. Enumeration of the m-partitions of a set must also be deferred. In fact, $|\mathbb{P}_m(U)|$ and $|\text{sur}(Y^X)|$ are closely related as we see in the next result.

C24 Proposition. *If M is an m-set, then*

$$|\mathbb{P}_m(U)| = \frac{|\text{sur}(M^U)|}{m!}.$$

Proof. Let φ: $\text{sur}(M^U) \to \mathbb{P}_m(U)$ by defining $\varphi(f)$ to be the partition of f. By Proposition B8, $\varphi(f)$ is a $|f[U]|$-partition. Since f is a surjection, $\varphi(f)$ is an m-partition. Since φ is clearly a surjection, we also have from B8 that $\{\varphi^{-1}[\mathscr{Q}]: \mathscr{Q} \in \mathbb{P}_m(U)\}$ is a partition of $\text{sur}(M^U)$. Thus

$$|\text{sur}(M^U)| = \sum_{\mathscr{Q} \in \mathbb{P}_m(U)} |\varphi^{-1}[\mathscr{Q}]|.$$

It remains only to show that $|\varphi^{-1}[\mathscr{Q}]| = m!$ for all $\mathscr{Q} \in \mathbb{P}_m(U)$.

Fix $\mathscr{Q} \in \mathbb{P}_m(U)$ and $g \in \varphi^{-1}[\mathscr{Q}]$. If $h \in \Pi(M)$, then clearly $\varphi(hg) = \varphi(g)$, i.e., $hg \in \varphi^{-1}[\mathscr{Q}]$. Hence we have a function γ: $\Pi(M) \to \varphi^{-1}[\mathscr{Q}]$ defined by $\gamma(h) = hg$. Since g is a surjection, we have by A23 that if $h_1 g = h_2 g$ then $h_1 = h_2$. Hence γ is an injection. Finally, it follows from B11 that for any $f \in \varphi^{-1}[\mathscr{Q}]$, there exists $h \in \Pi(M)$ such that $f = hg$. We conclude that γ is a bijection, and $|\varphi^{-1}[\mathscr{Q}]| = |\Pi(M)| = m!$. □

In order that the reader may become aware of the difficulties in counting surjections, he is asked in the next exercise to work out the two easiest non-trivial cases.

C25 *Exercise.* Compute $|\text{sur}(Y^X)|$ where $|Y| = |X| - i$ for $i = 1, 2$.

Of the fundamental objects that we have introduced, only the selections remain to be considered.

C26 Proposition. $|\mathbb{S}_m(U)| = \binom{|U|+m-1}{m}$, *except that* $|\mathbb{S}_0(\varnothing)| = 1$.

PROOF. Let $U = \{u_1, \ldots, u_n\}$ and let $X = \{1, 2, \ldots, n + m - 1\}$. We construct a function $\varphi: \mathscr{P}_{n-1}(X) \to \mathbb{S}_m(U)$ as follows. Let $Y = \{y_1, \ldots, y_{n-1}\} \in \mathscr{P}_{n-1}(X)$ where the elements of Y are indexed so that $y_1 < y_2 < \ldots < y_{n-1}$. Letting $y_0 = 0$ and $y_n = n + m$, we define $\varphi(Y)$ to be the selection $s \in \mathbb{S}(U)$ given by

$$s(u_i) = y_i - y_{i-1} - 1, \quad \text{for } i = 1, \ldots, n.$$

Note that

$$|s| = \sum_{i=1}^{n} s(u_i) = \sum_{i=1}^{n} (y_i - y_{i-1} - 1) = m,$$

i.e., $\varphi(Y) \in \mathbb{S}_m(U)$.

It suffices to show that φ is a bijection, since

$$|\mathscr{P}_{n-1}(X)| = \binom{|X|}{n-1} = \binom{n+m-1}{n-1} = \binom{|U|+m-1}{m}.$$

φ is an injection. Suppose that $\varphi(Y) = s = \varphi(W)$. We have $Y = \{y_1, \ldots, y_{n-1}\}$, $W = \{w_1, \ldots, w_{n-1}\} \in \mathscr{P}_{n-1}(X)$, and

$$y_i - y_{i-1} - 1 = s(u_i) = w_i - w_{i-1} - 1, \quad \text{for } i = 1, \ldots, n.$$

By induction on i one readily verifies that the system of equations $y_i - y_{i-1} - 1 = w_i - w_{i-1} - 1$ for $i = 1, \ldots, n$, and $y_0 = w_0$, $y_n = w_n$ has exactly one solution: $y_i = w_i$ for $i = 0, 1, \ldots, n$. Hence $Y = W$.

φ is a surjection. Let $s \in \mathbb{S}_m(U)$ and define $y_i = i + \sum_{j=1}^{i} s(u_j)$. One may easily verify that $y_0 = 0$, $y_n = n + m$, and $0 < y_1 < y_2 < \ldots < y_{n-1} < n + m$. Thus $\{y_1, \ldots, y_{n-1}\} \in \mathscr{P}_{n-1}(X)$, and $\varphi(\{y_1, \ldots, y_{n-1}\}) = s$. ☐

C27 *Exercise.* Compute $\sum_{m=0}^{r} |\mathbb{S}_m(U)|$ where r is any positive integer. (*Hint:* use Corollary C20.)

C28 *Exercise.* How many elements of $\mathbb{S}_m(U)$ select all elements of U at least once? How many select all elements an even (respectively, odd) number of times?

The last counting problem that we wish to discuss at this time is the following: how many functions in Y^X are distinct up to isomorphism? In other words, given that function-isomorphism is an equivalence relation on Y^X (B15), how many equivalence classes are there? Generally speaking, the equivalence classes will be of varying sizes. For instance, the set of bijections, if any, will form a single equivalence class of size $|X|!$. On the other hand, the constant functions form an equivalence class of size $|Y|$. Because these equivalence classes are not of uniform cardinality, we are unable to use that old "cowboy" technique applied in C24; in effect to "count their legs and divide by 4". However, it is clear that isomorphic functions will have isomorphic partitions and vice versa (B17). While the same can be said of the selections of isomorphic functions, it is more fruitful to consider partitions.

We have then that $\varphi\colon Y^X \to \mathbb{P}(X)$, where $\varphi(f)$ is the partition of f, is an injection which maps isomorphism classes onto isomorphism classes. What is the image of φ? Clearly a partition $\mathcal{Q} = \varphi(f)$ for some f if and only if $|\mathcal{Q}| \leq |Y|$. Hence the image of φ is $\mathbb{P}_1(X) \cup \mathbb{P}_2(X) \cup \ldots \cup \mathbb{P}_q(X)$ where $q = \min\{|X|, |Y|\}$. It is also clear that isomorphic partitions are of equal cardinality. Hence the problem reduces to counting the isomorphism classes of $\mathbb{P}_m(X)$ for each m. In fact, each isomorphism class can be uniquely represented by a selection s from $\mathbb{N} + \{0\}$ where $s(i)$ is the number of cells of cardinality i. This leads us to define a **partition of the positive integer** n to be a selection $s \in \mathbb{S}(\mathbb{N} + \{0\})$ such that $\sum_{i=1}^{\infty} is(i) = n$. If $|s| = m$, then s is called an m**-partition of** n.

As an example, let X be a 19-set, and suppose $\mathcal{Q} \in \mathbb{P}_7(X)$ has two single element cells, a 2-cell, three 3-cells, and a 6-cell. The selection corresponding to $\mathcal{Q}$ is a 7-selection with $s(1) = 2$, $s(2) = 1$, $s(3) = 3$, $s(4) = s(5) = 0$, $s(6) = 1$, and $s(i) = 0$ for $i > 6$.

We combine the results in this discussion in the following proposition.

C29 Proposition. *Let $p_m(n)$ denote the number of m-partitions of the positive integer n while $p(n)$ denotes the total number of partitions of n. Let X be an n-set. Then the number of isomorphism classes in $\mathbb{P}_m(X)$ is $p_m(n)$. The number of isomorphism classes in $\mathbb{P}(X)$ is $p(n)$. If $|Y| \leq n$, the number of isomorphism classes in Y^X is $\sum_{m=1}^{|Y|} p_m(n)$; if $|Y| \geq n$, the number of isomorphism classes in Y^X is $p(n)$.*

C30 *Exercise.* Show that the number of isomorphism classes in $\mathbb{S}_m(X)$ is $p_n(m)$, where X is an n-set.

We close this section with a small but representative assortment of problems analogous to the "word problems" of high school algebra or elementary calculus, insofar as their difficulty lies in translating the language of the stated problem into the abstract terminology of the theory. Observe that in some of these problems, the question "how many" does not always make precise a unique answer which is sought. When such ambiguity arises, the reader should investigate all alternative interpretations of the question.

C31 *Problem.* Prove the identity $\binom{n}{m}\binom{m}{k} = \binom{n}{k}\binom{n-k}{m-k}$ where $k \leq m \leq n$ by enumeration of appropriate sets rather than by direct computation (cf. the comment following C20).

C32 *Problem.* From a list of his party's n most generous contributors, the newly-elected President was expected to appoint three ambassadors (to different countries), a Commissioner of Indian Affairs, and a Fundraising Committee of five people. In how many ways could he have made his appointments?

C33 *Problem.* A candy company manufactures sour balls in tangy orange, refreshing lemon, cool lime, artificial cherry, and imitation grape flavors. They are randomly packaged in cellophane bags each containing a dozen sour balls. What is the probability of a bag containing at least one sour ball of each of the U.S. certified flavors?

C34 *Problem.* Let $m, k \in \mathbb{Z}$. How many solutions $(x_1, \ldots, x_n)$ are there to the equation

$$x_1 + \ldots + x_n = m$$

where x_i is an integer and $x_i \geq k$ $(i = 1, \ldots, n)$?

C35 *Problem.* A word is a sequence of letters. How many four-letter-words from the Latin alphabet have four distinct letters, at least one of which is a vowel? (An exhaustive list is beyond the scope of this book.)

C36 *Problem.* How many ways can the numbers $\{1, 2, \ldots, n\}$ be arranged on a "roulette" wheel? How many ways can alternate numbers lie in black (as opposed to red) sectors?

C37 *Problem.* Compute $p_3(n)$.

C38 *Problem.* What fraction of all 5-card poker hands have 4-of-a-kind? a full-house? 3-of-a-kind? 2-of-a-kind? a straight flush? a flush? a straight? none of these?

Two good sources for more problems of this type are C. L. Liu [ℓ.2, pp. 19–23] and Kemeny, Snell, and Thompson [k.2, pp. 97–99, 102–104, 106–108, 111–113, 136–139].

ID Systems

A **system** Λ is a triple (V, f, E) where V and E are disjoint sets and $f: E \to \mathscr{P}(V)$. The elements of E are called the **blocks** of Λ and the elements of V are called the **vertices** of Λ. If $x \in f(e)$, we say that the block e "contains" the vertex x, or that x and e are **incident** with each other. If $S \in \mathscr{P}(V)$, we say that the block e "contains" S ("is contained in" S) if $S \subseteq f(e)$ $(f(e) \subseteq S)$. Similarly we say that the block e "is contained in" the block e' if $f(e) \subseteq f(e')$. The **size** of a block e is the natural number $|f(e)|$. If all the blocks of Λ have size k, we say Λ has **blocksize** k.

The systems $\Lambda = (V, f, E)$ and $\Omega = (W, g, F)$ are **isomorphic** if there exist bijections $p: E \to F$, $q: V \to W$ such that $q[f(e)] = g(p(e))$ for all $e \in E$ (see Figure D1). The pair (p, q) is then called a **system-isomorphism.**

D1

$$\begin{array}{ccccc} V & & \mathscr{P}(V) & \xleftarrow{\quad f \quad} & E \\ q\Big\downarrow\cong & & q\Big\downarrow\cong & & p\Big\downarrow\cong \\ W & & \mathscr{P}(W) & \xleftarrow[\quad g \quad]{} & F \end{array}$$

D2 *Exercise.* Show that system-isomorphism is an equivalence relation.

Whenever (V, f, E) and (W, g, F) are isomorphic systems, then f and g are isomorphic functions. The converse of this statement is false, since a bijection from $\mathscr{P}(V)$ onto $\mathscr{P}(W)$ need not be induced by a bijection from V onto W.

If $\Lambda = (V, f, E)$ is a system and if f is an injection, then Λ is called a **set system.** For example, if $\mathscr{E} \subseteq \mathscr{P}(V)$ and if the "inclusion function" $j: \mathscr{E} \to \mathscr{P}(V)$ is defined by $j(S) = S$ for each $S \in \mathscr{E}$, then $(V, j, \mathscr{E})$ is a set system. In this case, the function j is suppressed and the set system is denoted simply by the pair $(V, \mathscr{E})$.

Let (V, f, E) be any set system. Let $\mathscr{E} = f[E]$ and let $j: \mathscr{E} \to \mathscr{P}(V)$ be the inclusion function. Since f is an injection, $f': E \to \mathscr{E}$ given by $f'(e) = f(e)$ for all $e \in E$ is a bijection. Then the pair $(f', 1_V)$ is a system-isomorphism between (V, f, E) and $(V, \mathscr{E}) = (V, f[E])$.

If V and E are sets and $f: E \to \mathscr{P}(V)$, the function $f^*: V \to \mathscr{P}(E)$ given by $f^*(x) = \{e \in E: x \in f(e)\}$ is called the **transpose** of f. Since

D3
$$x \in f(e) \Leftrightarrow e \in f^*(x), \quad \text{for all } x \in V, e \in E,$$

we have $f^{**} = f$. If $\Lambda = (V, f, E)$, then the system $\Lambda^* = (E, f^*, V)$ is called the **transpose** of Λ. Since $f^{**} = f$, $\Lambda^{**} = \Lambda$.

D4 Proposition. *If (V, f, E) is isomorphic to (W, g, F), then (E, f^*, V) is isomorphic to (F, g^*, W).*

Proof. Assume that (p, q) is a system-isomorphism from (V, f, E) to (W, g, F). We assert that (q, p) is a system-isomorphism from (E, f^*, V) to (F, g^*, W). Let $x \in V$. Then

$$\begin{aligned} p[f^*(x)] &= p[\{e \in E: x \in f(e)\}], \\ &= p[\{e \in E: q(x) \in q[f(e)]\}], \\ &= p[\{e \in E: q(x) \in g(p(e))\}], \\ &= \{p(e): q(x) \in g(p(e))\}, \\ &= \{d \in F: q(x) \in g(d)\}, \\ &= g^*(q(x)), \quad \text{as required.} \end{aligned}$$

□

For $\Lambda = (V, f, E)$ and $x, y \in V$, one has $f^*(x) = f^*(y)$ if and only if x and y are incident with precisely the same blocks. This motivates the following definition: a system Λ **distinguishes vertices** if for every two distinct vertices there is a block which contains exactly one of them. In this terminology,

D5 *Λ^* is a set system if and only if Λ distinguishes vertices.*

It is interesting to note that this property is analogous to the topological property T_0 (given a pair of distinct points in a T_0-topological space there is an open set containing one but not the other). This analogy may be extended. We could say that a system is "T_1" if given any two distinct vertices x and y, there is a block containing x but not y and vice versa.

D6 *Exercise.* Show that a system Λ is "T_1" if and only if Λ^* has the property that no block contains any other block.

A system of blocksize 2 is called a **multigraph.** If it is also a set system, it is called a **graph.** The blocks of a multigraph are called **edges.** Some mathematicians, taking the reverse approach from the one adopted here, have begun with a study of multigraphs and subsequently treated systems as generalizations of multigraphs. In particular, Berge [b.5] has defined the term **hypergraph** to denote a system (V, f, E) with the two additional properties that $f(e) \neq \varnothing$ for all $e \in E$ and that $f^*(x) \neq \varnothing$ for all $x \in V$.

A graph $(V, \mathscr{E})$ is said to be **bipartite** if $|V| \leq 1$ or if there exists a partition $\{V_1, V_2\} \in \mathbb{P}_2(V)$ such that $|E \cap V_1| = |E \cap V_2| = 1$ for all $E \in \mathscr{E}$. If $(V, \mathscr{E})$ is a bipartite graph, the partition $\{V_1, V_2\}$ need not be unique. When we wish to specify the partition we shall write: $(V, \mathscr{E})$ is a bipartite graph with respect to $\{V_1, V_2\}$, or $(\{V_1, V_2\}, \mathscr{E})$ is a bipartite graph.

There is a natural correlation between systems and bipartite graphs. If (U, f, D) is a system, we may define $V = U \cup D$ and let $\mathscr{E} = \{\{x, d\}: x \in f(d)\}$. Since $U \cap D = \varnothing$, $(V, \mathscr{E})$ is a bipartite graph called the **bipartite graph of** (U, f, D). From D3 it follows that (U, f, D) and (D, f^*, U) have the same bipartite graph. Conversely, if $(V, \mathscr{E})$ is a bipartite graph with respect to $\{V_1, V_2\}$, then $(V, \mathscr{E})$ is the bipartite graph of (at least) two systems, namely: (V_1, f, V_2) where $f(v_2) = \{v: \{v, v_2\} \in \mathscr{E}\}$ and (V_2, g, V_1) where $g(v_1) = \{v: \{v_1, v\} \in \mathscr{E}\}$. In fact, $g = f^*$.

Another method for representing a system (V, f, E) is obtained by indexing both V and E; thus $V = \{x_1, \ldots, x_v\}$, $E = \{e_1, e_2, \ldots, e_b\}$. We then construct the $v \times b$ matrix M where 1 is the (i, j)-entry if $x_i \in f(e_j)$; otherwise the (i, j)-entry is 0. M is called an **incidence matrix** of the system (V, f, E). It is not difficult to see that any $v \times b$ $\{0, 1\}$-matrix is an incidence matrix of some system. Furthermore, systems (V, f, E) and (W, g, F) are isomorphic if and only if for some indexing of $V, E, W,$ and F the corresponding incidence matrices are identical. Two $\{0, 1\}$-matrices M_1 and M_2 are incidence matrices for isomorphic systems if and only if M_2 may be obtained from

M_1 by row- or column-permutations. Clearly if M is an incidence matrix for Λ, then the transpose of M (denoted by M^*) is an incidence matrix for Λ^*.

A system $\Omega = (W, g, F)$ is a **subsystem** of the system $\Lambda = (V, f, E)$ if: $W \subseteq V$, $F \subseteq E$, and for all $e \in F$ it holds that $g(e) = f(e) \subseteq W$. For example, let $\Lambda = (V, f, E)$ and suppose $F \subseteq E$. Then $\Lambda_F = (W, g, F)$, where $W = \bigcup_{d \in F} f(d)$ and $g = f_{|F}$, is a subsystem of Λ. Λ_F is called the **subsystem induced by** F. We let $\Lambda_{(F)} = (V, f_{|E+F}, E + F)$. If $W \subseteq V$, the **subsystem induced by** W is the subsystem $\Lambda_W = (W, g, F)$ where $F = \{e \in E : f(e) \subseteq W\}$ and $g = f_{|F}$. We let $\Lambda_{(W)} = \Lambda_{V+W}$.

D7 Exercise. *Let M be the incidence matrix for the system $\Lambda = (V, f, E)$ corresponding to the indexing $V = \{x_1, \ldots, x_v\}$ and $E = \{e_1, \ldots, e_b\}$. If* $M^*M = [m_{ij}]$, show that $m_{ij} = |f(e_i) \cap f(e_j)|$ *for all* $i, j \in \{1, \ldots, b\}$. Interpret the entries of MM^*.

IE Parameters of Systems

If $\Lambda = (V, f, E)$ is a system, recall that the selection of the function f is

$$s : \mathscr{P}(V) \to \mathbb{N}$$

given by

$$s(S) = |f^{-1}[S]| \quad \text{for all } S \in \mathscr{P}(V).$$

For convenience, this selection will also be called the **selection of the system** Λ. When the symbol s is used to denote the selection of Λ, the selection of Λ^* will be denoted by s^*. If Λ is the set system $(V, \mathscr{E})$, then $s = s_{\mathscr{E}}$, the characteristic selection of $\mathscr{E}$.

We shall presently see that if two systems have the same selection, then they are isomorphic; however, two systems having isomorphic selections can still be nonisomorphic. This is consistent with the fact that two systems (V, f, E) and (W, g, F) need not be system-isomorphic even though f and g may be function-isomorphic. (See the discussion following D2.) The next proposition makes these remarks precise.

E1 Proposition. *Let (V, f, E) and (W, g, F) be systems with selections s and t, respectively. The following three statements are equivalent:*

(a) *(V, f, E) and (W, g, F) are system-isomorphic.*

(b) *There exists a bijection $q : V \to W$ such that $s(S) = t(q[S])$ for all $S \subseteq V$.* (See Figure E2a.)

(c) *There exists a bijection $p : E \to F$ such that $s^*(S) = t^*(p[A])$ for all $A \subseteq E$.* (See Figure E2b.)

E2

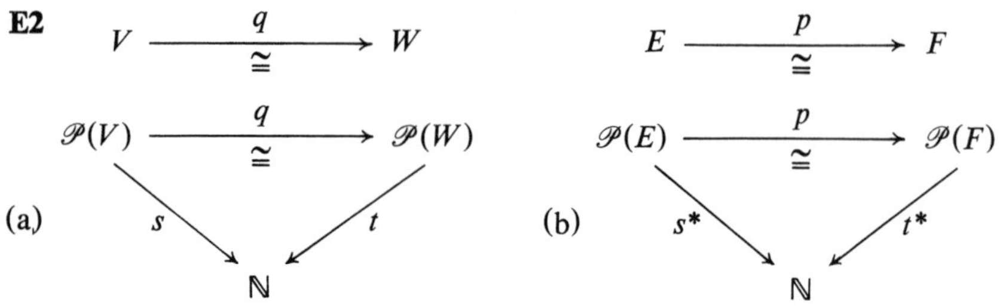

Proof. We need only demonstrate the equivalence of (a) and (b); the equivalence of (a) and (c) will then follow from this result and Proposition D4.

Assume that (a) holds. There exist bijections $p: E \to F$ and $q: V \to W$ such that

$$q[f(e)] = g(p(e)), \quad \text{for all } e \in E.$$

Thus (p, q) is a function-isomorphism from f to g. By B16a, q is a selection-isomorphism from s to t.

Conversely, assume that (b) holds. By B16b there exists a bijection $p': E \to F$ such that (p', q) is a function-isomorphism from f to g. The result follows from the definition of system-isomorphism. □

For any given selection $s \in \mathbb{S}(\mathscr{P}(V))$, a system (V, f, E) having s as its selection can always be constructed. For each $S \in \mathscr{P}(V)$, let E_S be an $s(S)$-set, and let all the sets E_S be disjoint from V and from each other. Let

$$E = \bigcup_{S \in \mathscr{P}(V)} E_S.$$

Now define $f: E \to \mathscr{P}(V)$ by $f(e) = S$ if $e \in E_S$. The selection of this system is obviously s. Moreover, this system is unique up to isomorphism. We may therefore identify systems having vertex set V with elements of $\mathbb{S}(\mathscr{P}(V))$.

From another point of view, the selection s of a system $\Lambda = (V, f, E)$ may be regarded as a list of parameters. For each $S \in \mathscr{P}(V)$, each value $s(S)$ is a parameter in the list, namely, the number of blocks which "coincide" with S. This list of parameters is a "complete list," inasmuch as Λ is uniquely determined (up to isomorphism) by the selection s. In the same way, the selection s^* determines the transpose Λ^* (up to isomorphism), and therefore by D4, the values of s^* on $\mathscr{P}(E)$ form another complete list of parameters determining Λ.

We now consider a third complete set of parameters which determines Λ (up to isomorphism, continuing to be understood). Unlike s and s^*, each of which tells the number of blocks "coinciding" with a given set, the function we are about to define will tell the number of blocks containing each given set.

For subsets $S, T \in \mathscr{P}(V)$, let $[S, T] \in \mathbb{N}$ be given by

$$[S, T] = \begin{cases} 1 & \text{if } S \subseteq T; \\ 0 & \text{otherwise.} \end{cases}$$

E3 Exercise. Show that *for any* $S, T, W \in \mathscr{P}(V)$,

$$[S, T][T, W] = [S + T, S + W][S, W].$$

For any selection $s \in \mathbb{S}(\mathscr{P}(V))$, we define $\bar{s} \in \mathbb{S}(\mathscr{P}(V))$ in terms of s by:

E4
$$\bar{s}(S) = \sum_{T \in \mathscr{P}(V)} [S, T]s(T), \quad \text{for all } S \in \mathscr{P}(V).$$

E5 Lemma. *For* $S, W \in \mathscr{P}(V)$,

$$\sum_{T \in \mathscr{P}(V)} (-1)^{|S+T|}[S, T][T, W] = 0^{|S+W|}.$$

PROOF.

$$\begin{aligned}
\sum_{T \in \mathscr{P}(V)} (-1)^{|S+T|}[S, T][T, W] &= \sum_{T \in \mathscr{P}(V)} (-1)^{|S+T|}[S + T, S + W][S, W] \\
&= \sum_{R \in \mathscr{P}(V)} (-1)^{|R|}[R, S + W][S, W] \\
&= \left[\sum_{R \in \mathscr{P}(S+W)} (-1)^{|R|}\right][S, W] \\
&= 0^{|S+W|}[S, W] \qquad \text{by C22} \\
&= 0^{|S+W|}.
\end{aligned}$$

□

The next result is the inverse of E4; it allows us to recover s when $\bar{s}$ is given.

E6 Proposition. *Let* $s \in \mathbb{S}(\mathscr{P}(V))$. *Then*

$$s(S) = \sum_{T \in \mathscr{P}(V)} (-1)^{|S+T|}[S, T]\bar{s}(T), \quad \textit{for all } S \in \mathscr{P}(V).$$

PROOF. Let $S \in \mathscr{P}(V)$. Then by definition of $\bar{s}$,

$$\begin{aligned}
\sum_{T \in \mathscr{P}(V)} (-1)^{|S+T|}[S, T]\bar{s}(T) &= \sum_{T \in \mathscr{P}(V)} (-1)^{|S+T|}[S, T] \sum_{W \in \mathscr{P}(V)} [T, W]s(W) \\
&= \sum_{W \in \mathscr{P}(V)} \left[\sum_{T \in \mathscr{P}(V)} (-1)^{|S+T|}[S, T][T, W]\right] s(W) \\
&= \sum_{W \in \mathscr{P}(V)} 0^{|S+W|}s(W) = s(S).
\end{aligned}$$

□

Since a system Λ is determined by the values of its selection s and since, by the above proposition, the values of s are in turn determined by the values of $\bar{s}$, it follows that the values of $\bar{s}$ form another "complete list of parameters" for Λ, as promised. Similarly, the values of $\overline{s^*}$ form a complete list of parameters for Λ^*, and hence also for Λ.

E7 *Exercise*. Show that the function Φ from $\mathbb{S}(\mathscr{P}(V))$ to itself given by $\Phi(s) = \bar{s}$ is an injection and satisfies the "linearity" condition:

$$\Phi(ms + nt) = m\Phi(s) + n\Phi(t)$$

for all $s, t \in \mathbb{S}(\mathscr{P}(V))$ and $m, n \in \mathbb{Z}$. Show further that Φ is never a surjection when $|V| \geq 2$.

When s is the selection of a system Λ, the values of the four selections s, s^*, $\bar{s}$, and $\overline{s^*}$ have important set-theoretical interpretations in terms of Λ, as summarized by the next result.

E8 Proposition. *Let s be the selection of the system (V, f, E). Let $S \in \mathscr{P}(V)$, and $A \in \mathscr{P}(E)$. Then:*

(a) $s(S) = |\{e \in E : f(e) = S\}|$;

(b) $\bar{s}(S) = |\{e \in E : f(e) \supseteq S\}|$;

(c) $s^*(A) = |(\bigcap_{e \in A} f(e)) \cap (\bigcap_{e \in E+A} (V + f(e)))|$;

(d) $\overline{s^*}(A) = |\bigcap_{e \in A} f(e)|$.

PROOF. (a) is, of course, the definition of s.

(b) represents the underlying motivation for defining $\bar{s}$ as we have. From the definitions of $\bar{s}$ and s,

$$\bar{s}(S) = \sum_{V \supseteq T \supseteq S} s(T) = \sum_{V \supseteq T \supseteq S} |\{e \in E : f(e) = T\}|,$$

whence the result.

(c) By definition,

$$\begin{aligned} s^*(A) &= |\{x \in V : f^*(x) = A\}| \\ &= |\{x \in V : \{e \in E : x \in f(e)\} = A\}| \\ &= |\{x \in V : x \in f(e) \Leftrightarrow e \in A\}| \\ &= |\{x \in V : e \in A \Rightarrow x \in f(e); e \in E + A \Rightarrow x \in V + f(e)\}| \\ &= |\{x \in V : x \in f(e) \text{ for all } e \in A\} \\ &\qquad \cap \{x \in V : x \in V + f(e) \text{ for all } e \in E + A\}| \\ &= \left|\left(\bigcap_{e \in A} f(e)\right) \cap \left(\bigcap_{e \in E+A} (V + f(e))\right)\right|. \end{aligned}$$

(d) Again by definition,

$$\overline{s^*}(A) = \sum_{A \subseteq C \subseteq E} s^*(C) = \sum_{A \subseteq C \subseteq E} |\{x \in V : \{e \in E : x \in f(e)\} = C\}|.$$

Let $C_1, C_2 \in \mathscr{P}(E)$ and let $V_i = \{x \in V : \{e \in E : x \in f(e)\} = C_i\}$ for $i = 1, 2$. Then $V_1 \cap V_2 = \varnothing$ if $C_1 \neq C_2$. Hence by C2,

$$\begin{aligned} &\sum_{A \subseteq C \subseteq E} |\{x \in V : \{e \in E : x \in f(e)\} = C\}| \\ &\qquad = \left|\bigcup_{A \subseteq C \subseteq E} \{x \in V : \{e \in E : x \in f(e)\} = C\}\right| \\ &\qquad = |\{x \in V : \{e \in E : x \in f(e)\} \supseteq A| \\ &\qquad = |\{x \in V : x \in f(e) \text{ for all } e \in A\}|, \end{aligned}$$

whence the result. □

We are now prepared, at least mathematically, to state and prove a major theorem in combinatorial theory. Since the statement of this result in the generality in which it will be given is probably less than transparent, we insert here an example and two exercises which should better familiarize the reader with the four selections considered in this section.

E9 *Example.* Consider the system $\Lambda = (V, f, E)$ where $E = \{e_1, e_2, e_3\}$. Let $S_i = f(e_i)$ $(i = 1, 2, 3)$, and let V, S_1, S_2, S_3 be represented by the Venn diagram E10, where $n_0, \ldots, n_{123}$ represent the cardinalities of the subsets corresponding to the regions in which they have been written.

E10

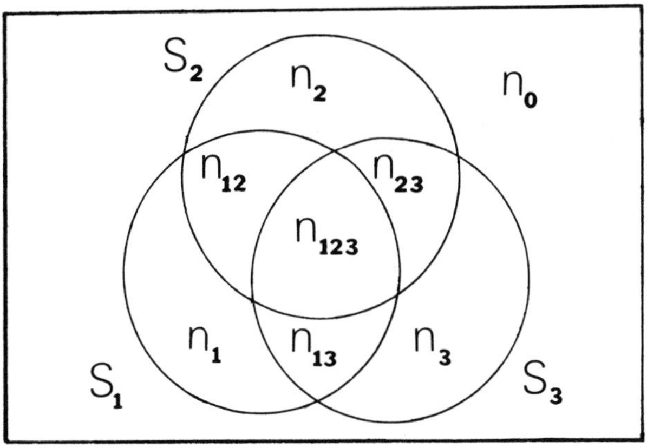

If s is the selection of Λ, then with i and j being distinct indices,

$$s^*(\varnothing) = n_0 \qquad \overline{s^*}(\varnothing) = |V|$$

$$s^*(\{e_i\}) = n_i \qquad \overline{s^*}(\{e_i\}) = |S_i|$$

$$s^*(\{e_i, e_j\}) = n_{ij} \qquad \overline{s^*}(\{e_i, e_j\}) = |S_i \cap S_j|$$

$$s^*(E) = \overline{s^*}(E) = n_{123} = |S_1 \cap S_2 \cap S_3|.$$

E11 *Exercise.* If Λ is the set system $(V, \mathscr{P}_k(V))$, determine the selection s of Λ and show that for all $S \in \mathscr{P}(V)$,

$$\bar{s}(S) = \begin{cases} 0 & \text{if } |S| > k; \\ \dbinom{|V| - |S|}{k - |S|} & \text{if } |S| \le k. \end{cases}$$

E12 *Exercise.* Let $\Lambda = (V, f, E)$ have selection s. Show that $\bar{s}(S) < m$ for all $S \in \mathscr{P}_t(V)$ if and only if $\overline{s^*}(A) < t$ for all $A \in \mathscr{P}_m(E)$.

We now present a fundamental counting theorem. Observe that its second statement is dual to the first.

E13 Theorem (The Principle of Inclusion–Exclusion). *Let* (V, f, E) *be a system with selection* s.

(a) *For* $r = 0, 1, \ldots, |E|$, *the number of vertices belonging to precisely* r *blocks is*

$$\sum_{i=0}^{|E|-r} (-1)^i \binom{r+i}{i} \sum_{|A|=r+i} \overline{s^*}(A).$$

(b) *For* $k = 0, 1, \ldots, |V|$, *the number of blocks of size* k *is*

$$\sum_{i=0}^{|V|-k} (-1)^i \binom{k+i}{i} \sum_{|S|=k+i} \bar{s}(S).$$

PROOF. As remarked above, it suffices to prove (a) alone.

If $A \in \mathscr{P}(E)$, then by E8c, $s^*(A)$ represents the number of vertices which belong to every block in A but to no other block. Thus the number of vertices which belong to precisely r blocks is $\sum_{|A|=r} s^*(A)$. By applying Proposition E6 to s^*, we get

$$\begin{aligned}
\sum_{|A|=r} s^*(A) &= \sum_{|A|=r} \sum_{C \in \mathscr{P}(E)} (-1)^{|C+A|}[A, C]\overline{s^*}(C) \\
&= \sum_{|C| \geq r} \left(\sum_{|A|=r} (-1)^{|C+A|}[A, C] \right) \overline{s^*}(C) \\
&= \sum_{|C| \geq r} (-1)^{|C|-r} \binom{|C|}{r} \overline{s^*}(C) \\
&= \sum_{i=0}^{|E|-r} \sum_{|C|=i+r} (-1)^{|C|-r} \binom{|C|}{r} \overline{s^*}(C) \\
&= \sum_{i=0}^{|E|-r} (-1)^i \binom{r+i}{i} \sum_{|C|=r+i} \overline{s^*}(C).
\end{aligned}$$

□

Returning to Example E9, let us now apply the Principle of Inclusion–Exclusion. The number of vertices belonging to precisely $r = 1$ block is

$$\sum_{i=0}^{2} (-1)^i \binom{1+i}{i} \sum_{|A|=1+i} \overline{s^*}(A) = \binom{1}{0}(|S_1| + |S_2| + |S_3|)$$

$$- \binom{2}{1}(|S_1 \cap S_2| + |S_2 \cap S_3| + |S_1 \cap S_3|) + \binom{3}{2}|S_1 \cap S_2 \cap S_3|,$$

which after substitution reduces to $n_1 + n_2 + n_3$.

Since $\overline{s^*}(A)$ is the number of vertices belonging to every block in A (see E8d), part (a) of the Principle of Inclusion–Exclusion gives the number of vertices contained in precisely r blocks in terms of the number of vertices contained in sets of r or more blocks. First we "include" the vertices belonging to at least r blocks, but because we have counted some of these more than once, we then "exclude" those belonging to at least $r + 1$ blocks. Having now excluded too much, we "reinclude" those vertices belonging to at least $r + 2$ blocks, and so on. Dually, since $\bar{s}(S)$ is simply the number of blocks containing S, part (b) gives the number of blocks of fixed size k in terms of the number of blocks containing subset S of V of size at least k.

The Principle of Inclusion–Exclusion has a wide range of applications. The remainder of this chapter is devoted to some of them. We begin by completing the answer to the question raised just after C12.

E14 Proposition. *For sets X and Y,*

$$|\mathrm{sur}(Y^X)| = (-1)^{|Y|} \sum_{i=0}^{|Y|} (-1)^i \binom{|Y|}{i} i^{|X|}.$$

PROOF. Let the function $\Phi: Y^X \to \mathscr{P}(Y)$ be given by $\Phi(f) = Y + f[X]$ for all $f \in Y^X$. Then (Y, Φ, Y^X) is a system. Let s denote its selection. Note that $f \in Y^X$ is a surjection if and only if $\Phi(f) = \varnothing$. Hence $|\mathrm{sur}(Y^X)|$ is the number of blocks of size $k = 0$. By Theorem E13b,

$$|\mathrm{sur}(Y^X)| = \sum_{j=0}^{|Y|} (-1)^j \sum_{|S|=j} \bar{s}(S).$$

By E8b,

$$\begin{aligned} \bar{s}(S) &= |\{f \in Y^X : \Phi(f) \supseteq S\}| \\ &= |\{f \in Y^X : f[X] \subseteq Y + S\}| \\ &= |Y + S|^{|X|}, \end{aligned}$$

by C7. Thus

$$|\mathrm{sur}(Y^X)| = \sum_{j=0}^{|Y|} (-1)^j \sum_{|S|=j} |Y + S|^{|X|} = \sum_{j=0}^{|Y|} (-1)^j \binom{|Y|}{j} (|Y| - j)^{|X|}.$$

Substituting i for $|Y| - j$ completes the proof. □

Combining this result with C24, we have

E15 Corollary.

$$|\mathbb{P}_m(V)| = \frac{(-1)^m}{m!} \sum_{i=0}^{m} (-1)^i \binom{m}{i} i^{|V|}.$$

In the literature the numbers $|\mathbb{P}_m(V)|$, usually denoted by $S(|V|, m)$, are called the **Stirling numbers of the second kind.** Another sequence of numbers well enough known to have been given a name is the sequence $\{D_n : n \in \mathbb{N}\}$ of **derangement numbers.** For each $n \in \mathbb{N}$, D_n is the number of **derangements**, i.e., permutations with no fixed points, of an n-set. The derangement numbers arise as a special case of the following result.

E16 Proposition. *The number of permutations of an n-set which have precisely r fixed-points is*

$$\frac{n!}{r!} \sum_{i=0}^{n-r} \frac{(-1)^i}{i!}.$$

PROOF. Let B be an n-set and let the function $f: B \to \mathscr{P}(\Pi(B))$ be given by $f(b) = \{\varphi \in \Pi(B): \varphi(b) = b\}$. Thus $(\Pi(B), f, B)$ is a system, and a given permutation φ belongs to a block b if and only if b is a fixed-point of φ. Hence we seek the number of vertices (i.e., permutations) which lie in exactly r blocks. This number is given in E13a; we now compute its value.

First we deduce from E8d that if $A \subseteq B$, then $\overline{s^*(A)}$ is the number of permutations of B which fix A pointwise. Clearly this is $(|B| - |A|)!$. Hence

$$\sum_{|A|=r+i} \overline{s^*(A)} = \binom{n}{r+i}(n-(r+i))! = \frac{n!}{(r+i)!}.$$

Substituting this into E13a yields

$$\sum_{i=0}^{n-r} (-1)^i \binom{r+i}{i} \frac{n!}{(r+i)!} = \frac{n!}{r!} \sum_{i=0}^{n-r} \frac{(-1)^i}{i!}. \qquad \square$$

Letting $r = 0$ in E16 we obtain

E17 Corollary.

$$D_n = n! \sum_{i=0}^{n} \frac{(-1)^i}{i!}.$$

By convention, $D_0 = 1$, and this corroborates the corollary. Observe that, $D_n/n!$ is the $(n + 1)$-st partial sum of the power series expansion of e^{-1}, and so $\lim_{n\to\infty} (D_n/n!) = e^{-1}$. In other words, and perhaps contrary to intuition, when n is large, approximately $1/e$ of all permutations of an n-set are derangements.

The next three exercises are concerned with derangements.

E18 *Exercise.* Prove that for $n \geq 2$, at least one-third of the permutations of an n-set are derangements.

E19 *Exercise.* Prove the following identities by set enumeration (cf. C31):

(a) $D_n = (n-1)(D_{n-1} + D_{n-2})$ for $n \geq 2$;

(b) $\sum_{i=0}^{n} \binom{n}{i} D_i = n!$.

E20 *Exercise.* Prove that for any n-set V, $n > 0$,

$$|\mathbb{P}(V)| = \frac{1}{n!} \sum_{i=0}^{n} (n-i)^n \binom{n}{i} D_i.$$

Our final application of the Principle of Inclusion–Exclusion is to derive a classical result from number theory. The function $\varphi: \mathbb{N} + \{0\} \to \mathbb{N}$ given by

$$\varphi(n) = |\{b \in \mathbb{N}: 0 < b \leq n; b \text{ is relatively prime to } n\}|,$$

for all $n \in \mathbb{N} + \{0\}$, is known as the "Euler φ-function". For example, $\varphi(n) = n - 1$ whenever n is prime.

E21 Theorem. *Let $n \in \mathbb{N}$, and let V be the set of prime divisors of n. Then*

$$\varphi(n) = n \prod_{p \in V} \left(1 - \frac{1}{p}\right).$$

PROOF. Clearly $\varphi(1) = 1$. If $n \geq 2$, let $B = \{1, 2, \ldots, n\}$, and let the function $f: B \to \mathscr{P}(V)$ be given by $f(b) = \{p \in V: p \text{ divides } b\}$, for each $b \in B$. Thus, (V, f, B) is a system, and $\varphi(n)$ is the number of blocks of size 0. Let s be the selection of (V, f, B). By E8b, for each $S \in \mathscr{P}(V)$, $\bar{s}(S)$ is the number of blocks divisible by every prime in S. Thus $\bar{s}(S) = n/\prod_{p \in S} p$. Substituting this into E13b with $k = 0$, we have

$$\varphi(n) = \sum_{i=0}^{|V|} (-1)^i \sum_{|S|=i} \frac{n}{\prod_{p \in S} p} = n \sum_{S \in \mathscr{P}(V)} \prod_{p \in S} \frac{-1}{p} = n \prod_{p \in V} \left(1 - \frac{1}{p}\right).$$

this last step requiring only algebraic manipulation. □

We close with two exercises of a general nature.

E22 *Exercise.* Verify that if $(V, \mathscr{E})$ is a set system, then

$$\left| \bigcup_{E \in \mathscr{E}} E \right| = \sum_{i=1}^{|\mathscr{E}|} (-1)^{i+1} \sum_{\mathscr{A} \in \mathscr{P}_i(\mathscr{E})} \left| \bigcap_{E \in \mathscr{A}} E \right|.$$

E23 *Exercise.* Let $s \in \mathbb{S}(\mathscr{P}(V))$ and let Φ be the function from $\mathbb{S}(\mathscr{P}(V))$ to itself given by $\Phi(s) = \bar{s}$ where

$$\bar{s}(S) = \sum_{T \in \mathscr{P}(V)} [T, S] s(T), \quad \text{for all } S \in \mathscr{P}(V).$$

State and prove results analogous to E6, E7, E8b, and E13b.

CHAPTER II

Algebraic Structures on Finite Sets

IIA Vector Spaces of Finite Sets

In *I*B we introduced the characteristic functions c_S for subsets S of a set U and proved (*I*B2) that the function $c_S \mapsto S$ is a bijection between $\mathbb{K}^U$ and $\mathscr{P}(U)$. Subsequently it was to be verified (Exercise *I*B3) that this same bijection made the assignments $c_S + c_T \mapsto S + T$ and $c_S c_T \mapsto S \cap T$. We have thereby that $(\mathscr{P}(U), +, \cap)$ is "algebra-isomorphic" to the commutative algebra $(\mathbb{K}^U, +, \cdot)$, and hence $(\mathscr{P}(U), +, \cap)$ is a commutative algebra over the field $\mathbb{K}$. In particular, $(\mathscr{P}(U), +)$ is a vector space over $\mathbb{K}$, while $(\mathscr{P}(U), +, \cap)$ is a commutative ring; $\varnothing$ is the additive identity and U itself is the multiplicative identity. For the present we shall be concerned only with the vector space structure.

For the reader who has studied vector spaces only over real or complex fields, we should remark that most of the results concerning such concepts as independence, spanning sets, bases, and dimension are not dependent upon the particular field in question but only upon the axioms common to all fields. These results are valid for $(\mathscr{P}(U), +)$ over $\mathbb{K}$, too. However, some results involving the inner product often not only involve properties characteristic of the real or complex numbers, but explicitly preclude the field $\mathbb{K}$.

We denote the dimension of a finite-dimensional vector space $\mathscr{V}$ by $\dim(\mathscr{V})$. For any set U, $\dim(\mathscr{P}(U)) = |U|$. This follows since $\dim(\mathbb{K}^U) = |U|$, but may also be seen directly by observing that the subcollection $\mathscr{P}_1(U)$ is a basis for $\mathscr{P}(U)$.

For each $S \subseteq U$, $\mathscr{P}(S)$ is a subspace of $\mathscr{P}(U)$. A subspace of this form is called a **coordinate subspace.**

Another subspace of interest consists of all the subsets of even cardinality:

$$\mathscr{E}(U) = \{S \in \mathscr{P}(U): |S| \text{ is even}\}.$$

A1 Proposition. *If $U \neq \varnothing$, then $\mathscr{E}(U)$ is a subspace of $\mathscr{P}(U)$ and* $\dim(\mathscr{E}(U)) = |U| - 1$. *If $U = \varnothing$, then $\mathscr{E}(U) = \mathscr{P}(U)$.*

PROOF. Clearly $\varnothing \in \mathscr{E}(U)$. If $U = \varnothing$, then $\mathscr{E}(U) = \{\varnothing\} = \mathscr{P}(U)$. We suppose $U \neq \varnothing$.

By *I*C5, if $S, T \in \mathscr{E}(U)$, then $S + T \in \mathscr{E}(U)$; also $0S = \varnothing$ and $1S = S$ belong to $\mathscr{E}(U)$. Since $\mathscr{E}(U)$ is closed with respect to $+$ and scalar multiplication, $\mathscr{E}(U)$ is a subspace of $\mathscr{P}(U)$.

Select $x_0 \in U$. One can easily verify that if $\mathscr{B} = \{\{x_0, x\}: x \in U + \{x_0\}\}$, then $\mathscr{B}$ is independent and $|\mathscr{B}| = |U| - 1$. Hence $\dim(\mathscr{E}(U)) \geq |U| - 1$. However, since $U \neq \varnothing, \mathscr{E}(U) \neq \mathscr{P}(U)$, so $\dim(\mathscr{E}(U)) < \dim(\mathscr{P}(U)) = |U|$. □

The following corollary offers a different approach to Exercise *I*C23.

A2 Corollary. *If $U \neq \varnothing$, then $|\mathscr{E}(U)| = |\mathscr{P}(U) + \mathscr{E}(U)| = 2^{|U|-1}$.*

PROOF. Let W be a $(|U| - 1)$-set. By the proposition, *I*C7, and *I*C8,

$$|\mathscr{E}(U)| = |\mathbb{K}^W| = |\mathbb{K}|^{|W|} = 2^{|U|-1} = |\mathscr{P}(U)|/2.$$

So $|\mathscr{P}(U) + \mathscr{E}(U)| = |\mathscr{P}(U)|/2$. □

A3 Exercise. *Let $\mathscr{S}_m = \langle \mathscr{P}_m(U) \rangle$,* i.e., *the subspace of $\mathscr{P}(U)$ spanned by $\mathscr{P}_m(U)$.* Show that

$$\mathscr{S}_m = \begin{cases} \mathscr{P}(U) & \text{if } 0 < m < |U| \text{ and } m \text{ is odd;} \\ \mathscr{E}(U) & \text{if } 0 < m < |U| \text{ and } m \text{ is even;} \\ \{\varnothing\} & \text{if } m = 0; \\ \{\varnothing, U\} & \text{if } m = |U|. \end{cases}$$

A4 *Exercise.* Let $f: U \to V$. Recall (§*I*A) the functions $f: \mathscr{P}(U) \to \mathscr{P}(V)$ and $f^{-1}: \mathscr{P}(V) \to \mathscr{P}(U)$ induced by f. First consider the vector spaces $(\mathscr{P}(U), +)$ and $(\mathscr{P}(V), +)$ and determine when f and f^{-1} are linear transformations and, in particular, nonsingular linear transformations. Then determine when f and f^{-1} are algebra-homomorphisms and, in particular, algebra-isomorphisms between the algebras $(\mathscr{P}(U), +, \cap)$ and $(\mathscr{P}(V), +, \cap)$. Finally, show that there can be no other algebra-isomorphisms.

A5 *Exercise.* Determine all subspaces of $\mathscr{P}(U)$ left invariant under the set of linear transformations induced by elements of $\Pi(U)$.

For any set U we define a function $\mathscr{P}(U) \times \mathscr{P}(U) \to \mathbb{K}$, called the **inner product** on $\mathscr{P}(U)$, as follows: if $S, T \in \mathscr{P}(U)$, then

$$S \cdot T = \begin{cases} 0 & \text{if } |S \cap T| \text{ is even;} \\ 1 & \text{if } |S \cap T| \text{ is odd.} \end{cases}$$

It follows at once from the definition, *I*A7, and *I*C5, that for all $R, S, T \in \mathscr{P}(U)$,

$$R \cdot S = S \cdot R;$$

$$R \cdot (S + T) = (R \cdot S) + (R \cdot T).$$

Furthermore,

$$R \cdot S = 0 \quad \text{for all } S \in \mathscr{P}(U) \Leftrightarrow R = \varnothing.$$

We say S is **orthogonal** to T if $S \cdot T = 0$. If $\mathscr{S} \subseteq \mathscr{P}(U)$, then the **orthogonal complement** of $\mathscr{S}$ is

$$\mathscr{S}^{\perp} = \{T \in \mathscr{P}(U) \colon S \cdot T = 0 \text{ for all } S \in \mathscr{S}\}.$$

Observe that $\mathscr{S}^{\perp}$ is always a subspace of $\mathscr{P}(U)$ and that $(\mathscr{S}^{\perp})^{\perp}$ is a subspace of $\mathscr{P}(U)$ which contains $\mathscr{S}$. In fact, $\mathscr{S}$ is a subspace of $\mathscr{P}(U)$ if and only if $(\mathscr{S}^{\perp})^{\perp} = \mathscr{S}$. Another important result concerning orthogonal complements (regardless of the characteristic of the field) is

A6 $\quad \dim(\mathscr{A}) + \dim(\mathscr{A}^{\perp}) = |U|, \quad$ for all subspaces $\mathscr{A} \subseteq \mathscr{P}(U)$.

The foregoing properties of inner products are all that will be required in this text. In a vector space over a subfield of the real numbers with the standard inner product, a subspace and its orthogonal complement have only the 0-vector in common. This is certainly not the case in the following example where the underlying field is $\mathbb{K}$.

Example. Let $|U| = n$ and let $\mathscr{A}$ be the subspace of $\mathscr{P}(U)$ spanned by a $[n/2]$-collection of pairwise-disjoint elements of $\mathscr{P}_2(U)$. Then $\dim(\mathscr{A}) = [n/2]$. Observe that if n is even, then $\mathscr{A} = \mathscr{A}^{\perp}$. If n is odd, then $\mathscr{A}^{\perp}$ is spanned by $\mathscr{A}$ together with the 1-set contained in no element of $\mathscr{A}$, and so $\mathscr{A} \subset \mathscr{A}^{\perp}$.

A7 Exercise. Show that $(\{\varnothing, U\})^{\perp} = \mathscr{E}(U)$.

A subspace of $\mathscr{P}(U)$ is **even** if all its elements are sets of even cardinality. This concept allows us to state the important algebraic result which underlies the classical Euler Theorem for graphs. This latter theorem will be encountered in its more traditional setting in §*III*A.

A8 Proposition. *Let $\mathscr{A}$ be any subspace of $\mathscr{P}(U)$. Then $U \in \mathscr{A}$ if and only if $\mathscr{A}^{\perp}$ is even.*

Proof. Since $\mathscr{A}$ is a subspace, $\mathscr{A} = (\mathscr{A}^\perp)^\perp$. Hence:

$$\begin{aligned} U \in \mathscr{A} &\Leftrightarrow S \cdot U = 0 \quad \text{for all } S \in \mathscr{A}^\perp \\ &\Leftrightarrow |S \cap U| \equiv 0 \text{ (modulo 2)} \quad \text{for all } S \in \mathscr{A}^\perp \\ &\Leftrightarrow |S| \quad \text{is even for all } S \in \mathscr{A}^\perp. \end{aligned}$$

□

If $\mathscr{S} \subseteq \mathscr{P}(U)$, the **foundation** of $\mathscr{S}$ is the set $\text{Fnd}(\mathscr{S}) = \bigcup_{S \in \mathscr{S}} S$. If $\mathscr{A}$ and $\mathscr{B}$ are subspace of $\mathscr{P}(U)$, then clearly their intersection $\mathscr{A} \cap \mathscr{B}$ is a subspace of $\mathscr{P}(U)$. Their **join** (also referred to as "sum") given by

$$\mathscr{A} \vee \mathscr{B} = \{S + T : S \in \mathscr{A}; T \in \mathscr{B}\}$$

is also a subspace of $\mathscr{P}(U)$. Clearly $\vee$ is a commutative and associative operation on the collection of subspaces of $\mathscr{P}(U)$. In the special case where the foundations of $\mathscr{A}$ and $\mathscr{B}$ are disjoint, the subspace $\mathscr{A} \vee \mathscr{B}$ is called the **coordinate sum** of $\mathscr{A}$ and $\mathscr{B}$ and is denoted by $\mathscr{A} \oplus \mathscr{B}$.

The following is a standard result from linear algebra:

A9 $$\dim(\mathscr{A} \vee \mathscr{B}) + \dim(\mathscr{A} \cap \mathscr{B}) = \dim(\mathscr{A}) + \dim(\mathscr{B}).$$

In particular,

$$\dim(\mathscr{A} \oplus \mathscr{B}) = \dim(\mathscr{A}) + \dim(\mathscr{B}).$$

We use the shorthand notation

$$\bigoplus_{i=1}^{n} \mathscr{A}_i$$

to represent $\mathscr{A}_1 \oplus \mathscr{A}_2 \oplus \ldots \oplus \mathscr{A}_n$.

It is easy to see that for each $S \subseteq U$,

$$\mathscr{P}(U) = \mathscr{P}(S) \oplus \mathscr{P}(U + S).$$

There is a function $\pi_S : \mathscr{P}(U) \to \mathscr{P}(S)$ given by $\pi_S(T) = T \cap S$ for all $T \in \mathscr{P}(U)$. It can be readily demonstrated using *I*A7 that π_S is a linear transformation. It is a surjection, its kernel is $\mathscr{P}(U + S)$, and it fixes $\mathscr{P}(S)$ pointwise. In vector space terminology, "π_S projects $\mathscr{P}(U)$ onto $\mathscr{P}(S)$ along $\mathscr{P}(U + S)$."

If $\mathscr{A}$ is a subspace of $\mathscr{P}(U)$ and if $S \subseteq U$, then $\pi_S[\mathscr{A}]$ is a subspace of $\mathscr{P}(S)$. Since the kernel of the restriction $\pi_{S|\mathscr{A}}$ is $\mathscr{P}(U + S) \cap \mathscr{A}$, we have

A10 $$\dim(\mathscr{P}(U + S) \cap \mathscr{A}) + \dim(\pi_S[\mathscr{A}]) = \dim(\mathscr{A}).$$

Since π_S fixes $\mathscr{P}(S)$ pointwise,

A11 $$\mathscr{P}(S) \cap \mathscr{A} \subseteq \pi_S[\mathscr{A}].$$

Of interest are those sets $S \in \mathscr{P}(U)$ for which equality holds in A11. The following result shows them to correspond to "summands" in a coordinate sum.

A12 Proposition. *Let* $\mathscr{A}$ *be a subspace of* $\mathscr{P}(U)$.
(a) *If* $\mathscr{A} = \mathscr{B} \oplus \mathscr{C}$ *and if* $B = \operatorname{Fnd}(\mathscr{B})$, *then* $\pi_B[\mathscr{A}] = \mathscr{P}(B) \cap \mathscr{A} = \mathscr{B}$.
(b) *Conversely, if* $B \in \mathscr{P}(U)$ *and if* $\pi_B[\mathscr{A}] = \mathscr{P}(B) \cap \mathscr{A}$, *then*

A13
$$\pi_{U+B}[\mathscr{A}] = \mathscr{P}(U + B) \cap \mathscr{A},$$

and

$$\mathscr{A} = (\mathscr{P}(B) \cap \mathscr{A}) \oplus (\mathscr{P}(U + B) \cap \mathscr{A}) = \pi_B[\mathscr{A}] \oplus \pi_{U+B}[\mathscr{A}].$$

Proof. (a) Let $\mathscr{A} = \mathscr{B} \oplus \mathscr{C}$ and $B = \operatorname{Fnd}(\mathscr{B})$. By these assumptions and A11, $\mathscr{B} \subseteq \mathscr{P}(B) \cap \mathscr{A} \subseteq \pi_B[\mathscr{A}]$. It suffices to show that $\pi_B[\mathscr{A}] \subseteq \mathscr{B}$.

Let $S \in \mathscr{A}$. Then $S = S_1 + S_2$ for some $S_1 \in \mathscr{B}$ and $S_2 \in \mathscr{C}$. By definition of $\oplus$, $B \cap S_2 = \varnothing$. Hence

$$\pi_B(S) = S \cap B = S_1 \in \mathscr{B}.$$

(b) Let $B \in \mathscr{P}(U)$ and assume that $\pi_B[\mathscr{A}] = \mathscr{P}(B) \cap \mathscr{A}$.
By A11, $\mathscr{P}(U + B) \cap \mathscr{A} \subseteq \pi_{U+B}[\mathscr{A}]$. By A10 and our assumption,

$$\begin{aligned} \dim(\mathscr{P}(U + B) \cap \mathscr{A}) &= \dim(\mathscr{A}) - \dim(\pi_B[\mathscr{A}]) \\ &= \dim(\mathscr{A}) - \dim(\mathscr{P}(B) \cap \mathscr{A}) \\ &= \dim(\pi_{U+B}[\mathscr{A}]), \end{aligned}$$

and A13 follows.

The coordinate sum $(\mathscr{P}(B) \cap \mathscr{A}) \oplus (\mathscr{P}(U + B) \cap \mathscr{A})$ is clearly a subspace of $\mathscr{A}$. Moreover, by A9, A13, and A10,

$$\begin{aligned} &\dim(\mathscr{P}(B) \cap \mathscr{A}) \oplus (\mathscr{P}(U + B) \cap \mathscr{A})) \\ &\qquad = \dim(\mathscr{P}(B) \cap \mathscr{A}) + \dim(\mathscr{P}(U + B) \cap \mathscr{A}) \\ &\qquad = \dim(\mathscr{P}(B) \cap \mathscr{A}) + \dim(\pi_{U+B}[\mathscr{A}]) \\ &\qquad = \dim(\mathscr{A}). \end{aligned}$$

Hence $\mathscr{A} = (\mathscr{P}(B) \cap \mathscr{A}) \oplus (\mathscr{P}(U + B) \cap \mathscr{A})$, which in turn equals $\pi_B[\mathscr{A}] \oplus \pi_{U+B}[\mathscr{A}]$ by A13. □

A14 Corollary. *Let* $\mathscr{A}$ *be a subspace of* $\mathscr{P}(U)$ *and let* $\mathscr{B}$ *be a subspace of* $\mathscr{A}$ *with* $B = \operatorname{Fnd}(\mathscr{B})$. *Then* $\mathscr{A} = \mathscr{B} \oplus \mathscr{C}$ *for some subspace* $\mathscr{C}$ *of* $\mathscr{P}(U)$ *if and only if* $\pi_B[\mathscr{A}] \subseteq \mathscr{B}$.

Proof. By A11, $\mathscr{B} \subseteq \mathscr{P}(B) \cap \mathscr{A} \subseteq \pi_B[\mathscr{A}]$. If $\pi_B[\mathscr{A}] \subseteq \mathscr{B}$, then by part (b) of the Proposition, we may let $\mathscr{C} = \pi_{U+B}[\mathscr{A}]$. The converse follows from part (a) of the Proposition. □

A15 Exercise. *Let* $V \subseteq U$ *and let* $\mathscr{S}$ *be a subspace of* $\mathscr{P}(U)$. Prove that $\pi_V[\mathscr{S}^{\perp}]$ *is the orthogonal complement in* $\mathscr{P}(V)$ *of* $\mathscr{P}(V) \cap \mathscr{S}$. (*Hint:* use A10 and A6.)

IIB Ordering

A **partial order** on a finite or infinite set U is a relation R on U which is reflexive, transitive, and **antisymmetric**, i.e.,

$$(x, y) \in R \quad \text{and} \quad (y, x) \in R \quad \text{imply} \quad x = y.$$

A partial order is frequently designated by a symbol such as $\leq$ which will be used in the following way. Instead of writing: $(x, y) \in \leq$, one writes: $x \leq y$. In this context, the symbol $<$ will be used to mean: $x \leq y$ and $x \neq y$. (Compare the use of $\subseteq$ and $\subset$.) Clearly $<$ is also a relation on U.

A pair $(U, \leq)$, where U is a finite or infinite set and $\leq$ is a partial order on U, is called a **partially-ordered set**. $(U, \leq)$ is a **totally-ordered set** and $\leq$ is a **total order** if either $x \leq y$ or $y \leq x$ for all $x, y \in U$. Isomorphism between sets with relations was defined in §*I*A.

B1 *Exercise.* If $(U, \leq)$ is a partially-ordered set, show that $<$ is antisymmetric and transitive on U.

Certainly if $(U, \leq)$ is a partially-ordered set, and if $S \in \mathscr{P}(U)$, then the intersection of $\leq$ with $S \times S$ is a partial order on S. We abuse notation and designate such a partially-ordered set by $(S, \leq)$.

The structures we have been considering readily provide examples of partial orderings.

Example. $(\mathscr{P}(U), \subseteq)$ and $(\mathscr{P}(U), \supseteq)$ are partially-ordered sets.

B2 *Example.* A partially-ordered set $(Y, \leq)$ determines a partially-ordered set $(Y^X, \leq)$ for any set X: if $f, g \in Y^X$, we say $f \leq g$ if $f(x) \leq g(x)$ for all $x \in X$. In particular, for any set U, $\mathbb{S}(U)$ will be regarded as a partially-ordered set, the partial order being determined by the total order on $\mathbb{N}$.

Let $\mathscr{S}, \mathscr{T} \subseteq \mathscr{P}(U)$. We say $\mathscr{S}$ **refines** $\mathscr{T}$, or $\mathscr{S}$ is a **refinement** of $\mathscr{T}$, if for every $S \in \mathscr{S}$ and $T \in \mathscr{T}$, either $S \cap T = \varnothing$ or $S \cap T = S$. Observe that refinement as a relation on $\mathscr{P}(\mathscr{P}(U))$ is generally not reflexive. In fact,

B3 *$\mathscr{S}$ refines itself if and only if the elements of $\mathscr{S}$ are pairwise-disjoint. Thus refinement is reflexive on $\mathbb{P}(U)$.*

We say that $\mathscr{S}$ **covers** U, or $\mathscr{S}$ is a **covering** of U, if $U \subseteq \text{Fnd}(\mathscr{S})$.

B4 Exercise. Show that *if $\mathscr{S}$ and $\mathscr{T}$ are coverings of U, and if each refines the other, then $\mathscr{S} = \mathscr{T} \in \mathbb{P}(U)$.*

Thus refinement is antisymmetric on the set of coverings (and hence on the set of partitions) of any set.

B5 Exercise. *Suppose $\mathscr{R}, \mathscr{S}, \mathscr{T} \subseteq \mathscr{P}(U)$ and that $\mathscr{S}$ covers U.* Show that *if $\mathscr{R}$ refines $\mathscr{S}$ and $\mathscr{S}$ refines $\mathscr{T}$, then $\mathscr{R}$ refines $\mathscr{T}$.*

We conclude that refinement is transitive on the set of coverings (and hence on the set of partitions) of any set, but one can readily verify that refinement need not be transitive on $\mathscr{P}(\mathscr{P}(U))$. From B3, B4, and B5 we conclude:

B6 Proposition. *For any set U, refinement is a partial order on $\mathbb{P}(U)$.*

Thus each of $\mathscr{P}(U)$, $\mathbb{S}(U)$, and $\mathbb{P}(U)$ admits a partial order in a rather natural way. Having defined various operations on these objects in the first chapter, let us observe how they relate to these objects as partially-ordered sets.

B7 *For all $R, S, T \in \mathscr{P}(U)$,*

$$S \cap T \subseteq T,$$

and

$$\text{if } S \subseteq T, \quad \text{then } S \cap R \subseteq T \cap R.$$

B8 *For all $\mathscr{R}, \mathscr{S}, \mathscr{T} \in \mathbb{P}(U)$, with $\leq$ denoting refinement,*

$$\mathscr{S}\mathscr{T} \leq \mathscr{T},$$

and

$$\text{if } \mathscr{S} \leq \mathscr{T}, \quad \text{then } \mathscr{S}\mathscr{R} \leq \mathscr{T}\mathscr{R}.$$

B9 *For all $r, s, t \in \mathbb{S}(U)$,*

$$s + t \geq t$$

and

$$\text{if } s \geq t, \quad \text{then } s + r \geq t + r.$$

B10 *Exercise.* Reconsider B7 with the operation $\cap$ replaced by $\cup$ (respectively, $+$), and reconsider B9 with addition replaced by multiplication. In each case prove or disprove the analogous assertions.

Let $(U, \leq)$ be a finite or infinite partially-ordered set. We say an element x of $(U, \leq)$ is **minimal** (respectively, **maximal**) if there exists no $x' \in U$ such that $x' < x$ (respectively, $x' > x$). An element x of $(U, \leq)$ is the **minimum** (respectively, the **maximum**) element of $(U, \leq)$ if $x \leq x'$ (respectively, $x \geq x'$) for all $x' \in U$. Two facts are immediate: first, a partially-ordered set need not have a minimum (respectively, maximum) element; second, if a partially-ordered set does have a minimum (respectively, maximum) element, then that element is the unique minimal (respectively, maximal) element of the partially-ordered set. The converse of this second remark is also true (see B12 below). If $f: Y \to \mathbb{N}$, we say $y \in Y$ is a **smallest** (respectively, **largest) element** of Y (relative to f being understood) if $f(y) \leq f(y')$ (respectively, $f(y) \geq f(y')$) for all $y' \in Y$.

Example. Let $U = \{a, b, c, d\}$ and let $\mathscr{S}$ consist of the sets $\{a\}$, $\{b\}$, $\{a, b\}$, $\{b, c\}$, $\{c, d\}$, $\{a, b, c\}$, $\{b, c, d\}$, $\{a, c, d\}$. Then $(\mathscr{S}, \subseteq)$ has neither a minimum nor a maximum element. The 3-sets in $\mathscr{S}$ are all maximal (with respect to inclusion) and largest (with respect to cardinality). Similarly the 1-sets in $\mathscr{S}$ are minimal and smallest. However, the set $\{c, d\}$ is also minimal but it is not smallest.

We return to the convention that all sets are presumed to be finite. If $(U, \leq)$ is a partially-ordered set, a totally ordered m-subset S of U is called a **chain** or, more specifically, an **m-chain**. The collection of all chains on $(U, \leq)$ is partially-ordered by the usual set-inclusion, and the term **maximal chain** denotes a maximal element of this ordered collection.

Clearly a maximal chain of $(U, \leq)$ contains both a unique maximal element of $(U, \leq)$ and a unique minimal element of $(U, \leq)$. For let $S = \{x_1, x_2, \ldots, x_m\}$ and suppose $x_1 < x_2 < \ldots < x_m$. If x_m is not maximal in $(U, \leq)$, then $x_m < w$ for some $w \in U$. It follows that $S + \{w\}$ is also a chain and $S \subset S + \{w\}$. Similarly one shows that x_1 is minimal in $(U, \leq)$. Now let $x \in U$. Since $\{x\}$ itself is a chain, the collection of all chains containing x is not empty and hence contains a largest member S. Clearly S is also a maximal chain. We have proved:

B11 Proposition. *Let $(U, \leq)$ be a partially-ordered set. If $x \in U$, then x is an element of some maximal chain in $(U, \leq)$. Moreover, $y \leq x \leq z$ for some minimal element $y \in U$ and some maximal element $z \in U$.*

B12 Exercise. *If the partially-ordered set $(U, \leq)$ has a unique minimal (respectively, maximal) element x,* prove *x is the minimum (respectively, maximum) element of $(U, \leq)$.*

The next proposition gives a further connection between partial orders and algebraic structures.

B13 Proposition. *Let $(U, \leq)$ be a partially-ordered set and let $\odot$ be a commutative, associative operation on U such that $x \odot x' \leq x$ for all $x, x' \in U$. If S is a nonempty subset of U closed under $\odot$, then $(S, \leq)$ has a minimum element.*

Proof. Let $S \subseteq U$ be closed under $\odot$ and let x_0 be the "product" of all elements in S. (Since $\odot$ is commutative and associative, x_0 is well-defined.) Clearly $x_0 \leq x$ for all $x \in S$, and of course $x_0 \in S$. □

Note that the finiteness of S is essential to the above proof. As an immediate application, we have:

B14 Corollary. *Any nonempty subcollection of* $(\mathscr{P}(U), \subseteq)$ *which is closed under* $\cap$ *(respectively,* $\cup$*) contains a minimum (respectively, maximum) subset.*

B15 Corollary. *Any nonempty subcollection of* $\mathbb{P}(U)$ *which is closed under multiplication of partitions contains a minimum partition (with respect to refinement).*

B16 Proposition. *Let* $\mathscr{S} \subseteq \mathscr{P}(U)$ *and let* $\leq$ *denote refinement on* $\mathscr{P}(\mathscr{P}(U))$. *Then* $\{\mathscr{Q} \in \mathbb{P}(U): \mathscr{S} \leq \mathscr{Q}\}$ *has a minimum element and* $\{\mathscr{Q} \in \mathbb{P}(U): \mathscr{Q} \leq \mathscr{S}\}$ *has a maximum element.*

PROOF. Let $\mathscr{R}_1, \mathscr{R}_2 \in \mathbb{P}(U)$ and suppose $\mathscr{S} \leq \mathscr{R}_1$ and $\mathscr{S} \leq \mathscr{R}_2$. We show that $\mathscr{S} \leq \mathscr{R}_1\mathscr{R}_2$. An arbitrary cell of $\mathscr{R}_1\mathscr{R}_2$ is of the form $R_1 \cap R_2$ where $R_i \in \mathscr{R}_i$. Let $S \in \mathscr{S}$. If $S \cap R_i \neq \varnothing$, then $S \subseteq R_i$ by the definition of refinement. It is immediate that either $S \subseteq R_1 \cap R_2$ or $S \cap R_1 \cap R_2 = \varnothing$. Hence $\{\mathscr{Q} \in \mathbb{P}(U): \mathscr{S} \leq \mathscr{Q}\}$ is closed under multiplication. By B15 it contains a minimum element.

To complete the proof, we define a relation $\sim$ on U whereby $x \sim y$ if

B17 $$x \in S \Leftrightarrow y \in S, \quad \text{for all } S \in \mathscr{S}.$$

Obviously $\sim$ is an equivalence relation on U, the equivalence classes of which form a partition $\mathscr{Q}_0 \in \mathbb{P}(U)$. Moreover $\mathscr{Q}_0 \leq \mathscr{S}$. If $\mathscr{Q} \in \mathbb{P}(U)$ and $\mathscr{Q} \leq \mathscr{S}$, and if x, y belong to the same cell of $\mathscr{Q}$, then indeed B17 holds, and $x \sim y$. Hence x and y belong to the same cell of $\mathscr{Q}_0$. It follows that $\mathscr{Q} \leq \mathscr{Q}_0$, and $\mathscr{Q}_0$ is the required maximum element. □

If $\mathscr{S} \subseteq \mathscr{P}(U)$, the minimum and maximum partitions guaranteed by B16 are called the **coarse partition** of $\mathscr{S}$ and the **fine partition** of $\mathscr{S}$, respectively. Thus the coarse partition of $\mathscr{S}$ is the "finest" partition refined by $\mathscr{S}$, and the fine partition of $\mathscr{S}$ is the "coarsest" of the partitions which refine $\mathscr{S}$. This is just fine, of course. If $\mathscr{S}$ has fine partition $\mathscr{Q}_0$ and coarse partition $\mathscr{Q}_1$, it follows from B5 that

B18 $$\mathscr{Q}_0 \leq \mathscr{Q}_1, \quad \text{whenever } \mathscr{S} \text{ is a covering of } U,$$

and from *B*4 and B5 that

B19 $$\mathscr{Q}_0 = \mathscr{Q}_1 \Leftrightarrow \mathscr{S} \in \mathbb{P}(U), \quad \text{whenever } \mathscr{S} \text{ is a covering of } U.$$

B20 *Exercise.* Show that the condition that $\mathscr{S}$ be a covering of U is essential in both B18 and B19.

B21 *Exercise.* Let $\mathscr{S}$ be a covering of U.

(a) If $\mathscr{S}$ has the property that

$$S_1, S_2 \in \mathscr{S} \Rightarrow S_1 \cap S_2 \neq \varnothing,$$

what is the coarse partition of $\mathscr{S}$?

(b) Let $\mathscr{S}$ have the property that the system $(U, \mathscr{S})$ distinguishes vertices. What is the fine partition of $\mathscr{S}$?

(c) Give examples of collections $\mathscr{S}$ having both of the above properties.

A collection $\mathscr{S} \subseteq \mathscr{P}(U)$ is said to be **incommensurable** if no element of $\mathscr{S}$ is a subset of any other element of $\mathscr{S}$. (Cf. *I*D6.)

B22 *Exercise.* Let $\mathscr{Q}_0, \mathscr{Q}_1 \in \mathbb{P}(U)$ be given with $\mathscr{Q}_0 \leq \mathscr{Q}_1$. Determine when there exists $\mathscr{S} \subseteq \mathscr{P}(U)$ such that $\mathscr{Q}_0$ is the fine partition of $\mathscr{S}$ and $\mathscr{Q}_1$ is the coarse partition of $\mathscr{S}$. What is the answer if it is imposed further that $\mathscr{S}$ be incommensurable?

B23 *Exercise* (Sperner [s.7]). Show that any largest incommensurable subcollection of $\mathscr{P}(U)$ has cardinality

$$\binom{|U|}{\left[\frac{|U|}{2}\right]}$$

(*Hint:* Let $\mathscr{C}$ be the set of $(|U| + 1)$-chains in $(\mathscr{P}(U), \subseteq)$. For each $S \in \mathscr{P}(U)$, let $\mathscr{C}_S$ consist of those chains in $\mathscr{C}$ of which S is an element. Begin by showing that if $\mathscr{S} \subseteq \mathscr{P}(U)$ is incommensurable, then $|\mathscr{C}| \geq \sum_{S \in \mathscr{S}} |\mathscr{C}_S|$.)

A pair (V, D), where V is a set and $D \subseteq (V \times V) + \{(v, v): v \in V\}$, is called a **directed graph.** The elements of V are called **vertices** and the elements of D are called **edges.** A sequence of vertices and edges of the form

$$v_0, (v_0, v_1), v_1, (v_1, v_2), v_2, \ldots, v_{k-1}, (v_{k-1}, v_k), v_k$$

from (V, D) is called a v_0v_k**-path.** The **length** of a path is the number of edges it contains. In particular, a single vertex constitutes a path of length 0. If $v \in V$, a vv-path is called a **directed circuit.** A directed graph is **acyclic** if all its directed circuits have length 0.

B24 Proposition. *Let (V, D) be a directed graph. Define the relation $\leq$ on V by: $u \leq v$ if (V, D) admits a uv-path. Then $\leq$ is reflexive and transitive. Moreover, (V, D) is acyclic if and only if $(V, \leq)$ is a partially-ordered set.*

PROOF. Trivially $\leq$ is reflexive.

If $u \leq v$ and $v \leq w$, then (V, D) admits a uv-path:

$$u = u_0, (u_0, u_1), u_1, \ldots, u_{j-1}, (u_{j-1}, u_j), u_j = v$$

and a vw-path:

$$v = v_0, (v_0, v_1), v_1, \ldots, v_{k-1}, (v_{k-1}, v_k), v_k = w.$$

The following sequence is a uw-path:

$$u, (u, u_1), u_1, \ldots, u_{j-1}, (u_{j-1}, u_j), v, (v_0, v_1), v_1, \ldots, v_{k-1}, (v_{k-1}, v_k), w.$$

Hence $\leq$ is transitive.

Now suppose $v \leq w$ and $w \leq v$. A repetition of the above construction yields a vv-path, and if $v \neq w$, this path will have positive length. Hence if (V, D) is acyclic, then $\leq$ is antisymmetric. Conversely, if there exists a directed circuit through distinct vertices v and w, then $v \leq w$ and $w \leq v$. □

From the previous result we have that every acyclic directed graph uniquely determines a partially-ordered set. Conversely, every partially-ordered set can be obtained in this way. However, a given partially-ordered set may be determined by many different directed graphs. For let $(V, \leq)$ be a partially-ordered set and let $D = \{(v, w) \in V \times V : v \leq w\}$. Clearly (V, D) is an acyclic directed graph which yields $(V, \leq)$ in the manner of the proof of the previous proposition. In this case, (V, D) is the directed graph of $(V, \leq)$ which has the largest possible number of edges.

Given $(V, \leq)$, we say that w is a **successor** of v if $v < w$ and if $v \leq x \leq w$ implies $x = v$ or $x = w$.

B25 *Exercise.* Given $(V, \leq)$, let $D_0 = \{(v, w): w \text{ is a successor of } v\}$. Show that the partial order determined by the directed graph (V, D_0) is precisely $(V, \leq)$. Moreover, $D_0 \subseteq D_1 \subseteq D$ if and only if (V, D_1) determines $(V, \leq)$.

Let $(U, \leq)$ be a finite or infinite partially-ordered set. If $x, y \in U$, we define the **meet** of x and y, denoted by $x \wedge y$, to be the maximum element of $\{z \in U: z \leq x; z \leq y\}$, if it exists. We define the **join** of x and y, denoted by $x \vee y$, to be the minimum element of $\{z \in U: x \leq z; y \leq z\}$, if it exists. A partially-ordered set $(U, \leq)$ is called a **lattice** if $x \wedge y$ and $x \vee y$ exist for all $x, y \in U$.

B26 *Example.* For any set U, the partially-ordered set $(\mathscr{P}(U), \subseteq)$ is a lattice, where the usual set-theoretic intersection and union are the two lattice operations of meet and join, respectively. Any lattice which is isomorphic to $(\mathscr{P}(U), \subseteq)$ for some set U is called a **Boolean lattice.**

B27 *Example.* Consider the partially-ordered set $(U, |)$ of all positive integral divisors of the positive integer n, where $|$ means "divides." The join of any two elements of U is their least common multiple and their meet is their greatest common divisor. It is not difficult to see that $(U, |)$ is a lattice, and it is Boolean if and only if n is divisible by no perfect square greater than 1.

We shall have frequent recourse to the following two examples.

B28 *Example.* Let $\mathsf{S}(\mathscr{V})$ denote the set of all subspaces of the vector space $\mathscr{V}$. If $\mathscr{A}, \mathscr{B} \in \mathsf{S}(\mathscr{V})$, let $\mathscr{A} \leq \mathscr{B}$ mean that $\mathscr{A}$ is a subspace of $\mathscr{B}$. With the join of $\mathscr{A}$ and $\mathscr{B}$ as defined in the previous section and their meet defined to be their intersection, $(\mathsf{S}(\mathscr{V}), \leq)$ becomes a lattice. The verification involves only elementary linear algebra. It will be shown subsequently that these lattices are not Boolean when $|\mathscr{V}| \geq 2$.

B29 *Example.* Let $\mathsf{S}(\Lambda)$ denote the set of all subsystems of the system $\Lambda = (V, f, E)$. If $\Omega_1, \Omega_2 \in \mathsf{S}(\Lambda)$, then $\Omega_1 \leq \Omega_2$ means that Ω_1 is a subsystem of Ω_2. Clearly $(\mathsf{S}(\Lambda), \leq)$ is a partially-ordered set. Let $\Omega_i = (W_i, g_i, F_i)$ for $i = 1, 2$, and define

$$\Omega_1 \wedge \Omega_2 = (W_1 \cap W_2, f_{|F_1 \cap F_2}, F_1 \cap F_2);$$

$$\Omega_1 \vee \Omega_2 = (W_1 \cup W_2, f_{|F_1 \cup F_2}, F_1 \cup F_2).$$

It is straightforward to verify that $(\mathsf{S}(\Lambda), \leq)$ now becomes a lattice.

B30 *Exercise.* Show that for any set U, the partially-ordered set $(\mathbb{P}(U), \leq)$ is a lattice.

If $(U, \leq)$ is a finite or infinite partially-ordered set, we define the dual order $\geq$ on U by: $x \geq y$ if and only if $y \leq x$, for all $x, y \in U$. Then $(U, \geq)$ is also a partially-ordered set. In particular, if $(U, \leq)$ is a lattice with meet and join denoted by $\wedge$ and $\vee$, respectively, then $(U, \geq)$ is its **dual lattice**, with meet and join given by $\vee$ and $\wedge$, respectively, as can be easily verified from the definitions. For example, $(\mathscr{P}(U), \supseteq)$ is dual to $(\mathscr{P}(U), \subseteq)$ in Example B26, where the roles of union and intersection have been interchanged. Clearly the dual of the dual of a lattice is the original lattice.

The next exercise is a list of algebraic properties to be verified for all lattices. They follow from basic definitions. The concept of duality can be used to substantially shorten the work.

B31 Exercise. *Let* $(U, \leq)$ *be a lattice.* Show that *for all* $x, y, z \in U$,

(a) $\wedge$ *and* $\vee$ *are idempotent* (i.e., $x \wedge x = x \vee x = x$);
(b) $\wedge$ *and* $\vee$ *are commutative and associative*;
(c) $x \wedge (x \vee y) = x = x \vee (x \wedge y)$;
(d) $x \leq y$ *implies both* $x \wedge z \leq y \wedge z$ and $x \vee z \leq y \vee z$;
(e) $(x \wedge y) \vee (x \wedge z) \leq x \wedge (y \vee z)$;
(f) $x \vee (y \wedge z) \leq (x \vee y) \wedge (x \vee z)$.

If W is any finite subset of U, then part (b) above yields that the meet of all the elements of W, denoted $\bigwedge_{x \in W} x$, is well-defined. Analogously, we write $\bigvee_{x \in W} x$. *Henceforth, all lattices are assumed to be finite.* We may now define two distinguished elements $0 = \bigwedge_{x \in U} x$ and $1 = \bigvee_{x \in U} x$. Thus $0 \leq x \leq 1$ for all $x \in U$. *Every (finite) lattice has a minimum element and a maximum element.* In Examples B26, B27, and B28, the minimum elements are $\varnothing$, 1, and $\{0\}$, respectively, while the maximum elements are U, n, and $\mathscr{V}$, respectively.

B32 Proposition. *The following statements are equivalent for any lattice* $(U, \leq)$:

(a) $(x \wedge y) \vee (x \wedge z) = x \wedge (y \vee z)$ *for all* $x, y, z \in U$;
(b) $(x \vee y) \wedge (x \vee z) = x \vee (y \wedge z)$ *for all* $x, y, z \in U$;
(c) $(x \vee y) \wedge z \leq x \vee (y \wedge z)$ *for all* $x, y, z \in U$.

PROOF. (a) $\Rightarrow$ (b). Assume (a) to hold and substitute $x \vee y$ for x and x for y, obtaining

$$[(x \vee y) \wedge x] \vee [(x \vee y) \wedge z] = (x \vee y) \wedge (x \vee z).$$

The left-hand member becomes

$x \vee [(x \vee y) \wedge z]$, by B31b and c;
$= x \vee [(x \wedge z) \vee (y \wedge z)]$, by assumption (with x and z interchanged);
$= x \vee (y \wedge z)$, by B31b and c.

(b) $\Rightarrow$ (c). Since $z \leq x \vee z$, B31d followed by our assumption (b) yields $(x \vee y) \wedge z \leq (x \vee y) \wedge (x \vee z) = x \vee (y \wedge z)$.

(c) $\Rightarrow$ (a). We need only prove that (c) implies the reverse inequality of B31f. With appropriate substitutions, two successive applications of assumption (c) yield

$$\begin{aligned}(x \vee y) \wedge (x \vee z) &\leq x \vee [y \wedge (x \vee z)]\\ &\leq x \vee [x \vee (y \wedge z)]\\ &= x \vee (y \wedge z), \quad \text{as required.}\end{aligned}$$ □

A lattice which satisfies any one (and hence all three) of the conditions of Proposition B32 is called a **distributive lattice.**

B33 *Example.* $(\mathscr{P}(U), \subseteq)$ is a distributive lattice for any set U. Hence *all Boolean lattices are distributive.*

B34 Exercise. Prove that

(a) *If U is a set with at least two elements, then the lattice* $(\mathsf{S}(\mathscr{P}(U)), \leq)$ *is not distributive (and hence not Boolean).*

(b) $(\mathsf{S}(\Lambda), \leq)$ *is a distributive lattice for any system* Λ.

B35 *Exercise.* Determine whether the lattice $(\mathbb{P}(U), \leq)$ is distributive.

A lattice $(U, \leq)$ is said to be **complemented** if for each $x \in U$ there exists $x' \in U$ such that $x \wedge x' = 0$ and $x \vee x' = 1$. In this case x' is called a **complement** of x.

B36 *Example.* In the lattice $(\mathscr{P}(U), \subseteq)$, the complement in the lattice of any set $S \in \mathscr{P}(U)$ is its set-theoretic complement $U + S$. Hence *all Boolean lattices are complemented.*

When $(U, \leq)$ is distributive, one may speak of *the* complement of x, for if both y and z were complements of x, one would have by B32c and b that $y = y \wedge 1 = y \wedge (x \vee z) \leq (x \wedge y) \vee z = 0 \vee z = z$. By symmetry $z \leq y$, and so $y = z$. Hence $x'' = x$ for all complemented elements x.

B37 Exercise. *Let* $(U, \leq)$ *be a distributive lattice. Let* $a, b \in U$ *and let* $a \leq b$. *Let* $W = \{x \in U: a \leq x \leq b\}$. Show that

(a) $(W, \leq)$ *is a distributive lattice.*

(b) *For each* $x \in W$ *there exists at most one element* $y \in U$ *such that* $x \wedge y = a$ *and* $x \vee y = b$.

(c) *If* $(U, \leq)$ *is complemented, then* $(W, \leq)$ *is complemented.*

If $(U, \leq)$ is a partially-ordered set and if $(W, \leq)$ is a lattice where $W \subseteq U$, then $(W, \leq)$ is a **sublattice** of $(U, \leq)$.

B38 Lemma. *Let* $(U, \leq)$ *be a lattice and let* $W \subseteq U$. *If* $x \wedge y, x \vee y \in W$ *for all* $x, y \in W$, *then* $(W, \leq)$ *is a sublattice of* $(U, \leq)$.

PROOF. By definition,

$$\begin{aligned} x \wedge y &= \max\{z \in U: z \leq x; z \leq y\} \\ &\geq \max\{z \in W: z \leq x; z \leq y\} \geq x \wedge y, \end{aligned}$$

since $x \wedge y \in W$. The argument for $x \vee y$ is analogous. □

When $(U, \leq)$ is a distributive lattice, we shall write $x_1 \oplus x_2 \oplus \ldots \oplus x_m$ for $x_1 \vee x_2 \vee \ldots \vee x_m$ if $x_i \wedge x_j = 0$ for $1 \leq i < j \leq m$. Clearly if $x \oplus y = 1$, then x and y are complements.

B39 Exercise. Show that *for any elements* x, y, z *of a distributive lattice,* $(x \oplus y) \oplus z = x \oplus y \oplus z$.

B40 Proposition. Let $(U, \leq)$ *be a distributive lattice. Then the set of complemented elements of* U *forms a complemented distributive sublattice of* $(U, \leq)$.

PROOF. Denote the set of complemented elements of U by W. Surely if $(W, \leq)$ is a sublattice of $(U, \leq)$, then it is distributive and complemented. It suffices, therefore, to show that the meet of any two elements of W is complemented and so belongs to W. (The analogous proof for the join follows by duality.) Specifically, we show that for $x, y \in W$, $(x \wedge y)' = x' \vee y'$. We use B32a:

$$\begin{aligned} (x \wedge y) \wedge (x' \vee y') &= x \wedge [y \wedge (x' \vee y')] \\ &= x \wedge [(y \wedge x') \vee (y \wedge y')] \\ &= x \wedge [(y \wedge x') \vee 0] \\ &= x \wedge y \wedge x' = 0. \end{aligned}$$

We next use B32b:

$$\begin{aligned} (x \wedge y) \vee (x' \vee y') &= [(x \wedge y) \vee x'] \vee y' \\ &= [(x \vee x') \wedge (y \vee x')] \vee y' \\ &= [1 \wedge (y \vee x')] \vee y' \\ &= y \vee x' \vee y' = 1 \end{aligned}$$

as required. □

An **atom** of a lattice $(U, \leq)$ is a minimal element of the partially-ordered set $(U + \{0\}, \leq)$. In Examples B26, B27, and B28, the atoms are, respectively, the 1-subsets of U, the prime divisors of n, and the 1-dimensional subspaces of $\mathscr{V}$. Note that if a and b are distinct atoms of $(U, \leq)$, then $a \wedge b = 0$, $a < a \vee b$, and $b < a \vee b$.

B41 Proposition. *Let $(U, \leq)$ be a complemented distributive lattice, and let A be the set of atoms of $(U, \leq)$. Let $f: \mathscr{P}(A) \to U$ be given by*

$$f(B) = \bigoplus_{a \in B} a \quad \textit{for each } B \in \mathscr{P}(A),$$

where the join of an empty collection is understood to be 0. Then f is a bijection.

Proof. Since the meet of any two atoms is 0, f is a well-defined function. To prove that f is surjective, we first prove that 1 is the join of all the atoms in U. For suppose that $\bigoplus_{a \in A} a = x$ for some $x < 1$. Then $a \leq x$ for all $a \in A$. Since $(U, \leq)$ is complemented, x' exists, and $x' > 0$. Hence the set $\{z \in U: z \leq x'\}$ contains at least one atom a. Thus $a \leq x'$, and since $a \leq x$, we have $a \leq x \wedge x' = 0$, which is absurd. Hence $\bigoplus_{a \in A} a = 1$.

Now let $x \in U$ and consider the set $W = \{z \in U: z \leq x\}$. By Exercise B37, $(W, \leq)$ is a complemented distributive sublattice of $(U, \leq)$. Moreover, if B is the set of atoms of W, then $B \subseteq A$. Applying the argument of the preceding paragraph to $(W, \leq)$, we obtain $x = \bigoplus_{a \in B} a = f(B)$.

To prove that f is injective, let B and C be distinct subsets of A such that $f(B) = f(C)$. We may pick $z \in B + C$; say $z \in B$ and $z \notin C$. Then

$$z \wedge \bigoplus_{a \in B} a = \bigvee_{a \in B} (z \wedge a) = z$$

while

$$z \wedge \bigoplus_{a \in C} a = \bigvee_{a \in C} (z \wedge a) = 0. \qquad \square$$

B42 Corollary. *The function $f: (\mathscr{P}(A), \subseteq) \to (U, \leq)$ of the above proposition is a lattice-isomorphism.*

Proof. It has already been established that f is a bijection. We need only verify that f is an isomorphism of partially-ordered sets.

Let $B, C \in \mathscr{P}(A)$ such that $C \subseteq B$. Clearly $\bigoplus_{a \in C} a \leq \bigoplus_{a \in B} a$.

Conversely, let $x, y \in U$ such that $y \leq x$. Then $y = \bigoplus_{c \in C} c$ and $x = \bigoplus_{b \in B} b$ for some subsets $C, B \in \mathscr{P}(A)$. Since $y \wedge x = y$, we have by "distributivity", $\bigvee_{c \in C, b \in B} (c \wedge b) = (\bigvee_{c \in C} c) \wedge (\bigvee_{b \in B} b) = \bigoplus_{c \in C} c$. Each term $c \wedge b$ clearly equals either c or 0. Specifically, $\bigvee_{c \in C, b \in B} (c \wedge b) = \bigoplus_{a \in C \cap B} a$. Thus $f(C \cap B) = f(C)$, and since f is injective, $C \cap B = C$. Hence $C \subseteq B$. $\square$

B43 Corollary. *A lattice is Boolean if and only if it is complemented and distributive.*

IIC Connectedness and Components

In the previous section we considered minimal, nonzero elements of a lattice ("atoms"); in this section we begin by considering the collection $\mathscr{M}(\mathscr{A})$ of minimal, nonempty subsets belonging to a collection $\mathscr{A}$ of sets. Like the set of atoms of a lattice, $\mathscr{M}(\mathscr{A})$ is an incommensurable collection. These subsets are called the **elementary sets** in $\mathscr{A}$. For example, $\mathscr{M}(\mathscr{P}(U)) = \mathscr{P}_1(U)$ and $\mathscr{M}(\mathscr{E}(U)) = \mathscr{P}_2(U)$. It holds not only in these two examples, but in general, that if $\mathscr{A}$ is a subspace of $\mathscr{P}(U)$, then $\mathscr{M}(\mathscr{A})$ spans $\mathscr{A}$. We shall look to $\mathscr{M}(\mathscr{A})$ to yield further properties about $\mathscr{A}$. Throughout this section $\mathscr{A}$ will denote a subspace of $(\mathscr{P}(U), +)$.

C1 Lemma. *Every subset in $\mathscr{A}$ is the sum of pairwise-disjoint elementary subsets in $\mathscr{A}$.*

PROOF. Let $S \in \mathscr{A}$. We proceed by induction on $|S|$. If $S = \varnothing$, then S is the sum over the empty collection. Let n be a positive integer, and assume the conclusion holds for T whenever $T \in \mathscr{A}$ and $|T| < n$. Now assume $|S| = n$. By Proposition B11, there exists an elementary set $M \subseteq S$. Thus $|S + M| = |S| - |M| < n$, and by the induction hypothesis, there exist pairwise-disjoint subsets $M_1, \ldots, M_k \in \mathscr{M}(\mathscr{A})$ such that $S + M = M_1 + \ldots + M_k$. Hence $S = M_1 + \ldots + M_k + M$ is the required sum. □

The incommensurability of $\mathscr{M}(\mathscr{A})$ plays an important role in proving the next result.

C2 Lemma. *Let $M_1, M_2 \in \mathscr{M}(\mathscr{A})$ such that $M_1 \cap M_2 \neq \varnothing$. Given $x_1 \in M_1$ and $x_2 \in M_2$, there exists $M \in \mathscr{M}(\mathscr{A})$ such that $\{x_1, x_2\} \subseteq M$.*

PROOF. We proceed by induction on $|M_1 \cup M_2|$. If $|M_1 \cup M_2| \leq 2$, the result is obviously true. Let $m > 2$ be given and suppose the lemma holds whenever $|M_1 \cup M_2| < m$.

Suppose $|M_1 \cup M_2| = m$, and let x_1, x_2 be given. Clearly if $x_i \in M_j$ for some $i \neq j$, there remains nothing to prove. We suppose therefore that $x_1, x_2 \in M_1 + M_2$.

By Lemma C1, $M_1 + M_2 = N_1 + \ldots + N_k$ for some pairwise-disjoint elementary subsets $N_1, \ldots, N_k$. If one of these sets N_i contains $\{x_1, x_2\}$, then set $M = N_i$. Hence we may assume without loss of generality that $x_1 \in N_1$ and $x_2 \in N_2$. Because $\mathscr{M}(\mathscr{A})$ is incommensurable, $[M_i + (M_1 \cap M_2)] \cap N_j \neq \varnothing$ for all $i = 1, 2$; $j = 1, \ldots, k$. Hence $N_1 \cup M_2 \subset M_1 \cup M_2$. We may therefore apply the induction hypothesis to $N_1 \cup M_2$, since $x_1 \in N_1$, $x_2 \in M_2$, and $N_1 \cap M_2 \neq \varnothing$. Hence $\{x_1, x_2\} \subseteq M$ for some set $M \in \mathscr{M}(\mathscr{A})$. □

C3 Lemma. *If $\mathscr{A} = \bigoplus_{i=1}^{k} \mathscr{B}_i$ and $\mathscr{B}_i \neq \{\varnothing\}$ for all i, then*

$$\{\mathscr{M}(\mathscr{B}_i)\colon i = 1, \ldots, k\} \in \mathbb{P}_k(\mathscr{M}(\mathscr{A})).$$

PROOF. We give a proof when $k = 2$; the general case then follows easily by induction. That $\mathscr{M}(\mathscr{B}_1) \cap \mathscr{M}(\mathscr{B}_2) = \varnothing$, is immediate. Hence we wish to prove $\mathscr{M}(\mathscr{A}) = \mathscr{M}(\mathscr{B}_1) \cup \mathscr{M}(\mathscr{B}_2)$. Let $B_i = \text{Fnd}(\mathscr{B}_i)$ for $i = 1, 2$. Let $A \in \mathscr{M}(\mathscr{A})$, and suppose that $A \cap B_i \neq \varnothing$ for some i. By A12a, $A \cap B_i \in \pi_{B_i}[\mathscr{A}] = \mathscr{B}_i \subseteq \mathscr{A}$. By the minimality of A, we must have $A \cap B_i = A$, and so $A \subseteq B_i$. Hence $A \in \mathscr{M}(\mathscr{B}_i)$. Conversely, let $A \in \mathscr{M}(\mathscr{B}_i)$ for some i. Then $A \in \mathscr{A}$ since $\mathscr{B}_i \subseteq \mathscr{A}$. If $A \notin \mathscr{M}(\mathscr{A})$, then $\varnothing \subset C \subset A$ for some $C \in \mathscr{A}$. But then $C \in \mathscr{A} \cap \mathscr{P}(B_i) \subseteq \pi_{B_i}[\mathscr{A}] = \mathscr{B}_i$ by A11 and A12a, contrary to the minimality of A. □

We are now ready to define some basic concepts of this chapter.

A subspace $\mathscr{B}$ is a **direct summand** of $\mathscr{A}$ if $\mathscr{A} = \mathscr{B} \oplus \mathscr{C}$ for some subspace $\mathscr{C}$. Clearly $\{\varnothing\}$ and $\mathscr{A}$ are always direct summands of $\mathscr{A}$. The subspace $\mathscr{A}$ is said to be **connected** if these are the only direct summands of $\mathscr{A}$. Finally, we define a **component** of $\mathscr{A}$ to be a connected direct summand of $\mathscr{A}$ other than $\{\varnothing\}$.

C4 *Example.* If $\varnothing \subset S \subset U$, then $\mathscr{P}(U) = \mathscr{P}(S) \oplus \mathscr{P}(U + S)$. Hence $\mathscr{P}(S)$ and $\mathscr{P}(U + S)$ are direct summands of $\mathscr{P}(U)$. It follows that $\mathscr{P}(U)$ is connected if and only if $|U| \leq 1$. Therefore the components of $\mathscr{P}(U)$ are all the subspaces of the form $\mathscr{P}(\{x\})$ where $x \in U$. It is true not only in this example, but in general, that the foundations of disjoint direct summands are disjoint.

C5 Lemma. *If* $\mathscr{A} = \bigoplus_{i=1}^{k} \mathscr{B}_i$ *and* $\mathscr{B}_i \neq \{\varnothing\}$ *for all* i, *then*

$$\{\text{Fnd}(\mathscr{B}_i): i = 1, \ldots, k\} \in \mathbb{P}_k(\text{Fnd}(\mathscr{A})).$$

PROOF. For $1 \leq i < j \leq k$, we have $\mathscr{B}_i \cap \mathscr{B}_j = \{\varnothing\}$, and so $\text{Fnd}(\mathscr{B}_i) \cap \text{Fnd}(\mathscr{B}_j) = \varnothing$. Since $\mathscr{B}_i \neq \{\varnothing\}$, $\text{Fnd}(\mathscr{B}_i) \neq \varnothing$. Finally let $x \in \text{Fnd}(\mathscr{A})$. Then $x \in A$ for some $A \in \mathscr{M}(\mathscr{A})$. By Lemma C3, $A \in \mathscr{B}_i$ for some i, whence $x \in \text{Fnd}(\mathscr{B}_i)$ as required. □

Given two systems $\Lambda_i = (V_i, f_i, E_i)$ for $i = 1, 2$ where $V_1 \cap V_2 = \varnothing = E_1 \cap E_2$, the system $\Lambda = (V_1 \cup V_2, f, E_1 \cup E_2)$ where $f(e) = f_i(e)$ for $e \in E_i$ is called the **direct sum** of Λ_1 and Λ_2 and is denoted by $\Lambda_1 \oplus \Lambda_2$. Since the operation $\oplus$ on systems is commutative and associative, this definition may again be extended to any finite number of systems $\Lambda_i = (V_i, f_i, E_i)$, $i = 1, \ldots, k$, as long as $V_i \cap V_j = E_i \cap E_j = \varnothing$ for all $i \neq j$. The resulting system Λ is called the **direct sum** of $\Lambda_1, \ldots, \Lambda_k$ and is denoted by $\bigoplus_{i=1}^{k} \Lambda_i$. Note that each Λ_i is a subsystem of $\bigoplus_{i=1}^{k} \Lambda_i$. Each Λ_i is called a **direct summand** of Λ. The system $(\varnothing, f, \varnothing)$ is called the **trivial system.** Clearly Λ itself and the trivial system are always direct summands of Λ. Hence we say that Λ is **connected** if these are its only direct summands. Finally, a connected nontrivial direct summand of Λ is called a **component** of Λ.

Direct summands for subspaces are closely related to those of systems, as we shall now see. If $\mathscr{A}$ is a subspace of $\mathscr{P}(V)$, we define $\Lambda(\mathscr{A})$ to be $(\mathrm{Fnd}(\mathscr{A}), \mathscr{M}(\mathscr{A}))$.

C6 Proposition. *Let $\mathscr{A}$ be a subspace of $\mathscr{P}(V)$.*

(a) *If $\mathscr{A} = \bigoplus_{i=1}^{k} \mathscr{B}_i$, then $\Lambda(\mathscr{A}) = \bigoplus_{i=1}^{k} \Lambda(\mathscr{B}_i)$.*

(b) *If $\Lambda(\mathscr{A}) = \bigoplus_{i=1}^{k} \Lambda_i$ where $\Lambda_i = (V_i, \mathscr{E}_i)$, then $\Lambda_i = \Lambda(\mathscr{A} \cap \mathscr{P}(V_i))$ for $i = 1, \ldots, k$. Furthermore $\mathscr{A} = \bigoplus_{i=1}^{k} (\mathscr{A} \cap \mathscr{P}(V_i))$.*

PROOF. (a) This follows at once from Lemmas C3 and C5 and the definitions.

(b) Let $\mathscr{B}_i = \mathscr{A} \cap \mathscr{P}(V_i)$ for $i = 1, \ldots, k$. Clearly $\mathrm{Fnd}(\mathscr{B}_i) \subseteq V_i$. Hence $\bigoplus_{i=1}^{k} \mathscr{B}_i$ is a well-defined subspace of $\mathscr{A}$. Now let $A \in \mathscr{A}$. By Lemma C1, $A = \sum_{i=1}^{k} B_i$, where B_i is a sum of sets in $\mathscr{E}_i \cap \mathscr{M}(\mathscr{A})$. Since for $i = 1, \ldots, k$, $B_i \in \mathscr{A}$ and $B_i \subseteq V_i$, we have $B_i \in \mathscr{B}_i$. Thus $A \in \bigoplus_{i=1}^{k} \mathscr{B}_i$, and we conclude that $\mathscr{A} = \bigoplus_{i=1}^{k} \mathscr{B}_i$. By part (a) above, $\Lambda(\mathscr{A}) = \bigoplus_{i=1}^{k} \Lambda(\mathscr{B}_i)$. Since $\mathrm{Fnd}(\mathscr{B}_i) \subseteq V_i$ for $i = 1, \ldots, k$, we must have $V_i = \mathrm{Fnd}(\mathscr{B}_i)$ and $\mathscr{E}_i = \mathscr{M}(\mathscr{B}_i)$. □

C7 Proposition. *Let $\Lambda_1, \ldots, \Lambda_k$ be the components of the system Λ. A subsystem Ω of Λ is a direct summand of Λ if and only if $\Omega = \bigoplus_{i \in S} \Lambda_i$ for some subset $S \subseteq \{1, \ldots, k\}$. In particular, $\Lambda = \bigoplus_{i=1}^{k} \Lambda_i$.*

PROOF. Let $\mathsf{D}(\Lambda)$ denote the collection of direct summands of Λ. This is precisely the set of complemented elements of the lattice $(\mathsf{S}(\Lambda), \leq)$ presented in Example B29. By B34b and B40, $(\mathsf{D}(\Lambda), \leq)$ is a distributive complemented sublattice of $(\mathsf{S}(\Lambda), \leq)$. The atoms of $(\mathsf{D}(\Lambda), \leq)$ are precisely the components of Λ, and the result follows from Corollary B42. □

This result combined with C6 yields:

C8 Corollary. *Let $\mathscr{C}_1, \ldots, \mathscr{C}_k$ be the components of $\mathscr{A}$. A subspace $\mathscr{B}$ of $\mathscr{A}$ is a direct summand of $\mathscr{A}$ if and only if $\mathscr{B} = \bigoplus_{i \in S} \mathscr{C}_i$ for some subset $S \subseteq \{1, \ldots, k\}$. In particular $\mathscr{A} = \bigoplus_{i=1}^{k} \mathscr{C}_i$.*

It follows from this Corollary and Lemma C5 that $\{\mathrm{Fnd}(\mathscr{C}) : \mathscr{C}$ is a component of $\mathscr{A}\}$ is a partition of $\mathrm{Fnd}(\mathscr{A})$. This partition is called the **component partition** of the subspace $\mathscr{A}$.

C9 *Example.* Let $U = \{s, t, u, v, w, x, y, z\}$ and let $\mathscr{A}$ be spanned by $S_1 = \{s, t, u, v\}$, $S_2 = \{s, t, w\}$, $S_3 = \{u, v, w, x, y, z\}$. Then $\mathscr{A}$ consists of all possible sums of these three sets. The remaining sets in $\mathscr{A}$ are: $\varnothing$, $S_1 + S_2 = \{u, v, w\}$, $S_1 + S_3 = \{s, t, w, x, y, z\}$, $S_2 + S_3 = \{s, t, u, v, x, y, z\}$, and $S_1 + S_2 + S_3 = \{x, y, z\}$. Since the sets in the list are all distinct, $|\mathscr{A}| = 2^3$, and so $\dim(\mathscr{A}) = 3$. Let $\mathscr{Q} = \{Q_1, Q_2\}$ where $Q_1 = \{s, t, u, v, w\}$ and $Q_2 = \{x, y, z\}$. Then $\mathscr{P}(Q_1) \cap \mathscr{A} = \{\varnothing, S_1, S_2, S_1 + S_2\}$ has dimension 2 and $\mathscr{P}(Q_2) \cap \mathscr{A} = \{\varnothing, S_1 + S_2 + S_3\}$ has dimension 1. Each of these two subspaces is connected.

Also, $\mathscr{A} = (\mathscr{P}(Q_1) \cap \mathscr{A}) \oplus (\mathscr{P}(Q_2) \cap \mathscr{A})$. Thus $\mathscr{P}(Q_1) \cap \mathscr{A}$ and $\mathscr{P}(Q_2) \cap \mathscr{A}$ are the components of $\mathscr{A}$, and $\mathscr{Q}$ is the component partition of $\mathscr{A}$.

C10 Exercise. Prove:
(**a**) The component partition of $\mathscr{A}$ is the coarse partition of $\mathscr{M}(\mathscr{A})$.
(**b**) *If* $\{x\} \in \mathscr{A}$ *for some* $x \in \mathrm{Fnd}(\mathscr{A})$, *then* $\mathscr{P}(\{x\})$ *is a component of* $\mathscr{A}$.
(**c**) *If* $x \in U + \mathrm{Fnd}(\mathscr{A})$, *then* $\mathscr{P}(\{x\})$ *is a component of* $\mathscr{A}^\perp$.

C11 Lemma. *Let* $|U| \geq 2$. $\mathrm{Fnd}(\mathscr{A}) = U$ *and* $\mathscr{A}$ *is connected if and only if* $\mathrm{Fnd}(\mathscr{A}^\perp) = U$ *and* $\mathscr{A}^\perp$ *is connected.*

Proof. Clearly by duality it suffices to prove this lemma in just one direction. Suppose $\mathrm{Fnd}(\mathscr{A}) = U$ and that $\mathscr{A}$ is connected, and let $x \in U + \mathrm{Fnd}(\mathscr{A}^\perp)$. By Exercise C10c, $\mathscr{P}(\{x\})$ is a component of $\mathscr{A}$. Since $|U| \geq 2$, $\mathscr{A}$ is not connected, contrary to assumption. Hence $\mathrm{Fnd}(\mathscr{A}^\perp) = U$.

Suppose $\mathscr{A}^\perp = \mathscr{B}_1 \oplus \mathscr{B}_2$. Let $\mathscr{A}_i$ be the orthogonal complement of $\mathscr{B}_i$ in $\mathscr{P}(\mathrm{Fnd}(\mathscr{B}_i))$ for $i = 1, 2$. For all $A \in \mathscr{A}_i$ and $A' \in \mathscr{A}^\perp$,

$$\begin{aligned}|A \cap A'| &= |A \cap (A' \cap \mathrm{Fnd}(\mathscr{B}_1))| + |A \cap (A' \cap \mathrm{Fnd}(\mathscr{B}_2))| \\ &= |A \cap (A' \cap \mathrm{Fnd}(\mathscr{B}_i))| \equiv 0 \pmod 2).\end{aligned}$$

Hence $A \in \mathscr{A}$, whence $\mathscr{A}_i \leq \mathscr{A}$. Since $\mathrm{Fnd}(\mathscr{A}_1) \cap \mathrm{Fnd}(\mathscr{A}_2) = \varnothing$, $\mathscr{A}_1 \oplus \mathscr{A}_2 \leq \mathscr{A}$. To prove the reverse inequality, we use A6:

$$\dim(\mathscr{A}_i) = |\mathrm{Fnd}(\mathscr{B}_i)| - \dim(\mathscr{B}_i) \quad i = 1, 2.$$

Since $\mathrm{Fnd}(\mathscr{A}^\perp) = U$, we have $\dim(\mathscr{A}_1 \oplus \mathscr{A}_2) = |U| - \dim(\mathscr{B}_1 \oplus \mathscr{B}_2) = |U| - \dim(\mathscr{A}^\perp) = \dim(\mathscr{A})$, and so $\mathscr{A}_1 \oplus \mathscr{A}_2 = \mathscr{A}$. If $\mathscr{A}_i = \{\varnothing\}$, then for $j \in \{1, 2\} + \{i\}$, $U = \mathrm{Fnd}(\mathscr{A}) = \mathrm{Fnd}(\mathscr{A}_j) \subseteq \mathrm{Fnd}(\mathscr{B}_j) = U + \mathrm{Fnd}(\mathscr{B}_i)$, and so $\mathrm{Fnd}(\mathscr{B}_i) = \varnothing$, whence $\mathscr{B}_i = \{\varnothing\}$. This proves that if $\mathscr{B}_1$ and $\mathscr{B}_2$ are nontrivial, then $\mathscr{A}$ is not connected, which completes the proof. □

C12 Proposition. *Let* $\mathscr{C}_1, \ldots, \mathscr{C}_k$ *be the components of* $\mathscr{A}$ *and let* $U + \mathrm{Fnd}(\mathscr{A}) = \{y_1, \ldots, y_p\}$. *Let* $\mathscr{D}_i$ *be the orthogonal complement of* $\mathscr{C}_i$ *in* $\mathscr{P}(\mathrm{Fnd}(\mathscr{C}_i))$, $i = 1, \ldots, k$. *Then the components of* $\mathscr{A}^\perp$ *are* $\mathscr{P}(\{y_i\})$ *for* $i = 1, \ldots, p$ *and the nontrivial spaces among* $\mathscr{D}_i$ *for* $i = 1, \ldots, k$. *In particular, if* $\mathrm{Fnd}(\mathscr{A}) = \mathrm{Fnd}(\mathscr{A}^\perp)$, *then* $\mathscr{A}$ *is connected if and only if* $\mathscr{A}^\perp$ *is connected.*

Proof. If $A \in \mathscr{A}$ and $D \in \mathscr{D}_i$, then $|A \cap D| = |\pi_{\mathrm{Fnd}(\mathscr{C}_i)}(A) \cap D| \equiv 0 \pmod 2$, since the projection of A belongs to $\mathscr{C}_i$ ($i = 1, \ldots, k$). Hence $D \in \mathscr{A}^\perp$, and so $\mathscr{D}_i \leq \mathscr{A}^\perp$ for $i = 1, \ldots, k$. By C10c, $\mathscr{P}(\{y_i\})$ is a component of $\mathscr{A}^\perp$ and so $\mathscr{B} = [\bigoplus_{i=1}^k \mathscr{D}_i] \oplus [\bigoplus_{i=1}^k \mathscr{P}(\{y_i\})] \leq \mathscr{A}^\perp$. Since $\dim(\mathscr{C}_i) = |\mathrm{Fnd}(\mathscr{C}_i)| - \dim(\mathscr{D}_i)$ for $i = 1, \ldots, k$, we have

$$\begin{aligned}\dim(\mathscr{A}) = \dim\left(\bigoplus_{i=1}^{k} \mathscr{C}_i\right) &= |U| - p - \dim\left(\bigoplus_{i=1}^{p} \mathscr{D}_i\right) \\ &= |U| - \dim(\mathscr{B}) \geq |U| - \dim(\mathscr{A}^\perp),\end{aligned}$$

and so $\mathscr{B} = \mathscr{A}^{\perp}$. Hence each subspace $\mathscr{D}_i$ is a direct summand of $\mathscr{A}^{\perp}$, and by the lemma, it is a component whenever it is nontrivial. The rest is immediate. □

C13 Exercise. Prove

(a) *$\mathscr{E}(U)$ is connected for any set U.*

(b) *If $|U| \geq 3$, then $\mathscr{E}(U)$ is the only connected ($|U| - 1$)-dimensional subspace of $\mathscr{P}(U)$.*

Continuing our notation, let M_i be an incidence matrix for the direct summand Λ_i for $i = 1, \ldots, k$. Then the matrix

$$M = \begin{bmatrix} M_1 & & & 0 \\ & M_2 & & \\ & & \ddots & \\ 0 & & & M_k \end{bmatrix}$$

is clearly an incidence matrix of $\bigoplus_{i=1}^{k} \Lambda_i$, provided none of the systems Λ_i has an empty vertex set or empty block set. Its transpose M^* has the same form except that M_i^* replaces the submatrix M_i for $i = 1, \ldots, k$. From this argument the following is clear:

C14 Proposition. *Let $\Lambda_1, \ldots, \Lambda_k$ be hypergraphs. If their direct sum is defined, then the direct sum of their transposes is defined, and*

$$\left(\bigoplus_{i=1}^{k} \Lambda_i\right)^* = \bigoplus_{i=1}^{k} \Lambda_i^*.$$

Since a system is trivial if and only if its transpose is trivial, we have

C15 Corollary. *A system is connected if and only if its transpose is connected.*

C16 Exercise. Show that *if $\Omega = (W, f_{|F}, F)$ is a direct summand of $\Lambda = (V, f, E)$ then $\Omega = \Lambda_W = \Lambda_F$ whenever $W \neq \varnothing \neq F$.*

C17 Exercise. *Let $\Lambda = (V, f, E)$ be a system such that $f(e) = \varnothing$ for some $e \in E$.* Prove *Λ is connected if and only if $V = \varnothing$ and $|E| = 1$.*

Let $\Lambda_1, \ldots, \Lambda_k$ be the components of the system $\Lambda = (V, f, E)$ and write $\Lambda_i = (V_i, f_i, E_i)$ for $i = 1, \ldots, k$. By C7, $\Lambda = \bigoplus_{i=1}^{k} \Lambda_i$, and so if $\Lambda_{i_1}, \ldots, \Lambda_{i_m}$ are the components with nonempty vertex sets, we have $\{V_{i_1}, \ldots, V_{i_m}\} \in \mathbb{P}(V)$. This partition is called the **component partition** of Λ.

C18 *Exercise.* Continuing this notation, find the component partition of Λ^*.

Let $\Lambda = (V, f, E)$ be a system, and let $s, t \in V \cup E$. An *st*-**path** is a sequence $s = s_0, s_1, \ldots, s_n = t$ of elements of $V \cup E$ such that:

(a) any three consecutive terms of the sequence are distinct;

(b) $\{s_{j-1}, s_j\}$ is an edge of the bipartite graph of Λ for $j = 1, \ldots, n$.

For example, if $x, y \in V$, an *xy*-path is an alternating sequence $x = x_0, e_1, x_2, e_3, \ldots, e_{n-1}, x_n = y$ of vertices and blocks such that $\{x_i, x_{i+2}\} \subseteq f(e_{i+1})$ for $i = 0, 2, \ldots, n-2$. Note that a single vertex or block is itself a path; such a path is said to be **trivial.** A path is said to be **elementary** if all of its terms are distinct.

C19 Exercise. (a) *Let $\Lambda = (V, f, E)$ be a system and let $s, t \in V \cup E$.* Show that *if Λ admits an st-path, then it admits an elementary st-path.*

(b) *Define the relation $\sim$ on $V \cup E$ by: $s \sim t$ if and only if there exists an st-path in Λ.* Show that *$\sim$ is an equivalence relation.*

C20 Proposition. *The component partition of $\Lambda = (V, f, E)$ is the partition of the equivalence relation $\sim$ of* C19b *restricted to V.*

Proof. Assume $\Lambda_{V_1}, \ldots, \Lambda_{V_k}$ are the components of $\Lambda = (V, f, E)$ and let $F_i = \{e \in E: f(e) \subseteq V_i\}$. Let $s \in V_i$ and $t \in V_j$ for some $i \neq j$. Suppose $s = s_0, s_1, \ldots, s_n = t$ is an *st*-path, and let s_k be the last term in the path in $V_i \cup F_i$. If s_k is a vertex, then $s_k \in f(s_{k+1})$ where $s_k \in V_i$ and $s_{k+1} \notin F_i$. Since $\Lambda_{V_1}, \ldots, \Lambda_{V_k}$ are the components of Λ, $s_{k+1} \in F_q$ for some $q \neq i$ and $f(s_{k+1}) \subseteq V_q$, i.e., $f(s_{k+1}) \cap V_i = \varnothing$. This is clearly impossible. If s_k is a block, then $s_{k+1} \in f(s_k)$, but $f(s_k) \subseteq V_i$ while $s_k \notin F_i$ which is impossible. We conclude that there exists no *st*-path. Hence the partition defined by $\sim$ refines the component partition.

Now suppose $s, t \in V_i$ for some i. Let $S = \{r \in V_i \cup F_i$: there is an *sr*-path$\}$. Observe that if $r \in S \cap F_i$, then $f(r) \subseteq S$ and hence $f(r) \subseteq S \cap V_i$. On the other hand if $r \in F_i + (S \cap F_i)$, then $f(r) \cap S = \varnothing$, i.e., $f(r) \in V_i + (S \cap V_i)$. We conclude that $\Omega_1 = (S \cap V_i, f_{|S \cap F_i}, S \cap F_i)$ and $\Omega_2 = (V_i + (S \cap V_i), f_{|F_i + (S \cap F_i)}, F_i + (S \cap F_i))$ are both well-defined subsystems of Λ. Furthermore, $\Omega_1 \wedge \Omega_2 = \{\varnothing, f, \varnothing\}$ and $\Omega_1 \vee \Omega_2 = \Lambda_{V_i}$, i.e., $\Omega_1 \oplus \Omega_2 = \Lambda_{V_i}$. However Λ_{V_i}, being a component, is connected. Hence Ω_1 or Ω_2 is trivial. Since $s \in S$, Ω_1 is not trivial. Thus $t \in V_i \subseteq S$, and $s \sim t$. □

C21 Corollary. *Let $\Lambda = (V, f, E)$. The following three conditions are equivalent:*

(a) *Λ is connected.*

(b) *$f(e) \neq \varnothing$ for all $e \in F$, and for every $s, t \in V$ there is an st-path.*

(c) *For every $s, t \in V \cup E$ there is an st-path.*

C22 Proposition. **(a)** *A necessary and sufficient condition for a subspace* $\mathscr{A}$ *to be connected is that given* $x_1, x_2 \in \mathrm{Fnd}(\mathscr{A})$, *there exists* $M \in \mathscr{M}(\mathscr{A})$ *such that* $\{x_1, x_2\} \subseteq M$.

(b) *The relation* $\sim$ *on* $\mathrm{Fnd}(\mathscr{A})$ *given by*

$$x \sim y \Leftrightarrow \{x, y\} \subseteq A \quad \textit{for some } A \in \mathscr{M}(\mathscr{A})$$

is an equivalence relation. The equivalence classes are precisely the cells of the component partition.

PROOF. The sufficiency of the condition in (a) is immediate.

By repeated application of Lemma C2, we see that there exists an x_1x_2-path in $\Lambda(\mathscr{A})$ if and only if $\{x_1, x_2\} \subseteq M$ for some $M \in \mathscr{M}(\mathscr{A})$, whence the necessity follows. Part (b) is merely a restatement of this principle. □

Let Γ be the bipartite graph of the system Λ. The terms of a path in Λ are precisely the vertices of a path in the system Γ, and conversely. Consequently by C17 we have:

C23 Proposition. *Let* Γ *be the bipartite graph of the system* Λ. Γ *is connected if and only if* Λ *is connected.*

C24 *Exercise.* Prove that if Γ_i is the bipartite graph of the system Λ_i for $i = 0, 1, \ldots, n$, then $\Lambda_1, \ldots, \Lambda_n$ are the components of Λ_0 if and only if $\Gamma_1, \ldots, \Gamma_n$ are the components of Γ_0.

C25 *Exercise.* Show that a bipartite graph is connected if and only if it is bipartite with respect to a unique partition.

C26 *Exercise.* Show that a bipartite graph with k components is the bipartite graph of precisely 2^k systems (but of at most 2^k nonisomorphic systems).

C27 *Exercise.* Let Ω be a subsystem of a system Λ. Show that Ω is a component of Λ if and only if Ω^* is a component of Λ^*, thereby extending C14 to all systems.

IID The Spaces of a System

Let $\Lambda = (V, f, E)$ be a system. The function f, when extended by linearity, yields a linear transformation $\bar{f}\colon \mathscr{P}(E) \to \mathscr{P}(V)$ given by

$$\bar{f}(A) = \sum_{e \in A} f(e) \quad \text{for all } A \subseteq E.$$

As a linear transformation, $\bar{f}$ determines two important subspaces. The image of $\bar{f}$, denoted by $\mathscr{Y}(\Lambda)$, is called the **space** of Λ, and the kernel of $\bar{f}$, denoted by $\mathscr{Z}(\Lambda)$, is called the **cycle space** of Λ. The space $\mathscr{Y}(\Lambda)$ is, of course, the subspace of $\mathscr{P}(V)$ spanned by $f[E]$. The space $\mathscr{Z}(\Lambda)$, on the other hand, is a subspace of $\mathscr{P}(E)$. Let $A \subseteq E$. Then

D1 $$A \in \mathscr{Z}(\Lambda) \quad \text{if and only if} \sum_{e \in A} f(e) = \varnothing.$$

The orthogonal complements $\mathscr{Y}^{\perp}(\Lambda)$ and $\mathscr{Z}^{\perp}(\Lambda)$ of $\mathscr{Y}(\Lambda)$ and $\mathscr{Z}(\Lambda)$ are called the **cospace** of Λ and the **cocycle space** of Λ, respectively. An element of $\mathscr{Z}(\Lambda)$ is called a **cycle** of Λ and an element of $\mathscr{Z}^{\perp}(\Lambda)$ is called a **cocycle** of Λ.

D2 *Exercise.* Let $R = \{a, b, c, d, e, f\}$. Let $S = \{a, b\}$, $T = \{a, c\}$, $U = \{b, d\}$, $V = \{c, d\}$, $W = \{b, c\}$, $X = \{d, e\}$, $Y = \{e, f\}$, $Z = \{d, f\}$. Let $\mathscr{E} = \{S, T, U, V, W, X, Y, Z\}$, and let Γ be the graph $(R, \mathscr{E})$. Determine $\mathscr{Y}(\Gamma)$, $\mathscr{Z}(\Gamma)$, $\mathscr{Y}^{\perp}(\Gamma)$, and $\mathscr{Z}^{\perp}(\Gamma)$. Compare these findings with your results in Example C9.

We display some immediate consequences of the definitions of these spaces. Since the dimension of the domain of $\bar{f}$ is $|E|$, we have

D3 $$\dim(\mathscr{Z}(\Lambda)) + \dim(\mathscr{Y}(\Lambda)) = |E|.$$

From A6, we have

D4 $$\dim(\mathscr{Z}(\Lambda)) + \dim(\mathscr{Z}^{\perp}(\Lambda)) = |E|,$$

and

D5 $$\dim(\mathscr{Y}(\Lambda)) + \dim(\mathscr{Y}^{\perp}(\Lambda)) = |V|.$$

Combining D3, D4, and D5, we have

D6 $$\dim(\mathscr{Y}(\Lambda)) = \dim(\mathscr{Z}^{\perp}(\Lambda)) = |E| - \dim(\mathscr{Z}(\Lambda)) = |V| - \dim(\mathscr{Y}^{\perp}(\Lambda)).$$

D7 Proposition. *Let Λ be a system. Then*

$$\mathscr{Y}(\Lambda^*) = \mathscr{Z}^{\perp}(\Lambda) \quad \textit{and} \quad \mathscr{Z}(\Lambda^*) = \mathscr{Y}^{\perp}(\Lambda).$$

Proof. Let $\Lambda = (V, f, E)$ and $A \in \mathscr{Z}(\Lambda)$. We show that A is orthogonal to each element of $\mathscr{Y}(\Lambda^*)$ by showing that A is orthogonal to each element of its spanning set $\{f^*(x): x \in V\}$. By D1, $\sum_{e \in A} f(e) = \varnothing$. That is to say, for each $x \in V$, $x \in f(e)$ for an even number of blocks $e \in A$. Thus $|f^*(x) \cap A| = |\{e \in A: x \in f(e)\}|$ is even, i.e., $A \cdot f^*(x) = 0$. Hence $\mathscr{Z}(\Lambda) \subseteq \mathscr{Y}^{\perp}(\Lambda^*)$, whence $\mathscr{Y}(\Lambda^*) \subseteq \mathscr{Z}^{\perp}(\Lambda)$. Dually we have $\mathscr{Y}(\Lambda) \subseteq \mathscr{Z}^{\perp}(\Lambda^*)$. By these two inclusions and D6, we have:

$$\dim(\mathscr{Y}(\Lambda)) \leq \dim(\mathscr{Z}^{\perp}(\Lambda^*)) = \dim(\mathscr{Y}(\Lambda^*)) \leq \dim(\mathscr{Z}^{\perp}(\Lambda)) = \dim(\mathscr{Y}(\Lambda)).$$

Equality must then hold throughout and in the above inclusions. □

The following example ties together many of these notions.

D8 *Example.* Let $n \geq 3$ be an odd integer, and let $E = \{e_1, \ldots, e_n\}$ and $V = \{x_1, \ldots, x_n\}$ be disjoint n-sets. Let $\Lambda = (V, f, E)$ be the set system where

$f(e_i) = V + \{x_i\}$ for $i = 1, \ldots, n$. Since $f[E] = \mathscr{P}_{n-1}(V)$, $\mathscr{Y}(\Lambda) = \mathscr{E}(V)$ by A3. Hence $\dim(\mathscr{Y}(\Lambda)) = n - 1$, by A1. By D3, $\dim(\mathscr{Z}(\Lambda)) = n - (n - 1) = 1$. Since $x_i \in f(e_j)$ if and only if $i \neq j$, each vertex is incident with an even number (namely, $n - 1$) of the blocks. Hence $\sum_{i=1}^{n} f(e_i) = \varnothing$, and $E \in \mathscr{Z}(\Lambda)$. Since $\dim(\mathscr{Z}(\Lambda)) = 1$, $\mathscr{Z}(\Lambda) = \{\varnothing, E\}$. By Exercise A7, if you were diligent, the cospace $\mathscr{Y}^{\perp}(\Lambda) = \{\varnothing, V\}$ while $\mathscr{Z}^{\perp}(\Lambda) = \mathscr{E}(E)$. (If you were not diligent, you could still obtain these two results. By A8, $V \in \mathscr{Y}^{\perp}(\Lambda)$, and by D5 $\dim(\mathscr{Y}^{\perp}(\Lambda)) = 1$. By A8, $\mathscr{Z}^{\perp}(\Lambda)$ is even, and by D4, $\dim(\mathscr{Z}^{\perp}(\Lambda)) = n - 1$. By C11, $\mathscr{Z}^{\perp}(\Lambda)$ is connected, and the result follows by C13b.) Now consider $\Lambda^* = (E, f^*, V)$. Thus $f^*(x_i) = E + \{e_i\}$ for $i = 1, \ldots, n$. Let $p: E \to V$ be given by $p(e_i) = x_i$ $(i = 1, \ldots, n)$. Clearly (p, p^{-1}) is a system-isomorphism from Λ to Λ^*. By the above discussion with the roles of V and E interchanged, $\mathscr{Y}(\Lambda^*) = \mathscr{E}(E)$ while $\mathscr{Z}(\Lambda^*) = \{\varnothing, V\}$. We have verified Proposition D7 for this example directly.

D9 Proposition. *Let $\Lambda = (V, f, E)$ be a hypergraph, and suppose $\Lambda = \Lambda_1 \oplus \Lambda_2$. Then $\mathscr{Y}(\Lambda) = \mathscr{Y}(\Lambda_1) \oplus \mathscr{Y}(\Lambda_2)$, $\mathscr{Z}(\Lambda) = \mathscr{Z}(\Lambda_1) \oplus \mathscr{Z}(\Lambda_2)$, $\mathscr{Y}^{\perp}(\Lambda) = \mathscr{Y}^{\perp}(\Lambda_1) \oplus \mathscr{Y}^{\perp}(\Lambda_2)$, and $\mathscr{Z}^{\perp}(\Lambda) = \mathscr{Z}^{\perp}(\Lambda_1) \oplus \mathscr{Z}^{\perp}(\Lambda_2)$.*

Proof. Let $\Lambda_i = (V_i, f_i, E_i)$ $(i = 1, 2)$. Since $f(e) = f_i(e) \subseteq V_i$ for all $e \in E_i$, and since $\mathscr{Y}(\Lambda_i)$ is spanned by $\{f(e): e \in E_i\}$, we have that $\operatorname{Fnd}(\mathscr{Y}(\Lambda_i)) \subseteq V_i$. Thus $\mathscr{Y}(\Lambda_1)$ and $\mathscr{Y}(\Lambda_2)$ have disjoint foundations, and $\mathscr{Y}(\Lambda_1) \oplus \mathscr{Y}(\Lambda_2)$ is well-defined. That this equals $\mathscr{Y}(\Lambda)$ follows when we observe that $\{f(e): e \in E_1\} + \{f(e): e \in E_2\}$ spans $\mathscr{Y}(\Lambda)$.

Since $\operatorname{Fnd}(\mathscr{Z}(\Lambda_i)) \subseteq E_i$, $\mathscr{Z}(\Lambda_1) \oplus \mathscr{Z}(\Lambda_2)$ is well-defined. Since $\mathscr{Z}(\Lambda_i)$ is clearly a subspace of $\mathscr{Z}(\Lambda)$,

D10
$$\mathscr{Z}(\Lambda_1) \oplus \mathscr{Z}(\Lambda_2) \subseteq \mathscr{Z}(\Lambda).$$

By D3,

$$\begin{aligned}\dim(\mathscr{Z}(\Lambda_1) \oplus \mathscr{Z}(\Lambda_2)) &= |E_1| - \dim(\mathscr{Y}(\Lambda_1)) + |E_2| - \dim(\mathscr{Y}(\Lambda_2)) \\ &= |E_1 + E_2| - \dim(\mathscr{Y}(\Lambda)) \\ &= \dim(\mathscr{Z}(\Lambda)).\end{aligned}$$

Thus equality holds in D10. The last two parts of the proposition follow from the first two parts, C14, and D7. □

D11 Corollary. *Let Λ be a hypergraph. If any one of the spaces $\mathscr{Y}(\Lambda)$, $\mathscr{Z}(\Lambda)$, $\mathscr{Y}^{\perp}(\Lambda)$, or $\mathscr{Z}^{\perp}(\Lambda)$ is connected, then Λ is connected.*

D12 *Exercise.* (a) Determine $\mathscr{Y}(\Gamma)$ for the graph $\Gamma = (V, \mathscr{P}_2(V))$.

(b) Fix $x \in V$, and for each $\{y, z\} \in \mathscr{P}_2(V + \{x\})$, let

$$\mathscr{S}_{yz} = \{\{x, y\}, \{x, z\}, \{y, z\}\}.$$

Show that $\{\mathscr{S}_{yz}: \{y, z\} \in \mathscr{P}_2(V + \{x\})\}$ is a basis for $\mathscr{Z}(\Gamma)$.

(c) In what remains of your youth, determine a basis for $\mathscr{Z}^{\perp}(\Gamma)$.

D13 Exercise. *Let* $\Omega = (W, g, F)$ *be a subsystem of a system* Λ. Prove:
(**a**) $\mathscr{Z}(\Omega) = \mathscr{P}(F) \cap \mathscr{Z}(\Lambda)$.
(**b**) $\mathscr{Z}^{\perp}(\Omega) = \pi_F[\mathscr{Z}^{\perp}(\Lambda)]$. (*Hint:* see A15.)
(**c**) $\dim(\mathscr{Z}(\Omega)) \leq \dim(\mathscr{Z}(\Lambda))$ *and* $\dim(\mathscr{Z}^{\perp}(\Omega)) \leq \dim(\mathscr{Z}^{\perp}(\Lambda))$.

IIE The Automorphism Groups of Systems

In *I*D we considered isomorphisms between two systems. In the present section we turn our attention to the isomorphisms between a system $\Lambda = (V, f, E)$ and itself. (It will always be assumed that $V \neq \varnothing$ or $E \neq \varnothing$.) Such a system-isomorphism is called an **automorphism** of Λ. The set of automorphisms is precisely:

$$G(\Lambda) = \{(p, q): p \in \Pi(E); q \in \Pi(V); q[f(e)] = f(p(e)) \text{ for all } e \in E\}.$$

Under the operation of componentwise composition

$$(p_2, q_2)(p_1, q_1) = (p_2 p_1, q_2 q_1),$$

it is immediate that $G(\Lambda)$ is a group, and we call $G(\Lambda)$ the **automorphism group** of Λ. Clearly $G(\Lambda) \cong G(\Lambda^*)$.

Note: the isomorphism indicated here as well as the isomorphisms below are to be interpreted in terms of the abstract group structure, and not necessarily of the permutation group structure.

Let

$$G_0(\Lambda) = \{q \in \Pi(V): (p, q) \in G(\Lambda) \text{ for some } p \in \Pi(E)\}$$

and

$$G_1(\Lambda) = \{p \in \Pi(E): (p, q) \in G(\Lambda) \text{ for some } q \in \Pi(V)\}.$$

Under composition $G_0(\Lambda)$ is a subgroup of $\Pi(V)$ and $G_1(\Lambda)$ is a subgroup of $\Pi(E)$. $G_0(\Lambda)$ is the **vertex group** of Λ, and $G_1(\Lambda)$ is the **block group** of Λ. Their elements are, respectively, **vertex-automorphisms** and **block-automorphisms** of Λ. Observe that

E1 $$G_0(\Lambda) = G_1(\Lambda^*); \qquad G_1(\Lambda) = G_0(\Lambda^*).$$

E2 Proposition. *Let* Λ *be a system.* Λ *is a set system if and only if* $G_0(\Lambda) \cong G(\Lambda)$.

Proof. Define $\pi: G(\Lambda) \to G_0(\Lambda)$ by $\pi(p, q) = q$ for all $(p, q) \in G(\Lambda)$. It is immediate that π is an epimorphism. The groups $G(\Lambda)$ and $G_0(\Lambda)$ will be isomorphic if and only if π is injective. If (p, q) is in the kernel of π, then $q = 1_V$. Hence $G(\Lambda) \cong G_0(\Lambda)$ if and only if

$$(p, 1_V) \in \ker(\pi) \Rightarrow p = 1_E.$$

Suppose Λ is a set system, and let $(p, 1_V) \in \ker(\pi)$. By definition of $G(\Lambda)$,

$$\textbf{E3} \qquad f(e) = 1_V[f(e)] = f(p(e)), \quad \text{for all } e \in E.$$

Since f is an injection, $p(e) = e$ for all $e \in E$. Hence $p = 1_E$.

If Λ is not a set system, there exists a set $\{e_1, e_2\} \in \mathscr{P}_2(E)$ such that $f(e_1) = f(e_2)$. We define

$$p(e) = \begin{cases} e_2 & \text{if } e = e_1; \\ e_1 & \text{if } e = e_2; \\ e & \text{if } e \in E + \{e_1, e_2\}. \end{cases}$$

Clearly p satisfies E3, and so $(p, 1_V) \in \ker(\pi)$. But $p \neq 1_E$. □

E4 Corollary. *Let Λ be a system. Λ distinguishes vertices if and only if $G(\Lambda) \cong G_1(\Lambda)$.*

Proof. By E2, Λ^* is a set system if and only if $G(\Lambda^*) = G_0(\Lambda^*)$. The corollary follows from E1 and *I*D5. □

E5 Corollary [w.5]. *Let Λ be a set system. Λ distinguishes vertices if and only if $G_0(\Lambda) \cong G_1(\Lambda)$.*

Proof. Apply E2 and E4. □

It can happen that $G_0(\Lambda) \cong G_1(\Lambda)$, where the groups are isomorphic even as permutation groups while Λ neither is a set system nor distinguishes vertices. Suppose, for example, that $V = \{x_1, \ldots, x_5\}$ and $E = \{e_1, \ldots, e_4\}$, and let

$$f(e_1) = f(e_2) = \{x_1, x_2, x_5\} \quad \text{and} \quad f(e_3) = f(e_4) = \{x_3, x_4, x_5\}.$$

In this case Λ is connected. One straightforwardly verifies that $G_0(\Lambda)$ is generated by the cyclic permutations $q_1 = (x_1, x_3, x_2, x_4)$ and $q_2 = (x_1, x_2)$, which satisfy the relations $q_1^4 = q_2^2 = (q_1q_2)^2 = 1_V$. Thus $G_0(\Lambda)$ is isomorphic to the dihedral group D_4. Similarly, $G_1(\Lambda)$ is generated by $p_1 = (e_1, e_3, e_2, e_4)$ and $p_2 = (e_1, e_2)$, satisfying $p_1^4 = p_2^2 = (p_1p_2)^2 = 1_E$. We see that $G_0(\Lambda)$ and $G_1(\Lambda)$ are isomorphic as abstract groups. In fact, if the vertex x_5 were to be removed, they would be isomorphic permutation groups. However, Λ neither is a set system nor does Λ distinguish vertices. Of course, neither group is isomorphic to $G(\Lambda)$.

E6 *Exercise.* Determine $G(\Lambda)$ in the above example.

E7 *Exercise.* Let Λ be a system and let $\eta: G_0(\Lambda) \to G_1(\Lambda)$ be a function which satisfies

$$(\eta(q), q) \in G(\Lambda) \quad \text{for all } q \in G_0(\Lambda).$$

Such a function clearly exists.

Prove:

(a) If Λ distinguishes vertices, then η is an injection.

(b) If Λ is a set system, then η is uniquely determined and is a (group) epimorphism.

(c) For any Λ, there exists a homomorphism $\eta: G_0(\Lambda) \to G_1(\Lambda)$ such that $(\eta(q), q) \in G(\Lambda)$ for all $q \in G_0(\Lambda)$.

E8 *Exercise.* Show that if Λ is allowed to be infinite, then the two-way implication in Corollary E5 need hold in only one sense [L. Babai, L. Lovász].

Consider an st-path $s = s_0, s_1, \ldots, s_n = t$ in Λ, and let us say, for definiteness, that $s \in V$. One easily verifies that for any $(p, q) \in G(\Lambda)$, the sequence $q(s_0), p(s_1), q(s_2), \ldots, p(s_{2i-1}), q(s_{2i}), \ldots, p(t)$ or $q(t)$ is also a path in Λ. From this together with C20 we immediately deduce:

E9 Proposition. *Let* $\Lambda = \bigoplus_{i=1}^{n} \Lambda_i$ *with components* $\Lambda_i = (V_i, f_i, E_i)$ *for* $i = 1, \ldots, n$. *If* $(p, q) \in G(\Lambda)$, *then to each* $i \in \{1, \ldots, n\}$ *there corresponds some* $j \in \{1, \ldots, n\}$ *such that* $q[V_i] = V_j$ *and* $p[E_i] = E_j$. *Moreover,* $(p_{|E_i}, q_{|V_i})$ *is a system-isomorphism from* Λ_i *to* Λ_j.

If X is a set and if G is a subgroup of $\Pi(X)$, then G is said to be **transitive** on X if for each $x, y \in X$, there exists some $p \in G$ such that $p(x) = y$. The system $\Lambda = (V, f, E)$ is **vertex-transitive** if $G_0(\Lambda)$ is transitive on V; Λ is **block-transitive** if $G_1(\Lambda)$ is transitive on E.

E10 *Exercise.* Let $\Lambda = \bigoplus_{i=1}^{n} \Lambda_i$ with components $\Lambda_i = (V_i, f_i, E_i)$ for $i = 1, \ldots, n$. Show that if Λ block-transitive (respectively, vertex-transitive) and if $E_i \neq \varnothing \neq E_j$ (respectively, $V_i \neq \varnothing \neq V_j$), then Λ_i and Λ_j are isomorphic subsystems.

E11 Proposition. *Let* Λ_1 *be a component of the system* Λ. *If* Λ *is vertex-transitive (respectively, block-transitive), then so is* Λ_1.

PROOF. Suppose Λ is vertex-transitive, and let $\Lambda_1 = (V_1, f_1, E_1)$. If $x, y \in V_1$, then $q(x) = y$ for some $q \in G_0(\Lambda)$, whence $(p, q) \in G(\Lambda)$ for some $p \in G_1(\Lambda)$. By E9, $(p_{|E_1}, q_{|V_1})$ is a system-automorphism of Λ_1. Hence $q_{|V_1} \in G_0(\Lambda_1)$ and maps x onto y.

If Λ is block-transitive, then Λ^* is vertex-transitive, and by what we have just proved, $\Lambda_1{}^*$ is also vertex-transitive. By C14, Λ_1 is block-transitive. □

We may suspect when studying vertex-transitive or block-transitive systems, that one could assume for all practical purposes that they are connected systems. The next proposition bears this out. It is essentially a theorem on permutation groups. As such, it is not truly in the domain of this book. It is therefore stated without proof. Let $\Pi(n)$ denote the permutation group

$\Pi(\{1, \ldots, n\})$. If G and G' are permutation groups, let G wreath G' denote the wreath product of G by G'.

E12 Proposition. *Denote the set of components of a system Λ by*

$$\{\Lambda_{ij}: i = 1, \ldots, n_j; j = 1, \ldots, r\}$$

where Λ_{ij} and $\Lambda_{i'j'}$ are system-isomorphic if and only if $j = j'$. Let $\Lambda_{ij} = (V_{ij}, f_{ij}, E_{ij})$ for each i, j. Let $m_{0j} = |\{i: V_{ij} \neq \varnothing\}|$ and $m_{1j} = |\{i: E_{ij} \neq \varnothing\}|$. Then for $h = 0, 1$,

$$G_h(\Lambda) = \bigoplus_{j=1}^{r} (\Pi(m_{hj}) \text{ wreath } G_h(\Lambda_{1j}))$$

(Here $\bigoplus$ denotes the direct product of permutation groups.)

If $G_0(\Lambda) = \{1_V\}$, then Λ is said to be **asymmetric**. If Λ is not asymmetric but if every element of $G_0(\Lambda)$ is a derangement of V, then Λ is said to be **fixed-point free**.

The next result was first proved by G. Sabidussi [s.3] for the special case of graphs.

E13 Proposition. *Let Λ be fixed-point free. If Λ is not connected, then at most two components Λ_1 and Λ_2 have nonempty vertex sets, and Λ_1 and Λ_2 are each asymmetric.*

PROOF. Let $\Lambda = \bigoplus_{i=1}^{k} \Lambda_i$, where $\Lambda_i = (V_i, f_i, E_i)$ for $i = 1, \ldots, k$ are the components of Λ, and $V_1, V_2 \neq \varnothing$. Suppose some component, say Λ_1, is not asymmetric. Then $G_0(\Lambda_1)$ contains some vertex-automorphism $q_1 \neq 1_{V_1}$. Now define $q \in \Pi(V)$ by

$$q(x) = \begin{cases} q_1(x) & \text{if } x \in V_1; \\ x & \text{if } x \in V + V_1. \end{cases}$$

To show that $q \in G_0(\Lambda)$, define $p \in \Pi(E)$ by letting $p(e) = e$ for $e \in E + E_1$ and $p(e) = p_1(e)$ if $e \in E_1$, where $(p_1, q_1) \in G(\Lambda_1)$. Then $(p, q) \in G(\Lambda)$. Thus $q \neq 1_V$ but q has a fixed-point, contrary to hypothesis. Hence each component of Λ is asymmetric.

Now suppose $k \geq 3$, that $V_1, V_2, V_3 \neq \varnothing$, and that (p_1, q_1) is an isomorphism from Λ_1 to Λ_2. Define p and q by

$$p(e) = \begin{cases} p_1(e) & \text{if } e \in E_1; \\ p_1^{-1}(e) & \text{if } e \in E_2; \\ e & \text{if } e \in E + (E_1 \cup E_2); \end{cases}$$

$$q(x) = \begin{cases} q_1(x) & \text{if } x \in V_1; \\ q_1^{-1}(x) & \text{if } x \in V_2; \\ x & \text{if } x \in V + (V_1 \cup V_2). \end{cases}$$

Clearly $(p, q) \in G(\Lambda)$, and so $q \in G_0(\Lambda)$. But $q \neq 1_V$ while $q(x) = x$ for some $x \in V_3$, contrary to hypothesis. □

E14 *Exercise.* Let Λ be fixed-point free and let Λ_1 and Λ_2 be distinct components with nonempty vertex sets. Show that if $G_0(\Lambda)$ is nontrivial, then (a) Λ_1 and Λ_2 are isomorphic; (b) $|G_0(\Lambda)| = 2$; (c) if Λ is a set system with $|V| \geq 3$, then $|G_1(\Lambda)| = 2$.

CHAPTER III

Multigraphs

Throughout this chapter the symbol $\Gamma = (V, f, E)$ will be used exclusively to denote a multigraph. Multigraphs have been studied far more than any other kind of system. They are the simplest interesting systems, since those with blocksize ≤ 1 have only trivial components. Another and perhaps more important reason for the extensive research in multigraphs is that they are the abstract mathematical objects which lie behind the many diagrams one often draws. Historically, multigraphs were first studied as topological objects.

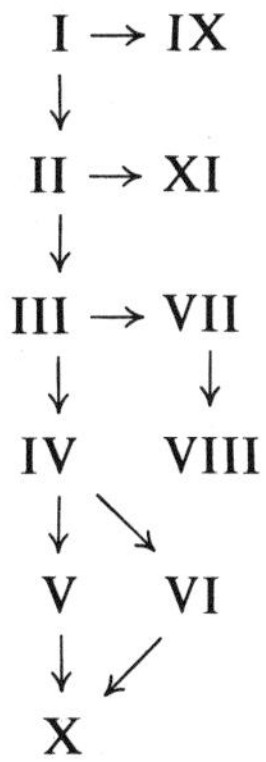

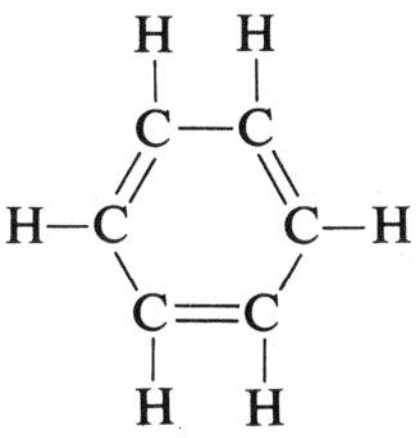

The vertices were points in the plane or 3-space, and the edges were simple arcs joining the vertices. As a result of these "graphic" beginnings, much of the terminology is geometric in spirit, and most of the results can be geometrically motivated. The reader is encouraged to draw pictures and to use them as an aid in following the proofs and doing the exercises.

In this chapter we are mainly interested in the particular results of graph theory which arise as a consequence of considering multigraphs as systems.

Some of our results can be stated and proved for more general systems; others cannot. The reader is encouraged to make appropriate generalizations whenever possible.

IIIA The Spaces of a Multigraph

The number of components of the system Γ will be denoted by $\nu_{-1}(\Gamma)$. We shall let $\nu_0(\Gamma)$ denote the number of vertices of Γ and $\nu_1(\Gamma)$ the number of edges. When there is no risk of confusion, we shall write briefly ν_{-1}, ν_0, and ν_1.

If s is the selection of Γ and x is a vertex of Γ, then $\bar{s}(\{x\})$ is called the **valence** of x and is denoted by $\rho(x)$. If $\rho(x) = 0$, x is an **isolated vertex**; if $\rho(x) = 1$, x is a **pendant vertex**. If for some $k \in \mathbb{N}$, $\rho(x) = k$ for all $x \in V$, then Γ is k-**valent**. Γ is **isovalent** if it is k-valent for some k. (In the literature, the word "regular" is often used in place of "isovalent".)

Let M be an incidence matrix for Γ. Counting the 1's in M by rows, we get $\sum_{x \in V} \rho(x)$. Since there are precisely two 1's in each column, this number is also $2\nu_1(\Gamma)$. Hence

A1
$$\nu_1(\Gamma) = \frac{1}{2} \sum_{x \in V} \rho(x).$$

It follows from A1 that *the number of vertices with odd valence is even.*

In view of A1, it is reasonable to define the **average valence** of Γ to be

$$\rho(\Gamma) = \frac{2\nu_1(\Gamma)}{\nu_0(\Gamma)}.$$

A2 *Exercise.* Prove that if Γ is a graph with $\nu_0(\Gamma) \geq 2$, then there exist at least two vertices with the same valence. Describe all graphs which have exactly two vertices of the same valence.

A3 *Exercise.* Prove that for each $n > 2$ there exists a multigraph Γ with $\nu_0(\Gamma) = n$ which has no two vertices of the same valence. For each n, find a multigraph Γ satisfying these properties for which $\nu_1(\Gamma)$ is as small as possible.

In §*II*C we defined a path for an arbitrary system. For a multigraph we use **path** exclusively for paths which have vertices as both initial and final terms and wherein *all edges are distinct.*

We now show that the space of a multigraph $\Gamma = (V, f, E)$ has a particularly simple structure. From *II*C20 and the definition of "component," we see that $x, y \in V$ lie in the same component of Γ if and only if Γ admits an xy-path. If $x \neq y$ and if $x, e_1, x_1, e_2, \ldots, x_{k-1}, e_k, x_k = y$ is such a path, then $\sum_{i=1}^{k} f(e_i) = \{x, y\}$. Hence if Γ is connected, then $\mathscr{E}(V) \subseteq \mathscr{Y}(\Gamma)$. On the other hand, $\mathscr{Y}(\Gamma)$ is spanned by $\{f(e) : e \in E\}$, and $|f(e)|$ is even for all $e \in E$. Hence $\mathscr{Y}(\Gamma) \subseteq \mathscr{E}(V)$, and equality holds when Γ is connected. Thus by *II*D9, we have

A4 Proposition. *If* Γ *is a multigraph with components* $\Gamma_i = (V_i, f_i, E_i)$, $(i = 1, \ldots, k)$, *then* $\mathscr{Y}(\Gamma) = \bigoplus_{i=1}^{k} \mathscr{E}(V_i)$.

If Λ is a system, a subsystem Γ which is a multigraph (respectively, graph) is called a **submultigraph** (respectively, **subgraph**) of Λ. By a **circuit** we shall mean a nontrivial path $x_0, e_1, x_1, \ldots, e_k, x_k = x_0$ where all of the edges are distinct. Such a circuit is said to be **elementary** if $x_0, x_1, \ldots, x_{k-1}$ are all distinct. The **length** of a path (and hence of a circuit) is the number of edges in the path.

Observe that the vertices and edges of a path (elementary path) in Γ form a submultigraph (subgraph) of Γ. We will often identify the path with this corresponding submultigraph. Note that many paths may correspond to the submultigraph of a given path. To say that one path or circuit is contained in another path or circuit means that the submultigraph of the one is contained in the submultigraph of the other. The graph consisting of a single elementary circuit of length k will be denoted by Δ_k.

A5 Exercise. Prove that *every circuit of length* k *contains* Δ_n *for some* $n \leq k$.

A6 Exercise. Prove:

(a) *If* $\rho(x) \geq 2$ *for every vertex* x *of the multigraph* Γ, *then* Γ *contains a circuit.*

(b) If $\rho(x) \geq 2$ for every vertex x of the multigraph Γ, and if there exists a vertex of Γ in no circuit, then Γ contains two disjoint circuits.

(c) If $\rho(\Gamma) \geq 2$, then Γ contains a circuit.

(d) Let Γ be a connected multigraph. Then Γ contains a circuit if and only if $\rho(\Gamma) \geq 2$.

(e) Let Γ be a connected multigraph. Then Γ contains exactly one circuit if and only if $\rho(\Gamma) = 2$.

A7 Exercise. *Let* (V_i, f_i, E_i) *be a circuit of* Γ *for* $i = 1, 2$. *Suppose* $E_1 \cap E_2 = \varnothing$, *but* $V_1 \cap V_2 \neq \varnothing$. Show that $(V_1 \cup V_2, f, E_1 + E_2)$, *where* $f(e) = f_i(e)$ *for all* $e \in E_i$, *is a circuit.*

A8 Exercise. Prove that *in a multigraph, every circuit has even length if and only if every elementary circuit has even length.*

A9 Proposition. *If* $x_0, e_1, x_1, \ldots, e_k, x_0$ *is a circuit (respectively, an elementary circuit) of* Γ, *then* $\{e_1, \ldots, e_k\}$ *is a cycle (respectively, an elementary cycle) of* Γ. *Conversely, all elementary cycles of* Γ *are the edge sets of elementary circuits.*

Proof. Let $x_0, e_1, \ldots, e_k, x_k = x_0$ be a circuit. Since $f(e_j) = \{x_{j-1}, x_j\}$ for each $j = 1, \ldots, k$, we have $\sum_{j=1}^{k} f(e_j) = \varnothing$. Hence by *II*D1, $\{e_1, \ldots, e_k\}$ is a cycle. Assume the circuit is elementary, and assume that $f(e_i)$ belongs to a subset of $\{f(e_1), \ldots, f(e_k)\}$ whose sum is zero. Then that subset must also include $f(e_{i+1})$ in order to "cancel out" x_i (the indices being read modulo k). We conclude that $\{e_1, \ldots, e_k\}$ is elementary.

Conversely, assume that Z is an elementary cycle of Γ. Since $\sum_{e \in Z} f(e) = \varnothing$, the vertices of Γ_Z all have valence 2 or more. By Exercise A6a, Γ_Z contains a circuit, the edges of which form a cycle $Z' \in \mathscr{Z}(\Gamma_Z) \subseteq \mathscr{Z}(\Gamma)$. Clearly $Z' \subseteq Z$. Since Z is elementary, $Z' = Z$, and Γ_Z is a circuit. □

Note that the "converse" in this proposition is only a "partial converse." It is not in fact true that *every* cycle is the set of edges of some circuit. For example, the edges of two disjoint circuits taken together form such a cycle. The strongest possible "converse" is given in the next proposition.

A10 Proposition. *If $Z \in \mathscr{Z}(\Gamma) + \{\varnothing\}$ and if the submultigraph Γ_Z is connected, then Γ_Z is a circuit.*

Proof. By *II*C1, Z is the sum of pairwise-disjoint elementary cycles. Hence by A9, Γ_Z contains a circuit. If Γ_Z is not itself a circuit, let Δ be a largest circuit in Γ_Z. The set Z'' of the edges of Δ is a cycle by A9. Clearly $Z'' \subset Z$, and Z'' is disjoint from the cycle $Z' = Z + Z''$. Let $\Delta' = \Gamma_{Z'}$. If Δ and Δ' have no common vertex, then $\Gamma_Z = \Delta \oplus \Delta'$, contrary to hypothesis. Thus there exists a vertex x common to Δ and Δ'. By *II*C1, $Z' = Z_1' + \ldots + Z_k'$ where each cycle Z_i' is elementary. Let $\Delta_i' = \Gamma_{Z_i'}$ for $i = 1, \ldots, k$. We may assume without loss of generality that x is a vertex of Δ_1'. Hence by A7, Δ and Δ_1' yield a circuit in Γ_Z larger than Δ. □

With the relationship between cycles and circuits on a firm footing, we may now explore the graph-theoretical significance of the cocycle space. By *II*D7, the cocycle space $\mathscr{Z}^{\perp}(\Gamma) = \mathscr{Y}(\Gamma^*)$. Recall that $\mathscr{Y}(\Gamma^*)$ is spanned by the images under f^* of the blocks of Γ^*, i.e., of the vertices of Γ. For each $x \in V$, the set $f^*(x) = \{e \in E: x \in f(e)\}$ is called a **vertex cocycle**. Since Γ is a multigraph, each edge is an element of precisely two vertex cocycles. Hence

A11
$$\sum_{x \in V} f^*(x) = \varnothing.$$

This relation shows that while the collection of vertex cocycles spans $\mathscr{Z}^{\perp}(\Gamma)$, it is too large to be a basis. Two more observations of use are that $\text{Fnd}(\mathscr{Z}^{\perp}(\Gamma)) = \bigcup_{x \in V} f^*(x) = E$, and that $|f^*(x)| = \rho(x)$ for all $x \in V$.

We now use these algebraic tools to prove perhaps the oldest theorem in graph theory. A circuit in a multigraph is called an **Euler circuit** if it includes every edge and every vertex. A multigraph which contains an Euler circuit is said to be **Eulerian**. Intuitively, an Eulerian graph can be "drawn" completely without having either to retrace any "edge" or withdraw one's quill from one's parchment. For the interesting historical background of the next result, see [b.8].

A12 Theorem (L. Euler [e.8], 1736). *A multigraph is Eulerian if and only if it is connected and every vertex has even valence.*

PROOF. Let $\Gamma = (V, f, E)$ be a multigraph. By *II*A8, $E \in \mathscr{Z}(\Gamma)$ if and only if $\mathscr{Z}^{\perp}(\Gamma)$ is even, i.e., if and only if each vertex has even valence. If Γ is connected and if E is a cycle, then by Proposition A10, Γ is an Euler circuit. Conversely, if Γ has an Euler circuit, it is connected. Moreover, the set E of the edges of the Euler circuit is a cycle. □

The traditional proof of Euler's Theorem is constructive. The algebraic proof, however, has the advantage that it shows that Euler's Theorem and the following well-known result are really "dual" to one another.

A13 Theorem. *A multigraph is bipartite if and only if every circuit has even length.*

PROOF. Let $\Gamma = (V, f, E)$. To say that every circuit has even length is by A8 and A9 to say that $\mathscr{Z}(\Gamma)$ is an even space, which is equivalent (*II*A8 again) to saying that $E \in \mathscr{Z}^{\perp}(\Gamma)$. Equivalently, $E = \sum_{x \in U} f^*(x)$ for some $U \subseteq V$. This means that each edge is incident with exactly one vertex in U and that $\{U, V + U\}$ is the required partition of V. Conversely, if Γ is bipartite with partition $\{U, V + U\}$, then $E = \sum_{x \in U} f^*(x)$, and one pursues the chain of equivalent statements in the reverse direction. □

A14 *Exercise.* A path in a multigraph is said to be an **Euler path** if it contains each edge exactly once and each vertex at least once. State and prove a proposition about Euler paths analogous to Euler's Theorem.

A15 Proposition. *For any multigraph with space $\mathscr{Y}$ and cycle space $\mathscr{Z}$:*
- (a) $\dim(\mathscr{Y}) = \dim(\mathscr{Z}^{\perp}) = \nu_0 - \nu_{-1}$;
- (b) $\dim(\mathscr{Z}) = \nu_1 - \nu_0 + \nu_{-1}$;
- (c) $\dim(\mathscr{Y}^{\perp}) = \nu_{-1}$.

PROOF. (a) is an immediate consequence of *II*D6, A4, and *II*A1. (b) follows from (a) and *II*D3. (c) follows from (a) and *II*D5. □

A16 Corollary. *If $\Gamma = (V, f, E)$ is connected and $x_0 \in V$, then $\mathscr{B} = \{f^*(x) : x \in V + \{x_0\}\}$ is a basis for $\mathscr{Z}^{\perp}(\Gamma)$.*

PROOF. By A11, $\mathscr{B}$ spans $\mathscr{Z}^{\perp}(\Gamma)$. Since $\nu_{-1}(\Gamma) = 1$, we have $|\mathscr{B}| = \nu_0 - 1 = \dim(\mathscr{Z}^{\perp}(\Gamma))$. □

A17 *Exercise.* Determine $\mathscr{Y}^{\perp}(\Gamma)$.

Propositions A9 and A10 gave precise graph-theoretical interpretations of the algebraic notion of a cycle. The following proposition attempts to give a corresponding interpretation for cocycles. However, we must first introduce some notation.

Let $F \subseteq E$ and recall that $\Gamma_{(F)}$ is the submultigraph $(V, f_{|E+F}, E + F)$. If

$F = \{e\}$ we write $\Gamma_{(e)}$ for $\Gamma_{(\{e\})}$. Note that $\Gamma_{(F)}$ and Γ_{E+F} differ only in that $\Gamma_{(F)}$ may contain some vertices in addition to those in Γ_{E+F}. Since none of these vertices is incident with edges in $E + F$, they are isolated vertices in $\Gamma_{(F)}$. There are precisely $\nu_{-1}(\Gamma_{(F)}) - \nu_{-1}(\Gamma_{E+F})$ of them. It follows that $\mathscr{Z}(\Gamma_{(F)}) = \mathscr{Z}(\Gamma_{E+F})$ and $\mathscr{Z}^{\perp}(\Gamma_{(F)}) = \mathscr{Z}^{\perp}(\Gamma_{E+F})$.

A18 Proposition. *Let F be a set of edges of the multigraph* Γ.

(a) *F contains a nonempty cocycle if and only if* $\nu_{-1}(\Gamma_{(F)}) > \nu_{-1}(\Gamma)$.

(b) *F is an elementary cocycle if and only if* $\nu_{-1}(\Gamma_{(F)}) > \nu_{-1}(\Gamma)$ *and* $\nu_{-1}(\Gamma_{(F+\{e\})}) = \nu_{-1}(\Gamma)$ *for every* $e \in F$, *i.e., F is minimal with respect to the property given in* (a).

PROOF. (a) If C is an arbitrary vertex cocycle of Γ, then $\pi_{E+F}[C] = C \cap (E + F)$ is a vertex cocycle of $\Gamma_{(F)}$. Moreover, all vertex cocycles of $\Gamma_{(F)}$ are of this form. Thus $\pi_{E+F}\colon \mathscr{Z}^{\perp}(\Gamma) \to \mathscr{Z}^{\perp}(\Gamma_{(F)})$ is a surjection. Hence $\dim(\mathscr{Z}^{\perp}(\Gamma)) = \dim(\ker \pi_{E+F}) + \dim(\mathscr{Z}^{\perp}(\Gamma_{(F)})) = \dim(\mathscr{Z}^{\perp}(\Gamma) \cap \mathscr{P}(F)) + \dim(\mathscr{Z}^{\perp}(\Gamma_{(F)}))$. Thus by Proposition A15a,

$$\begin{aligned}\dim(\mathscr{Z}^{\perp}(\Gamma) \cap \mathscr{P}(F)) &= (\nu_0(\Gamma) - \nu_{-1}(\Gamma)) - (\nu_0(\Gamma_{(F)}) - \nu_{-1}(\Gamma_{(F)}))\\ &= \nu_{-1}(\Gamma_{(F)}) - \nu_{-1}(\Gamma).\end{aligned}$$

Hence $\nu_{-1}(\Gamma_{(F)}) > \nu_{-1}(\Gamma)$ if and only if $\dim(\mathscr{Z}^{\perp}(\Gamma) \cap \mathscr{P}(F)) > 0$, i.e., if and only if F contains a nonempty cocycle.

(b) If $\nu_{-1}(\Gamma_{(F)}) > \nu_{-1}(\Gamma)$, then F contains a nonempty cocycle (by (a) above). Also, if $\nu_{-1}(\Gamma_{(F+\{e\})}) = \nu_{-1}(\Gamma)$, then $F + \{e\}$ contains no nonempty cocycle, for each $e \in F$. Hence F is an elementary cocycle. Conversely, if F is a nonempty cocycle, $\nu_{-1}(\Gamma_{(F)}) > \nu_{-1}(\Gamma)$. If in addition, F is elementary, then $F + \{e\}$ contains no nonempty cocycle and $\nu_{-1}(\Gamma_{(F+\{e\})}) = \nu_{-1}(\Gamma)$ for all $e \in F$. □

An edge $e \in F$ is called an **isthmus** of Γ if the 1-set $\{e\}$ is a cocycle. A vertex x of Γ is called an **articulation vertex** if the vertex cocycle $f^*(x)$ is not an elementary cocycle.

Example. In Figure A19, the vertices x, y, and z are articulation vertices.

A19

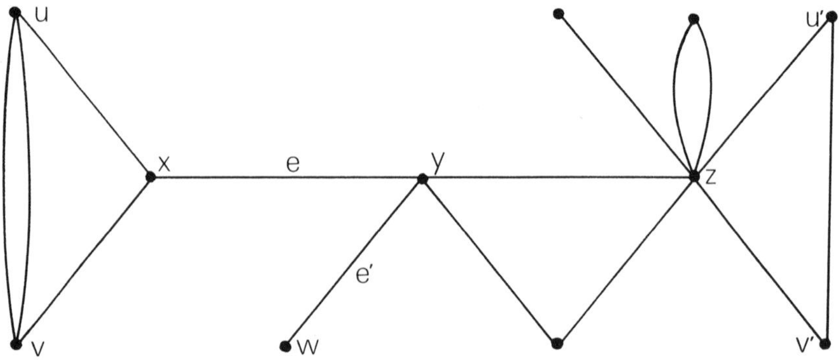

To see that z is an articulation vertex, note that the cocycle $f^*(u') + f^*(v')$ is a proper subset of $f^*(z)$. We may also observe that e is an isthmus, for $\{e\} = f^*(u) + f^*(v) + f^*(x)$. The edge e' is also an isthmus; in this case w is a pendant vertex and $\{e'\} = f^*(w)$.

A20 *Exercise.* If a vertex x is incident with an isthmus in a multigraph, show that x is either an articulation vertex or a pendant vertex.

A21 Exercise. Show that $\mathrm{Fnd}(\mathscr{Z}(\Gamma)) = E + I$, *where I is the set of isthmuses of* Γ.

IIIB Biconnectedness

When a connected multigraph with at least two edges has no articulation vertex, it is said to be **biconnected**. Thus, *a multigraph containing two or more edges is biconnected if and only if it is connected and every vertex cocycle is elementary*. We shall see that there is a close relationship between the structure of the cycle space and biconnectedness. Also there is an interesting parallel between the relationship of paths to connectedness (cf. §*II*C) and the relationship of elementary circuits to biconnectedness.

B1 Lemma. *If $\mathscr{Z}^\perp(\Gamma)$ is connected, then Γ has no articulation vertex.*

Proof. There is no loss of generality in assuming that Γ has no isolated vertices and hence is a hypergraph. Thus by *II*D11, Γ is connected. Suppose that x_0 is an articulation vertex of Γ. Let C_1 be an elementary cocycle such that $C_1 \subset f^*(x_0)$, and let $C_2 = f^*(x_0) + C_1$. By A16, $C_1 = \sum_{x \in U_1} f^*(x)$ for some $U_1 \subset V + \{x_0\}$. Let $U_2 = V + \{x_0\} + U_1$. Thus

B2
$$C_1 + \sum_{x\in U_1} f^*(x) = \varnothing, \quad \text{and} \quad C_2 + \sum_{x \in U_2} f^*(x) = \varnothing.$$

(The second equation is obtained by addition of the first equation to A11.) Let $\mathscr{B}_i$ be the subspace of $\mathscr{Z}^\perp(\Gamma)$ spanned by $\{f^*(x)\colon x \in U_i\}$, for $i = 1, 2$. Since each edge belongs to exactly two vertex cocycles, B2 implies that $\mathscr{B}_1$ and $\mathscr{B}_2$ have disjoint foundations. Hence $\mathscr{B}_1 \oplus \mathscr{B}_2$ is well-defined. It follows from A16 that $\mathscr{B}_1 \oplus \mathscr{B}_2 = \mathscr{Z}^\perp(\Gamma)$. Finally, since $C_i \in \mathscr{B}_i$ for $i = 1, 2$, $\mathscr{B}_i$ is not trivial, and $\mathscr{Z}^\perp(\Gamma)$ is not connected. □

B3 Proposition. *Let Γ have at least two edges but no isolated vertices. The following are equivalent*:

(a) Γ *is biconnected*;
(b) $\mathscr{Z}^\perp(\Gamma)$ *is connected*;
(c) $\mathscr{Z}(\Gamma)$ *is connected and* $\mathrm{Fnd}(\mathscr{Z}(\Gamma)) = E$;
(d) *every two edges of* Γ *are in a common elementary circuit.*

Proof. (a) $\Rightarrow$ (b). Assume Γ is biconnected and that $\mathscr{Z}^\perp(\Gamma) = \mathscr{B}_1 \oplus \mathscr{B}_2$. Since Γ is biconnected, the vertex cocycles are the elementary cocycles. By

*II*C3, each vertex cocycle lies in exactly one of $\mathscr{B}_1$ and $\mathscr{B}_2$. Let $U_i = \{x \in V : f^*(x) \in \mathscr{B}_i\}$ for $i = 1, 2$. Thus $U_1 \cap U_2 = \varnothing$. Therefore no edge is incident with both a vertex from U_1 and a vertex from U_2. Hence all the vertices of any path lie entirely in U_1 or entirely in U_2. Since Γ is connected, we conclude that either U_1 or U_2 is empty, and hence that either $\mathscr{B}_1$ or $\mathscr{B}_2$ is trivial. Thus $\mathscr{Z}^\perp(\Gamma)$ is connected.

(b) $\Rightarrow$ (c). Since $\mathscr{Z}^\perp(\Gamma)$ is connected and $|E| \geq 2$, Γ has no isthmuses by *II*C10b. By Exercise A21, $\operatorname{Fnd}(\mathscr{Z}(\Gamma)) = E = \operatorname{Fnd}(\mathscr{Z}^\perp(\Gamma))$. By *II*C12, $\mathscr{Z}(\Gamma)$ is connected.

(c) $\Rightarrow$ (b). Since $\operatorname{Fnd}(\mathscr{Z}^\perp(\Gamma)) = E$, the result follows from *II*C12.

(c) $\Leftrightarrow$ (d). This follows from *II*C22a.

(b) $\Rightarrow$ (a). Assume that $\mathscr{Z}^\perp(\Gamma)$ is connected. It follows from B1 that Γ contains no articulation vertex. Now let x and x' be any two vertices of Γ; we must show that there exists an xx'-path. Since Γ has no isolated vertices, we may choose edges e, e' incident with x and x', respectively. Since (b) implies (d), e and e' (and hence x and x') are in a common elementary circuit. Thus there exists an xx'-path, and Γ is connected. □

B4 Lemma. **(a)** $\mathscr{Z}(\Gamma_F) = \mathscr{Z}(\Gamma_{(E+F)}) = \mathscr{Z}(\Gamma) \cap \mathscr{P}(F)$, *for all* $F \subseteq E$;
(b) $\mathscr{Z}^\perp(\Gamma_F) = \mathscr{Z}^\perp(\Gamma_{(E+F)}) = \pi_F[\mathscr{Z}^\perp(\Gamma)]$, *for all* $F \subseteq E$.

Proof. Let Z be any elementary cycle of Γ. By A9, Γ_Z is an elementary circuit of Γ. Thus

$$\begin{aligned} Z \in \mathscr{Z}(\Gamma_F) &\Leftrightarrow \Gamma_Z \text{ is a submultigraph of } \Gamma_F, \\ &\Leftrightarrow Z \subseteq F, \\ &\Leftrightarrow Z \in \mathscr{Z}(\Gamma) \cap \mathscr{P}(F), \end{aligned}$$

as required to prove (a). Taking the orthogonal complement of each term in (a) and applying *II*A15, we get (b). □

Let $\mathscr{Q}$ be the component partition of the space $\mathscr{Z}^\perp(\Gamma)$. For each $Q \in \mathscr{Q}$, Γ_Q is called a **lobe** of Γ; a subgraph consisting of an isolated vertex of Γ is also called a **lobe** of Γ. By B4a, *II*C12, and the definition of component of a space, we have

B5 $$\mathscr{Z}(\Gamma) = \bigoplus_{Q \in \mathscr{Q}} \mathscr{Z}(\Gamma_Q) \quad \text{and} \quad \mathscr{Z}^\perp(\Gamma) = \bigoplus_{Q \in \mathscr{Q}} \mathscr{Z}^\perp(\Gamma_Q).$$

Let us consider the internal structure of the lobes Γ_Q of Γ. First note that distinct lobes have no common edges. If $Q = \{e\}$, then e is an isthmus. Conversely, if an edge e is an isthmus, then by *II*C10b and *II*C22b, $\{e\}$ is a cell of the component partition of $\mathscr{Z}^\perp(\Gamma)$. We have shown:

B6 Lemma. *An edge e is an isthmus of* Γ *if and only if* $\{e\}$ *induces a lobe of* Γ.

B7 Lemma. *A biconnected submultigraph* $\Gamma' = (V', f', E')$ *of* Γ *is a submultigraph of some lobe of* Γ.

PROOF. By B3, any two edges of Γ' belong to a common elementary circuit. By *II*C3, $E' \subseteq Q$ for some cell Q of the component partition of $\mathscr{Z}(\Gamma)$. It follows that Γ' is a submultigraph of the lobe induced by Q. □

B8 Lemma. *Let Γ_Q be a lobe of Γ with $|Q| > 1$. Then Γ_Q is biconnected.*

PROOF. Since Q is a cell of the component partition of $\mathscr{Z}(\Gamma)$, we have by *II*C22a, that every two edges in Q belong to a common elementary cycle. By A9, every two edges in Q belong to a common elementary circuit. Finally by B3, Γ_Q is biconnected. □

B9 Proposition. *A submultigraph Γ' of Γ is a lobe of Γ if and only if Γ' is induced by an isolated vertex or Γ' is induced by an isthmus or Γ' is a maximal biconnected submultigraph of Γ.*

PROOF. Let Γ' be a lobe of Γ which is not induced by an isolated vertex, i.e., $\Gamma' = \Gamma_Q$ for some cell Q of the component partition of $\mathscr{Z}^{\perp}(\Gamma)$. If $|Q| = 1$ then by B6, Γ' is induced by an isthmus. If $|Q| > 1$, then by B8, Γ' is biconnected, and by B7, Γ' is a maximal biconnected submultigraph of Γ.

Conversely, if Γ' is induced by an isthmus, then it is a lobe by B6. If Γ' is a maximal biconnected submultigraph of Γ, then it is a lobe by B7 and B8. □

Observe that the multigraph in Figure A19 has exactly six lobes, two of which are isthmuses and four of which are maximal biconnected submultigraphs.

B10 *Exercise.* Prove that a biconnected multigraph Γ with $\rho(\Gamma) = 2$ is an elementary circuit.

B11 Exercise. *Let Γ be a multigraph and let $x \in V$.* Prove that *the following three statements are equivalent*:

(**a**) *x is an articulation vertex;*

(**b**) *x is a vertex of more than one lobe of Γ;*

(**c**) *there exist vertices $y, z \in V + \{x\}$ such that there exists a yz-path and every yz-path contains x.*

B12 *Exercise.* Prove that two distinct lobes have at most one common vertex.

B13 Exercise. *Assuming that $\nu_1(\Gamma) \geq 2$*, prove that *the following seven statements are equivalent*:

(**a**) *Γ is biconnected.*

(**b**) *Γ consists of a single lobe.*

(**c**) *Every two vertices of Γ belong to a common elementary circuit.*

(**d**) *Each vertex and each edge from Γ belong to a common elementary circuit.*

(**e**) *Given any $x, y \in V$ and any $e \in E$, there exists an elementary xy-path containing e.*

(**f**) *Given distinct $x, y, z \in V$, there exists an elementary xy-path containing z.*
(**g**) *Given distinct $x, y, z \in V$, there exists an elementary xy-path avoiding z.*

B14 *Exercise.* Prove that if a multigraph has an odd circuit, then it has an odd elementary circuit.

B15 *Exercise.* Prove that a biconnected multigraph is either bipartite or has the property that each edge lies on an odd elementary circuit.

B16 *Exercise* [w.1]. Prove that a biconnected multigraph either is an odd circuit or contains an elementary even circuit.

B17 *Exercise* [o.1]. For each vertex x of Γ let $i(x) = \nu_{-1}(\Gamma_{V+\{x\}}) - \nu_{-1}(\Gamma)$. Prove that the number of lobes of Γ is $(\sum_{x \in V} i(x)) + 1$.

B18 *Exercise.* Let Γ be a biconnected multigraph with $\dim(\mathscr{Z}(\Gamma)) \geq 2$. Prove that every edge of Γ belongs to at least two distinct elementary circuits.

A graph is **geodetic** if given any two vertices x and y, there exists a unique xy-path of smallest length.

B19 *Exercise.* Prove that:
(a) Γ is geodetic if and only if every lobe of Γ is geodetic.
(b) If Γ is geodetic and biconnected, then $\dim(\mathscr{Z}(\Gamma)) \neq 2$.

IIIC Forests

A multigraph Γ is said to be a **forest** if $\mathscr{Z}(\Gamma) = \{\varnothing\}$. A connected forest is called a **tree**. Clearly every forest is a graph; for if $f(e) = f(e')$ for distinct $e, e' \in E$, then $f(e) + f(e') = \varnothing$, i.e., $\{e, e'\} \in \mathscr{Z}(\Gamma)$. We will show that every multigraph Γ contains certain special subgraphs which are forests (trees if the multigraph is connected), and that these subgraphs give particular information about the structure of Γ itself. These subgraphs will be constructed by removing, one at a time, edges which belong to circuits, thereby destroying all the circuits. We start our discussion by considering the algebraic consequences of deleting a single edge from a multigraph.

C1 Proposition. *Let e be an edge of the multigraph* Γ. *Then:*

(**a**) $$\dim(\mathscr{Z}(\Gamma)) - \dim(\mathscr{Z}(\Gamma_{(e)})) = \begin{cases} 0 & \text{if } e \text{ is an isthmus;} \\ 1 & \text{otherwise.} \end{cases}$$

(**b**) $$\dim(\mathscr{Z}^{\perp}(\Gamma)) - \dim(\mathscr{Z}^{\perp}(\Gamma_{(e)})) = \begin{cases} 1 & \text{if } e \text{ is an isthmus;} \\ 0 & \text{otherwise.} \end{cases}$$

(**c**) $$\nu_{-1}(\Gamma_{(e)}) - \nu_{-1}(\Gamma) = \begin{cases} 1 & \text{if } e \text{ is an isthmus;} \\ 0 & \text{otherwise.} \end{cases}$$

PROOF. By *II*A10, $\dim(\mathscr{Z}(\Gamma)) = \dim(\pi_{\{e\}}[\mathscr{Z}(\Gamma)]) + \dim(\mathscr{P}(E + \{e\}) \cap \mathscr{Z}(\Gamma))$. By B4a, $\mathscr{P}(E + \{e\}) \cap \mathscr{Z}(\Gamma) = \mathscr{Z}(\Gamma_{(e)})$. Combining these, we get

C2 $$\dim(\mathscr{Z}(\Gamma)) - \dim(\mathscr{Z}(\Gamma_{(e)})) = m,$$

where $m = \dim(\pi_{\{e\}}[\mathscr{Z}(\Gamma)])$. By *II*A6, $\dim(\mathscr{Z}(\Gamma)) + \dim(\mathscr{Z}^{\perp}(\Gamma)) = \nu_1(\Gamma)$ while $\dim(\mathscr{Z}(\Gamma_{(e)})) + \dim(\mathscr{Z}^{\perp}(\Gamma_{(e)})) = \nu_1(\Gamma) - 1$. Substituting these into C2 yields

C3 $$\dim(\mathscr{Z}^{\perp}(\Gamma)) - \dim(\mathscr{Z}^{\perp}(\Gamma_{(e)})) = 1 - m.$$

By A15a, $\dim(\mathscr{Z}^{\perp}(\Gamma)) = \nu_0(\Gamma) - \nu_{-1}(\Gamma)$ while $\dim(\mathscr{Z}^{\perp}(\Gamma_{(e)})) = \nu_0(\Gamma_{(e)}) - \nu_{-1}(\Gamma_{(e)}) = \nu_0(\Gamma) - \nu_{-1}(\Gamma_{(e)})$. Substituting these into C3 yields

C4 $$\nu_{-1}(\Gamma_{(e)}) - \nu_{-1}(\Gamma) = 1 - m.$$

If e is an isthmus, no cycle contains e. Hence $\pi_{\{e\}}[\mathscr{Z}(\Gamma)] = \{\varnothing\}$, and $m = 0$. If e is not an isthmus, some cycle contains e; hence $\pi_{\{e\}}[\mathscr{Z}(\Gamma)] = \mathscr{P}(\{e\})$ and $m = 1$. Thus C2, C3, and C4 give (a), (b), and (c), respectively. □

C5 Corollary. *For each integer* $j = 0, 1, \ldots, \dim(\mathscr{Z}(\Gamma))$, *there exists a subset* $F \subseteq E$ *such that* $\dim(\mathscr{Z}(\Gamma_{(F)})) = j$ *while* $\dim(\mathscr{Z}^{\perp}(\Gamma_{(F)})) = \dim(\mathscr{Z}^{\perp}(\Gamma))$ *and* $\nu_{-1}(\Gamma_{(F)}) = \nu_{-1}(\Gamma)$.

PROOF. We proceed by induction on $\dim(\mathscr{Z}(\Gamma))$. If $\dim(\mathscr{Z}(\Gamma)) = 0$, then $F = \varnothing$ and the result is valid. Let $n > 0$ and suppose the result is valid whenever $\dim(\mathscr{Z}(\Gamma)) < n$. If $\dim(\mathscr{Z}(\Gamma)) = n$, there exists a nonempty cycle and hence an edge e which is not an isthmus. By the proposition, $\dim(\mathscr{Z}(\Gamma_{(e)})) = \dim(\mathscr{Z}(\Gamma)) - 1$, while $\dim(\mathscr{Z}^{\perp}(\Gamma_{(e)})) = \dim(\mathscr{Z}^{\perp}(\Gamma))$ and $\nu_{-1}(\Gamma_{(e)}) = \nu_{-1}(\Gamma)$. Let j be an integer such that $0 \leq j \leq \dim(\mathscr{Z}(\Gamma))$. If $j = \dim(\mathscr{Z}(\Gamma))$, let $F = \varnothing$. If $j < \dim(\mathscr{Z}(\Gamma))$, we apply the induction hypothesis to $\Gamma_{(e)}$ to choose $F' \subseteq E + \{e\}$ so that with $F = F' + \{e\}$, $\dim(\mathscr{Z}(\Gamma_{(F)})) = j$, $\dim(\mathscr{Z}^{\perp}(\Gamma_{(F)})) = \dim(\mathscr{Z}^{\perp}(\Gamma_{(e)})) = \dim(\mathscr{Z}^{\perp}(\Gamma))$, and $\nu_{-1}(\Gamma_{(F)}) = \nu_{-1}(\Gamma_{(e)}) = \nu_{-1}(\Gamma)$. Since $F \subseteq E$, we are done. □

Implicit in the foregoing proof is the following result:

C6 Corollary. *If* $F \subseteq E$, *then*

(a) $\nu_{-1}(\Gamma_{(F)}) \geq \nu_{-1}(\Gamma)$;

(b) $\dim(\mathscr{Z}^{\perp}(\Gamma_{(F)})) \leq \dim(\mathscr{Z}^{\perp}(\Gamma))$;

(c) $\dim(\mathscr{Z}(\Gamma_{(F)})) \leq \dim(\mathscr{Z}(\Gamma))$.

Moreover, equality holds in (a) *if and only if it holds in* (b); *equality holds simultaneously in* (a), (b), *and* (c) *if and only if* $F = \varnothing$.

C7 Corollary. *Let* $C \subseteq E$. *C is an elementary cocycle of* Γ *if and only if* $\nu_{-1}(\Gamma_{(C)}) = \nu_{-1}(\Gamma) + 1$ *and* $\nu_{-1}(\Gamma_{(C+\{e\})}) = \nu_{-1}(\Gamma)$ *for every* $e \in C$; *i.e., the subset C is minimal with respect to the first equality.*

PROOF. Apply A18b and C1c. □

C8 *Exercise*. Show that if Γ is a multigraph with at least one edge, then $\dim(\mathscr{Z}) < \nu_1$. Characterize all multigraphs for which $\dim(\mathscr{Z}) = \nu_1 - 1$.

C9 Proposition. *For a multigraph* Γ, *conditions* (a), (b), *and* (c) *below are equivalent. If two of the conditions hold including at least one from among* (d) *and* (e), *then all five hold. In particular, if* Γ *is a tree, then all five hold.*

(a) Γ *is a forest.*

(b) *Every edge is an isthmus.*

(c) $\dim(\mathscr{Z}^\perp) = \nu_1$.

(d) Γ *is connected.*

(e) $\nu_1 = \nu_0 - 1$.

PROOF. It follows directly from the definitions that each of (a), (b), and (c) is equivalent to $\mathscr{Z}^\perp = \mathscr{P}(E)$. Hence they are equivalent to each other.

Note that (d) is equivalent to $\nu_{-1} = 1$. Thus by A15a, any two of the conditions (c), (d), (e) imply the third. If Γ is a tree, then (a) and (d) hold. $\square$

In Figure C10, graph (a) satisfies only conditions (a), (b), and (c) above; graph (b) satisfies only condition (d); graph (c) satisfies only condition (e).

C10

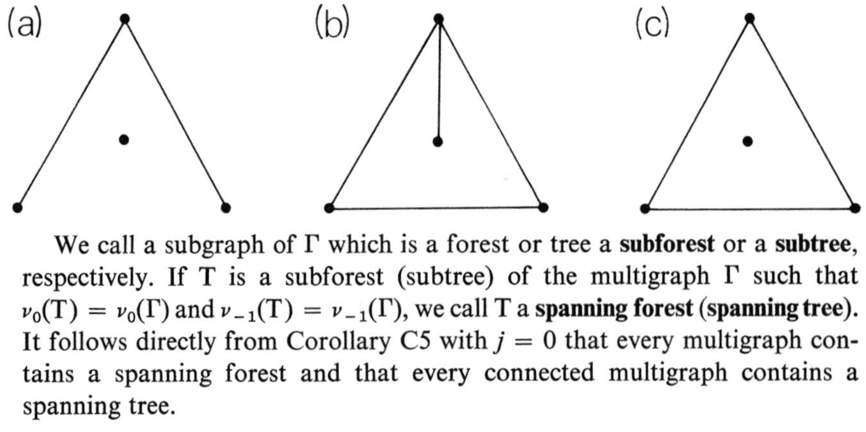

We call a subgraph of Γ which is a forest or tree a **subforest** or a **subtree**, respectively. If T is a subforest (subtree) of the multigraph Γ such that $\nu_0(\mathrm{T}) = \nu_0(\Gamma)$ and $\nu_{-1}(\mathrm{T}) = \nu_{-1}(\Gamma)$, we call T a **spanning forest (spanning tree)**. It follows directly from Corollary C5 with $j = 0$ that every multigraph contains a spanning forest and that every connected multigraph contains a spanning tree.

C11 Proposition. *Let* $F \subseteq E$ *and let* $\mathrm{T} = \Gamma_{(E+F)}$. *Then the following six statements are equivalent*:

(a) $|F| = \dim(\mathscr{Z}^\perp(\mathrm{T})) = \dim(\mathscr{Z}^\perp(\Gamma))$;

(b) *T is a spanning forest of* Γ;

(c) *F is a maximal subset of E which contains no nonempty cycle of* Γ;

(d) *F is a minimal subset of E which meets each nonempty cocycle of* Γ;

(e) $\mathscr{Z}(\Gamma)$ *admits a basis* $\{Z_e : e \in E + F\}$ *such that for each* $e \in E + F$, $Z_e \cap (E + F) = \{e\}$;

(f) $\mathscr{Z}^\perp(\Gamma)$ *admits a basis* $\{C_e : e \in F\}$ *such that for each* $e \in F$, $C_e \cap F = \{e\}$.

PROOF. Let $E_2 \subset E_1 \subseteq E$ and let $\Gamma_i = \Gamma_{(E+E_i)}$ ($i = 1, 2$.) By C6 we have

C12 $\quad \nu_{-1}(\Gamma_1) \geq \nu_{-1}(\Gamma); \qquad \dim(\mathscr{Z}^{\perp}(\Gamma_1)) \leq \dim(\mathscr{Z}^{\perp}(\Gamma));$

C13 $\quad \dim(\mathscr{Z}(\Gamma_1)) \leq \dim(\mathscr{Z}(\Gamma));$

C14 $\quad \nu_{-1}(\Gamma_2) \geq \nu_{-1}(\Gamma_1); \qquad \dim(\mathscr{Z}^{\perp}(\Gamma_2)) \leq \dim(\mathscr{Z}^{\perp}(\Gamma_1));$

C15 $\quad \dim(\mathscr{Z}(\Gamma_2)) \leq \dim(\mathscr{Z}(\Gamma_1)).$

Equality must hold or fail simultaneously for the two inequalities in each of C12 and C14. Furthermore, equality cannot hold in both C14 and C15. Finally, equality holds in both C12 and C13 if and only if $\Gamma_1 = \Gamma$.

(a) $\Rightarrow$ (b). Let $\Gamma_1 = \mathrm{T}$, and note that the second equality in (a) implies $\nu_{-1}(\mathrm{T}) = \nu_{-1}(\Gamma)$. By C9, the first equality in (a) implies T is a forest. Finally, since $\nu_0(\mathrm{T}) = \nu_0(\Gamma)$, T is a spanning forest.

(b) $\Rightarrow$ (c). If T is a spanning forest, then, of course, F contains no nonempty cycle. If $\mathrm{T} = \Gamma$, we are done. Otherwise, let $\Gamma_2 = \mathrm{T}$ and let E_1 satisfy $F \subset E_1 \subseteq E$. Since T is a spanning forest, $\nu_{-1}(\mathrm{T}) = \nu_{-1}(\Gamma)$, and so $\nu_{-1}(\mathrm{T}) = \nu_{-1}(\Gamma_1)$. Hence equality holds in C14 and fails in C15. We conclude that $\dim(\mathscr{Z}(\Gamma_1)) > 0$ and E_1 contains a nonempty cycle.

(b) $\Rightarrow$ (d). Let $\Gamma_1 = \mathrm{T}$ and let $E_2 \subset F$. Since T is a spanning tree, $\nu_{-1}(\mathrm{T}) = \nu_{-1}(\Gamma)$ and $\dim(\mathscr{Z}(\mathrm{T})) = 0$. Thus equality holds in C12 and C15 and fails in C14. Thus $\nu_{-1}(\Gamma_2) < \nu_{-1}(\Gamma)$. By A18a, $E + F$ contains no nonempty cocycle while $E + E_2$ contains at least one nonempty cocycle, i.e., F meets every nonempty cocycle, but no proper subset E_2 of F has this property.

(c) $\Rightarrow$ (e). For each $e \in E + F$ there exists $Z_e \in \mathscr{Z}(\Gamma) + \{\varnothing\}$ such that $Z_e \subseteq F + \{e\}$. Since F contains no nonempty cycle, $Z_e \cap (E + F) = \{e\}$. Clearly $\{Z_e : e \in E + F\}$ is an independent set. Let $Z \in \mathscr{Z}(\Gamma)$. Since the cycle $Z + \sum_{e \in Z \cap (E+F)} Z_e \subseteq F$, it must be empty, i.e., $Z = \sum_{e \in Z \cap (E+F)} Z_e$. It follows that $\{Z_e : e \in E + F\}$ spans $\mathscr{Z}(\Gamma)$ and hence is a basis.

(d) $\Rightarrow$ (f). For each $e \in F$, there exists $C_e \in \mathscr{Z}^{\perp}(\Gamma) + \{\varnothing\}$ such that $C_e \cap (F + \{e\}) = \varnothing$. Since $C_e \cap F \neq \varnothing$, $C_e \cap F = \{e\}$. Clearly $\{C_e : e \in F\}$ is an independent set. To show that this collection spans $\mathscr{Z}^{\perp}(\Gamma)$, let $C \in \mathscr{Z}^{\perp}(\Gamma)$. Then the cocycle $C + \sum_{e \in C+F} C_e$ is disjoint from F and hence must be empty, i.e., $C = \sum_{e \in C \cap F} C_e$. Thus $\{C_e : e \in F\}$ spans $\mathscr{Z}^{\perp}(\Gamma)$ and hence is a basis.

(e) $\Rightarrow$ (a). Since $\{Z_e : e \in E + F\}$ is a basis for $\mathscr{Z}(\Gamma)$, $\dim(\mathscr{Z}(\Gamma)) = |E + F|$, which implies $\dim(\mathscr{Z}^{\perp}(\Gamma)) = |F|$. Clearly $E + F$ meets every nonempty sum of cycles in this basis. Hence F contains no nonempty cycle. It follows from B4a that $\mathscr{Z}(\mathrm{T}) = \{\varnothing\}$ and hence $\dim(\mathscr{Z}^{\perp}(\mathrm{T})) = |F|$.

(f) $\Rightarrow$ (a). Since $\{C_e : e \in F\}$ is a basis for $\mathscr{Z}^{\perp}(\Gamma)$, $\dim(\mathscr{Z}^{\perp}(\Gamma)) = |F|$. To complete the proof we need only show that F contains no nonempty cycles. Let $A \subseteq F$ be any nonempty subset of F. Then $|A \cap C_e| = 1$ for each $e \in A$. Hence A is not orthogonal to C_e and cannot be a cycle. □

C16 Exercise. *Let Γ_F be a spanning forest of* Γ. Show that *for any $e \in E + F$, there exists $e' \in F$ such that $\Gamma_{F+\{e,e'\}}$ is a spanning forest of* Γ. (This is called the **Exchange Property** because a new spanning forest is obtained from an old one by the "exchange" of one edge for another. It will be generalized in §XA.)

C17 *Exercise.* Show that any spanning forest of Γ can be obtained from any other spanning forest by a finite number of applications of the Exchange Property.

If $\Gamma_{(F)}$ is a spanning forest of Γ, then $\Gamma_{(E+F)}$ is a **spanning coforest**. (The edge set of a spanning coforest is called a "dendroid" by Tutte [t.6].) The following result is dual to C11 and should be proved as an exercise.

C18 Proposition. *Let $F \subseteq E$. The following six statements are equivalent*:

(a) $|F| = \dim(\mathscr{Z}(\Gamma_{(E+F)})) = \dim(\mathscr{Z}(\Gamma))$;

(b) $\Gamma_{(E+F)}$ *is a spanning coforest*;

(c) *F is a maximal subset of E which contains no nonempty cocycle of Γ*;

(d) *F is a minimal subset of E which meets every nonempty cycle of Γ*;

(e) *$\mathscr{Z}(\Gamma)$ admits a basis $\{Z_e : e \in F\}$ such that for each $e \in F, Z_e \cap F = \{e\}$*;

(f) *$\mathscr{Z}^{\perp}(\Gamma)$ admits a basis $\{C_e : e \in E + F\}$ such that for each $e \in E + F$, $C_e \cap (E + F) = \{e\}$.*

C19 *Exercise.* State and prove an "Exchange Property" for coforests.

IIID Graphic Spaces

Since the concept of the cycle space of a multigraph is easy to grasp intuitively, it is a natural question to ask: under what conditions will an arbitrary subspace $\mathscr{A} \subseteq \mathscr{P}(E)$ be the cycle space of some multigraph $\Gamma = (V, f, E)$? Since the cycle space of Γ is the orthogonal complement of its cocycle space, this question is clearly equivalent to asking whether $\mathscr{A}^{\perp}$ can be realized as the cocycle space of Γ. It turns out that necessary and sufficient conditions are more easily stated in terms of the cocycle space.

When the term "vertex cocycle" was defined, it was noted that every cocycle of Γ is a sum of vertex cocycles $f^*(x)$ for $x \in V$. Also, every edge belongs to $f^*(x)$ for exactly two vertices $x \in V$. Note our cautious wording both here and in Proposition D3 below; we will *not* say, "Every edge belongs to exactly two vertex cocycles." Although an edge belongs to the vertex cocycle of two distinct vertices, it may belong to only one vertex cocycle. For example, we may have, for some $e \in E$, $f(e) = \{x, y\}$ where $f^*(x) = f^*(y)$. In this case Γ has a component $\Gamma' = (\{x, y\}, k, f^*(x))$ where k is the constant function onto $\{x, y\}$.

D1 Exercise. Prove that *for a multigraph $\Gamma = (V, f, E)$, the following statements are equivalent*:

(a) *f^* is an injection*;

(b) *no component of Γ contains exactly two vertices, and Γ has no more than one isolated vertex.*

Prove also that *these statements must hold if* Γ *is connected and* $\nu_0(\Gamma) \geq 3$.

D2 *Exercise.* Prove that the vertex group and the edge group of a finite graph are group-isomorphic if and only if the graph has at most one isolated vertex and no component has exactly one edge. [*Hint:* use *II*E5.] Show that this statement is false for infinite graphs. (The result is due to Sabidussi [s.2] and Harary and Palmer [h.6]. The suggested method of proof appears in [w.5].)

D3 Proposition. *Let E be a set, let* $C_1, \ldots, C_m \in \mathscr{P}(E)$, *and let* $\mathscr{A}$ *be a subspace of* $\mathscr{P}(E)$. $\mathscr{A}$ *is the cocycle space of a multigraph whose edge set is E and whose vertex cocycles are* $C_1, \ldots, C_m$ *if and only if*

(a) $\{C_1, \ldots, C_m\}$ *spans* $\mathscr{A}$; *and*

(b) *if* $e \in E$, *then* $e \in C_i$ *for exactly two indices i.*

Furthermore, if $\mathscr{A}$ *is the cocycle space of some multigraph, then* $\mathscr{A}$ *is the cocycle space of a connected multigraph.*

Proof. If $\mathscr{A} = \mathscr{Z}^{\perp}(\Gamma)$, we have seen that the sets $f^*(x)$ for $x \in V$ satisfy condition (a) and condition (b).

Conversely, assume the existence of $C_1, \ldots, C_m \in \mathscr{A}$ satisfying (a) and (b). Let $V = \{1, \ldots, m\}$ and define $f: E \to \mathscr{P}(V)$ by

$$f(e) = \{i \in V : e \in C_i\}, \quad \text{for all } e \in E.$$

By (b), the system $\Gamma = (V, f, E)$ is a multigraph. Since $f^*(i) = C_i$ for all $i \in V$, and since $\mathscr{Z}^{\perp}(\Gamma)$ is spanned by the set of all vertex cocycles, condition (a) implies $\mathscr{Z}^{\perp}(\Gamma) = \mathscr{A}$.

We may reindex $C_1, \ldots, C_m$ so that for some $k \leq m$, $\{C_1, \ldots, C_k\}$ is a basis for $\mathscr{A}$. Since $\mathrm{Fnd}(\mathscr{A}) = E$, each edge belongs to at least one set C_i with $1 \leq i \leq k$. Then the set $C' = C_1 + C_2 + \ldots + C_k$ consists of those edges which belong to exactly one of the sets $C_1, \ldots, C_k$. Hence $C_1, \ldots, C_k, C'$ satisfies condition (b). Since it clearly satisfies (a), we have shown that we may take $m = k + 1 = \dim(\mathscr{A}) + 1$. In this case $\dim(\mathscr{Z}^{\perp}(\Gamma)) = m - 1 = \nu_0(\Gamma) - 1$, and by A15a, Γ is connected. □

The subsets $C_1, \ldots, C_m$ in the above proposition form what is called a **graphical realization** of $\mathscr{A}$, and $\mathscr{A}$ is said to be a **graphic subspace**. It is natural to ask, under what conditions will the choice of $C_1, \ldots, C_m$ be uniquely determined (up to permutation)? In other words, under what conditions will a multigraph be determined by its cocycle space (or, equivalently, by its cycle space)? This, the second question of this section, was answered by H. Whitney [w.10] in 1933. However, his proof did not capitalize on the linear algebra. Rather, he posed the question in nonalgebraic terms: under what conditions can a bijection between the edge sets of two multigraphs which takes the edge set of an elementary circuit onto the edge set of an elementary circuit be "extended" to a system-isomorphism?

Before we answer these questions, let us consider some examples. First of all, if Γ' is an arbitrary multigraph and if $\mathscr{A} = \mathscr{L}^{\perp}(\Gamma')$, the constructions in the proof of Proposition D3 applied to $\mathscr{A}$ will yield anew a connected multigraph Γ with the same edge set, the same cocycle space, and the same cycle space as Γ'. We can restate our second question as, what are sufficient conditions on Γ' to assure that Γ will be system-isomorphic to Γ'? Since Γ is connected, it is clearly necessary that Γ' be connected. Connectivity alone, however, is not sufficient. The first three multigraphs in Figure D4 are all nonisomorphic but have identical cocycle spaces. Observe that both Γ_2 and Γ_3 are connected.

D4

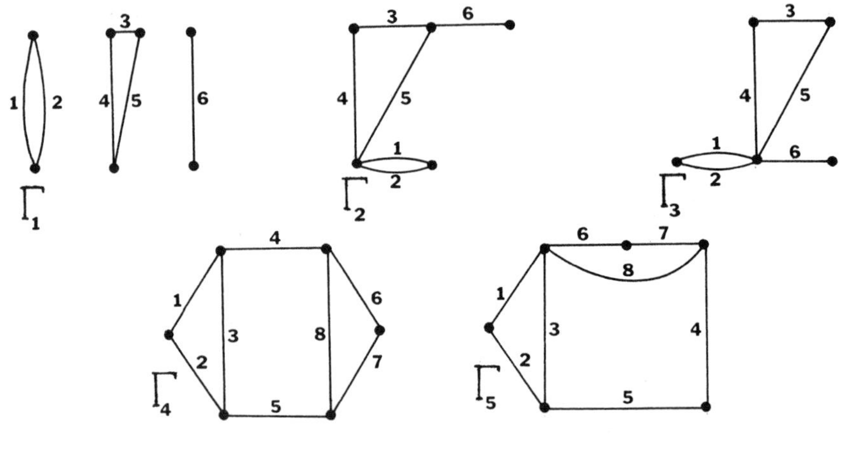

None of these graphs is biconnected, and we may inquire whether perhaps biconnectivity will suffice. It does not; Γ_4 and Γ_5 are biconnected, nonisomorphic, and have identical cycle and cocycle spaces. We observe in this example that while $\{1, 3, 4\}$ is a vertex cocycle of Γ_4, it is not a vertex cocycle of Γ_5. We wish then to find conditions on a multigraph so that its vertex cocycles may be distinguished *algebraically* from the other cocycles.

Let Γ be connected and let C be an elementary cocycle. By C7, $\Gamma_{(C)}$ has two components. If C is not a vertex cocycle, both of these components contain edges, and so $\mathscr{L}^{\perp}(\Gamma_{(C)})$ is not connected. (For example, let $C = \{4, 5\}$ in Γ_4 of the Figure.) We have shown:

D5 Proposition. *If C is an elementary cocycle of the biconnected multigraph Γ, and if $\mathscr{L}^{\perp}(\Gamma_{(C)})$ is connected, then C is a vertex cocycle.*

For example, if $C = \{1, 2\}$ in $\Gamma = \Gamma_4$ (Figure D4), then $\mathscr{L}^{\perp}(\Gamma_{(C)})$ is connected. By the proposition, $\{1, 2\}$ must be a vertex cocycle in both Γ_4 and Γ_5 as well as in any other graphical realization of $\mathscr{L}^{\perp}(\Gamma_4)$. The following corollary restates the above proposition in terms of the cocycle space of Γ rather than the cocycle space of $\Gamma_{(C)}$.

D6 Corollary. *If C is an elementary cocycle of the biconnected multigraph Γ, and if $\pi_{E+C}[\mathscr{Z}^{\perp}(\Gamma)]$ is connected, then C is a vertex cocycle of Γ.*

PROOF. By B4b, $\pi_{E+C}[\mathscr{Z}^{\perp}(\Gamma)] = \mathscr{Z}^{\perp}(\Gamma_{(C)})$. Now apply the proposition. □

A multigraph Γ is **triconnected** if for each $x \in V$, the submultigraph $\Gamma_{V+\{x\}}$ is biconnected.

D7 Exercise. Show that *every triconnected multigraph Γ* **(a)** *is biconnected and hence connected,* **(b)** *has at least three vertices, and* **(c)** *satisfies* $\dim(\mathscr{Z}^{\perp}(\Gamma)) \geq 2$.

D8 Proposition. *Let Γ be a triconnected multigraph, and let Γ' be a multigraph with no isolated vertices. If $\mathscr{Z}(\Gamma') = \mathscr{Z}(\Gamma)$, then Γ and Γ' are system-isomorphic.*

PROOF. Let $\Gamma = (V, f, E)$ and $\Gamma' = (U, g, E)$ satisfy the hypotheses. Then for any $x \in V$, $\Gamma_{V+\{x\}}$ is biconnected. Since the submultigraphs $\Gamma_{V+\{x\}}$ and $\Gamma_{(f^*(x))}$ differ only insofar as the latter includes the isolated vertex x, the two submultigraphs have equal cocycle spaces, which by B3 must be connected. Also $\mathscr{Z}^{\perp}(\Gamma'_{(f^*(x))})$ is connected, since by B4b it is equal to $\pi_{E+f^*(x)}[\mathscr{Z}^{\perp}(\Gamma')] = \mathscr{Z}^{\perp}(\Gamma_{(f^*(x))})$. Since $f^*(x)$ is an elementary cocycle of Γ, and hence of Γ', we may apply Proposition D5 to Γ' to deduce that $f^*(x)$ is also a vertex cocycle of Γ'.

Since $\nu_1(\Gamma') = \nu_1(\Gamma) \geq 2$ (Γ is triconnected and hence biconnected by Exercise D7), since Γ' was presumed to have no isolated vertices, and since $\mathscr{Z}^{\perp}(\Gamma')$ is connected, we infer from B3 that Γ' is connected. By D7, $\dim(\mathscr{Z}^{\perp}(\Gamma')) = \dim(\mathscr{Z}^{\perp}(\Gamma)) \geq 2$, and so by A15a, $\nu_0(\Gamma') \geq 3$. By D1, g^* is an injection. Therefore, the function $q: V \to U$ given by

$$g^*(q(x)) = f^*(x), \quad \text{for all } x \in V$$

is well-defined, and Figure D9 is a commutative diagram.

D9

$$\begin{array}{ccccc} V & \xrightarrow{\ f^*\ } & \mathscr{P}(E) & & E \\ q \downarrow \cong & & \cong \downarrow 1_E & & \cong \downarrow 1_E \\ U & \xrightarrow[\ g^*\ ]{} & \mathscr{P}(E) & & E \end{array}$$

It remains only to show that q is a bijection. It is clearly an injection, since both f^* and g^* are. Since every element $e \in E$ belongs to $f^*(x)$ for two distinct values of $x \in V$, e belongs to $g^*(q(x))$ for two distinct values of $x \in V$. Hence e can belong to no other vertex cocycle of Γ'. Since Γ' has no isolated vertices, $g^*(u) \neq \varnothing$ for all $u \in U$. It follows that q is a surjection. □

We recast the above proposition in the language of Whitney's original paper [w.10].

D10 Corollary. (H. Whitney). *Let* $\Gamma = (V, f, E)$ *be triconnected, and let* $\Theta = (U, h, F)$ *be a multigraph with no isolated vertices. Let there be a bijection* $p\colon E \to F$ *such that* Z *is the edge set of a circuit in* Γ *if and only if* $p[Z]$ *is the edge set of a circuit in* Θ. *Then* Γ *and* Θ *are system-isomorphic.*

Proof. Define $g = hp$, and observe that the lower rectangle in Figure D11 commutes.

D11

$$\begin{array}{ccccccc} V & & \mathscr{P}(V) & \xleftarrow{\quad f \quad} & E & & \\ q\big\downarrow\cong & & q\big\downarrow\cong & & 1_E\big\downarrow\cong & & \\ U & & \mathscr{P}(U) & \xleftarrow{\quad g \quad} & E & & p \\ 1_U\big\downarrow\cong & & 1_U\big\downarrow\cong & & p\big\downarrow\cong & & \\ U & & \mathscr{P}(U) & \xleftarrow{\quad h \quad} & E & & \end{array}$$

Thus $\Gamma' = (U, g, E)$ and Θ are system-isomorphic. By this and the hypothesis, $p[\mathscr{Z}(\Gamma')] = \mathscr{Z}(\Theta) = p[\mathscr{Z}(\Gamma)]$. Since p is a bijection, $\mathscr{Z}(\Gamma') = \mathscr{Z}(\Gamma)$. Hence $\mathscr{Z}^{\perp}(\Gamma') = \mathscr{Z}^{\perp}(\Gamma)$, and the result follows from the proposition. □

If V is an n-set for $n > 0$, any graph which is isomorphic to the graph $(V, \mathscr{P}_2(V))$ is called a **complete graph of order** n, and the symbol K_n is always used to denote such a graph. Thus, K_n is connected and $\dim(\mathscr{Z}(K_n)) = \binom{n}{2} - n + 1 = \binom{n-1}{2}$.

D12 Exercise. *Let* $K_n = (V, \mathscr{P}_2(V))$ *and let* $\mathscr{A}_n = \mathscr{Z}(K_n)$. Show that:

(a) $\mathscr{A}_n$ *is spanned by* $\{\mathscr{P}_2(W)\colon W \in \mathscr{P}_3(V)\}$. [*Hint*: use induction on n.]

(b) $\mathscr{A}_n$ *is connected for all* n, *and hence* K_n *is biconnected for all* $n \geq 3$.

(c) *If* Γ' *is a multigraph such that* $\mathscr{Z}^{\perp}(\Gamma') = \mathscr{A}_n$, *then* $\mathscr{Z}^{\perp}(\Gamma'_{(\mathscr{P}_2(W))})$ *is connected and* $\mathscr{P}_2(W)$ *is a vertex cocycle of* Γ' *for all* $W \in \mathscr{P}_3(V)$.

(d) *Each edge of* Γ' *belongs to* $\mathscr{P}_2(W)$ *for exactly* $n - 2$ *sets* $W \in \mathscr{P}_3(V)$.

(e) Combine parts (a)–(d) of this exercise to deduce:

D13 Proposition. $\mathscr{Z}(K_n)$ *is graphic if and only if* $n \leq 4$.

D14 *Exercise.* Prove that for any set E, the space $\mathscr{E}(E)$ is graphic.

D15 *Exercise.* Let $E = \{1, 2, \ldots, 7\}$. Let $\mathscr{A}$ be the subspace of $\mathscr{P}(E)$ spanned by

$$\{\{1, 2, 3, 4\}, \{2, 4, 5, 6\}, \{3, 4, 6, 7\}\}.$$

Show that neither $\mathscr{A}$ nor $\mathscr{A}^{\perp}$ is graphic.

IIIE Planar Multigraphs

Intuitively, a "planar multigraph" is a multigraph which can be represented in the plane in such a way that edges meet only at vertices. One can be more rigorous in topological language, but it requires regarding a multigraph as a topological object, namely as a 1-dimensional simplicial complex (except that two vertices may be joined by more than a single edge), an edge being regarded as a homeomorph of a closed real interval. Such a topological "multigraph" is called planar if it can be homeomorphically embedded into 2-dimensional Euclidean space, or equivalently, into the 2-sphere. Our approach here, however, will be combinatorial and ultimately algebraic. Demonstrating the equivalence of various mathematical approaches to planarity is no easy or elegant matter. Since one cannot seem to exploit the best from all possible mathematical worlds simultaneously, we will confine our rigor to combinatorics (except in §*VII*D and §*VII*E). Nonetheless we will freely use more pictorial language for both motivation and reinforcement.

Working unrigorously, the reader may observe by trial and error that K_n, for example, can be drawn in the plane if and only if $n \leq 4$. In particular, K_5 must be drawn with some edges meeting other than at common vertices (Figure E1a). With care the number of these "cross-overs" can be reduced to just one (Figure E1b).

After the reader has spent some time on this trial-and-error method, the difficulties of demonstrating for instance, that K_5 is "nonplanar" become

E1

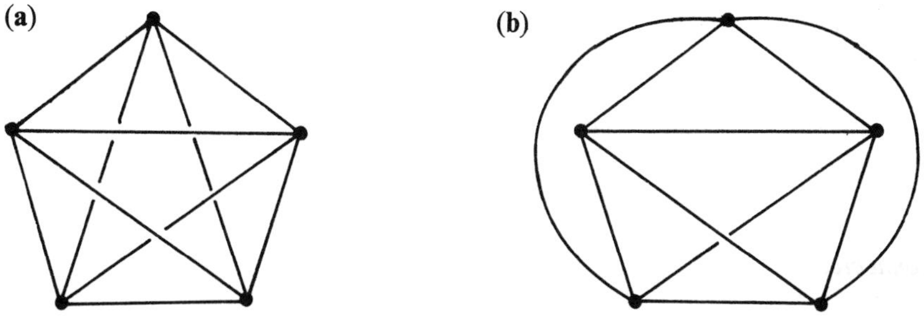

apparent, and the lack of a rigorous terminology should no doubt contribute to the frustration. Among the various combinatorial developments of planarity, ours will most closely parallel the work of S. MacLane [m.1] and will be closely related also to the approach of H. Whitney [w.9].

The first observation related to MacLane's approach to planarity is that when a multigraph Γ is realized in the plane (without "cross-overs"), there are certain cycles which play a special role. The set-theoretic complement in the plane of the realization of Γ has connected components, which are usually called "regions." The boundaries of these "regions," except when they contain isthmuses, correspond to circuits of Γ. The edge sets of these circuits, as noted in §A, belong to the cycle space of Γ and have algebraic significance in this space. These topological regions will be identified with their bounding cycles. MacLane used these cycles to characterize planarity.

We illustrate for a specific multigraph Γ, MacLane's "bounding cycles." Let Γ be the multigraph drawn in Figure E2. Its "regions" are:

$$\{e_1, e_2\}, \{e_2, e_3, e_4\}, \{e_4, e_5\}, \{e_5, e_6, e_7\}, \{e_1, e_3, e_6, e_7\}.$$

E2

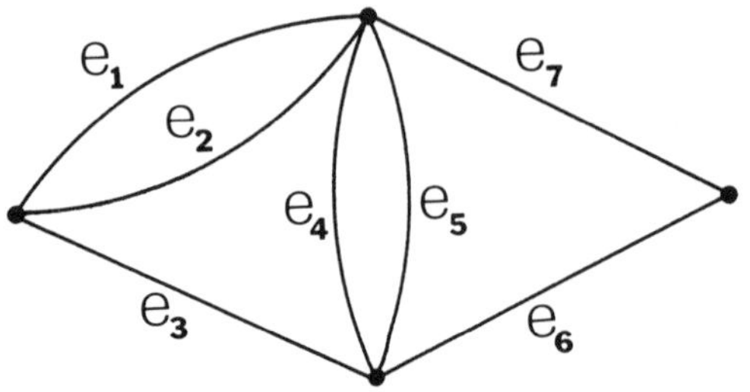

We observe that

(a) this set of cycles spans $\mathscr{Z}(\Gamma)$, and
(b) each edge of Γ is an element of exactly two of these cycles.

Observe how these properties are precisely dual to the properties of vertex cocycles in Proposition D3. What MacLane proved is that, in general, the existence of a list of cycles of Γ satisfying conditions (a) and (b) is equivalent to the existence of a *topological* realization of the graph in the plane. The approach of this text, however, is combinatorial rather than topological. What we will do is to use MacLane's combinatorial characterization of planarity as the *definition* of planarity.

A sequence $Z_1, \ldots, Z_k$ of cycles of the multigraph Γ is called a **planar imbedding** of Γ if:

(a) $\{Z_1, \ldots, Z_k\}$ spans $\mathscr{Z}(\Gamma)$, and
(b) if $e \in \text{Fnd}(\mathscr{Z}(\Gamma))$, then $e \in Z_i$ for precisely two indices $i \in \{1, \ldots, k\}$.

We say that Γ is **planar** if it admits a planar imbedding.

An isthmus of a multigraph of course belongs to no cycle while (cf. A21) all other edges belong to $\text{Fnd}(\mathscr{Z}(\Gamma))$. By condition (b),

E3
$$\sum_{i=1}^{k} Z_i = \varnothing$$

holds for any planar imbedding $Z_1, \ldots, Z_k$. We return to this development after a brief look at Whitney's approach.

The motivation behind Whitney's development is that when a multigraph Γ without isthmuses is realized in the plane, there is a "natural" construction which leads to a realization of a second multigraph Θ (whose edges are those of Γ). Whitney called Θ a "combinatorial dual" to Γ. The algebraic relation between Γ and Θ is that $\mathscr{Z}^{\perp}(\Theta) = \mathscr{Z}(\Gamma)$. For this reason we prefer to say that Θ is "orthogonal" to Γ, rather than "dual" to Γ. Having characterized orthogonal multigraphs in purely combinatorial terms, Whitney then proved that a realization of a multigraph is planar in the topological sense if and only if some multigraph is orthogonal to it.

An orthogonal multigraph Θ is shown in Figure E4 in solid lines superimposed on Γ (cf. Figure E2) in broken lines. Solid and broken edges crossing each other receive the same label. Note how the "regions" of Γ are precisely

E4

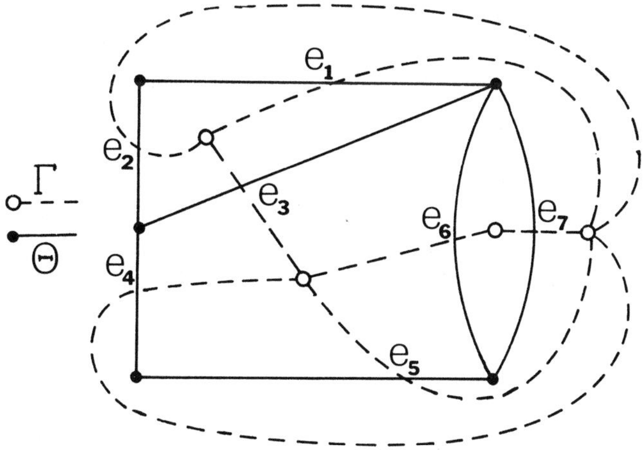

the vertex cocycles of Θ, and vice versa. Surely the relationship between Γ and Θ here is intuitively apparent. To obtain a realization of Θ from a realization of Γ, place a vertex in the interior of each "region" of Γ. These are the vertices of Θ. Then across each edge *e* of Γ draw an edge of Θ joining the two vertices of Θ in the two regions of Γ having *e* on their boundary. Note that at the same time one vertex of Γ appears in each "region" of Θ. Had this construction been carried out beginning with Θ, one would thereby have obtained Γ. We may rightly perceive at this juncture as did Whitney the equivalence between the planarity of Γ and the existence of a multigraph Θ "orthogonal" to it.

When this same multigraph Γ is realized in the plane with different "regions" as in Figure E5a, however, the multigraph Θ obtained by the above method has instead the form of Figure E5b, which is obviously not isomorphic to the Θ of Figure E4. Comparing Figures E4 and E5b with Γ_4 and Γ_5 in Figure D4, one observes that the two constructions of Θ have the same cycle space and the same cocycle space. Whitney's result D10 suggests that triconnectedness may be required to insure the uniqueness of the construction of Θ. Observe also that each vertex cocycle of Θ in each of the two constructions consists of edges corresponding to edges of a "region" of Γ, and each vertex cocycle of Γ determines a "region" of the appropriately constructed Θ.

E5

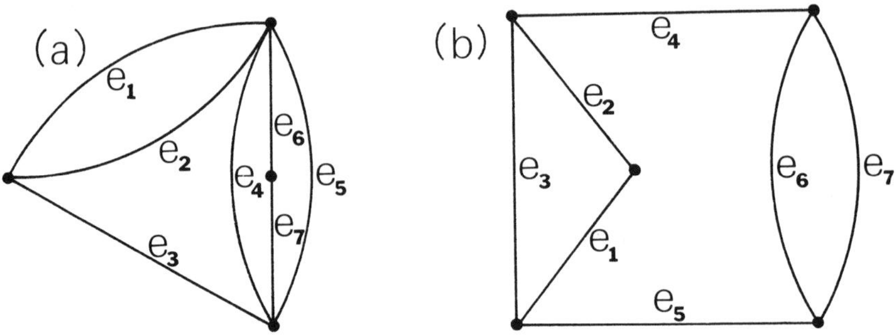

Let the multigraphs Γ and Θ have a common edge set E. We say that Γ is **orthogonal** to Θ, and write $\Gamma \perp \Theta$ if $\mathscr{Z}(\Gamma) = \mathscr{Z}^{\perp}(\Theta)$.

Some consequences of the assertion $\Gamma \perp \Theta$ are immediate. First of all, $\Theta \perp \Gamma$. Secondly, $E = \text{Fnd}(\mathscr{Z}^{\perp}(\Gamma)) = \text{Fnd}(\mathscr{Z}(\Gamma))$, and so neither Γ nor Θ may have an isthmus. Thirdly, if neither graph has isolated vertices, it then follows from B3 that Γ is biconnected if and only if Θ is biconnected. Fourthly, if $\Gamma \perp \Theta'$ also holds, then Θ and Θ' have identical cycle space and cocycle space.

Because we have given a *combinatorial* definition for planarity, we obtain a short proof of Whitney's characterization of planar multigraphs.

E6 Theorem. (H. Whitney). *Let the multigraph Γ contain no isthmus. Γ is planar if and only if there exists a multigraph Θ such that $\Gamma \perp \Theta$.*

Proof. The theorem is an immediate consequence of the definition of a planar imbedding, the fact that $\text{Fnd}(\mathscr{Z}(\Gamma)) = \text{Fnd}(\mathscr{Z}^{\perp}(\Gamma)) = E$, and Proposition D3. □

If $Z_1, \ldots, Z_m$ is a planar imbedding of a connected multigraph Γ without isthmus, then the multigraph Θ whose vertex cocycles are $Z_1, \ldots, Z_m$ is called the **multigraph orthogonal to Γ with respect to (the planar imbedding)** $Z_1, \ldots, Z_n$. One observes that Θ is already furnished with an imbedding $C_1, \ldots, C_n$ (which are the vertex cocycles of Γ) and that Γ is the multigraph orthogonal to Θ with respect to this imbedding.

A planar imbedding $Z_1, \ldots, Z_k$ is **elementary** if each cycle Z_i is an elementary cycle; it is a **simple** imbedding if E3 is the only relation among $Z_1, \ldots, Z_k$; i.e., $\dim(\mathscr{Z}(\Gamma)) = k - 1$.

Example. Let Γ be represented by Figure E7. Then $Z = \{e_1, e_2\}$ and $Z' = \{e_3, e_4, e_5\}$ are the only elementary cycles of Γ. The list $Z, Z', Z + Z'$ is a simple imbedding but it is not elementary, while Z, Z, Z', Z' is an elementary imbedding which is not simple. Up to reordering the cycles, these

E7

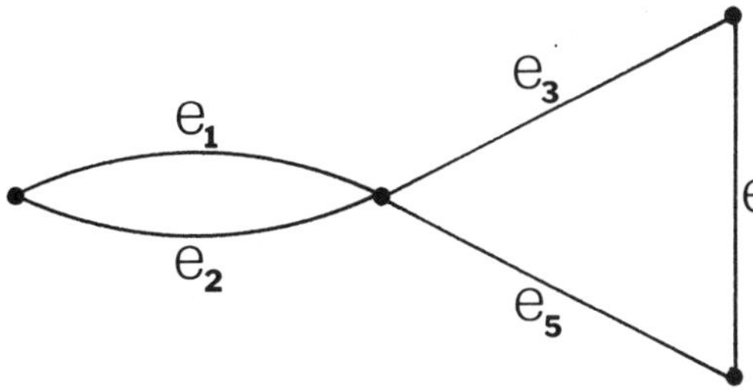

are the only planar imbeddings of Γ. Observe (in anticipation of Proposition E9) that $\mathscr{Z}(\Gamma)$ is not connected.

E8 Exercise. Prove that *every planar multigraph admits an elementary imbedding and a simple imbedding.*

E9 Proposition. *Let Γ be a planar multigraph. Every planar imbedding of Γ is both simple and elementary if and only if $\mathscr{Z}(\Gamma)$ is connected.*

PROOF. Suppose that $Z_1, \ldots, Z_k$ is a planar imbedding of Γ which is not simple, in which case, for some reordering of the cycles,

E10
$$\sum_{i=1}^{h} Z_i = \varnothing \quad \text{for some } h < k.$$

Let $\mathscr{B}_1$ and $\mathscr{B}_2$ be the subspaces spanned by $\{Z_1, \ldots, Z_h\}$ and $\{Z_{h+1}, \ldots, Z_k\}$, respectively, and let $B_j = \text{Fnd}(\mathscr{B}_j)$ for $j = 1, 2$. These subspaces are not trivial. If $e \in B_1 \cap B_2$, then by the definition of a planar imbedding, e belongs to only one of the cycles $Z_1, \ldots, Z_h$, contrary to E10. Hence $B_1 \cap B_2 = \varnothing$, and $\mathscr{B}_1 \oplus \mathscr{B}_2$ is a well-defined subspace of $\mathscr{Z}(\Gamma)$. On the other hand, if $Z \in \mathscr{Z}(\Gamma)$, then for some $a_1, \ldots, a_k \in \mathbb{K}$, we have $Z = \sum_{i=1}^{k} a_i Z_i = \sum_{i=1}^{h} a_i Z_i + \sum_{i=h+1}^{k} a_i Z_i \in \mathscr{B}_1 \oplus \mathscr{B}_2$. Hence $\mathscr{Z}(\Gamma)$ is not connected.

If we suppose instead that the planar imbedding $Z_1, \ldots, Z_k$ is not elementary, then some cycle, say Z_k, is not elementary, and by *II*C1, $Z_k = \sum_{i=1}^{n} W_i$ where $W_1, \ldots, W_n$ are pairwise-disjoint elementary cycles, and $n > 1$. Hence $Z_1, \ldots, Z_{k-1}, W_1, \ldots, W_n$ is a planar imbedding of Γ with more than $\dim(\mathscr{Z}(\Gamma)) - 1$ cycles. Thus the existence of a planar imbedding which is not elementary implies the existence of one which is not simple, and we proceed as in the previous paragraph.

Conversely, if $\mathscr{Z}(\Gamma)$ is a nontrivial direct sum $\mathscr{B}_1 \oplus \mathscr{B}_2$, we suppose that the planar imbedding $Z_1, \ldots, Z_k$ of Γ is elementary. By *II*C3, we may reindex this imbedding so that $Z_1, \ldots, Z_h \in \mathscr{B}_1$ while $Z_{h+1}, \ldots, Z_k \in \mathscr{B}_2$ for some h satisfying $1 \leq h \leq k - 1$. But since $\mathscr{B}_1$ and $\mathscr{B}_2$ have disjoint foundations, the definition of a planar imbedding implies that E10 must hold. Therefore, the imbedding is not simple. □

One of the important and topologically obvious properties of planar multigraphs is that their submultigraphs are also planar, as are the multigraphs obtainable therefrom by identifying a pair of vertices incident with a common edge.

A system $\Gamma' = (V', f', E')$ is a **contraction** of the multigraph $\Gamma = (V, f, E)$ if

(a) $V' \in \mathbb{P}(V)$ and Γ_W is connected for all $W \in V'$;
(b) $E' = \{e \in E: f(e) \nsubseteq W \text{ for all } W \in V'\}$;
(c) $f'(e) = \{W \in V': f(e) \cap W \neq \varnothing\}$, for all $e \in E'$.

A contraction of a multigraph is clearly a multigraph, but a contraction of a graph need not itself be a graph. The contraction Γ' is uniquely determined by Γ and V'. Every multigraph is obviously a contraction of itself. In Figure E11, each of the last three multigraphs is a contraction of the first.

E11

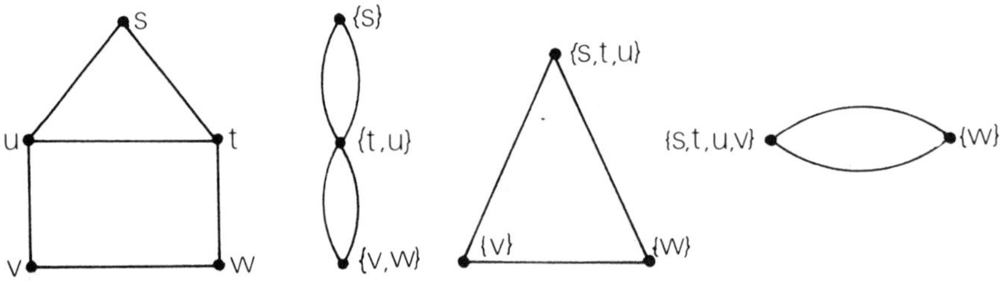

A **subcontraction** of a multigraph Γ is a submultigraph of a contraction of Γ. For example, every submultigraph of Γ is also a subcontraction of Γ.

The next two exercises cover most of the elementary but essential properties of contractions and subcontractions.

E12 Exercise. *Let* Γ_1 *be a subcontraction of* Γ. Prove:

(a) *If* $Z_1 \in \mathscr{Z}(\Gamma_1)$, *then* $Z_1 \subseteq Z$ *for some* $Z \in \mathscr{Z}(\Gamma)$.

(b) *If an edge e of* Γ_1 *is an isthmus of* Γ, *then e is an isthmus of* Γ_1.

E13 Exercise. Prove:

(a) *If* Γ_2 *is a contraction of* Γ_1 *and if* Γ_1 *is a contraction of* Γ, *then* Γ_2 *is a contraction of* Γ.

(b) *If* Γ_2 *is a subcontraction of* Γ, *then there exists a submultigraph* Γ_1 *of* Γ *such that* Γ_2 *is a contraction of* Γ_1.

(c) *Let* $\Gamma_1, \ldots, \Gamma_k$ *be multigraphs such that* Γ_i *is either a submultigraph or a contraction of* Γ_{i-1} $(i = 2, \ldots, k)$. *Then* Γ_k *is a subcontraction of* Γ_1.

(d) *If* Γ' *is a subcontraction of* Γ, *then there exists a sequence of multigraphs* $\Gamma = \Gamma_1, \ldots, \Gamma_k = \Gamma'$ *such that for each* $i = 2, \ldots, k$, $\Gamma_i = (V_i, f_i, E_i)$ *is one of the following*:

(i) $(\Gamma_{i-1})_{(x)}$ *where x is an isolated vertex of* Γ_{i-1};

(ii) $(\Gamma_{i-1})_{(e)}$ *for some* $e \in E_{i-1}$;
(iii) *the contraction of* Γ_{i-1} *obtained by identifying two vertices* $x, y \in V_{i-1}$ *incident with a common edge. Thus*

$$V_i = \{\{w\}: w \in V_{i-1} + \{x, y\}\} \cup \{\{x, y\}\}$$

and

$$E_i = E_{i-1} + f_{i-1}^{-1}[\{x, y\}].$$

E14 Lemma. *If* $\Gamma_1 = (V_1, f_1, E_1)$ *is a contraction of* Γ, *then*

$$\mathscr{Z}^{\perp}(\Gamma_1) = \mathscr{Z}^{\perp}(\Gamma) \cap \mathscr{P}(E_1) \quad \textit{and} \quad \mathscr{Z}(\Gamma_1) = \pi_{E_1}[\mathscr{Z}(\Gamma)].$$

PROOF. In the light of Exercise E13a and the fact that every contraction of Γ may be obtained by iterating the procedure of E13d(iii), we may assume that Γ_1 has been obtained from Γ by just one application of this procedure. Thus

$$V_1 = \{\{w\}: w \in V + \{x, y\}\} \cup \{\{x, y\}\}$$

and

$$E_1 = \{e \in E: f(e) \neq \{x, y\}\}.$$

$\mathscr{Z}^{\perp}(\Gamma_1)$ is spanned by the collection of its vertex cocycles, and these are

$$f_1^*(\{w\}) = f^*(w) \quad \text{for } w \in V + \{x, y\}$$

$$f_1^*(\{x, y\}) = f^*(x) + f^*(y).$$

We show that this very same collection spans $\mathscr{Z}^{\perp}(\Gamma) \cap \mathscr{P}(E_1)$. For if $C \in \mathscr{Z}^{\perp}(\Gamma) \cap \mathscr{P}(E_1)$, then $C = \sum_{w \in U} f^*(w)$ for some $U \subseteq V$. Since $C \subseteq E_1$, either $\{x, y\} \subseteq U$ or $\{x, y\} \cap U = \varnothing$. Hence $\mathscr{Z}^{\perp}(\Gamma_1) = \mathscr{Z}^{\perp}(\Gamma) \cap \mathscr{P}(E_1)$.

The second equality follows from the first and *II*A15. □

Comparing this lemma with *II*D13a and b, we see that the roles of the cycle space and cocycle space are interchanged, and one may infer a duality between contractions of Γ and submultigraphs of Γ. This principle recurs in Exercise E16.

E15 Proposition. *Every subcontraction of a planar multigraph is planar.*

PROOF. Let $\Gamma = (V, f, E)$ be a planar multigraph. By Exercise E13d, it suffices to prove that the multigraph Γ' is planar if Γ' is obtained from Γ by one of the following three operations: (i) deletion of an isolated vertex, (ii) deletion of an edge, (iii) identification of two vertices incident with a common edge.

Clearly if $\mathscr{Z}(\Gamma) = \mathscr{Z}(\Gamma')$, then any planar imbedding for Γ is also a planar imbedding for Γ'. This is indeed the case when Γ' is obtained from Γ by operation (i) or by operations (ii) or (iii) where the edge in question is an isthmus. Therefore, let $e \in \text{Fnd}(\mathscr{Z}(\Gamma))$—recall A21—and let $Z_1, \ldots, Z_k$ be a planar imbedding of Γ such that $e \in Z_1 \cap Z_2$.

To show that $Z_1 + Z_2, Z_3, \ldots, Z_k$ spans $\mathscr{Z}(\Gamma_{(e)})$, let $Z = \sum_{i=1}^{k} a_i Z_i$ for some $a_1, \ldots, a_k \in \mathbb{K}$. If $Z \in \mathscr{Z}(\Gamma_{(e)})$, then $a_1 = a_2$, and $Z = a_1(Z_1 + Z_2) + \sum_{i=3}^{k} a_i Z_i$ as required. It follows that each edge in $\text{Fnd}(\mathscr{Z}(\Gamma_{(e)}))$ belongs to at least one of the cycles $Z_1 + Z_2, Z_3, \ldots, Z_k$. Since $Z_1, \ldots, Z_k$ is a planar imbedding, each such edge is in at most two of $Z_1 + Z_2, Z_3, \ldots, Z_k$. Finally, since $(Z_1 + Z_2) + Z_3 + \ldots + Z_k = \varnothing$ (by E3), each edge in $\text{Fnd}(\mathscr{Z}(\Gamma_{(e)}))$ lies in exactly two of the cycles. Hence $Z_1 + Z_2, Z_3, \ldots, Z_k$ is a planar imbedding of $\Gamma_{(e)}$.

If Γ' is obtained by identification of the two vertices in $f(e)$, let $F = \text{Fnd}(\mathscr{Z}(\Gamma)) + f^{-1}[f(e)]$. By Lemma E14, $\mathscr{Z}(\Gamma') = \pi_F[\mathscr{Z}(\Gamma)]$, which implies that $Z_1 \cap F, \ldots, Z_k \cap F$ spans $\mathscr{Z}(\Gamma')$. As in the previous paragraph, one easily demonstrates that part (b) of the definition of planar imbedding is satisfied. □

E16 *Exercise.* Let $\Theta = (W, g, E)$.

(a) Let $\Gamma \perp \Theta$, and let e be in their (common) edge set. Let Γ' be obtained from Γ by identification of the vertices incident with e. Prove that $\Gamma' \perp \Theta_{(g^{-1}[g(e)])}$, thereby showing that the deletion of the edges having a given common image and the identification of their two incident vertices are dual operations.

(b) Let $\Gamma \perp \Theta$ and let Γ' be a subcontraction of Γ without isthmuses. Then $\Gamma' \perp \Theta'$ for some subcontraction Θ' of Θ.

E17 Corollary. *Γ is planar if and only if every lobe of Γ is planar.*

Proof. If Γ is planar, then by the proposition, every lobe is also planar. Conversely, let $\Gamma_1, \ldots, \Gamma_m$ be the lobes of Γ which are not isolated vertices or isthmuses, and let $Z_{i,1}, \ldots, Z_{i,k_i}$ be a planar imbedding of Γ_i $(i=1, \ldots, m)$. Since $\mathscr{Z}(\Gamma) = \bigoplus_{i=1}^{m} \mathscr{Z}(\Gamma_i)$ by B5, the list $Z_{1,1}, \ldots, Z_{1,k_1}, Z_{2,1}, \ldots, Z_{2,k_2}, \ldots, Z_{m,k_m}$ of cycles clearly spans $\mathscr{Z}(\Gamma)$. If $e \in \text{Fnd}(\mathscr{Z}(\Gamma))$, then $e \in \text{Fnd}(\mathscr{Z}(\Gamma_i))$ for exactly one lobe Γ_i. Hence $e \in Z_{i,j}$ for exactly two indices j, and e is in no other cycle of the planar imbedding. □

E18 Exercise. Show that *the same cycle Z can occur twice in a planar imbedding of Γ if and only if Γ_Z is a lobe of Γ.*

In the light of this exercise, whenever $\dim(\mathscr{Z}(\Theta)) > 1$ for each lobe Θ of Γ, one can treat every planar imbedding of Γ as a *set* of cycles rather than as a *list* of cycles.

The symbol $K_{m,n}$ will denote the bipartite graph $(\{V_1, V_2\}, \mathscr{E})$ where $|V_1| = m$, $|V_2| = n$, and $\mathscr{E} = \{\{x_1, x_2\}: x_i \in V_i\}$. Such a graph is called a **complete bipartite graph.**

E19 Exercise. Prove that:

(**a**) *If* $m \leq 2$ *or* $n \leq 2$, *then* $K_{m,n}$ *is planar;*

(**b**) *If* $n \leq 4$ *then* K_n *is planar.*

In the next section it will be shown (F9) that neither part of E19 can be sharpened.

E20 *Exercise.* In §*VI*E we will prove: if Γ is a triconnected graph and if $\Gamma \perp \Theta$, then Θ is likewise a triconnected graph. Assuming this result, show that if Γ is a triconnected planar graph, then the set of regions of a planar imbedding is unique and that there exists only one connected multigraph Θ such that $\Gamma \perp \Theta$.

Let $\mathbb{Z}_n$ denote the cyclic group with n elements. A **cyclic ordering** of an n-set U is a bijection $x: \mathbb{Z}_n \to U$; the image of i under x is then denoted by x_i.

E21 Lemma. *Let* $\Gamma = (V, f, E)$ *be a biconnected multigraph. Given any planar imbedding for* Γ, *for each vertex* x *of* Γ *there exists a cyclic ordering* $e_0, \ldots, e_{\rho(x)-1}$ *of the elements of* $f^*(x)$ *such that* e_i *and* e_{i+1} *lie in a common cycle of the imbedding for* $i = 0, \ldots, \rho(x) - 1$.

Proof. Let $Z_1, \ldots, Z_k$ be a planar imbedding for Γ, and let x be a vertex of Γ. By reordering if necessary, we may suppose that $Z_1, \ldots, Z_h$ are the cycles of the imbedding which contain edges incident with x. Since Γ is biconnected, the imbedding is elementary by B3 and E9. Hence for $i = 1, \ldots, h$, the cycle Z_i contains exactly two edges belonging to $f^*(x)$. Let $e_0 \in Z_1 \cap f^*(x)$. Then the edge $e_1 \in (Z_1 \cap f^*(x)) + \{e_0\}$ is uniquely determined. We suppose then that $e_0, e_1, \ldots, e_m$ have been selected and that $Z_1, \ldots, Z_h$ have been reordered so that $e_i \in Z_{i-1} \cap Z_i \cap f^*(x)$ for $i = 1, \ldots, m$, where $2 \leq m \leq h - 1$. It suffices merely to show that $e_0 \notin Z_m$.

If $e_0 \in Z_m$, then there exists an edge $e_{m+1} \in f^*(x)$ such that $e_{m+1} \notin \bigcup_{i=1}^{m} Z_i$. Since Γ is biconnected, we deduce from B3 that e_{m+1} and e_0 lie on some common elementary circuit. By A9, e_{m+1} and e_0 belong to a common elementary cycle Z. Let us write $Z = \sum_{i=1}^{k} a_i Z_i$, where $a_1, \ldots, a_k \in \mathbb{K}$. Since therefore $e_1, \ldots, e_{m-1} \notin Z$, we must have $a_1 = \ldots = a_m$. But $e_0 \in Z_m \cap Z_1$, which implies that $e_0 \notin Z$. □

Let Γ have a planar imbedding and let Δ and Δ' be elementary circuits with Z and Z' as corresponding cycles. We say Δ and Δ' **cross at a vertex** x if the two edges in $f^*(x) \cap Z$ and the two edges in $f^*(x) \cap Z'$ alternate in any cyclic ordering for $f^*(x)$ constructed as in Lemma E21. We may now state and prove:

E22 Theorem (Jordan Curve Theorem for Planar Multigraphs). *Let* Γ *be a planar multigraph. If two elementary circuits* Δ *and* Δ' *of* Γ *cross at a vertex* *in some planar imbedding of* Γ, *then they have at least one other vertex in common.*

PROOF. Let $Z_1, \ldots, Z_k$ be a planar imbedding in which Δ and Δ' cross at a vertex. We write $Z = \sum_{i=1}^{k} a_i Z_i$, where Z is the cycle of Δ. Each edge of Γ is then of one of three types: an edge is of type j ($j = 0, 1, 2$) if it belongs to exactly j cycles Z_i having coefficient $a_i = 1$. Clearly the edges of type 1 are the edges in Z. By the lemma, if the vertex x is not in Δ, then the edges in $f^*(x)$ are either all of type 0 or all of type 2.

Now let Δ' be given by $x_0, e_1, x_1, \ldots, e_n, x_n = x_0$, and suppose Δ and Δ' cross at x_0. By the lemma we may assume without loss of generality that e_1 is of type 0 and e_n is of type 2. Thus there is a first index $i > 1$ such that e_i is not of type 0. The vertex $x_{i-1} \neq x_0$ is then a vertex common to Δ and Δ'. □

To conclude this section, we give a heuristic proof of the result of MacLane stated at the beginning of the section, namely that our combinatorial definition of planarity is equivalent to the usual topological definition.

Consider first the case where Γ is an arbitrary biconnected multigraph and, by our definition, planar. Let $Z_1, \ldots, Z_k$ be the regions of a planar imbedding of Γ. By B3 and E9, $Z_1, \ldots, Z_k$ are all elementary. Hence Γ_i, defined to be Γ_{Z_i}, is an elementary circuit of Γ. Let D_i be a topological disk whose boundary is a topological realization of Γ_i. Identify D_i with D_j along each edge and each vertex that Γ_i and Γ_j have in common. Let K denote the resulting cell complex. Clearly each point of K which is not a vertex of Γ has a neighborhood homeomorphic with a disk. By Lemma E21, each vertex of Γ also has a neighborhood in K homeomorphic with a disk. Hence K is a surface. To see that K is in fact a sphere, observe that the topological imbedding of Γ into K yields a cell decomposition of K with Euler characteristic

$$k - \nu_1(\Gamma) + \nu_0(\Gamma) = \dim(\mathscr{Z}(\Gamma)) - \nu_1(\Gamma) + \nu_0(\Gamma) + 1,$$

which equals 2 by A15b.

Now assume Γ is an arbitrary planar multigraph (again by our definition). By E17 each lobe of Γ is planar, and by the above argument each lobe of Γ is planar in the topological sense. Since the statement of E17 is also valid when the terms are understood topologically, Γ is therefore planar in the topological sense.

Conversely, if Γ is any multigraph which is planar in the topological sense, we need only delete the isthmuses. The bounding cycles of the regions then form a planar imbedding by our definition.

It should be apparent now that our special treatment of isthmuses is necessitated by the fact that no 1-subset of edges can be a cycle in a multigraph—as we have defined multigraphs. In order to accommodate such cycles, we would have to introduce "loops," i.e., edges which join a vertex to itself. In Figure E23 we carry out, for a multigraph Γ with isthmuses, the topological construction of an orthogonal multigraph, illustrating that "loops" must then be included.

E23

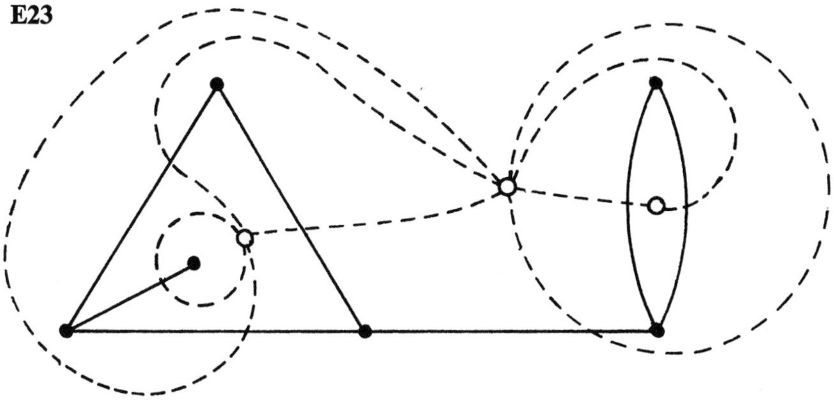

E24 Exercise. *Let Γ be a connected planar multigraph. Let Z be a cycle of Γ which is not a region of some planar imbedding of Γ, and let U be the set of vertices incident with edges in Z.* Show that *if Γ_U is an elementary circuit, then $\Gamma_{(U)}$ is not connected.*

IIIF Euler's Formula

We have seen by example in the previous section that a planar multigraph need not uniquely determine which cycles will be regions of a simple planar imbedding. It does, however, determine the *number* of regions in any simple planar imbedding.

F1 Proposition. *The number of regions in any simple planar imbedding of a multigraph Γ is*

$$\nu_1(\Gamma) - \nu_0(\Gamma) + \nu_{-1}(\Gamma) + 1.$$

PROOF. If $Z_1, \ldots, Z_k$ is a simple planar imbedding of Γ, then by the definition and A15b,

$$k = \dim(\mathscr{Z}(\Gamma)) + 1 = \nu_1(\Gamma) - \nu_0(\Gamma) + \nu_{-1}(\Gamma) + 1. \qquad \square$$

The number of regions in a simple planar imbedding of Γ is thus a parameter of Γ and is denoted by $\nu_2(\Gamma)$, abbreviated by ν_2 when there is no risk of confusion. We now state a result familiar to both graph theorists and topologists, according to how the symbols are interpreted.

F2 Corollary. (a) *For any planar multigraph,*

$$\nu_2 - \nu_1 + \nu_0 - \nu_{-1} = 1.$$

(b) (The Euler Formula). *For any connected planar multigraph,*

$$\nu_2 - \nu_1 + \nu_0 = 2.$$

For the remainder of this section it will be understood that $\Gamma = (V, f, E)$ denotes a planar multigraph. We define the **average covalence** of Γ to be the number

F3
$$\rho^{\perp}(\Gamma) = \frac{2\nu_1(\Gamma)}{\nu_2(\Gamma)},$$

writing simply $\rho^{\perp}$ when there is no risk of confusion. In terms of $\rho = \rho(\Gamma)$ and $\rho^{\perp}$, the Euler Formula has another useful form:

F4 Corollary. *For a connected planar multigraph with* $\nu_1 > 0$,

$$\frac{1}{\rho} + \frac{1}{\rho^{\perp}} = \frac{1}{2} + \frac{1}{\nu_1}.$$

PROOF. One merely substitutes into the Euler Formula the values $\nu_0 = 2\nu_1/\rho$ and $\nu_2 = 2\nu_1/\rho^{\perp}$ from A1 and F3, respectively, and then divides by $2\nu_1$. □

Suppose now that I is the set of isthmuses of Γ and that $Z_1, \ldots, Z_{\nu_2(\Gamma)}$ is a planar imbedding of Γ. The **covalence** of the region Z_i is the integer $\rho^{\perp}(Z_i) = |Z_i|$. Note that if $\Gamma_{(I)} \perp \Theta$ where $\Theta = (U, g, E + I)$ and $Z_i = g^*(u_i)$ for $u_i \in U$ $(i=1, \ldots, \nu_2(\Gamma_{(I)}))$, then the covalence of $g^*(u_i)$ is precisely the valence of u_i. Thus not only do we have

F5
$$\Gamma_{(I)} \perp \Theta \Rightarrow \nu_2(\Gamma) = \nu_0(\Theta),$$

but from A1,

F6
$$\nu_1(\Gamma) = \frac{1}{2}\left(\sum_{i=1}^{\nu_2(\Gamma)} \rho^{\perp}(Z_i)\right) + |I|.$$

F7 Proposition. *Let* $Z_1, \ldots, Z_{\nu_2}$ *be a planar imbedding for* Γ.

(a) *If* Γ *has no isthmuses, then*

$$\rho^{\perp}(\Gamma) = \frac{1}{\nu_2(\Gamma)} \sum_{i=1}^{\nu_2(\Gamma)} \rho^{\perp}(Z_i).$$

(b) *There exists some index* i *such that* $\rho^{\perp}(Z_i) \leq \rho^{\perp}(\Gamma)$.

PROOF. From F6 we have

$$\frac{2(\nu_1 - |I|)}{\nu_2} = \frac{1}{\nu_2} \sum_{i=1}^{\nu_2} \rho^{\perp}(Z_i).$$

Hence

$$\rho^{\perp}(\Gamma) - \frac{2|I|}{\nu_2} = \frac{1}{\nu_2} \sum_{i=1}^{\nu_2} \rho^{\perp}(Z_i),$$

and (a) follows when $I = \varnothing$. If $\rho^{\perp}(Z_i) > \rho^{\perp}(\Gamma)$ for all i, then

$$\rho^{\perp}(\Gamma) - \frac{2|I|}{\nu_2} = \frac{1}{\nu_2} \sum_{i=1}^{\nu_2} \rho^{\perp}(Z_i) > \rho^{\perp}(\Gamma),$$

which is impossible. □

Suppose that Γ is without isthmuses and that $Z_1, \ldots, Z_{\nu_2(\Gamma)}$ is a simple planar imbedding of Γ. If for some $k \in \mathbb{N}$, $\rho^{\perp}(Z_i) = k$ for all $i = 1, \ldots, \nu_2(\Gamma)$, then the planar imbedding is said to be **k-covalent**. If for some $k \in \mathbb{N}$, every simple planar imbedding of Γ is k-covalent, we say that Γ is **k-covalent**. We say that Γ is **isocovalent** if it is k-covalent for some $k \in \mathbb{N}$.

Examples. The circuit Δ_n is n-covalent. K_4 is both 3-valent and 3-covalent. A multigraph with $\nu_0 = 2$ and $\nu_1 \geq 2$ is ν_1-valent and 2-covalent. The multigraph shown in Figure F8 has a 3-covalent planar imbedding, as shown in F8a, but it is not a 3-covalent multigraph, as seen by F8b.

F8

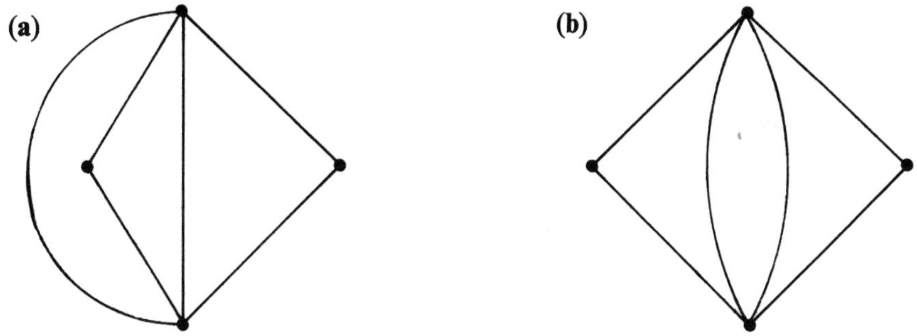

The following necessary condition for planarity is known as the "Kuratowski criterion." The proof of its sufficiency is much more difficult and will be given in the next section.

F9 Proposition. *Any graph having* K_5 *or* $K_{3,3}$ *as a subcontraction is not planar.*

PROOF. By E15 it suffices to prove only that neither K_5 nor $K_{3,3}$ is planar.

By D13, $\mathscr{Z}(K_5)$ is not graphic, and since K_5 has no isthmus, K_5 is not planar, by Proposition E6.

Now suppose that $K_{3,3}$ is planar. By substituting $\nu_1(K_{3,3}) = 9$ and $\rho(K_{3,3}) = 3$ into F4, we obtain $\rho^{\perp}(K_{3,3}) = \frac{18}{5} < 4$. Hence some cycle of $K_{3,3}$ consists of three or fewer edges, which is impossible since $K_{3,3}$ is a bipartite graph. □

F10 *Exercise.* Prove that K_5 is not planar directly from F4.

F11 Proposition. *Let* $\Gamma = (V, f, E)$ *be a planar multigraph.*

(a) *If* $\rho^{\perp}(\Gamma) \geq 3$, *then* $\rho(x) \leq 5$ *for some* $x \in V$.

(b) *If* $\rho(\Gamma) \geq 3$, *then every planar imbedding contains a region* Z *such that* $\rho^{\perp}(Z) \leq 5$.

PROOF. (a) Since $1/\rho^{\perp}(\Gamma) \le \frac{1}{3}$, F4 yields

$$\frac{1}{\rho(\Gamma)} \ge \frac{1}{2} + \frac{1}{\nu_1(\Gamma)} - \frac{1}{3} > \frac{1}{6}.$$

Hence $\rho(\Gamma) < 6$, and $\rho(x) < 6$ for at least one vertex $x \in V$.

(b) Similarly, if $\rho(\Gamma) \ge 3$, then $\rho^{\perp}(\Gamma) < 6$. The result then follows by Corollary F7b. □

F12 Corollary. *Every planar graph has a vertex of valence at most* 5.

PROOF. Every nonempty cycle of a graph Γ has covalence at least 3. By F7b, $\rho^{\perp}(\Gamma) \ge 3$. The result follows from F11a above. □

F13 Exercise. Show that *if a planar graph has smallest valence* 5, *then it has at least* 12 *vertices of valence* 5.

F14 *Exercise.* Let Γ be a planar graph with $\nu_0(\Gamma) \ge 4$. Then Γ has at least four vertices of valence at most 5.

F15 Proposition. *Let* Γ *be a planar isovalent multigraph without isthmuses or isolated vertices, and let* Γ *have an m-covalent imbedding for some m. Then the parameters* $\nu_0, \nu_1, \nu_2, \rho, \rho^{\perp}$ *of* Γ *must conform to one of the seven types in Table* F16, *where k is any integer greater than* 1.

F16

Type	ν_0	ν_1	ν_2	ρ	$\rho^{\perp}$
I	2	k	k	k	2
II	k	k	2	2	k
III	4	6	4	3	3
IV	6	12	8	4	3
V	8	12	6	3	4
VI	12	30	20	5	3
VII	20	30	12	3	5

PROOF. Since Γ is isovalent, ρ is an integer. Since Γ has no isolated vertices or isthmuses, $\rho > 1$. By F7a, $\rho^{\perp}$ is likewise an integer greater than 1. The parameters ν_1, ρ, and $\rho^{\perp}$ of Γ thus form an integral solution to the system of inequalities:

F17 $$\rho \ge 2, \qquad \rho^{\perp} \ge 2, \qquad \nu_1 \ge 2, \quad \text{and} \quad \frac{1}{\rho} + \frac{1}{\rho^{\perp}} = \frac{1}{2} + \frac{1}{\nu_1};$$

this last equation is from Corollary F4.

If $\rho \ge 4$ and $\rho^{\perp} \ge 4$, we have $(1/\rho) + (1/\rho^{\perp}) \le \frac{1}{2}$ and the last equality in F17 cannot hold. Hence either $\rho \le 3$ or $\rho^{\perp} \le 3$. We should also observe that the system F17 is symmetric in ρ and $\rho^\perp$. Thus if we find the solutions with $\rho \geq \rho^\perp$, we may obtain all other solutions by interchanging the values of ρ and $\rho^\perp$.

Case 1: $\rho^\perp = 2$. Then F17 becomes:

$$\rho \geq 2, \qquad \nu_1 \geq 2, \qquad \frac{1}{\rho} = \frac{1}{\nu_1}.$$

This yields the solutions $\rho = \nu_1 = k$, $\rho^\perp = 2$, for $k = 2, 3, \ldots$.

Case 2: $\rho^\perp = 3$. Then F17 becomes:

$$\rho \geq 3, \qquad \nu_1 \geq 2, \qquad \frac{1}{\rho} = \frac{1}{6} + \frac{1}{\nu_1}.$$

Clearly we must have $\rho < 6$, i.e., $\rho = 3$, 4, or 5, whence the solutions:

$$\rho = 3, \qquad \nu_1 = 6;$$
$$\rho = 4, \qquad \nu_1 = 12;$$
$$\rho = 5, \qquad \nu_1 = 30.$$

We have then a total of 4 solutions with $\rho \geq \rho^\perp$ and we get three more by symmetry. Using the definitions of average valence and average covalence, we may compute ν_0 and ν_2 and fill in the table. □

F18 Proposition. *For each integer $k \geq 2$, there is a unique planar multigraph of each of the Types I and II in Table* F16.

Proof. Suppose $\Gamma = (V, f, E)$ is of Type I. Since $\nu_0 = 2$, $f(e) = V$ for all $e \in E$. Since $\nu_1 = k$, Γ is uniquely determined. To prove existence, let $\Gamma = (V, f, E)$ where $\nu_0 = 2$, $E = \{e_1, \ldots, e_k\}$ and $f(e_i) = V$ for all $i = 1, \ldots, k$. It is easy to see that $\mathscr{Z}(\Gamma) = \mathscr{E}(E)$. It is also easy to see that

$$\{Z_i = \{e_i, e_{i+1}\} : i = 1, \ldots, k - 1\}$$

is a basis for $\mathscr{Z}(\Gamma)$. It follows at once that $Z_1, \ldots, Z_k$, where $Z_k = \{e_k, e_1\}$, is a planar imbedding of Γ and that Γ is of Type I.

The remainder of the proof (for Type II) is left as an **exercise** for the reader. □

A planar multigraph of one of the five remaining types is, in fact, a graph and is called a **Platonic graph**. This name comes from the fact that the 1-skeletons of the five Platonic solids are graphs of these five types. They are illustrated in Figures F19 through F23. Actually, these are the only multigraphs of these types, i.e., there are precisely five Platonic graphs. We will prove the uniqueness for Types III, IV, and V, leaving the "dirty" cases of Types VI and VII as exercises for the reader.

F19

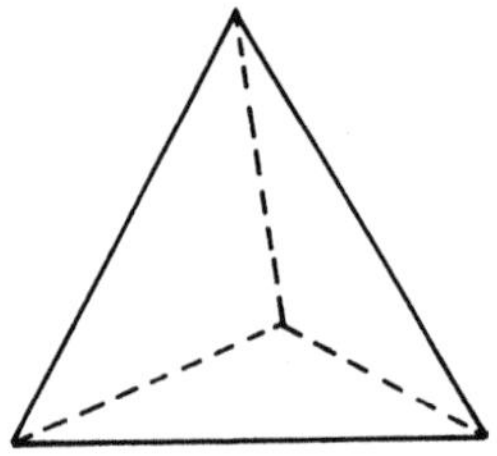

Tetrahedron

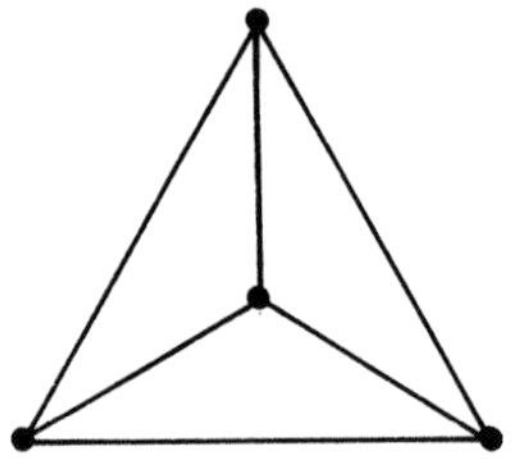

Type III

F20

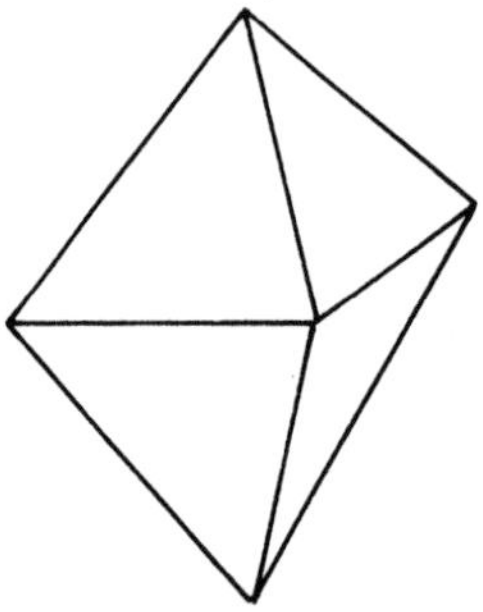

Octahedron

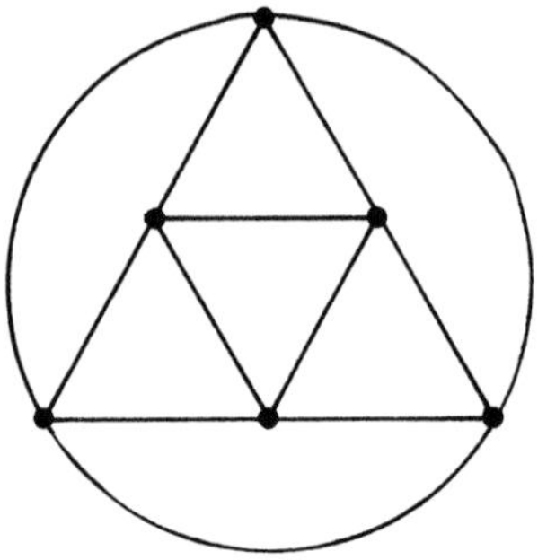

Type IV

F21

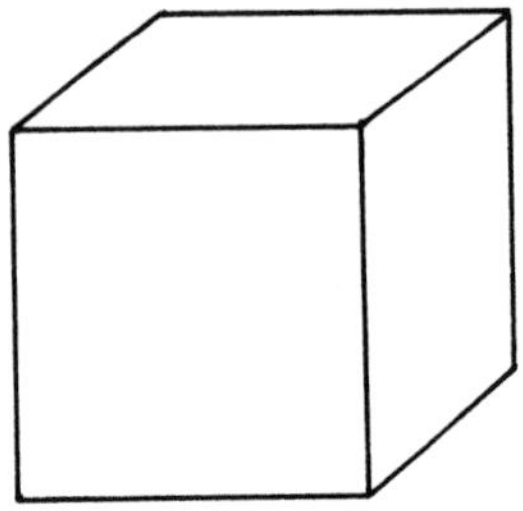

Cube

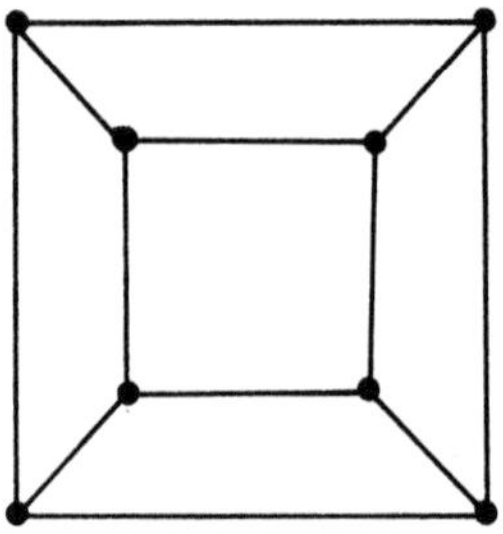

Type V

F22

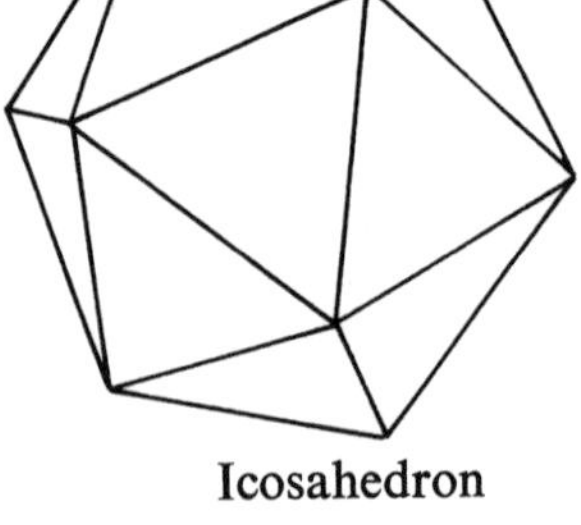

Icosahedron

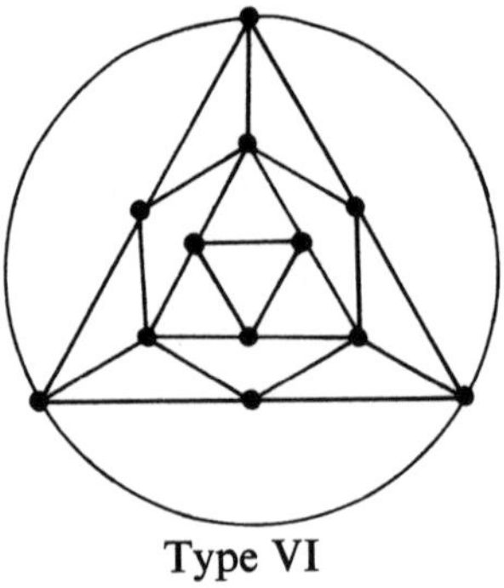

Type VI

F23

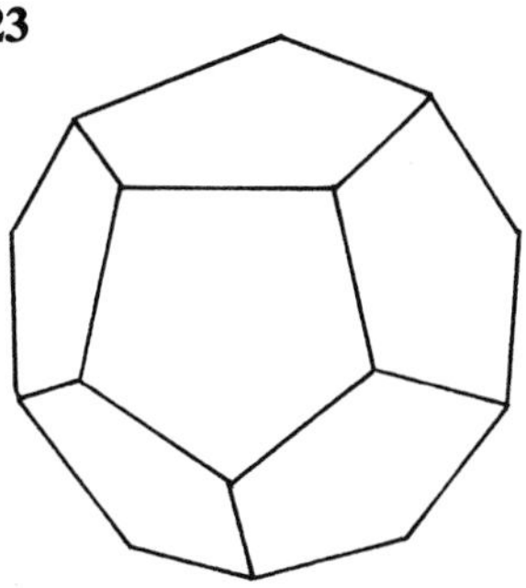

Dodecahedron

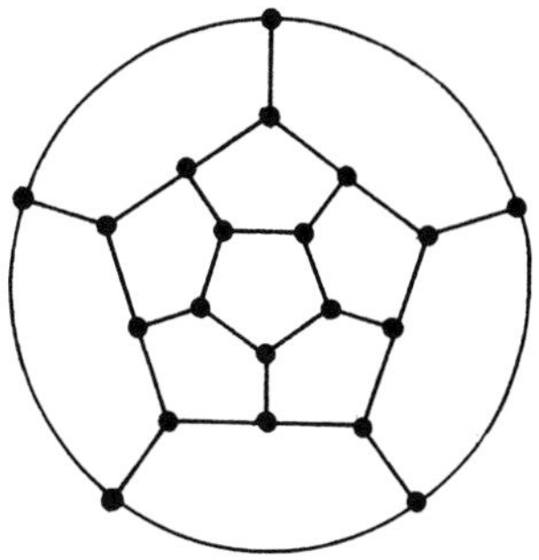

Type VII

F24 Lemma. *If* Γ *is a Platonic graph, there exists a connected Platonic graph* Θ *such that* $\Gamma \perp \Theta$.

PROOF. Let Γ be a Platonic graph and let $Z_1, \ldots, Z_k$ be the regions of an isocovalent imbedding of Γ. By the definition and D3, $Z_1, \ldots, Z_k$ form the vertex cocycles of a connected multigraph Θ orthogonal to Γ. Clearly Θ is isovalent without isthmuses or isolated vertices. The vertex cocycles of Γ form a $\rho(\Gamma)$-covalent imbedding of Θ. Hence by F15, Θ is of one of the types in Table F16. Since $\rho(\Theta) = \rho^{\perp}(\Gamma)$ and $\rho^{\perp}(\Theta) = \rho(\Gamma)$, Θ is Platonic. □

F25 Lemma. *Every Platonic graph is biconnected.*

PROOF. By two successive applications of the foregoing Lemma, we infer the existence of a connected Platonic graph Γ of each type.

If $\rho(\Gamma) = 3$, let x_0 be an articulation vertex of Γ. Since $f^*(x_0)$ is not elementary, while $|f^*(x_0)| = 3$, at least one of the edges in $f^*(x_0)$ is an isthmus. But by the definition of a Platonic graph, Γ contains no isthmuses. If $\rho(\Gamma) > 3$, then by the lemma, there exists a Platonic graph Θ such that $\Theta \perp \Gamma$. From F16 we see that $\rho(\Theta) = \rho^{\perp}(\Gamma) = 3$. Hence Θ is biconnected and it follows that Γ is biconnected. □

F26 Lemma. *If* $\Gamma = (V, f, E)$ *is Platonic and* $\rho(\Gamma) = 3$, *then* Γ *is a graph.*

PROOF. Suppose $f(e_1) = f(e_2) = \{x, y\}$. Since $\rho(x) = 3$, there is a third edge e_3 such that $x \in f(e_3)$. Hence $f^*(x) = \{e_1, e_2, e_3\}$. For any region Z of an isocovalent imbedding of Γ, $|Z \cap \{e_1, e_2, e_3\}|$ is even. Since, in addition, each of these edges belongs to two regions, there exist regions Z_1, Z_2, Z_3 such that $Z_i \cap f^*(x) = \{e_j, e_k\}$ for $\{i, j, k\} = \{1, 2, 3\}$. Hence Z_3 contains the cycle $\{e_1, e_2\}$. Since $\rho^{\perp}(\Gamma) \geq 3, Z_3$ is not elementary. But since Γ is biconnected, each region must be elementary by E9. □

F27 Proposition. *There is only one Platonic graph of each of the Types III, IV, and V. Furthermore, each is an isocovalent graph.*

Proof. Assume Γ is of Type III. By F26, it is a graph. Since it has 4 vertices each of valence 3, it must be K_4, which is isocovalent since it has a unique planar imbedding.

Assume Γ is of Type V. It is also a graph by F26. Furthermore, $\mathscr{Z}(\Gamma)$ is even, and hence by A13, Γ is bipartite, say with respect to $\{V_1, V_2\}$. Since Γ is isovalent, $|V_1| = |V_2|$. Thus Γ is a subgraph of $K_{4,4}$. Since $\nu_1(\Gamma) = 12 = \nu_1(K_{4,4}) - 4$, Γ must be obtained by deleting 4 edges from $K_{4,4}$ which reduces the valence of each vertex by 1. This can be done in essentially one way, yielding the cube (F21).

Assume Γ is of Type IV. Let Θ be Platonic and let $\Theta \perp \Gamma$. We have just shown that Θ is the cube. Since the cube has no cocycle consisting of two edges, Γ has no cycle of length 2 and is therefore a graph. Hence Γ is a subgraph of K_6. Since $\nu_1(\Gamma) = \nu_1(K_6) - 3$, Γ must be obtained by deleting three edges from K_6 in such a way as to reduce the valence of each vertex by 1. This can be done in only one way (up to system-isomorphism) yielding the octahedron (F20). One may easily verify that the cube and the octahedron have each only one planar imbedding and hence are isocovalent. □

IIIG Kuratowski's Theorem

This section is devoted to proving:

G1 Theorem (Kuratowski [k.6]). *A necessary and sufficient condition for a multigraph* Γ *to be planar is that neither* K_5 *nor* $K_{3,3}$ *is a subcontraction of* Γ.

Actually Kuratowski's original formulation was slightly different:

A necessary and sufficient condition for a graph to be planar is that it have no subgraph "homeomorphic" to K_5 *or* $K_{3,3}$.

Our combinatorial (and therefore nontopological) approach to graph theory precludes our proving the theorem in its original form. To understand better the relationship between Theorem G1 and Kuratowski's original statement, consider the graph obtained by replacing some of the edges of K_5 (or $K_{3,3}$) by elementary paths of length more than 1. The resulting graph is "homeomorphic" to K_5 (or $K_{3,3}$) but only the appropriate contraction of it is K_5 (or $K_{3,3}$). The closest we can come to Kuratowski's original statement is:

G2 Corollary. *If* Γ *is a nonplanar multigraph with* $\rho(x) \geq 3$ *for every vertex* x *and such that every subgraph (other than* Γ *itself) is planar, then* Γ *is* K_5 *or* $K_{3,3}$.

The proofs of these results will come at the end of this section.

Let $\Gamma = (V, f, E)$ be a connected multigraph. A set $W \subseteq V$ is called a **separating set** of Γ if $\nu_{-1}(\Gamma_{(W)}) > 1$. In E17, we proved that a graph is planar if and only if its lobes are planar. In the next proposition we extend this result to biconnected and triconnected multigraphs.

G3 Proposition. *Let* $\Gamma = (V, f, E)$ *be a connected multigraph, let* W *be a minimal separating set such that* $|W| \leq 3$. *Let* U_1 *be the vertex set of a component of* $\Gamma_{(W)}$, *and let* $U_2 = V + W + U_1$. *Then* Γ *is planar if and only if the contraction* Γ_j *obtained by contracting* U_j *to a single vertex is planar for* $j = 1, 2$.

Proof. If Γ is planar, then so are Γ_1 and Γ_2 by E15.

The converse follows from E17 in the case $|W| = 1$. We will prove the converse in the case $|W| = 3$, leaving the simpler case $|W| = 2$ to the reader.

Let $W = \{x_1, x_2, x_3\}$, and let $\{x_0, e_1, e_2, e_3\}$ be a set of four distinct elements disjoint from all sets in question. For $j = 1, 2$, let $V_j = U_j + W + \{x_0\}$; let F_1 be the edge set of $\Gamma_{W + U_1}$; let $F_2 = E + F_1$; let $E_j = F_j \cup \{e_1, e_2, e_3\}$; and finally let $\Theta_j = (V_j, f_j, E_j)$ where

$$f_j(e) = \begin{cases} \{x_0, x_i\} & \text{if } e = e_i \ (i = 1, 2, 3); \\ f(e) & \text{otherwise} \end{cases}$$

(see Figure G4). One easily sees that Θ_j is a subcontraction of Γ.

G4

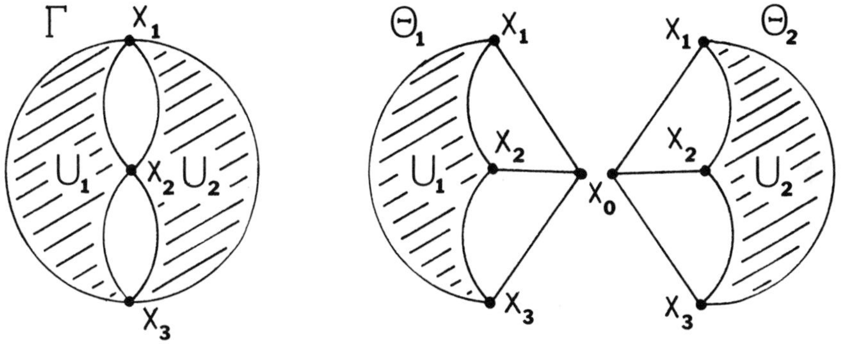

Γ_j and Θ_j may differ only in that Γ_j may admit more than one edge whose image is $\{x_0, x_i\}$. By assumption Γ_j is planar, and so by E15, Θ_j is planar.

Let $Z_1^j, \ldots, Z_{m_j}^j$ be a simple imbedding of Θ_j $(j = 1, 2)$. If e_i were an isthmus in either Θ_1 or Θ_2, then x_i would be an articulation vertex of Γ, contrary to the minimality of W. Hence e_i is contained in two of the cycles of the imbedding of Θ_j $(i = 1, 2, 3; j = 1, 2)$. Since $\rho(x_0) = 3$, any cycle containing one of e_1, e_2, e_3 contains exactly two of these edges. We may therefore assume without loss of generality that $\{e_2, e_3\} \subseteq Z_1^j$, $\{e_1, e_3\} \subseteq Z_2^j$, and $\{e_1, e_2\} \subseteq Z_3^j$ for $j = 1, 2$. Now consider the list

G5 $\quad Z_1^1 + Z_1^2, Z_2^1 + Z_2^2, Z_3^1 + Z_3^2, Z_4^1, \ldots, Z_{m_1}^1, Z_4^2, \ldots, Z_{m_2}^2.$

We assert that this list is a planar imbedding of Γ.

Part 1: *The cycles in* G5 *are all in* $\mathscr{Z}(\Gamma)$ *and each edge of* Γ *which is not an isthmus belongs to exactly two of these cycles.*

For $i \geq 4$, $j = 1, 2$, $Z_i^j \subseteq F_j$ and hence is a cycle of $(\Theta_j)_{U_j+W} = \Gamma_{U_j+W}$. But any cycle of Γ_{U_j+W} is a cycle of Γ. We observe that $Z_1^j + \{e_2, e_3\}$ is the edge set of an x_2x_3-path in Γ_{U_j+W}. Thus, $Z_1^1 + Z_1^2$ is the edge set of a circuit in Γ passing through x_2 and x_3, and by A9, $Z_1^1 + Z_1^2 \in \mathscr{Z}(\Gamma)$. Similarly $Z_2^1 + Z_2^2$ and $Z_3^1 + Z_3^3$ are cycles of Γ.

Observe that E is the disjoint union of F_1 and F_2, and that if $e \in F_j$ is an isthmus of Θ_j, then it is an isthmus of Γ. Thus if $e \in F_j$ and is not an isthmus of Γ, it belongs to two of the cycles $Z_1^j, \ldots, Z_{m_j}^j$, and hence to two of the cycles in G5.

Part 2: *The set of cycles in* G5 *spans* $\mathscr{Z}(\Gamma)$.

Since $Z_1^j, \ldots, Z_{m_j}^j$ is a simple imbedding of Θ_j, we have by Euler's Formula:

$$m_j - \nu_1(\Theta_j) + \nu_0(\Theta_j) = 2, \quad \text{for } j = 1, 2.$$

Adding these two equations together yields

$$m_1 + m_2 - (\nu_1(\Gamma) + 6) + (\nu_0(\Gamma) + 5) = 4,$$

which in turn yields by A15b,

$$m_1 + m_2 - 3 = \nu_1(\Gamma) - \nu_0(\Gamma) + 2 = \dim(\mathscr{Z}(\Gamma)) + 1.$$

Thus there are exactly $\dim(\mathscr{Z}(\Gamma)) + 1$ cycles in G5. It suffices to show that there is only one nontrivial relation over $\mathbb{K}$ among these cycles. Assume

G6
$$\sum_{i=1}^{3} a_i^1(Z_i^1 + Z_i^2) + \sum_{j=1}^{2} \sum_{i=4}^{m_j} a_i^j Z_i^j = \varnothing,$$

where $a_i^j \in \mathbb{K}$ for $j = 1, 2$; $i = 1, \ldots, m_j$. Then

$$\sum_{i=1}^{m_1} a_i^1 Z_i^1 + \sum_{i=1}^{m_2} a_i^2 Z_i^2 = \varnothing,$$

where $a_i^2 = a_i^1$ for $i = 1, 2, 3$. From this we conclude

$$\sum_{i=1}^{m_1} a_i^1 Z_i^1 = \sum_{i=1}^{m_2} a_i^2 Z_i^2 = Z,$$

where Z is a cycle in $E_1 \cap E_2 = \{e_1, e_2, e_3\}$. It follows that $Z = \varnothing$. Since $Z_1^j, \ldots, Z_{m_j}^j$ is a simple imbedding of Θ_j, it follows that $a_1^j = a_2^j = \ldots = a_{m_j}^j$ for $j = 1, 2$. But $a_1^1 = a_1^2$, hence all of the coefficients in G6 are equal. □

A subcontraction Θ of Γ is **proper** if Θ is not isomorphic to Γ.

G7 Exercise. *Let* Γ *be a nonplanar multigraph with the property that every proper subcontraction is planar.* Prove that *then*:

(a) Γ *is a triconnected graph*;

(b) *If W is any separating 3-set, then W is the set of vertices incident with a vertex of valence 3.*

G8 Lemma. *Let $\Gamma = (V, \mathscr{E})$ be a nonplanar graph such that every proper subcontraction is planar. If for some edge the subgraph obtained by deleting the two vertices incident with that edge is an elementary circuit, then Γ is K_5 or $K_{3,3}$.*

PROOF. Let $V = \{y_1, y_2, x_1, \ldots, x_k\}$, and suppose $\{y_1, y_2\}, \{x_i, x_{i+1}\} \in \mathscr{E}$, the indices being read modulo k, and that there are no other edges of the form $\{x_i, x_j\}$.

Case 1: Three or more of the vertices $x_1, \ldots, x_k$ have valence 4 (i.e., *are incident with both y_1 and y_2*). Say x_1, x_p, and x_q all have valence 4 ($1 < p < q \le k$). Then the contraction defined by the partition

$$\{\{x_1, \ldots, x_{p-1}\}, \{x_p, \ldots, x_{q-1}\}, \{x_q, \ldots, x_k\}, \{y_1\}, \{y_2\}\}$$

of V contains K_5 as a subgraph. We conclude Γ is K_5.

Case 2: At most two of the vertices in $\{x_1, \ldots, x_k\}$ have valence 4. Since $|V| > 4$ by E19b, and since each vertex of Γ has valence at least 3 by G7a, we may assume x_1 is incident with y_1 and not with y_2 and that x_k is incident with y_2. Now let p be the least index such that x_p is incident with y_2, and let q be the first index in cyclic order after p such that x_q is incident with y_1. If $q = k$ or 1, we assert that Γ is planar, contrary to assumption. We have illustrated this fact in Figure G9, leaving the reader to list the cycles of this imbedding. We assume then that x_1, x_p, x_q, and x_k are distinct. Then the contraction defined by the partition

$$\{\{x_1, \ldots, x_{p-1}\}, \{x_p, \ldots, x_{q-1}\}, \{x_q, \ldots, x_{k-1}\}, \{x_k\}, \{y_1\}, \{y_2\}\}$$

has a subgraph isomorphic to $K_{3,3}$. We conclude Γ is $K_{3,3}$. □

G9

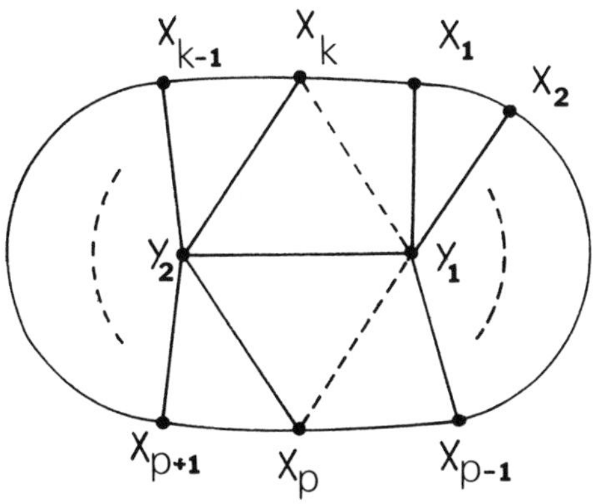

G10 Exercise. Prove that *if Γ is a triconnected graph such that for each edge the two vertices incident with that edge are also incident with a common vertex of valence* 3, *then Γ is K_4.*

G11 Proposition. *If* $\Gamma = (V, \mathscr{E})$ *is a nonplanar graph such that every proper subcontraction is planar, then* Γ *is* K_5 *or* $K_{3,3}$.

Proof. By Exercise G7a, Γ is triconnected. If $\{x, y\} = E \in \mathscr{E}$ and if Γ_{V+E} (the graph obtained by deleting the vertices x and y) is not biconnected, then by G7b, x and y are incident with a common vertex of valence 3. It follows from Exercise G10 that there exists at least one edge $E_0 = \{x_1, x_2\}$ such that Γ_{V+E_0} is biconnected.

Let $\{E_0, E_1^j, E_2^j, \ldots, E_{m_j}^j\}$ be the edges incident with x_j ($j = 1, 2$), and let $\mathscr{E}'$ be the set of edges incident with neither x_1 nor x_2. Let x_0 be an element distinct from all other objects under consideration, let $F_i^j = E_i^j + \{x_0, x_j\}$ for $j = 1, 2;\ i = 1, \ldots, m_j$, and let $\mathscr{F} = \{F_i^j : j = 1, 2; i = 1, \ldots, m_j\}$. Finally let $\Theta = (V + \{x_0, x_1, x_2\}, \mathscr{E}' \cup \mathscr{F}\}$. Thus Θ is isomorphic with a subgraph of the contraction of Γ obtained by contracting $\{x_1, x_2\}$ to a single vertex. By assumption Θ is planar.

Let $Z_1, \ldots, Z_m$ be a planar imbedding of Θ. Since Γ is biconnected, Θ is biconnected, and by E9 and B3, this imbedding is both simple and elementary. Thus each region Z_i contains exactly two or none of the edges in $\mathscr{F}$. Assume that $Z_1, \ldots, Z_k$ are the cycles in the list avoiding $\mathscr{F}$, and let $Z = Z_{k+1} + \ldots + Z_m$. We assert that $Z_1, \ldots, Z_k, Z$ is a planar imbedding of $\Theta_{(x_0)} = (V + \{x_1 x_2\}, \mathscr{E}')$.

Since each edge of $\mathscr{F}$ is contained in two of the cycles $Z_{k+1}, \ldots, Z_m$ and since each of these cycles contains two of the edges in $\mathscr{F}$, $|\mathscr{F}| = m - k$. Hence

$$\begin{aligned}\dim(\mathscr{Z}(\Theta_{(x_0)})) &= \nu_1(\Theta_{(x_0)}) - \nu_0(\Theta_{(x_0)}) + 1\\ &= (\nu_1(\Theta) - |\mathscr{F}|) - (\nu_0(\Theta) - 1) + 1\\ &= \dim(\mathscr{Z}(\Theta)) + 1 - |\mathscr{F}| = k.\end{aligned}$$

Furthermore, one easily sees that $Z_1, \ldots, Z_k, Z$ satisfies only one nontrivial relation over $\mathbb{K}$. (Any relation among $Z_1, \ldots, Z_k, Z$ yields a relation among $Z_1, \ldots, Z_m$.) We conclude that $Z_1, \ldots, Z_k, Z$ spans $\mathscr{Z}(\Theta_{(x_0)})$ and that every edge of $\Theta_{(x_0)}$ belongs to a positive even number of the cycles in this list. Finally, no edge of $\Theta_{(x_0)}$ can belong to more than two of the cycles $Z_1, \ldots, Z_k, Z$. Thus $Z_1, \ldots, Z_k, Z$ is a planar imbedding of $\Theta_{(x_0)}$. Since by assumption $\Theta_{(x_0)} = \Gamma_{V+E_0}$ is biconnected, this imbedding is elementary, and in particular Z is an elementary cycle.

Let $\mathscr{E}'' = \mathscr{E} + \mathscr{E}'$ (i.e., the set of edges of Γ incident with x_1 or x_2) and consider the subgraph $\Gamma_{Z+\mathscr{E}''}$. If $\Gamma = \Gamma_{Z+\mathscr{E}''}$, then Γ satisfies the hypothesis of Lemma G8 and we conclude Γ is K_5 or $K_{3,3}$. We suppose then that $\Gamma_{Z+\mathscr{E}''}$ is a proper subgraph of Γ and show that this supposition leads to a contradiction.

By hypothesis, $\Gamma_{Z+\mathscr{E}''}$ is planar. There exists a planar imbedding $Z_1, \ldots, Z_{n+1}$ of $\Gamma_{Z+\mathscr{E}''}$. Since $\Gamma_{Z+\mathscr{E}''}$ is clearly biconnected, this imbedding is both simple and elementary. Thus

$$\begin{aligned}n = \dim(\mathscr{Z}(\Gamma_{Z+\mathscr{E}''})) &= (|Z| + |\mathscr{E}''|) - (|Z| + 2) + 1 = |\mathscr{E}''| - 1\\ &= (\rho(x_1) + \rho(x_2) - 1) - 1.\end{aligned}$$

By direct computation, exactly $\rho(x_1) + \rho(x_2) - 2$ cycles contain edges in $\mathscr{E}''$. Thus one cycle in the list, say Z_{n+1}, is contained entirely in Z, i.e., $Z_{n+1} = Z$.

We wish to show that $Z_1, \ldots, Z_k, Z_1', \ldots, Z_n'$ is a planar imbedding of Γ. These are clearly cycles of Γ. Every edge in $Z + \mathscr{E}'$ belongs to two of the cycles $Z_1, \ldots, Z_k$. Every edge in $\mathscr{E}''$ belongs to two of the cycles $Z_1', \ldots, Z_n'$, and every edge in Z belongs to one of the cycles from each list. Finally, $Z_1, \ldots, Z_k, Z_1', \ldots, Z_n'$ spans $\mathscr{Z}(\Gamma)$ since these cycles clearly satisfy only one relation and since

$$\begin{aligned}\dim(\mathscr{Z}(\Gamma)) &= \nu_1(\Gamma) - \nu_0(\Gamma) + 1 \\ &= \nu_1(\Theta_{(x_0)} + |\mathscr{E}''|) - (\nu_0(\Theta_{(x_0)}) + 2) + 1 \\ &= (\dim \mathscr{Z}(\Theta_{(x_0)})) + |\mathscr{E}''| - 2 = k + n - 1. \qquad \square\end{aligned}$$

The necessity of Kuratowski's Theorem as we have stated it follows from Proposition F9. The sufficiency follows from Proposition G11. We leave the proof of Corollary G2 as an exercise for the reader.

CHAPTER IV

Networks

IVA Algebraic Preliminaries

Let X be a set. In §*II*A we discussed the algebraic structure of $\mathbb{K}^X$, and we demonstrated an isomorphism from $\mathbb{K}^X$ onto $\mathscr{P}(X)$. In this section, we develop an analogous theory for $\mathbb{Q}^X$. Many of the following results admit immediate generalizations to $\mathbb{F}^X$ where $\mathbb{F}$ is an arbitrary field or at least an arbitrary ordered field.

The set $\mathbb{Q}^X$ is a commutative algebra under the usual operations. For $h_1, h_2 \in \mathbb{Q}^X$, we have addition of functions:

$$(h_1 + h_2)(x) = h_1(x) + h_2(x) \quad \text{for all } x \in X;$$

multiplication of functions:

$$(h_1h_2)(x) = h_1(x)h_2(x) \quad \text{for all } x \in X;$$

and scalar multiplication:

$$(\eta h_1)(x) = \eta h_1(x) \quad \text{for all } \eta \in \mathbb{Q},\ x \in X.$$

Characteristic functions c_S for $S \subseteq X$ as defined in *I*B1 acquire a different meaning since the symbols 0 and 1 are now understood to be in $\mathbb{Q}$ instead of in $\mathbb{K}$. When $S = \{x\}$, we shall supress the braces and write c_x for $c_{\{x\}}$. The reader should verify that the set $\{c_x: x \in X\}$ is a basis for $\mathbb{Q}^X$ (as it was for $\mathbb{K}^X$), that $c_\varnothing$ is the additive identity of the vector space $\mathbb{Q}^X$, and that c_X is the multiplicative identity of the algebra $\mathbb{Q}^X$. No confusion should arise from our use of the symbols 0 and 1 also to designate the identities $c_\varnothing$ and c_X, respectively.

The algebra $\mathbb{Q}^X$ yields a rather natural inner product:

A1 $$h_1 \cdot h_2 = \sum_{x \in X} h_1(x)h_2(x), \quad \text{for all } h_1, h_2 \in \mathbb{Q}^X.$$

Observe that when $\mathbb{Q}$ is replaced by $\mathbb{K}$, then A1 reduces to the inner product on $\mathscr{P}(X)$ defined in §*II*A. The **support function** $\sigma: \mathbb{Q}^X \to \mathscr{P}(X)$ is given by

$$\sigma(h) = \{x \in X: h(x) \in \mathbb{Q} + \{0\}\}, \quad \text{for all } h \in \mathbb{Q}^X,$$

and the set $\sigma(h)$ is called the **support** of h. Thus when $\mathbb{Q}$ is replaced by $\mathbb{K}$, σ is the algebra isomorphism from $(\mathbb{K}^X, +, \cdot)$ onto $(\mathscr{P}(X), +, \cap)$ discussed at the beginning of §*II*A. However, in the present case, σ need not be an injection. Even still there is a relationship between $\mathbb{Q}^X$ and $\mathscr{P}(X)$, and this is the subject matter of the present section.

Let L be a subspace of $\mathbb{Q}^X$ and let

$$\mathscr{N}(L) = \{\sigma(h): h \in L\}.$$

As a subcollection of $\mathscr{P}(X)$, $\mathscr{N}(L)$ inherits the partial order $\subseteq$. We let $\mathscr{M}(L)$ denote the collection of elementary sets in $\mathscr{N}(L)$. If $h \in L$ and $\sigma(h) \in \mathscr{M}(L)$, we say that h is a **minimal function** of L.

A2 Proposition. *Let L be a subspace of $\mathbb{Q}^X$. Let $h_1, h_2 \in L$, where h_1 is a minimal function of L and $\sigma(h_2) \subseteq \sigma(h_1)$. Then $h_2 = \eta h_1$ for some $\eta \in \mathbb{Q}$.*

PROOF. By hypothesis, $\sigma(h_1) \neq \varnothing$, and so we may fix $x \in \sigma(h_1)$. Let

$$h_3 = h_2 - \frac{h_2(x)}{h_1(x)} h_1.$$

Clearly $h_3 \in L$ and $\sigma(h_3) \subseteq \sigma(h_1) \cup \sigma(h_2) = \sigma(h_1)$. In fact, $\sigma(h_3) \subset \sigma(h_1)$ since $x \in \sigma(h_1)$ but $h_3(x) = 0$. But since $\sigma(h_1) \in \mathscr{M}(L)$, $\sigma(h_3) = \varnothing$. Hence $h_3 = 0$ and

$$h_2 = \frac{h_2(x)}{h_1(x)} h_1. \qquad \square$$

Let $Y \subseteq X$ and consider the injection $j: \mathbb{Q}^Y \to \mathbb{Q}^X$ given by

$$(j(h))(x) = \begin{cases} h(x) & \text{if } x \in Y; \\ 0 & \text{if } x \in X + Y, \end{cases}$$

for $h \in \mathbb{Q}^Y$. We say that $j(h)$ is the "extension by 0" of h to a function on X. Clearly j is a nonsingular linear transformation. Hence $\mathbb{Q}^Y$ is isomorphic to $j[\mathbb{Q}^Y]$, which is the subspace of $\mathbb{Q}^X$ consisting of all functions $h \in \mathbb{Q}^X$ with $\sigma(h) \subseteq Y$. Identifying $\mathbb{Q}^Y$ with $j[\mathbb{Q}^Y]$, we henceforth consider $\mathbb{Q}^Y$ as a subalgebra of $\mathbb{Q}^X$. It is in fact an ideal of $\mathbb{Q}^X$.

The above identification is analogous to the fact that $\mathscr{P}(Y)$ is a subspace of $\mathscr{P}(X)$. While such "coordinate subspaces" $\mathbb{Q}^Y$ could provide a theory of connectedness for subspaces of $\mathbb{Q}^X$ analogous to the theory in §*II*C, we shall have different aims and emphases in the present chapter. Nonetheless, many of our techniques will be reminiscent of earlier ones, and a perusal of §*II*A and §*II*C is recommended, with an eye toward comparing present and past results as we proceed.

For each subset $Y \subseteq X$, we define the projection $\pi_Y\colon \mathbb{Q}^X \to \mathbb{Q}^Y$ where $\pi_Y(h)$ is the product hc_Y for each $h \in \mathbb{Q}^X$. Thus $\pi_Y(h)$ is the restriction $h_{|Y}$ extended by 0 to all of X. The image of π_Y has been identified with $\mathbb{Q}^Y$. Its kernel is $\mathbb{Q}^{X+Y}$. Thus $\pi_Y[L]$ is a subspace of $\mathbb{Q}^Y$ whenever L is a subspace of $\mathbb{Q}^X$. Since $L \cap \mathbb{Q}^{X+Y}$ is the kernel of the restriction $\pi_{Y|L}$, we have

A3 $$\dim(L \cap \mathbb{Q}^{X+Y}) + \dim(\pi_Y[L]) = \dim(L).$$

A4 Exercise. Prove that *for any subspace $L \subseteq \mathbb{Q}^X$ and any $Y \subseteq X$,*

$$(L \cap \mathbb{Q}^Y)^{\perp_Y} = \pi_Y[L^\perp]$$

and

$$(\pi_Y[L])^{\perp_Y} = L^\perp \cap \mathbb{Q}^Y,$$

where $\perp_Y$ indicates the orthogonal complement in $\mathbb{Q}^Y$.

A5 *Exercise.* For $Y \subseteq X$ and L a subspace of $\mathbb{Q}^X$, prove that the minimal functions of $L \cap \mathbb{Q}^Y$ are precisely the minimal functions of L with support contained in Y.

The next proposition, which is parallel to Proposition *II*C1, is the central result of this section.

A6 Proposition. *Let L be a subspace of $\mathbb{Q}^X$, and let $h \in L$. Then $h = \sum_{i=1}^m h_i$ where $h_i \in L$ for $i = 1, \ldots, m$, and*

- **(a)** $\sigma(h_i) \in \mathcal{M}(L)$;
- **(b)** $\sigma(h_i) \subseteq \sigma(h)$;
- **(c)** $h_i(x)h(x) \geq 0$ *for all* $x \in X$.

Proof. We proceed by induction on $|\sigma(h)|$. If $\sigma(h) = \varnothing$, then $h = 0$, and the proposition is trivially satisfied with $m = 0$.

Now assume that $|\sigma(h)| = n > 0$ and, as the induction hypothesis, that the conclusion holds for all $g \in L$ with $|\sigma(g)| < n$.

If $\sigma(h) \in \mathcal{M}(L)$, we are done; so assume $\sigma(h) \notin \mathcal{M}(L)$. We shall demonstrate the existence of a function $g \in L$ having simultaneously the following three properties:

- (i) $\sigma(g) \subseteq \sigma(h)$;
- (ii) $\sigma(g) \in \mathcal{M}(L)$;
- (iii) $g(x)h(x) \geq 0$ for all $x \in X$.

By *II*B11, we may select a function g'' satisfying (i) and (ii). Assuming (iii) to fail for g'', let

$$Y = \{x \in X\colon g''(x)h(x) < 0\}.$$

Hence $Y \neq \varnothing$, and one may let

$$\mu = \max\left\{\frac{h(x)}{g''(x)}\colon x \in Y\right\}.$$

Then $\mu < 0$. Let $g' = h - \mu g''$. Since $\sigma(h) \notin \mathcal{M}(L)$, g'' is not a scalar multiple of h, and so $g' \neq 0$. Clearly g' satisfies (i).

We next show that g' satisfies (iii). If $x \in X$, then $g'(x)h(x) = [h(x)]^2 - \mu g''(x)h(x)$, which is clearly nonnegative when $x \in X + Y$. If $x \in Y$, then $h(x)/g''(x) \leq \mu < 0$, and so

$$g'(x)h(x) = [h(x)]^2 - \frac{\mu g''(x)}{h(x)}[h(x)]^2$$

$$= [h(x)]^2\left(1 - \frac{\mu g''(x)}{h(x)}\right) \geq 0$$

as required.

For some point $y \in Y$, $\mu = h(y)/g''(y)$, and this point y lies in $\sigma(h) + \sigma(g')$. Hence $|\sigma(g')| < |\sigma(h)|$, and we may apply the induction hypothesis to g' to obtain

$$g' = \sum_{i=1}^{p} g_i,$$

where g_i satisfies (i) and (ii) and $g_i(x)g'(x) \geq 0$ for each $i = 1, \ldots, p$. Thus for each $x \in X$, $g_i(x)$ and $g'(x)$ never have opposite sign. We have already shown that $g'(x)$ and $h(x)$ never have opposite sign. It follows that $g_1, \ldots, g_p$ also satisfy (iii). For definiteness, let $g = g_1$, and we have the desired function.

Let $\nu = \min\{h(x)/g(x): x \in \sigma(g)\}$, and let

$$h_1 = \nu g \quad \text{and} \quad h' = h - h_1.$$

Since g satisfies (iii), $\nu \geq 0$; but in fact since g also satisfies (i), $\nu > 0$. Hence

$$h_1(x)h(x) = \nu g(x)h(x) \geq 0, \quad \text{for all } x \in X.$$

If $x \in \sigma(g)$, then clearly $\nu \leq h(x)/g(x)$, whence

A7 $$\nu g(x)h(x) \leq [h(x)]^2.$$

When $x \in X + \sigma(g)$, the left-hand member of A7 is 0, and so A7 holds for all $x \in X$. Hence for all $x \in X$,

A8 $$0 \leq [h(x)]^2 - \nu g(x)h(x) = h(x)(h(x) - \nu g(x)) = h(x)h'(x).$$

For some point $x_0 \in \sigma(g)$, $h(x_0) = \nu g(x_0)$, and hence $h'(x_0) = 0$. Thus $\sigma(h') \subset \sigma(h)$. Applying the induction hypothesis to h', we obtain

$$h' = \sum_{i=2}^{m} h_i$$

where $h_i \in \mathcal{M}(L)$, $\sigma(h_i) \subseteq \sigma(h') \subset \sigma(h)$, and

A9 $$h_i(x)h'(x) \geq 0, \quad \text{for all } x \in X;\ i = 2, \ldots, m.$$

Clearly

$$h = \sum_{i=1}^{m} h_i$$

as required. The condition required in (c) now follows from the definition of h_1, A8, and A9. □

The decomposition of h guaranteed by Proposition A6 is called an **$\mathscr{M}$-decomposition** of h.

The following is immediate.

A10 Corollary. *If L is a subspace of $\mathbb{Q}^X$ and if $\{h_1, \ldots, h_m\}$ is an $\mathscr{M}$-decomposition of $h \in L$, then*

$$|h(x)| = \sum_{i=1}^{m} |h_i(x)|, \quad \textit{for all } x \in X.$$

A11 Corollary. *Let L be a subspace of $\mathbb{Q}^X$, let $Y \subseteq X$, and let g be a minimal function of $\pi_Y[L]$. Then $g = \pi_Y(h)$ for some minimal function h of L.*

PROOF. Since $g \in \pi_Y[L]$, $g = h'c_Y$ for some function $h' \in L$. Applying the proposition, let $\{h_1, \ldots, h_m\}$ be an $\mathscr{M}$-decomposition of h'. Clearly,

$$g = h'c_Y = \sum_{i=1}^{m} h_i c_Y,$$

and $\sigma(h_i c_Y) \subseteq \sigma(g)$. By Proposition A2, there exists $\eta_i \in \mathbb{Q}$ such that $h_i c_Y = \eta_i g$ $(i = 1, \ldots, m)$. Since $g \neq 0$, $\eta_i \neq 0$ for some index i; say $\eta_1 \neq 0$. Let

$$h = \frac{1}{\eta_1} h_1.$$

Then h is a minimal function of L and $g = hc_Y = \pi_Y(h)$. □

If L is a subspace of $\mathbb{Q}^X$ and if h is a minimal function of L, then there exists a smallest positive number $\theta \in \mathbb{Q}$ such that the function $g = \theta h$ is integer-valued. Such a function g is called an **elementary function** of L. We list some immediate consequences of this definition:

A12 A function $g \in L$ is an elementary function of L if and only if g is an integer-valued minimal function of L with the property that 1 is the greatest common divisor of the set $\{g(x): x \in \sigma(g)\}$.

A13 For each set $S \in \mathscr{M}(L)$, there are precisely two elementary functions of L whose support is S. If g is one such function, then $-g$ is the other.

A14 There are finitely many elementary functions of L.

With the notion of "elementary function," Proposition A6 yields:

A15 Corollary. *Let L be a subspace of $\mathbb{Q}^X$ and let $h \in L$. Then*

$$h = \sum_{i=1}^{m} \eta_i g_i$$

where η_i is a positive rational number, g_i is an elementary function of L, and $g_i(x)h(x) \geq 0$ for all $x \in X$ and all $i = 1, \ldots, m$.

PROOF. Apply the proposition to h. By A2 and A13, each minimal function h_i may be replaced by $\eta_i g_i$ where $\eta_i > 0$ and g_i is an elementary function $(i = 1, \ldots, m)$.

A subspace L of $\mathbb{Q}^X$ is **unimodular** if $g[X] \subseteq \{0, 1, -1\}$ for every elementary function g of L.

A16 *Example*. Let $L = \{f \in \mathbb{Q}^X : \sum_{x \in X} f(x) = 0\}$. One easily verifies that L is a subspace of $\mathbb{Q}^X$ and that $\mathscr{M}(L) = \mathscr{P}_2(X)$. An elementary function f with $\sigma(f) = \{x, y\}$ has the form

$$f(u) = \begin{cases} \pm 1 & \text{if } u = x \\ \mp 1 & \text{if } u = y \\ 0 & \text{if } u \in X + \{x, y\}. \end{cases}$$

Thus L is unimodular. Observe that $L^\perp$ is the subspace of the constant functions in $\mathbb{Q}^X$. Note that $\{\sigma(f) : f \in L\} = \mathscr{E}(X)$, and $\{\sigma(f) : f \in L^\perp\} = \{\varnothing, X\}$ (cf. *IIA7*).

A17 Exercise. *Let L be a subspace of $\mathbb{Q}^X$ and let $h \in L \cap \mathbb{Z}^X$.* Prove that $h = \sum_{i=1}^m g_i$ *where g_i is an elementary function of L, $\sigma(g_i) \subseteq \sigma(h)$, and $g_i(x)h(x) \geq 0$ for all $x \in X$ and all $i = 1, \ldots, m$.* [*Hint:* Prove by induction on $\sum_{x \in X} |h(x)|$ using A6.]

A18 Exercise. Prove that *if L is a unimodular subspace of $\mathbb{Q}^X$ and if $Y \subseteq X$, then both $L \cap \mathbb{Q}^Y$ and $\pi_Y[L]$ are unimodular subspaces.*

A19 Proposition. *Let L be a subspace of $\mathbb{Q}^X$. If L is unimodular, then $L^\perp$ is unimodular.*

PROOF. We shall assume that $|X| \geq 2$ since otherwise the result is trivial.

Assuming L to be unimodular, let us first resolve the special case where $L^\perp$ has the properties:

A20 $$\dim(L^\perp) = 1; \qquad \mathrm{Fnd}(L^\perp) = X.$$

Let $x_0 \in X$. It is evident that $L^\perp$ has a basis $\{h\}$ where $\sigma(h) = X$ and $h(x_0) = 1$, and it suffices to prove that $h[X] \subseteq \{1, -1\}$.

If $g \in L$, then $g \cdot h = 0$, and if $g \neq 0$, one must have $|\sigma(g)| \neq 1$. Arbitrarily select $x_1 \in X + \{x_0\}$, and define

$$g_1(x) = \begin{cases} h(x_1) & \text{if } x = x_0; \\ -1 & \text{if } x = x_1; \\ 0 & \text{if } x \in X + \{x_0, x_1\}. \end{cases}$$

Since $g_1 \cdot h = h(x_1) - h(x_1) = 0$, $g_1 \in L$. Furthermore, since $|\sigma(g_1)| = 2$, g_1 is a minimal function of L and hence by A2, g_1 is a multiple of some elementary function with the same support. Since L is unimodular, the fact that $g_1(x_1) = -1$ implies that $h(x_1) = g_1(x_0) = \pm 1$. Since x_1 was arbitrarily chosen and $h(x_0) = 1$, we conclude that $h[X] \subseteq \{1, -1\}$. Hence $L^\perp$ is unimodular.

Now let the conditions A20 on $L^\perp$ be relaxed, and let h be any minimal function of $L^\perp$. (We assume $L^\perp \neq \{0\}$, since trivial subspaces are trivially unimodular.) Let $S = \sigma(h)$ and let $M = \pi_S[L]$. By Exercise A4, $M^{\perp_S} = L^\perp \cap \mathbb{Q}^S$, where $\perp_S$ denotes orthogonal complements in $\mathbb{Q}^S$. Since $S \in \mathscr{M}(L^\perp)$, it follows from this and Proposition A2 that $M^{\perp_S} = \{\eta h : \eta \in \mathbb{Q}\}$. By Exercise A18, M is unimodular and hence we may invoke the special case above (with M in place of L and S in place of X) to conclude that $M^{\perp_S}$ is unimodular. Hence $h[S] \subseteq \{\theta, -\theta\}$ for some $\theta \in \mathbb{Q}$, and so $h[X] \subseteq \{0, \theta, -\theta\}$. $L^\perp$ is therefore unimodular. □

IVB The Flow Space

Let V be a set, and fix the letter $W = (V \times V) + \{(x, x) : x \in V\}$ throughout this chapter. Recall that a basis for the vector space $\mathbb{Q}^V$ is $\{c_x : x \in V\}$, and so a basis for $\mathbb{Q}^W$ is $\{c_{(x,y)} : (x, y) \in W\}$.

We define $\partial(c_{(x,y)}) = c_y - c_x$ for all $(x, y) \in W$ and extend by linearity, i.e.,

$$\text{if } h = \sum_{(x,y) \in W} a_{(x,y)} c_{(x,y)}, \quad \text{then} \quad \partial(h) = \sum_{(x,y) \in W} a_{(x,y)}(c_y - c_x).$$

We thus obtain a linear transformation $\partial : \mathbb{Q}^W \to \mathbb{Q}^V$, called the **boundary operator** on W.

B1 Exercise. *For $h \in \mathbb{Q}^W$ and $x_0 \in V$, verify that*

$$(\partial(h))(x_0) = \sum_{x \in V + \{x_0\}} (h(x, x_0) - h(x_0, x)).$$

The kernel of ∂ is called the **flow space**, and is denoted by $F(V)$ or simply by F. By B1,

B2 $$\sum_{x \in V + \{x_0\}} h(x, x_0) = \sum_{x \in V + \{x_0\}} h(x_0, x), \quad \text{for all } h \in F \text{ and all } x_0 \in V.$$

An intuitive description of a flow space $F(V)$ may be given as follows. Consider the directed graph (V, W), and let $h \in \mathbb{Q}^W$. For each $(x, y) \in W$, imagine $h(x, y)$ to measure a "flow" of fluid or current or a commodity through the edge (x, y) in the direction from x to y. If $h \in F$, Equation B2 can be interpreted to mean that the total "flow" into any vertex x_0 equals the total "flow" out of x_0. This is a sort of principle of conservation of matter or energy or money. It is this situation which has historically motivated the abstract notion of a flow space, an element of which is called a **flow**. A **minimal (elementary) flow** is a minimal (elementary) function in F.

For the remainder of this chapter, Γ will denote the multigraph (V, f, W), where for each $(x, y) \in W$, we define $f(x, y) = \{x, y\}$.

B3 *Exercise.* The three diagrams B4 below all depict the directed graph (V, W) or equally well the multigraph $\Gamma = (V, f, W)$, where the arrow on an "edge" from, say, u to v is to designate the edge $(u, v) \in W$ rather than the edge (v, u). The number beside an "edge" in the ith diagram indicates the value of h_i on that "edge" ($i = 1, 2, 3$). Compute the value of $\partial(h_i)$ for $i = 1, 2, 3$. When is $h_i \in F(V)$?

B4

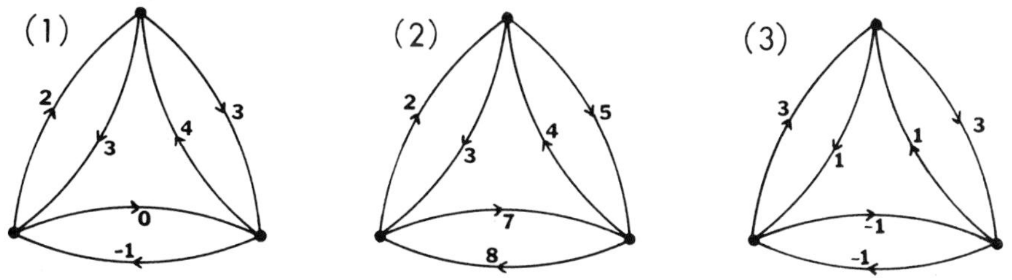

B5 Proposition. **(a)** $\dim(F) = (|V| - 1)^2$; **(b)** $\dim(F^\perp) = |V| - 1$.

PROOF. (a) From the fundamental result

$$\dim(\mathbb{Q}^W) = \dim(\ker \partial) + \dim(\partial[\mathbb{Q}^W]),$$

it follows that

$$\dim(F) = |W| - \dim(\partial[\mathbb{Q}^W]) = |V|^2 - |V| - \dim(\partial[\mathbb{Q}^W]).$$

Let $L = \{g \in \mathbb{Q}^V : \sum_{x \in V} g(x) = 0\}$. Clearly L is the kernel of the transformation $h \mapsto \sum_{x \in V} h(x)$ from $\mathbb{Q}^V$ onto $\mathbb{Q}$. Hence

B6 $$\dim(L) = \dim(\mathbb{Q}^V) - \dim(\mathbb{Q}) = |V| - 1,$$

and to prove (a) it will suffice to prove that $L = \partial[\mathbb{Q}^W]$.

From B1 we have that for each $h \in \mathbb{Q}^W$,

$$\sum_{x \in V} (\partial h)(x) = \sum_{x \in V} \sum_{y \in V + \{x\}} (h(y, x) - h(x, y)) = 0.$$

Hence $\partial[\mathbb{Q}^W] \subseteq L$. On the other hand, for any fixed $x_0 \in V$,

$$\{\partial(c_{(x_0, y)}) : y \in V + \{x_0\}\}$$

is an independent $(|V| - 1)$-set contained in L. By B6, it is a basis for L which is contained in $\partial[\mathbb{Q}^W]$. Hence $\partial[\mathbb{Q}^W] = L$.

(b) $$\dim(F^\perp) = \dim(\mathbb{Q}^W) - \dim(F)$$

$$= (|V|^2 - |V|) - (|V| - 1)^2 = |V| - 1. \qquad \square$$

Let $x_0, e_1, x_1, e_2, \ldots, e_m, x_m = x_0$ be a circuit of Γ. In the present context we need to distinguish between different cyclic orderings corresponding to the same submultigraph. Therefore, we shall denote this cyclically-ordered circuit by the symbol D instead of by the usual capital Greek letter used for submultigraphs. We define the function $h_{\mathsf{D}} \in \mathbb{Q}^W$ by

$$h_{\mathsf{D}}(e) = \begin{cases} 1 & \text{if } e = (x_{i-1}, x_i) \text{ for some } i = 1, \ldots, m; \\ -1 & \text{if } e = (x_i, x_{i-1}) \text{ for some } i = 1, \ldots, m; \\ 0 & \text{if } e \in W + \{e_1, \ldots, e_m\}. \end{cases}$$

B7 Lemma. *If D is a circuit in Γ, then $h_{\mathsf{D}} \in F(V)$.*

PROOF. As above, represent D by the list: $x_0, e_1, x_1, e_2, \ldots, e_m, x_m = x_0$ and let $\{I, J\}$ be a 2-partition of $\{1, \ldots, m\}$ where

$$I = \{i: e_i = (x_{i-1}, x_i)\} \quad \text{and} \quad J = \{i: e_i = (x_i, x_{i-1})\}.$$

By definition, $h_{\mathsf{D}} = \sum_{i \in I} c_{e_i} - \sum_{i \in J} c_{e_i}$. Hence,

$$\begin{aligned} \partial(h_{\mathsf{D}}) &= \sum_{i \in I} \partial(c_{e_i}) - \sum_{i \in J} \partial(c_{e_i}) \\ &= \sum_{i \in I} (c_{x_i} - c_{x_{i-1}}) - \sum_{i \in J} (c_{x_{i-1}} - c_{x_i}) \\ &= \sum_{i=1}^{m} (c_{x_i} - c_{x_{i-1}}) = 0, \end{aligned}$$

since $x_0 = x_m$. □

B8 Proposition. *$F(V)$ is a unimodular subspace of $\mathbb{Q}^W$. Moreover, a function $h \in \mathbb{Q}^W$ is an elementary flow if and only if $h = h_{\mathsf{D}}$ for some elementary circuit D of Γ.*

PROOF. Let h be a minimal flow and let x_0 be a vertex of the subgraph $\Gamma_{\sigma(h)}$. By B2, $\sum_{x \in V + \{x_0\}} (h(x, x_0) - h(x_0, x)) = 0$. Since some term $h(x, x_0)$ or $h(x_0, x)$ is nonzero, there must be at least two such nonzero terms. Hence x_0 has valence $\rho(x_0) \geq 2$. By Exercise *III*A6a, $\Gamma_{\sigma(h)}$ contains an elementary circuit, which we can represent by the list D. Hence $\sigma(h_{\mathsf{D}}) \subseteq \sigma(h)$. By B7 and A2, $h_{\mathsf{D}} = \eta h$ for some $\eta \in \mathbb{Q}$. Since $h_{\mathsf{D}} \neq 0$, $\eta \neq 0$, and we observe that $h[W] \subseteq \{0, 1/\eta, -1/\eta\}$. This proves that F is unimodular.

If, moreover, h is an elementary flow, then necessarily $\eta = \pm 1$, and so $h = \pm h_{\mathsf{D}}$. If $h = -h_{\mathsf{D}}$, then $h = h_{\mathsf{D}'}$, where D' is the list D in the reverse order.

Conversely, let D be an elementary circuit of Γ. By Lemma B7, h_{D} is a flow. In order to show that h_{D} is an elementary flow, it suffices to prove that h_{D} is a minimal flow, since by definition, $h_{\mathsf{D}}[W] \subseteq \{0, 1, -1\}$. We select a minimal flow h such that $\sigma(h) \subseteq \sigma(h_{\mathsf{D}})$. By the first part of this same proof, there exists an elementary circuit D'' such that $\sigma(h_{\mathsf{D}''}) \subseteq \sigma(h)$, and so $\sigma(h_{\mathsf{D}''}) \subseteq \sigma(h_{\mathsf{D}})$. By *III*A9, both $\sigma(h_{\mathsf{D}''})$ and $\sigma(h_{\mathsf{D}})$ are elementary cycles of Γ. Hence

$$\sigma(h_{\mathsf{D}}) = \sigma(h) = \sigma(h_{\mathsf{D}''}) \in \mathscr{M}(F). \qquad \square$$

As an immediate consequence of this proposition and *III*A9 we have

B9 Corollary. *$\mathscr{M}(F)$ is the set of elementary cycles of Γ.*

The close relationship between flows and cycles of Γ delineated by B7, B8, and B9 suggests that $F^{\perp}$ may be related to the cocycle space of Γ. This is indeed the case.

For each vertex $x \in V$, define $g_x \in \mathbb{Q}^W$ by

$$g_x(u, v) = \begin{cases} 1 & \text{if } u = x; \\ -1 & \text{if } v = x; \\ 0 & \text{otherwise.} \end{cases}$$

Clearly $\sigma(g_x) = f^*(x)$.

B10 Exercise. Show that $g_x \in F^{\perp}$ *for all* $x \in V$. [*Hint*: Show that $g_x \cdot h_{\mathsf{D}} = 0$ for any elementary circuit D of Γ; then use Proposition A6.]

B11 Lemma. *Let $g = \sum_{x \in U} \eta_x g_x$ where $U \subseteq V$ and $\eta_x \in \mathbb{Q} + \{0\}$. Then $\sum_{x \in U} f^*(x) \subseteq \sigma(g)$. Furthermore, equality holds if and only if $\eta_x = \eta_y$ for all $x, y \in U$.*

Proof. Let $e = (x, y)$ and let $Y = \{x, y\} \cap U$. Then $e \in \sum_{u \in U} f^*(u)$ if and only if $|Y| = 1$. On the other hand,

$$g(e) = \begin{cases} 0 & \text{if } Y = \varnothing; \\ \eta_x & \text{if } Y = \{x\}; \\ -\eta_y & \text{if } Y = \{y\}; \\ \eta_x - \eta_y & \text{if } Y = \{x, y\}. \end{cases}$$

Hence $e \in \sigma(g)$ if and only if either $|Y| = 1$ or $|Y| = 2$ with $\eta_x \neq \eta_y$. $\square$

B12 Proposition.

(**a**) *$F^{\perp}$ is spanned by $\{g_x : x \in V\}$.*

(**b**) *$\sum_{x \in V} g_x = 0$ is the only relation among the functions g_x for $x \in V$.*

(**c**) *$\sum_{x \in U} g_x$ is an elementary function of $F^{\perp}$ if and only if $\sum_{x \in U} f^*(x)$ is an elementary cocycle of Γ.*

(d) $\mathscr{M}(F^\perp)$ *is the set of elementary cocycles of* Γ.
(e) $F^\perp$ *is unimodular*.

Proof. (a) and (b). Let G be the subspace of $\mathbb{Q}^W$ spanned by $\{g_x: x \in V\}$. Suppose that $g = \sum_{x \in V} \eta_x g_x = 0$. By the lemma, $\varnothing = \sigma(g) \supseteq \sum_{x \in U} f^*(x)$, where $U = \{x: \eta_x \neq 0\}$. Furthermore, $\sigma(g) = \sum_{x \in U} f^*(x)$ and $g = \sum_{x \in U} \eta g_x$ for some fixed $\eta \in \mathbb{Q} + \{0\}$. Since (V, f, W) is connected, we have from *III*A11 and *III*A16 that $\sum_{x \in U} f^*(x) = \varnothing$ if and only if $U = \varnothing$ or $U = V$. Hence the only (nontrivial) relation among the functions in the set $\{g_x: x \in V\}$ is $\eta \sum_{x \in V} g_x = 0$, or $\sum_{x \in U} g_x = 0$. We conclude that $\dim(G) = |V| - 1$. But by Exercise B10, $G \leq F^\perp$ and by Proposition B5b, $\dim(F^\perp) = |V| - 1$. We conclude that $G = F^\perp$, and that (a) and (b) hold.

(c) and (d). It follows from the lemma and part (a) that if $g \in F^\perp + \{0\}$, then $\sigma(g)$ contains a nonempty cocycle of Γ and that if $C \in \mathscr{Z}^\perp(\Gamma)$, then $C = \sigma(g)$ for some $g \in F^\perp$. Hence $\mathscr{M}(F^\perp) = \{C \in \mathscr{Z}^\perp(\Gamma): C \text{ is elementary}\}$. On the other hand, if C is an elementary cocycle of Γ, then $C = \sum_{x \in U} f^*(x)$ for some $U \subseteq V$. Since $g = \sum_{x \in U} g_x$ takes on only the values 0, +1, −1, and since $\sigma(g) = C$, g is elementary. Furthermore, $-g = \sum_{x \in U + V} g_x$. Hence by A13, these are all of the elementary functions of $F^\perp$.

(e) This follows from B8 and A19. □

B13 *Exercise.* Prove that in Proposition B12, conclusion (e) follows directly from (a)–(d), i.e., without the use of A19.

IVC Max-Flow–Min-Cut

A **network** is a pair (V, k) where V is a set, and $k \in \mathbb{Q}^W$ with $k(e) \geq 0$ for all $e \in W$. The function k is called the **capacity**, and the value $k(e)$ is called the **capacity of** e. A flow $h \in F(V)$ is said to be **feasible** if

$$0 \leq h(e) \leq k(e) \quad \text{for all } e \in W.$$

If (V, k) is a network and $K = \sigma(k)$, then (V, K) is a directed graph. It is a "sub-directed graph" of the directed graph (V, W) discussed at the beginning of the previous section. In line with the interpretation developed there, the values of the capacity function represent the actual "capacities" of the various links in the highway system or pipeline, etc. When two "vertices" are joined by no road or pipe, we assign a capacity of 0 to the corresponding edge. The directed graph (V, K) is then an abstraction of the highway system or pipeline, etc. If h is a feasible flow, the numbers assigned to each edge of (V, K) are nonnegative but do not exceed the capacity of that edge. The sum of these numbers over the edges entering a vertex equals the sum of the numbers assigned to the edges leaving that vertex. Hence a feasible flow represents a possible flow of traffic or fluid or money, etc. through the system.

A **cut** of (V, k) is a cocycle of the multigraph $\Gamma = (V, f, W)$ and a **cut through** e will mean, of course, a cut containing the edge e.

Let C be any cut through $e_0 = (y_0, x_0)$. Then for some $U \subseteq V$, $C = \sum_{x \in U} f^*(x)$. Replacing U by $V + U$ if necessary, we may assume $x_0 \in U \subseteq V + \{y_0\}$. For all $(x, y) \in W$, define:

C1
$$g_U(x, y) = \begin{cases} 1 & \text{if } \{x, y\} \cap U = \{x\}; \\ -1 & \text{if } \{x, y\} \cap U = \{y\}; \\ 0 & \text{otherwise.} \end{cases}$$

Clearly $g_U = \sum_{x \in U} g_x$. Hence $g_U \in F^\perp$ by B12a, and $g_U(e_0) = -1$. Furthermore, since Γ is connected, U is uniquely determined by C and e_0. In terms of this function g_U, we define the **capacity of the cut** C through e_0 to be

$$k(C; e_0) = \sum_{g_U(e) = 1} k(e).$$

C2 Proposition. *Let $e_0 \in W$. Then $h(e_0) \leq k(C; e_0)$ for every feasible flow h and every cut C through e_0.*

PROOF. Let $h \in F$ be feasible, let C be a cut, and let g be the function g_U determined by C and e_0 as above. Since Since $g \in F^\perp$, we have

$$0 = h \cdot g = \sum_{e \in W} h(e) g(e) = \sum_{g(e) = 1} h(e) - \sum_{g(e) = -1} h(e).$$

Hence $\sum_{g(e)=1} h(e) = \sum_{g(e)=-1} h(e)$. Since h is feasible,

$$h(e_0) \leq \sum_{g(e) = -1} h(e) = \sum_{g(e) = 1} h(e) \leq \sum_{g(e) = 1} k(e) = k(C; e_0),$$

as required. □

If $e_0 \in W$, a feasible flow h is said to be a **maximum flow through** e_0 if $h(e_0) \geq h'(e_0)$ for any feasible flow h'. Clearly if h is a feasible flow and if $h(e_0) = k(e_0)$, then h is a maximum flow through e_0. A cut through e_0 is said to be a **minimum cut through** e_0 if $k(C; e_0) \leq k(C'; e_0)$ for any cut C' through e_0. The following result is immediate from these definitions.

C3 Corollary. *Let $e_0 \in W$, let h be a feasible flow, and let C be a cut through e_0. If $h(e_0) = k(C; e_0)$, then h is a maximum flow through e_0 and C is a minimum cut through e_0.*

Example. Let $V = \{1, 2, \ldots, 9\}$. Define $k \in \mathbb{Q}^W$ to be 0 except as follows: $k(1, 2) = 3$, $k(1, 3) = 1$, $k(3, 1) = 2$, $k(4, 1) = 1$, $k(5, 1) = 1$, $k(1, 7) = 5$, $k(1, 9) = 5$, $k(9, 8) = 6$, $k(8, 2) = 6$, $k(2, 3) = 7$, $k(3, 4) = 4$, $k(4, 5) = 3$, $k(5, 6) = 1$, $k(6, 7) = 1$ and $k(7, 8) = 1$. We have drawn the directed graph (V, K) in Figure C4, listing the capacity of each edge beside it.

C4

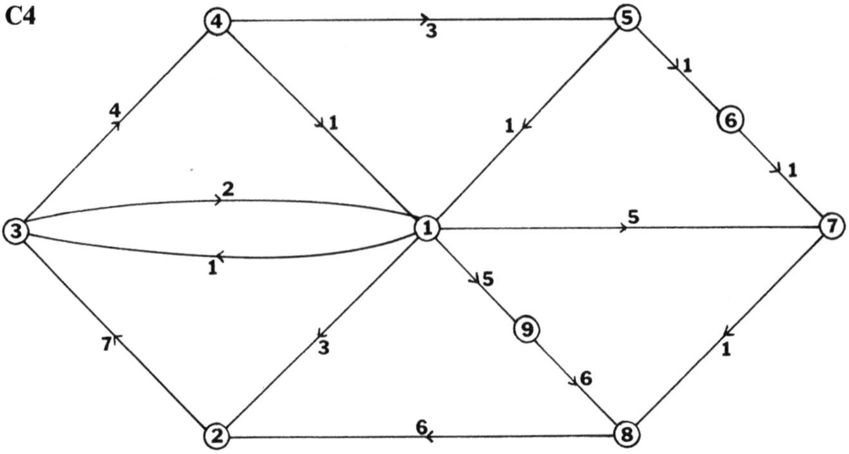

Define $h(i, j)$ to be 1 either if $j = i + 1$ and $i = 2, 3, \ldots, 7$ or if $(i, j) = (8, 2)$, and to be 0 otherwise. This may be visualized as a flow of one unit "around the outer rim" of the figure. It is clearly a feasible flow. If $e_0 = (8, 2)$, the cut C determined by $U = \{1, 2, 3, 4\}$ is a cut through e_0. The edges which appear in the figure and belong to C are (8, 2), (1, 9), (1, 7), (5, 1), and (4, 5). Furthermore, $k(C; e_0) = 13$.

C5 *Problem.* Find an integer-valued feasible flow for the network described in C4 which has support as large as is possible. Let

$$C = \{(8, 2), (1, 2), (1, 3), (3, 1), (3, 4)\}.$$

What is its capacity as a cut through (8, 2)? through (3, 1)? Verify Proposition C2 for this flow and these cuts. Find a maximum flow through (8, 2) and a minimum cut through (8, 2).

A function $h \in \mathbb{Q}^W$ is said to be **integral** if $h \in \mathbb{Z}^W$. We may thus speak of an integral capacity and an integral flow. A network (V, k) is an **integral network** if k is an integral capacity.

The next theorem is the main result of the present chapter. It is also the point of departure for both Chapters V and VI.

C6 The Max-Flow–Min-Cut Theorem. *Let (V, k) be a network and let $e_0 = (y_0, x_0) \in W$. Then there exists a maximum flow h through e_0 such that*

$$h(e_0) = \min(\{k(e_0)\} \cup \{k(C; e_0)\colon C \text{ is a cut through } e_0\}).$$

Furthermore, if (V, k) is an integral network, h may be taken to be an integral flow.

PROOF. Let n be the least positive integer such that nk is integral. Let

$$F' = \{h \in F\colon h \text{ is feasible; } nh \text{ is integral}\}.$$

Then for all $h' \in F'$ and all $e \in W$,

$$h'(e) \in \{0, 1/n, 2/n, \ldots, k(e)\}.$$

Moreover,

C7 **(a)** $h'(e) > 0 \Rightarrow h'(e) \geq 1/n$; **(b)** $h'(e) < k(e) \Rightarrow h'(e) \leq k(e) - 1/n$.

Clearly there exists a flow $h \in F'$ such that $h(e_0) \geq h'(e_0)$ for all $h' \in F'$. In particular, if k is integral, then $n = 1$ and h is integral. If $h(e_0) = k(e_0)$, then h is a maximum flow through e_0 and the theorem is proved. Hence suppose

C8 $$h(e_0) < k(e_0).$$

An x_0x-path $x_0, e_1, x_1, \ldots, e_m, x$ is said to be **unsaturated** with respect to h if

$$e_i \neq e_0 \quad \text{for } i = 1, \ldots, m;$$
$$e_i = (x_{i-1}, x_i) \Rightarrow h(e_i) < k(e_i);$$
$$e_i = (x_i, x_{i-1}) \Rightarrow h(e_i) > 0.$$

Let $U = \{x \in V$: there exists an unsaturated x_0x-path$\}$. Trivially $x_0 \in U$. Also if $x \in U$, then so is every vertex on an unsaturated x_0x-path.

Case 1: $y_0 \in U$. Let $x_0, e_1, x_1, \ldots, e_m, y_0$ be an unsaturated x_0y_0-path, and let D be the circuit $x_0, e_1, x_1, \ldots, e_m, y_0, e_0, x_0$. From the definition of "unsaturated", C8, and C7, it follows that $h + (1/n)h_{\mathsf{D}} \in F'$. This is a contradiction since $(h + (1/n)h_{\mathsf{D}})(e_0) = h(e_0) + 1/n > h(e_0)$.

Case 2: $y_0 \notin U$. The cocycle $C = \sum_{x \in U} f^*(x)$ is then a cut through e_0. Let g_U be defined as in C1. Since $g_U \in F^{\perp}$ and $h \in F$,

C9 $$0 = \sum_{e \in W} h(e)g_U(e) = \sum_{g_U(e)=1} h(e) - \sum_{g_U(e)=-1} h(e).$$

Let $e = (x, y)$ and suppose $g_U(x, y) = 1$. Thus $x \in U$ and $y \notin U$. Hence there exists an unsaturated x_0x-path $x_0, e_1, \ldots, e_m, x$. If $h(e) < k(e)$, the x_0y-path $x_0, e_1, \ldots, e_m, x, e, y$ would be unsaturated, which is impossible since $y \notin U$. We conclude that if $g_U(e) = 1$, then $h(e) = k(e)$. Similarly one can show that if $g_U(e) = -1$ and $e \neq e_0$, then $h(e) = 0$. Substituting these values into C9 we get

$$0 = \left(\sum_{g(e)=1} k(e)\right) - h(e_0) = k(C; e_0) - h(e_0).$$

Thus $h(e_0) = k(C; e_0)$. By C3, h is a maximum flow through e_0 and C is a minimum cut through e_0. □

C10 *Problem*. Let (V, k) be defined by Figure C4. Find a maximum flow Through (1, 9) and the corresponding minimum cut. (*Hint:* Start with the flow you produced in C5 and try to find an unsaturated path from 9 to 1. If you can find such a path, increase the flow by one unit as in Case 1 above. If there is no such path, construct the cut as in Case 2 above.)

IVD The Flow Algorithm

Let (V, k) be any network. As in the proof of the Max-Flow–Min-Cut Theorem, let n be the least positive integer such that nk is integral. Then (V, nk) is an integral network. Furthermore, h is a feasible flow in (V, k) if and only if nh is a feasible flow in (V, nk). Finally, for any $e_0 \in W$, h is a maximum flow through e_0 in (V, k) if and only if nh is a maximum flow through e_0 in (V, nk). It should thus be clear that for the purposes of actually computing maximum flows, it suffices to restrict oneself to integral networks.

D1 Proposition. *Let (V, k) be an integral network. Let h be a feasible flow which is not a maximum flow through e_0. Then there exists an elementary circuit D of Γ such that $h_{\mathsf{D}}(e_0) = 1$ and $h + h_{\mathsf{D}}$ is a feasible flow.*

PROOF. By the Max-Flow–Min-Cut Theorem (C6), there exists an integral maximum flow h' through e_0. By A17, the integral flow $h' - h$ admits an $\mathcal{M}$-decomposition $\{g_1, \ldots, g_m\}$ where g_i is elementary, $\sigma(g_i) \subseteq \sigma(h' - h)$, and $(h' - h)(e)g_i(e) \geq 0$ for all $e \in W$ and all $i = 1, 2, \ldots, m$. Since $h'(e_0) > h(e_0)$ by assumption, $g_j(e_0) = 1$ for some j. By B8, $g_j = h_{\mathsf{D}}$ for some elementary circuit D of Γ, and so $h_{\mathsf{D}}(e_0) = 1$. Since $h + h_{\mathsf{D}}$ is a flow, it remains only to prove that it is feasible. Let $e \in W$.

Case 1: $(h' - h)(e) = 0$. Since $\sigma(h_{\mathsf{D}}) \subseteq \sigma(h' - h)$, $h_{\mathsf{D}}(e) = 0$. It follows that $(h + h_{\mathsf{D}})(e) = h(e)$, which lies between 0 and $k(e)$.

Case 2: $(h' - h)(e) > 0$. Then $h'(e) \geq h(e) + 1$, and $h_{\mathsf{D}}(e) = 0$ or 1. Hence $0 \leq h(e) \leq (h + h_{\mathsf{D}})(e) \leq h'(e) \leq k(e)$.

Case 3: $(h' - h)(e) < 0$. Then $h'(e) \leq h(e) - 1$ and $h_{\mathsf{D}}(e) = 0$ or -1. Hence $0 \leq h'(e) \leq (h + h_{\mathsf{D}})(e) \leq h(e) \leq k(e)$. □

Both the Max-Flow–Min Cut Theorem and the foregoing proposition are assertions concerning existence. In particular, the former asserts the existence of a maximum flow through a given edge e_0 and gives the value of that flow through e_0. On the other hand, the Flow Algorithm below is actually a procedure for obtaining a maximum flow through e_0. It is basically an "improvement process" and is motivated by Proposition D1.

If h is a feasible flow for an integral network (V, k), a circuit D of Γ is said to be **unsaturated with respect to** h **and** e_0 if $h_{\mathsf{D}}(e_0) = 1$ and $h + h_{\mathsf{D}}$ is feasible.

D2 Exercise. Prove that *a circuit $x_0, e_1, x_1, \ldots, e_m, y_0, e_0, x_0$ is unsaturated with respect to h and e_0 if and only if $h(e_0) < k(e_0)$ and the path $x_0, e_1, x_1, \ldots, e_m, y_0$ is unsaturated with respect to h.*

We can now state:

D3 The Flow Algorithm. *Let (V, k) be an integral network and let $e_0 \in W$.*

Step 1: Select a feasible flow h for (V, k). (This is always possible, since the zero-function is always a feasible flow.)

Step 2: Search for an elementary circuit D *which is unsaturated with respect to h and e_0.*

Step 3: If such a circuit D *exists, replace h by $h + h_{\mathrm{D}}$ and return to Step* 2.

Step 4: If no such D *exists then h is a maximum flow through e_0 by* D1.

This algorithm is indicated schematically in Figure D4.

D4

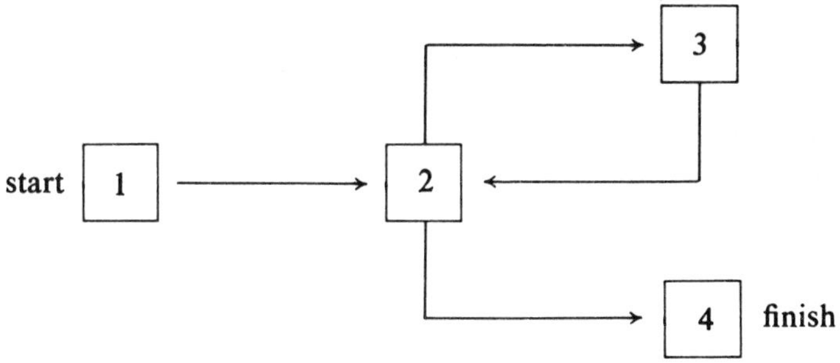

Before investigating the details of the Flow Algorithm we show:

D5 Proposition. *Let (V, k) be an integral network, $e_0 \in W$, and h a feasible flow. With h as initial feasible flow, the Flow Algorithm yields a maximum flow through e_0 within at most $k(e_0) - h(e_0)$ iterations.*

PROOF. Let $h_0 = h$, and let h_i be the feasible flow obtained in the ith iteration of Step 3 of the Flow Algorithm. By the definition of h_i, $h_i(e_0) = h_{i-1}(e_0) + 1$ for $i \geq 1$. Thus $h_i(e_0) = h(e_0) + i$. Since h_i is feasible, $k(e_0) \geq h_i(e_0)$ for all i, and so the maximum number of iterations possible is $k(e_0) - h(e_0)$. □

Evidently the efficiency of the Flow Algorithm depends upon the efficiency of the search technique used in implementing Step 2. There is a search technique implicit in the proof of the Max-Flow–Min-Cut Theorem; it is the one most often associated with the Flow Algorithm in the literature. This section concludes with a description of this search technique.

Let (V, k) be an integral network, $e_0 = (y_0, x_0) \in W$, and h an integral feasible flow. We seek either to construct an elementary circuit in Γ which is unsaturated with respect to h and e_0 or to show that such a circuit cannot exist.

If $h(e_0) = k(e_0)$, then there exists no unsaturated circuit. Assume then that $h(e_0) < k(e_0)$. By Exercise D2, it suffices either to find an unsaturated

x_0y_0-path (with respect to h, henceforth being understood) or to show that such a path cannot exist. For $i = 0, 1, 2, \ldots$, let

D6 $\quad U_i = \{x \in V\colon$ there exists an unsaturated x_0x-path of length i but none of shorter length$\}$.

Clearly $y_0 \in U_i$ for some i if and only if there is an unsaturated x_0y_0-path. Since $U_i \cap U_j = \varnothing$ for $i \neq j$, $U_i \neq \varnothing$ for only finitely many indices i. Moreover, if $U_i = \varnothing$, then $U_j = \varnothing$ for all $j \geq i$.

The sets U_i admit an inductive construction as follows. Let $U_0 = \{x_0\}$, and assume U_i has been constructed. Then U_{i+1} is the collection of all $x \in V + U_0 + U_1 + \ldots + U_i$ such that for some $y \in U_i$, one has either $h(y, x) < k(y, x)$ or $h(x, y) > 0$. One readily sees that this construction is consistent with D6.

Let m be the smallest integer such that either $U_m = \varnothing$ or $y_0 \in U_m$. If $U_m = \varnothing$, then no unsaturated x_0y_0-path exists. If $y_0 \in U_m$, then an unsaturated x_0y_0-path exists and one may be constructed as follows. Write $x_m = y_0$, and for $j = m - 1, m - 2, \ldots, 0$, one may inductively select $x_j \in U_j$ and $e_{j+1} \in W$ such that either

$$e_{j+1} = (x_j, x_{j+1}) \quad \text{and} \quad h(e_{j+1}) < k(e_{j+1}), \quad \text{or}$$

$$e_{j+1} = (x_{j+1}, x_j) \quad \text{and} \quad h(e_{j+1}) > 0.$$

In using the Flow Algorithm with this "subroutine" as the search technique, we terminate the process when a feasible flow has been constructed such that either $h(e_0) = k(e_0)$ or $h(e_0) = k(C; e_0)$ for some cut C through e_0. In the latter case such a cut C is actually constructible by means of this subroutine, since there exists no x_0y_0-path unsaturated with respect to h. Hence $U_m = \varnothing$ in the last iteration of the subroutine. The set $U = U_0 + U_1 + \ldots + U_{m-1}$ is then the very same set U defined in the proof of the Max-Flow–Min-Cut Theorem (C6). The required cut is $C = \sum_{x \in U} f^*(x)$.

Example. Consider (V, k) as in Figure C4. We will use the Flow Algorithm with the above subroutine to find a maximum flow through $e_0 = (8, 2)$. Start with h equal to the zero-function. Searching for an unsaturated 2,8-path, we have $U_0 = \{2\}$, $U_1 = \{3\}$, $U_2 = \{1, 4\}$, $U_3 = \{5, 7, 9\}$, $U_4 = \{6, 8\}$. This yields among others the elementary cycle $\mathsf{D} = 2, (2, 3), 3, (3, 1), 1, (1, 9), 9, (9, 8), 8, (8, 2), 2$, and we replace h by $h + h_{\mathsf{D}}$, which is a unit flow through D. Returning to Step 2, we start our search again. However, there is a short cut. We have just constructed D and we could check to see whether it is unsaturated with respect to the new h. It is, and we replace h by $h + h_{\mathsf{D}}$ again. As we return to Step 3 our flow is $h + 2h_{\mathsf{D}}$. Since D is saturated with respect to this flow ($h(3, 1) = 2 = k(3, 1)$), we must make another search:

$$U_0 = \{2\}, U_1 = \{3\}, U_2 = \{4\}, U_3 = \{1, 5\}, U_4 = \{6, 7, 9\}, U_5 = \{8\}.$$

A new circuit D is 2, (2, 3), 3, (3, 4), 4, (4, 1), 1, (1, 9), 9, (9, 8), 8, (8, 2), 2. Adding this flow to h yields the flow whose values are underlined in Figure D7.

D7

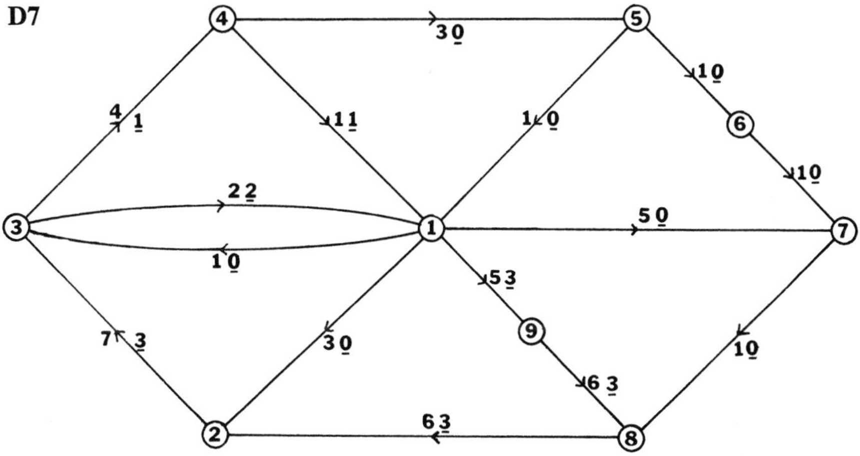

The next search yields: $U_0 = \{2\}$, $U_1 = \{3\}$, $U_2 = \{4\}$, $U_3 = \{5\}$, $U_4 = \{1, 6\}$, $U_5 = \{7, 9\}$, $U_6 = \{8\}$, which in turn yields a circuit D = 2, (2, 3), 3, (3, 4), 4, (4, 5), 5, (5, 1), 1, (1, 7), 7, (7, 8), 8, (8, 2), 2. The new flow $h + h_D$ is then underlined in Figure D8.

D8

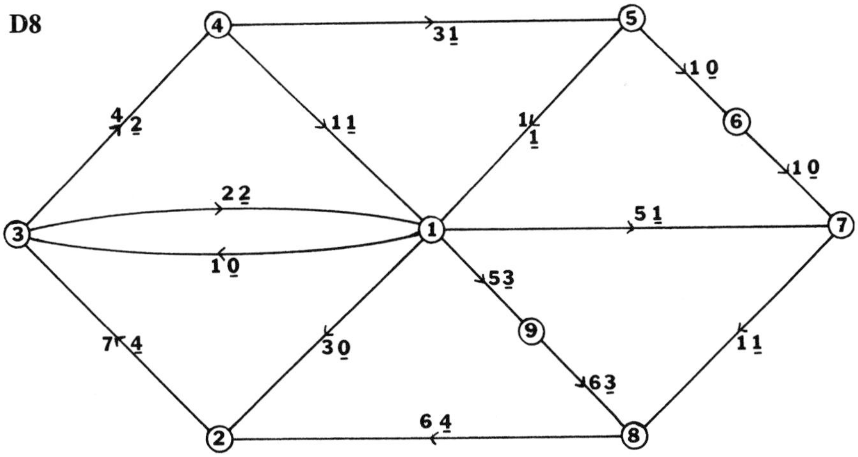

Repeating the "subroutine" again yields: $U_0 = \{2\}$, $U_1 = \{3\}$, $U_2 = \{4\}$, $U_3 = \{5\}$, $U_4 = \{6\}$, $U_5 = \{7\}$, $U_6 = \{1\}$, $U_7 = \{9\}$, and $U_8 = \{8\}$. This clearly determines uniquely the circuit

2, (2, 3), 3, . . . , 6, (6, 7), 7, (1, 7), 1, (1, 9), 9, . . . , 2.

Note the edge (1, 7); it is our first example of a "back flow." The new flow is depicted in Figure D9.

D9

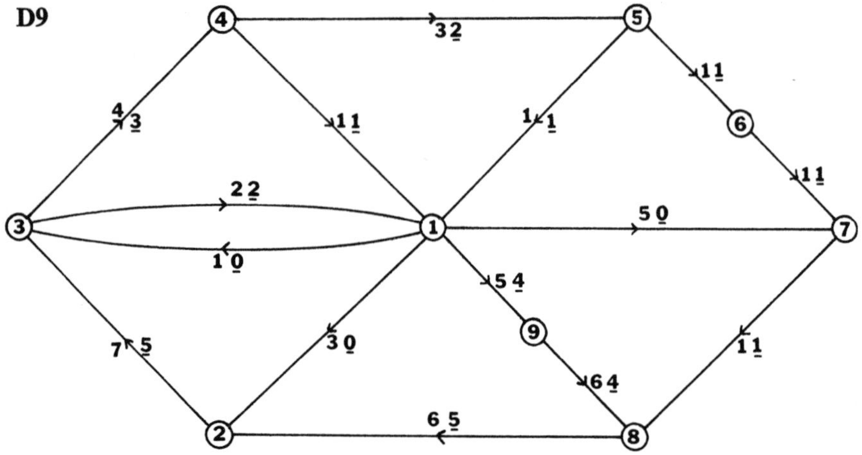

The final application of the subroutine now yields a minimum cut C. We have $U_0 = \{2\}$, $U_1 = \{3\}$, $U_2 = \{4\}$, $U_3 = \{5\}$, $U_4 = \varnothing$. Hence $U = \{2, 3, 4, 5\}$. The edges of C which appear in D9 are (8, 2), (1, 2), (1, 3), (3, 1), (4, 1), (5, 1), and (5, 6). One easily checks that $k(C; e_0) = 5 = h(e_0)$.

IVE The Classical Form of the Max-Flow–Min-Cut Theorem

The Max-Flow–Min-Cut Theorem was developed, stated, and proved in this chapter in such a way as to be consistent with the algebraic structure of this book. For completeness we reformulate this result in the traditional form. This is the form in which the reader is most likely to encounter it in the literature, especially in the context of optimization problems.

Rather than the multigraph $\Gamma = (V, f, W)$ considered in the previous sections, we now deal with an arbitrary directed graph (V, D) and use the term "a capacity for (V, D)" to indicate any function $k: D \to \mathbb{N}$. Two vertices $x_0, y_0 \in V$ are distinguished and are called the "source" and "sink", respectively. A "network" now denotes a 4-tuple $((V, D), k, x_0, y_0)$.

A function $f: D \to \mathbb{N}$ is called a "feasible flow" for this network if

(a) $0 \leq f(x, y) \leq k(x, y)$ for all $(x, y) \in D$, and

(b) $\sum_{y \in V, (x,y) \in D} f(x, y) = \sum_{y \in V, (y,x) \in D} f(y, x)$, for all $x \in V + \{x_0, y_0\}$.

As before, the zero-function is still a feasible flow. If f is a feasible flow, we define its "value" to be

$$v(f) = \sum_{\substack{y \in V \\ (x_0, y) \in D}} f(x_0, y) - \sum_{\substack{y \in V \\ (y, x_0) \in D}} f(y, x_0).$$

Of course, the zero-function has value 0.

E1 *Exercise.* Show that if f is a feasible flow, then

$$v(f) = \sum_{\substack{y \in V \\ (y, y_0) \in D}} f(y, y_0) - \sum_{\substack{y \in V \\ (y_0, y) \in D}} f(y_0, y).$$

Recall how a cut in a network was expressed as the sum of vertex cocycles over a subset of V. Classically, a cut is identified with that subset. Thus, by a "cut" we shall mean a subset $U \subseteq V$ such that $x_0 \in U \subseteq V + \{y_0\}$. The "capacity of U" is the number

$$k(U) = \sum_{\substack{(x, y) \in D \\ x \in U \\ y \in V + U}} k(x, y).$$

E2 Max-Flow–Min-Cut Theorem: Classical Form (L. R. Ford, Jr. and D. R. Fulkerson [f.3]). *In any network* $((V, D), k, x_0, y_0)$, $\max\{v(f): f$ *is a feasible flow*$\} = \min\{k(U)$: U *is a cut*$\}$.

E3 *Exercise.* Prove that E2 and C6 are equivalent for integral networks.

Often E2 is stated and proved in the case where the capacity function and the feasible flows are defined as functions from D into the nonnegative elements of $\mathbb{Q}$ or $\mathbb{R}$. The proof of C6 takes care of the rational case of E2, but not the real case. The difficulty in the real case is that the iteration process need not increase the value of the flow by a fixed minimum amount and hence may never terminate. If the process does not terminate, suppose $h_0, h_1, \ldots$ is a sequence of feasible flows constructed via this iteration process. Since the sequence is bounded above by k, there is a function $h \in \mathbb{R}^D$ such that $\lim_{i \to \infty} h_i(x, y) = h(x, y)$ for all $(x, y) \in D$. While it is not difficult to show that h is a feasible flow, it cannot be shown—in fact, it need not even be true—that h is a maximum flow. The existence of a maximum flow in the real case must be proved using the fact that (under the product topology on $\mathbb{R}^D$) the set of feasible flows is a closed and bounded subset and that the value function v is continuous. Once the existence of a maximum flow is established, the proof of C6 may be adapted to show that its value equals the capacity of a "minimum" cut.

IVF The Vertex Form of Max-Flow–Min-Cut

A variant of the Max-Flow–Min-Cut Theorem is obtained by assigning capacities to the vertices of a directed graph. One then considers cuts as consisting of sets of vertices which interrupt all directed circuits through a given vertex rather than as sets of edges which interrupt all directed circuits through a given edge. In this section we prove that the "edge form" C6 of the theorem implies the "vertex form." The latter is particularly important as it affords an elegant proof of Menger's Theorem in *VI*A below. The cycle of equivalence will be completed in the same section. There Menger's Theorem will in turn be used to give another proof of Theorem C6.

Let (V, D) be a directed graph and let $j: V \to \mathbb{N}$. The function j is called a **vertex capacity** for (V, D). By a **flow** in (V, D) we shall mean a flow h in $F(V)$ such that $\sigma(h) \subseteq D$. The **value of the flow** h **at the vertex** x is denoted by $h(x)$ and defined by

$$h(x) = \sum_{y \in V + \{x\}} h(y,x).$$

Since the "inflow" equals the "outflow" at x, we have by B2 that

F1
$$h(x) = \sum_{y \in V + \{x\}} h(x, y).$$

A flow h in (V, D) is said to be **feasible** if $0 \leq h(x, y)$ for each $(x, y) \in W$ while $h(x) \leq j(x)$ for all $x \in V$. Finally, h is said to be a **maximum flow through** x_0 if h is a feasible flow and if $h(x_0) \geq h'(x_0)$ for every feasible flow h' in (V, D). A **vertex-cut through** x_0 is a subset $U \subseteq V + \{x_0\}$ such that every directed circuit through x_0 contains an element of U. Clearly $V + \{x_0\}$ is a vertex-cut through x_0. If there are no directed circuits through x_0, then $\varnothing$ is a **vertex-cut through** x_0. The capacity of a vertex-cut U is

$$j(U) = \sum_{x \in U} j(x).$$

F2 Max-Flow–Min-Cut Theorem (Vertex Form). *Let (V, D) be a directed graph. Let j be a vertex capacity for (V, D), and let $x_0 \in V$. Then there exists a maximum flow h through x_0 such that*

$$h(x_0) = \min(\{j(x_0)\} \cup \{j(U)\colon U \text{ is a vertex-cut through } x_0\}).$$

PROOF. Let $X = V \times \{1, 2\}$. For $(x, i) \in X$ we write x^i. Let $Y = (X \times X) + \{(x^i, x^i)\colon x^i \in X\}$, and let μ denote some integer greater than $\sum_{x \in V} j(x)$. Let $k\colon Y \to \mathbb{N}$ be given by

$$k(x^i, y^i) = 0 \quad \text{for all } x, y \in V, i \in \{1, 2\};$$

$$k(x^1, y^2) = \begin{cases} \mu & \text{if } (x, y) \in D; \\ 0 & \text{if } (x, y) \notin D; \end{cases}$$

$$k(x^2, y^1) = \begin{cases} j(x) & \text{if } x = y; \\ 0 & \text{if } x \neq y. \end{cases}$$

The pair (X, k) is then a network. As was indicated in Figure C4, a network (X, k) may be viewed in terms of the directed graph $(X, \sigma(k))$. In Figure F3 we indicate how $(X, \sigma(k))$ may be constructed locally from (V, D). The numbers beside the edges in Figure F3b represent the values of k.

F3 (a) (V, D)

(b) $(X, \sigma(k))$

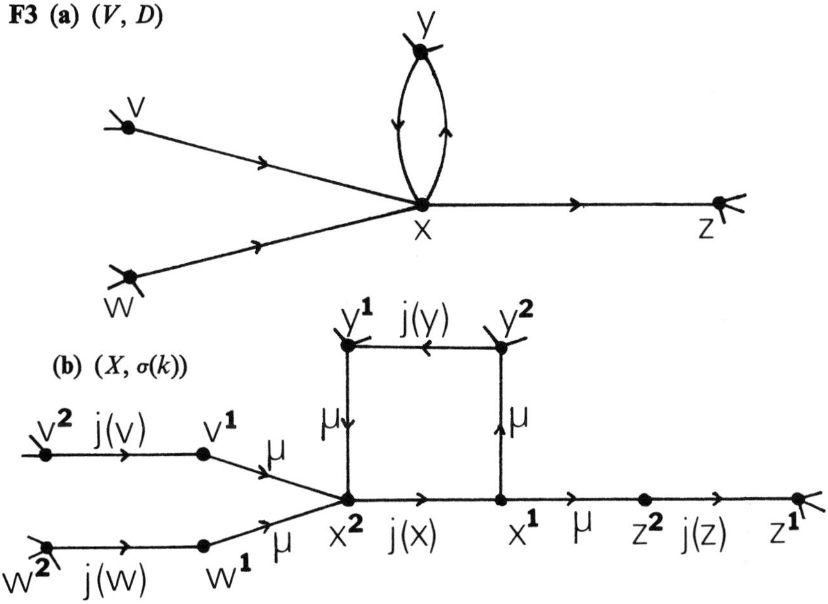

Let h be a feasible flow in (V, D). We define $\bar{h}\colon Y \to \mathbb{N}$ by

$$\bar{h}(x^i, y^i) = 0 \quad \text{for all } x, y \in V, i \in \{1, 2\};$$

$$\bar{h}(x^1, y^2) = h(x, y) \quad \text{for all } x, y \in V;$$

$$\bar{h}(x^2, y^1) = \begin{cases} h(x) & \text{if } x = y; \\ 0 & \text{if } x \neq y. \end{cases}$$

By this definition, the definition of feasible flow in (V, D), and F1, we have that $\bar{h}$ is a feasible flow in the network (X, k). In fact, one easily verifies that the mapping $h \mapsto \bar{h}$ is a bijection from the set of feasible flows on (V, D) onto the set of feasible flows on (X, k). Furthermore, since $h(x_0) = \bar{h}(e_0)$ where $e_0 = (x_0{}^2, x_0{}^1)$, we have

F4
$$\max\{h(x_0)\colon h \text{ is a feasible flow in } (V, D)\} = \max\{\bar{h}(e_0)\colon \bar{h} \text{ is a feasible flow in } (X, k)\}.$$

Define $g\colon Y \to \mathscr{P}_2(X)$ by $g(x^i, y^m) = \{x^i, y^m\}$, and let Θ denote the multigraph (X, g, Y). A cut C through e_0 in (X, k) is then of the form $C = \sum_{x^i \in S} g^*(x^i)$ for some subset $S \subseteq X$ such that $x_0{}^1 \in S$ and $x_0{}^2 \in X + S$. Let $T_i = \{x \in V\colon x^i \in S\}$ for $i = 1, 2$. The edges $(x^i, y^m) \in C$ which contribute to $k(C; e_0)$, the capacity of C, are those for which $x^i \in S$ and $y^m \in X + S$. These edges fall into two classes:

$$C_1 = \{(x^1, y^2)\colon x \in T_1;\, y \in V + T_2;\, (x, y) \in D\},$$

$$C_2 = \{(x^2, x^1)\colon x \in T_2 \cap (V + T_1)\}.$$

Each edge in C_1 contributes the quantity μ to $k(C; e_0)$, while each edge $(x^2, x^1) \in C_2$ contributes the quantity $j(x)$. Thus $k(C; e_0) \geq \mu$ if $C_1 \neq \varnothing$, while if $C_1 = \varnothing$ and $U = T_2 + (V + T_1)$, then

F5 $$k(C; e_0) = \sum_{x \in U} j(x) = j(U) < \mu.$$

If there exists at least one cut C through e_0 with $C_1 = \varnothing$, then any minimum cut must also have that property. Hence we may assume for the cuts under consideration that $C_1 = \varnothing$. We will show that cuts of this type do exist and that they in fact correspond to vertex-cuts of (V, D).

Let $C = \sum_{x^i \in S} g^*(x^i)$ and assume that $C_1 = \varnothing$. Let T_1, T_2 be defined as above in terms of S. We proceed to show that $U = T_2 \cap (V + T_1)$ is a vertex-cut of (V, D). Let $x_0, (x_0, x_1), x_1, \ldots, (x_m, x_0), x_0$ be an arbitrary directed circuit through x_0. We may assume $x_i \neq x_0$ for $i = 1, \ldots, m$, for if not we may replace the above circuit by a shorter one. This path induces an $x_0^1 x_0^2$-path in Θ:

F6 $$x_0^1, (x_0^1, x_1^2), x_1^2, (x_1^2, x_1^1), x_1^1, \ldots, (x_m^1, x_0^2), x_0^2.$$

Since $x_0^1 \in S$ while $x_0^2 \in V + S$, some edge in F6 has its "first vertex" in S and its "second vertex" in $V + S$; such an edge is in C. Since $C_1 = \varnothing$, this edge must be of the form (x_i^2, x_i^1) for some $i = 1, \ldots, m$. Thus $x_i \neq x_0$, $x_i \in T_2$, $x_i \notin T_1$. In other words, $U \neq \varnothing$.

We next show that for every vertex-cut $U \subseteq V + \{x_0\}$, there exists a cut C in (X, k) such that $k(C; e_0) = j(U)$. Let U be a vertex-cut of (V, D). Let $U_1 = \{x \in V$: there exists a directed x_0x-path containing no vertex of $U\}$, and let $U_2 = U + U_1 + \{x_0\}$. Thus $x_0 \in U_1$, $x_0 \notin U_2$, and $U = U_2 \cap (V + U_1)$. Let $S = (U_1 \times \{1\}) \cup (U_2 \times \{2\})$, and let $C = \sum_{x^i \in S} g^*(x^i)$. Since $x_0^1 \in S$ and $x_0^2 \in V + S$, C is a cut through e_0 in Θ. It suffices to show that $C_1 = \varnothing$. Suppose that $(x^1, y^2) \in C_1$. Since $x \in U_1$, there exists a directed x_0x-path "avoiding" U, which may clearly be extended via (x, y) to a directed x_0y-path. Since $y \notin U_2$, the extended path avoids U. Therefore $y \in U$, which is only possible if $y = x_0$. But this means that the extended path is a directed circuit through x_0 avoiding U, contrary to our assumption that U is a vertex-cut.

Combining these last two arguments and F5, we have

$$\begin{aligned} &\min\{k(C; e_0): C \text{ is a cut through } e_0 \text{ in } (X, k)\} \\ &\qquad = \min\{j(U): U \text{ is a vertex-cut through } x_0 \text{ in } (V, D)\}. \end{aligned}$$

Since $k(e_0) = j(x_0)$,

F7 $$\begin{aligned} &\min(\{k(e_0)\} \cup \{k(C; e_0): C \text{ is a cut through } e_0 \text{ in } (X, k)\}) \\ &\qquad = \min(\{j(x_0)\} \cup \{j(U): U \text{ is a vertex-cut through } x_0 \text{ in } (V, D)\}). \end{aligned}$$

By Theorem C6, the left-hand number of F7 equals the right-hand member of F4. This string of equalities proves the theorem. □

IVG Doubly-Capacitated Networks and Dilworth's Theorem

A **doubly-capacitated network** is a triple (V, k_1, k_2) where $k_i \in \mathbb{Z}^W$ for $i = 1, 2$, and $k_1(e) \leq k_2(e)$ for all $e \in W$. The functions k_1 and k_2 are called, respectively, the **lower** and **upper capacities**. An integral flow h is said to be **feasible** if

$$k_1(e) \leq h(e) \leq k_2(e) \quad \text{for all } e \in W.$$

If $e_0 \in W$, h is a **maximum flow through** e_0 if h is a feasible flow and $h(e_0) \geq h'(e_0)$ for any feasible flow h'. An integral network (V, k) as defined in §C can be regarded as the doubly-capacitated network $(V, 0, k)$ where 0 denotes the zero function in $\mathbb{Z}^W$. It follows from the discussion at the beginning of §D that the results of the present section hold as well with $\mathbb{Z}$ replaced by $\mathbb{Q}$.

Consider a system of pipelines where $k_2(e)$ represents the capacity of a link e as did $k(e)$ in §C. Suppose, however, that to prevent deterioration, each link e must carry a certain minimum flow $k_1(e)$. The doubly-capacitated network (V, k_1, k_2) can be regarded as an abstraction of just such a situation as this. While in this situation k_1 and k_2 are nonnegative functions, it is not required that they always be so.

It should be emphasized that in general a doubly-capacitated network need not always admit a feasible flow. In Figure G1, for example, where $|V| = 5$, suppose that $k_1(e) = 1$ and $k_2(e) = 2$ for every edge e represented in the figure, while $k_1(e) = k_2(e) = 0$ for every other edge. If there existed a feasible flow, the "inflow" into the vertex x could be at most 2 while its "outflow" must be at least 3.

G1

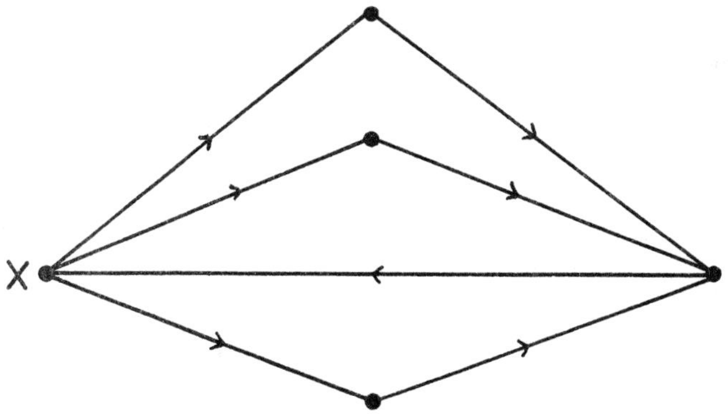

Let $e_0 = (y_0, x_0) \in W$. A **cut** C **through** e_0 in the doubly-capacitated network (V, k_1, k_2) is defined exactly as it is for a network; it is of the form $C = \sum_{x \in U} f^*(x)$ where $x_0 \in U \subseteq V + \{y_0\}$. With g_U as defined in C1, the **upper capacity of** C **through** e_0 is defined as

$$k_2(C; e_0) = \sum_{g_U(e)=1} k_2(e) - \sum_{\substack{e \neq e_0 \\ g_U(e)=-1}} k_1(e),$$

and the **lower capacity of C through e_0** as

$$k_1(C; e_0) = \sum_{g_U(e)=1} k_1(e) - \sum_{\substack{e \neq e_0 \\ g_U(e)=-1}} k_2(e).$$

If h is a feasible flow,

$$0 = h \cdot g_U = \sum_{g_U(e)=1} h(e) - \sum_{g_U(e)=-1} h(e) \leq k_2(C; e_0) - h(e_0).$$

Similarly, $0 \geq k_1(C; e_0) - h(e_0)$. Thus

G2 $$k_1(C; e_0) \leq h(e_0) \leq k_2(C; e_0)$$

for any feasible flow h and any cut C through e_0.

G3 Max-Flow–Min-Cut Theorem (Doubly-Capacitated Form). *If $e_0 \in W$ and if the doubly-capacitated network (V, k_1, k_2) admits a feasible flow, then it admits a maximum flow h through e_0 and*

$$h(e_0) = \min(\{k_2(e_0)\} \cup \{k_2(C; e_0): C \text{ is a cut through } e_0\}).$$

Proof. The proof of C6 may be adapted to prove this result with the following modest changes. First, replace k by k_2 throughout. Second, alter the third condition in the definition of "unsaturated path" to read: $e_i = (x_i, x_{i-1}) \Rightarrow h(e) > k_1(e)$. □

The first inequality in G2 gives a lower bound for flows through e_0. We are thus led to call a feasible flow h a **minimum flow through** e_0 if $h(e_0) \leq h'(e_0)$ for any feasible flow h'. Clearly if k_1 is a flow, then it is a minimum flow through each edge. However, when k_1 is not a flow, the following result is nontrivial.

G4 Min-Flow–Max-Cut Theorem. *If $e_0 \in W$ and if the doubly-capacitated network (V, k_1, k_2) admits a feasible flow, then it admits a minimum flow h through e_0, and*

$$h(e_0) = \max(\{k_1(e_0)\} \cup \{k_1(C; e_0): C \text{ is a cut through } e_0\}).$$

Proof. Let $k_1' = -k_2$ and $k_2' = -k_1$. Then (V, k_1', k_2') is also a doubly-capacitated network. Moreover, a flow h is feasible (minimum through e_0) in (V, k_1, k_2) if and only if $-h$ is feasible (maximum through e_0) in (V, k_1', k_2'). Finally, for any cut C through e_0, $k_1(C; e_0) = -k_2'(C; e_0)$. Thus

$$\begin{aligned}&\max(\{k_1(e_0)\} \cup \{k_1(C; e_0): C \text{ is a cut through } e_0\})\\ &\quad = -\min(\{k_2'(e_0)\} \cup \{k_2'(C; e_0): C \text{ is a cut through } e_0\}),\end{aligned}$$

and the result follows from Theorem G3. □

The *existence* of a feasible flow in a doubly-capacitated network (V, k_1, k_2) is the essential question, for without it, Theorems G3 and G4 add nothing to what has already been said. Let $e_0 \in W$ and let a cut $C = \sum_{x \in U} f^*(x)$ through e_0 be given. If h is a feasible flow, then by G2,

$$k_2(C; e_0) \geq h(e_0) \geq k_1(e_0),$$

and so

G5 $$0 \leq k_2(C; e_0) - k_1(e_0) = \sum_{g_U(e)=1} k_2(e) - \sum_{g_U(e)=-1} k_1(e).$$

If $k(U)$ denotes the right-hand quantity in G5, then the assertion: $k(U) \geq 0$ whenever $\varnothing \subset U \subset V$, is a necessary condition for the existence of a feasible flow. Indeed, it is also sufficient.

G6 *Exercise.* Prove that (V, k_1, k_2) admits a feasible flow if and only if $k(U) \geq 0$ for all $U \subseteq V$. [*Hint:* construct a new doubly-capacitated network as follows. Let x_0, y_0 be two vertices not in V and let $V' = V + \{x_0, y_0\}$. Let

$$k_1'(e) = \begin{cases} k_1(e) & \text{if } e \in W; \\ 0 & \text{otherwise}; \end{cases}$$

$$k_2'(e) = \begin{cases} k_2(e) & \text{if } e \in W; \\ \mu & \text{if } e = (y_0, x_0), (x_0, x), \text{ or } (x, y_0) \text{ for } x \in V; \\ 0 & \text{otherwise}; \end{cases}$$

where μ is a fixed integer greater than $\sum_{e \in V} k_2(e)$. Show that (V', k_1', k_2') has a feasible flow and hence a minimum flow h through $e_0 = (y_0. x_0)$. Observe that if $h(e_0) = 0$, then $h_{|W}$ is a feasible flow on (V, k_1, k_2), while if $h(e_0) > 0$, then there exists no feasible flow on (V, k_1, k_2).]

Let $(X, \leq)$ be a partially-ordered set and recall the definition of a chain from §*II*B. The notion of an **incommensurable set** defined prior to Exercise *II*B22 is generalized here to denote a set $S \subseteq X$ such that $x < y$ fails for all $x, y \in S$.

G7 Theorem (R. P. Dilworth [d.2], 1950). *Let $(X, \leq)$ be a partially-ordered set. Then*

$$\begin{aligned} &\max\{|S| : S \text{ is incommensurable in } (X, \leq)\} \\ &\quad = \min\{|\mathcal{Q}| : \mathcal{Q} \in \mathbb{P}(X);\ Q \text{ is a chain for all } Q \in \mathcal{Q}\}. \end{aligned}$$

Proof. Let $V = (X \times \{1, 2\}) \cup \{x_0, y_0\}$ where x_0 and y_0 are distinct objects which are not elements of $X \times \{1, 2\}$. As in §F, we write x^i for the element $(x, i) \in V$. We form the doubly-capacitated network (V, k_1, k_2), where if $e \in W$, then

$$k_1(e) = \begin{cases} 1 & \text{if } e = (x^1, x^2) \text{ for some } x \in X; \\ 0 & \text{otherwise}; \end{cases}$$

and

$$k_2(e) = \begin{cases} 1 & \text{if } k_1(e) = 1; \\ |X| & \text{if } e = (x_0, x^1) \text{ or } (x^2, y_0) \text{ for some } x \in X; \\ |X| & \text{if } e = (y_0, x_0); \\ |X| & \text{if } e = (x^2, y^1) \text{ for } x, y \in X \text{ and } x < y; \\ 0 & \text{otherwise.} \end{cases}$$

(For example, let $X = \{x, y, z\}$ and suppose $x < y < z$. In Figure G8 the directed graph $(V, \sigma(k_2))$ is shown where the integer beside the edge e represents $k_2(e)$.)

G8

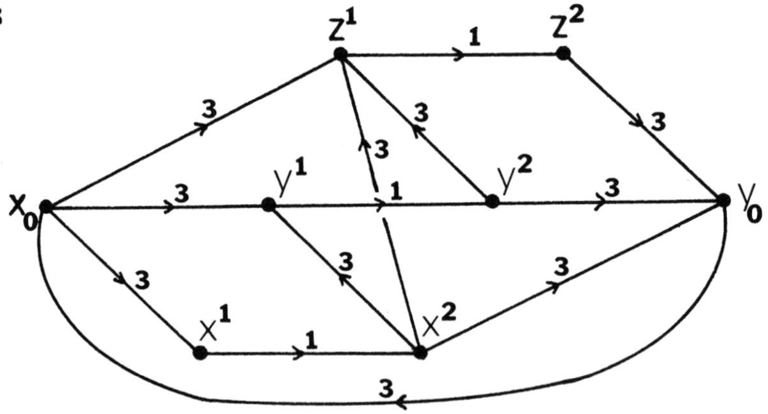

We assert that (V, k_1, k_2) admits a feasible flow. Indeed the function $\bar{h}$ given by

$$\bar{h}(e) = \begin{cases} 1 & \text{if } e = (x_0, x^1), (x^1, x^2), \text{ or } (x^2, y_0) \text{ for some } x \in X; \\ |X| & \text{if } e = (y_0, x_0); \\ 0 & \text{otherwise;} \end{cases}$$

is a feasible flow. Hence Theorem G4 applies to (V, k_1, k_2) nonvacuously.

Let h be any feasible flow in (V, k_1, k_2). Then h is nonnegative since k_1 is. By Exercise A17, we can write

$$h = \sum_{i=1}^{m} g_i$$

where g_i is a nonnegative elementary flow and $\sigma(g_i) \subseteq \sigma(h)$ for $i = 1, \ldots, m$. Let i now be fixed. By B8, $g_i = h_{\mathsf{D}}$ for some circuit D of $\Gamma = (V, f, W)$. Since h_{D} is nonnegative, D must assume the form

$$x_0, (x_0, x_1{}^1), x_1{}^1, (x_1{}^1, x_1{}^2), x_1{}^2, (x_1{}^2, x_2{}^1), x_2{}^1, (x_2{}^1, x_2{}^2), x_2{}^2, \ldots$$
$$\ldots, x_j{}^1, (x_j{}^1, x_j{}^2), x_j{}^2, (x_j{}^2, y_0), y_0, (y_0, x_0), x_0,$$

where $x_1 < x_2 < \ldots < x_j$ for some $j \geq 1$. Thus h is a sum of elementary flows, each of which corresponds to a chain in $(X, \leq)$. But since h is a feasible flow, $h(x^1, x^2) = 1$ for all $x \in X$. This implies that the set of chains corresponding to elementary flows $g_1, \ldots, g_m$ is an m-partition of X. Conversely, each partition $\mathcal{Q} \in \mathbb{P}(X)$ whose cells are nonempty chains corresponds to a feasible flow whose value at e_0 is $|\mathcal{Q}|$. We have proved:

G9 $$\min\{h(e_0)\colon h \text{ is a feasible flow}\} = \min\{|\mathcal{Q}|\colon \mathcal{Q} \in \mathbb{P}(X);\ Q \text{ is a chain, for all } Q \in \mathcal{Q}\}.$$

Consider an arbitrary cut $C = \sum_{x \in U} f^*(x)$ through e_0, where it may be assumed that $x_0 \in U \subseteq V + \{y_0\}$. Let

$$S = \{x \in X\colon x^1 \in U;\ x^2 \in V + U\},$$

$$T = \{x \in X\colon x^1 \in V + U;\ x^2 \in U\},$$

$$D = \{(x, y) \in X \times X\colon x < y;\ x^2 \in V + U;\ y^1 \in U\}.$$

By definition,

G10 $$k_1(C; e_0) = \sum_{\substack{u \in U \\ v \in V + U}} k_1(u, v) - \sum_{\substack{u \in V + U \\ v \in U}} k_2(u, v) = |S| - (|T| + |D||X|).$$

Since $|S| \leq |X|$, $k_1(C; e_0) \leq 0$ whenever $D \neq \varnothing$. If $\bar{x}, \bar{y} \in S$, then $\bar{x}^2 \in V + U$ and $\bar{y}^1 \in U$. Hence if $(\bar{x}, \bar{y}) \notin D$, then $\bar{x} < \bar{y}$ fails. Thus if $D = \varnothing$, then $x < y$ fails for all $x, y \in S$, i.e., S is incommensurable. Since $k_1(e_0) = 0$, we have proved:

G11 $$\max(\{k_1(e_0)\} \cup \{k_1(C; e_0)\colon C \text{ is a cut through } e_0\}) \leq \max\{|S|\colon S \text{ is incommensurable in } (X, \leq)\}.$$

On the other hand, let S_0 be any incommensurable set and define

$$U = \{x^1 \in X \times \{1\}\colon x \leq y \text{ for some } y \in S_0\} \cup \{x^2 \in X \times \{2\}\colon x < y \text{ for some } y \in S_0\} \cup \{x_0\}.$$

One readily observes that $S = S_0$ and $D = T = \varnothing$. By G10, $k_1(C; e_0) = |S_0|$. This proves that equality indeed holds in G11. This equality combined with G9 and Theorem G4 yield a string of equalities which proves the theorem. □

CHAPTER V

Matchings and Related Structures

The spaces of a system were studied in Chapter II; in Chapter III they were interpreted in the context of multigraphs and in Chapter IV in the context of networks. Once again we shall see how a single combinatorial notion transcends the peculiarities of the model which serves as the vehicle for its presentation. The Main Matching Theorem, to be presented and proved in the first section of this chapter, is the keystone for the rest of the chapter. Initially the vehicle for presentation is the bipartite graph. In the subsequent sections, the Main Matching Theorem will give information about many outwardly dissimilar yet nearly equivalent structures.

Since our proof of the Main Matching Theorem will be facilitated by the Max-Flow–Min-Cut Theorem, the reader is advised to refamiliarize himself with the definitions and the statements of results in §*IV*B and §*IV*C before proceeding.

VA Matchings in Bipartite Graphs

Let $(V, \mathscr{E})$ denote a graph. The notion of incidence introduced earlier was a "relation" between the sets V and $\mathscr{E}$. We now extend it to a reflexive and symmetric relation on $V \cup \mathscr{E}$ as follows. For $x_1, x_2 \in V$ and $E_1, E_2 \in \mathscr{E}$:

x_1 is **incident** with E_1 if $x_1 \in E_1$;

x_1 is **incident** with x_2 if either $x_1 = x_2$ or $\{x_1, x_2\} \in \mathscr{E}$;

E_1 is **incident** with E_2 if $E_1 \cap E_2 \neq \varnothing$.

A set $S \subseteq V \cup \mathscr{E}$ is a **vertex-covering set** (respectively, **edge-covering set**) if each vertex of positive valence (respectively, edge) is incident with some element of S. Observe that supersets of vertex- (respectively, edge-) covering sets are also vertex- (respectively, edge-) covering sets.

A set $S \subseteq V \cup \mathscr{E}$ is **independent** if no two distinct elements of S are incident. Clearly subsets of independent sets are also independent. An independent vertex set (also called an "internally stable set") is an independent subset of V. The cardinality of a largest independent vertex set in Γ is called the **vertex-independence number** (or "internal stability") of Γ and is denoted by $\alpha_0(\Gamma)$, or simply by α_0. An independent edge set is defined analogously, and the **edge-independence number** is denoted by $\alpha_1(\Gamma)$, or simply α_1. We define the function $N: \mathscr{P}(V) \to \mathscr{P}(V)$ whereby for each $U \in \mathscr{P}(V)$,

$$N(U) = \{y \in V + U: \{x, y\} \in \mathscr{E} \text{ for some } x \in U\}.$$

In this section we are concerned only with the bipartite graph $\mathrm{B} = (\{V_1, V_2\}, \mathscr{E})$, *the significance of these letters being hereby fixed for all of this section.* If $U \subseteq V_1$ or $U \subseteq V_2$, we define the **deficiency** of U to be the integer

$$\delta(U) = |U| - |N(U)|,$$

and we define

$$\delta_i(\mathrm{B}) = \max\{\delta(U): U \subseteq V_i\} \quad (i = 1, 2),$$

writing briefly δ_i when B is understood from the context.

Since $\delta(\varnothing) = 0$, $\delta_i \geq 0$ for any bipartite graph. A subset $U \subseteq V_i$ such that $\delta(U) = \delta_i(\mathrm{B})$ is called a **critical set**. An independent set $\mathscr{E}' \subseteq \mathscr{E}$ is called a **matching**. If $U_i \subseteq V_i$ for $i = 1, 2$, a matching $\mathscr{E}'$ is a **matching of U_1 into U_2** if

(a) $|\mathscr{E}'| = |U_1|$, and
(b) each edge in $\mathscr{E}'$ is incident with one vertex in U_1 and one vertex in U_2.

Of course, the condition $|U_1| \leq |U_2|$ is necessary though hardly sufficient for such a matching to exist. When we say briefly, "a matching of U_1," one should understand "a matching of U_1 into V_2."

Recall that to say that a matching $\mathscr{E}'$ in B is largest means that no other matching $\mathscr{E}''$ of any subset of V_1 whatever satisfies $|\mathscr{E}'| < |\mathscr{E}''|$.

A1 Main Matching Theorem. *In the bipartite graph* $\mathrm{B} = (\{V_1, V_2\}, \mathscr{E})$, $\alpha_1(\mathrm{B}) = |V_1| - \delta_1(\mathrm{B}) = |V_2| - \delta_2(\mathrm{B})$.

Proof. Form the set $V = V_1 \cup V_2 \cup \{a, z\}$, where a and z are "new" objects not in $V_1 \cup V_2$. Let μ be some integer greater than $|V|^2$. We then form the integral network (V, k) where

$$k(e) = \begin{cases} 1 & \text{if } e = (a, x_1) \text{ or } (x_2, z) \text{ for some } x_i \in V_i, \\ & \text{or if } e = (x_1, x_2) \text{ for } x_i \in V_i \text{ and } \{x_1, x_2\} \in \mathscr{E}; \\ \mu & \text{if } e = (z, a); \\ 0 & \text{otherwise.} \end{cases}$$

(In Figure A2 we show a bipartite graph B together with the directed graph $(V, \sigma(k))$ obtained as described. All edges shown except (z, a) have capacity 1.)

A2

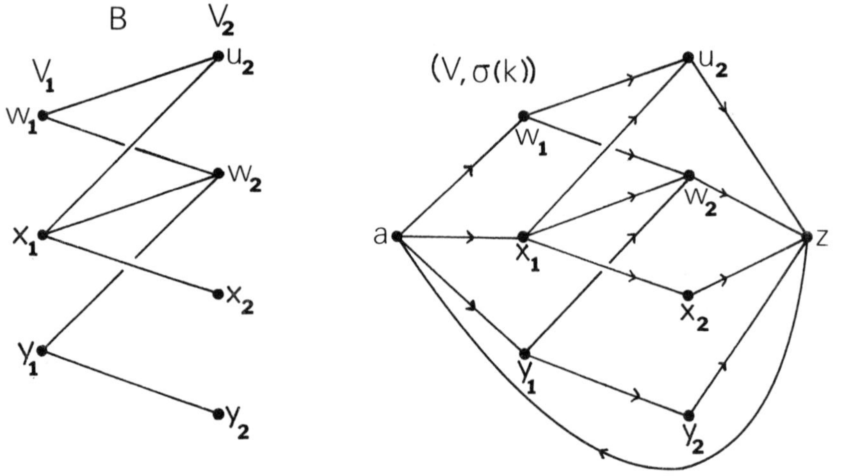

If h is an integer-valued feasible flow on (V, k), then clearly h assigns 0 or 1 to every element of $W = (V \times V) + \{(x, x): x \in V\}$ except perhaps to (z, a). Since h is a flow, one sees by the structure of (V, k) that

$$h(z, a) = |\{(x_1, x_2): x_i \in V_i; h(x_1, x_2) = 1\}|$$
$$= \sum_{x_1 \in V_1} \sum_{x_2 \in V_2} h(x_1, x_2).$$

Also the set $\mathscr{E}' = \{\{x_1, x_2\} \in \mathscr{E}: h(x_1, x_2) = 1\}$ is a matching in B of cardinality $h(z, a)$. Conversely, any matching in B of an m-subset of V_1 determines a feasible flow h in (V, k) with $h(z, a) = m$. This proves that

A3 $\qquad \alpha_1(\mathrm{B}) = \max\{h(z, a): h \text{ is a feasible flow in } (V, k)\}.$

Let the set U satisfy $a \in U \subseteq V + \{z\}$, and let $U_i = U \cap V_i$ $(i = 1, 2)$. Let C be the cut determined by U. Then C is a cut through (z, a), and its capacity satisfies

$$\begin{aligned} k(C; (z, a)) &= |V_1 + U_1| + \sum_{\substack{x_1 \in U_1 \\ x_2 \in V_2 + U_2}} k(x_1, x_2) + |U_2| \\ &\geq |V_1 + U_1| + |N(U_1) + (N(U_1) \cap U_2)| + |U_2| \\ &\geq |V_1 + U_1| + |N(U_1)| \\ &= |V_1| - (|U_1| - |N(U_1)|) \\ &= |V_1| - \delta(U_1) \geq |V_1| - \delta_1(\mathrm{B}). \end{aligned}$$

In particular, suppose U_1 is a critical subset of V_1. Then the set $U = \{a\} + U_1 + N(U_1)$ determines a cut C through (z, a) of capacity $|V_1| - \delta_1(\mathrm{B})$. This proves

A4 $\qquad |V_1| - \delta_1(\mathrm{B}) = \min\{k(C; (z, a)): C \text{ is a cut through } (z, a)\}.$

By the Max-Flow–Min-Cut Theorem (*IV*C6), A3, and A4, one has $\alpha_1(\mathrm{B}) = |V_1| - \delta_1(\mathrm{B})$. The other equality is proved symmetrically. □

A5 Corollary. *For any bipartite graph,* $\nu_0 = 2\alpha_1 + \delta_1 + \delta_2$.

A6 Corollary. *A necessary and sufficient condition for* B *to admit a matching of* V_i *is*

$$U \subseteq V_i \Rightarrow |U| \leq |N(U)| \quad (i = 1, 2).$$

PROOF: *Sufficiency.* The condition is clearly equivalent to the assertion that $\delta(U) \leq 0$ for all $U \subseteq V_i$, i.e., that $\delta_i(\mathrm{B}) = 0$. Now apply the theorem.

Necessity. Suppose that B admits a matching of V_i and let $U \subseteq V_i$. By the theorem,

$$0 = \delta_i(\mathrm{B}) \geq \delta(U) = |U| - |N(U)|,$$

whence the condition follows. □

A7 Corollary. *If in* B *one has* $\mathscr{E} \neq \varnothing$ *and* $\rho(x_1) \geq \rho(x_2)$ *for all* $x_i \in V_i$, *then there exists a matching of* V_1.

PROOF. Let $m = \min\{\rho(x): x \in V_1\}$, and let $U \subseteq V_1$. Since any edge incident with a vertex in U is also incident with a vertex in $N(U)$, one has

A8 $$m|U| \leq \sum_{x_1 \in U} \rho(x_1) \leq \sum_{x_2 \in N(U)} \rho(x_2) \leq m|N(U)|,$$

the last inequality in A8 holding since $\rho(x_2) \leq m$ for all $x_2 \in V_2$. Since $\mathscr{E} \neq \varnothing$, $\rho(x_2) > 0$ for some $x_2 \in V_2$. Hence $m > 0$, and A8 implies $\delta(U) \leq 0$. The result follows from the preceding corollary. □

The reader may wish to find other sufficient conditions on B for there to exist a matching of V_1. A particularly easy case presents itself when B is m-valent for some $m > 0$. Then $m|V_1| = |\mathscr{E}| = m|V_2|$, and every matching of V_1 into V_2 is also a matching of V_2 into V_1. Such a matching is called a **mutual matching**.

A9 Exercise. Prove that *if* B *is m-valent, then there exists an m-partition of* $\mathscr{E}$, *each cell of which is a mutual matching.*

A10 Exercise. *Given* $m, n \in \mathbb{N}$ *with* $m \leq n$, prove that *there exists an m-valent bipartite graph* $(\{V_1, V_2\}, \mathscr{E})$ *with* $|V_1| = |V_2| = n$.

For any multigraph Γ, the largest valence among the vertices in Γ will be denoted by $\hat{\rho}(\Gamma)$ or briefly by $\hat{\rho}$ when there is no risk of ambiguity. Similarly, $\check{\rho}(\Gamma)$ will denote the smallest valence in Γ. The next proposition shows how a bipartite graph B may be extended to an isovalent bipartite graph with the same largest valence without "joining by an edge" any vertices of B not already so joined.

A11 Proposition. *Given any bipartite graph* B, *there exists a* $\hat{\rho}(\mathrm{B})$*-valent bipartite graph* $\mathrm{A} = (\{W_1, W_2\}, \mathscr{F})$ *such that* $V_i \subseteq W_i$ *for* $i = 1, 2$ *and* $\mathrm{B} = \mathrm{A}_{V_1 \cup V_2}$.

Proof. Let V_i' be formed from V_i by adjoining sufficiently many new vertices to V_i ($i = 1, 2$) so that $|V_1'| = |V_2'|$ and that

A12 $$\hat{\rho}(\mathrm{B}) - 1 \leq \hat{\rho}(\mathrm{B})|V_1'| - |\mathscr{E}|.$$

Then $\mathrm{B}' = (\{V_1', V_2'\}, \mathscr{E})$ is a bipartite graph. Its valence function, being the extension by zero of ρ, will also be denoted by ρ.

Let n denote the right-hand member of A12. By Exercise A10, there exists a $(\hat{\rho}(\mathrm{B}) - 1)$-valent bipartite graph $\mathrm{B}'' = (\{V_1'', V_2''\}, \mathscr{E}'')$ with $|V_1''| = |V_2''| = n$. We may assume that B″ is disjoint from B′.

For each $i = 1, 2$,

$$\sum_{x \in V_i'} (\hat{\rho}(\mathrm{B}) - \rho(x)) = \hat{\rho}(\mathrm{B})|V_i'| - |\mathscr{E}| = n.$$

Hence for $j \in \{1, 2\}$, $j \neq i$, V_j'' admits a partition $\{U_x : x \in V_i';\ \rho(x) < \hat{\rho}(\mathrm{B})\}$ wherein each cell U_x has (positive) cardinality $|U_x| = \hat{\rho}(\mathrm{B}) - \rho(x)$. Now let

$$\mathscr{E}' = \{\{x, y\} : x \in V_1' \cup V_2';\ y \in U_x\},$$
$$\mathscr{F} = \mathscr{E} \cup \mathscr{E}' \cup \mathscr{E}'',$$

and $W_i = V_i' \cup V_i''$ for $i = 1, 2$.

Then $\mathrm{A} = (\{W_1, W_2\}, \mathscr{F})$ is the required bipartite graph. (See Figure A13.) □

A13

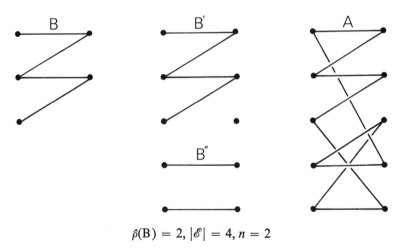

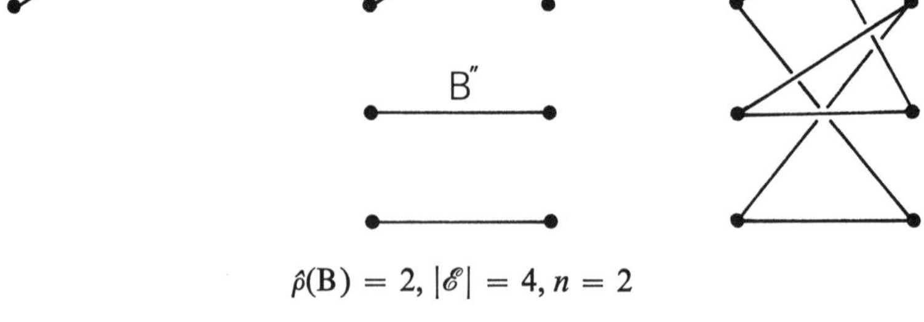

$\hat{\rho}(\mathrm{B}) = 2$, $|\mathscr{E}| = 4$, $n = 2$

The set $\mathscr{E}$ of edges of a bipartite graph B clearly admits a partition each of whose cells is a matching. Indeed, each matching could consist of a single edge, and we would have a $|\mathscr{E}|$-partition. At the other extreme, no such partition could have fewer than $\hat{\rho}(B)$ cells. The next result shows that this bound is "best possible" in general and not only in the case of an isovalent bipartite graph (cf. A9).

A14 Proposition. *For any bipartite graph* B *there exists a* $\hat{\rho}(B)$*-partition of* $\mathscr{E}$ *each of whose cells is a matching.*

PROOF. By Proposition A11 we may let $A = (\{W_1, W_2\}, \mathscr{F})$ be a $\hat{\rho}(B)$-valent bipartite graph such that $V_i \subseteq W_i$ $(i = 1, 2)$ and $B = A_{V_1 \cup V_2}$. By Exercise A9, there exists a partition $\{\mathscr{F}_1, \ldots, \mathscr{F}_{\hat{\rho}(B)}\}$ of $\mathscr{F}$, each of whose cells is a mutual matching of W_1. Since B contains a vertex incident with precisely $\hat{\rho}(B)$ edges in $\mathscr{E}$, it is readily seen that for each $i = 1, \ldots, \hat{\rho}(B)$, the set $\mathscr{E} \cap \mathscr{F}_i$ is a nonempty matching in B. Hence $\{\mathscr{E} \cap \mathscr{F}_1, \ldots, \mathscr{E} \cap \mathscr{F}_{\hat{\rho}(B)}\}$ is the required partition. □

The remainder of this section is concerned with the properties of critical sets and the role that they play in matchings.

First we observe that for any $U_1, U_2 \subseteq V_1$, one has

A15 $$N(U_1 \cup U_2) = N(U_1) \cup N(U_2)$$

while

A16 $$N(U_1 \cap U_2) \subseteq N(U_1) \cap N(U_2).$$

(Cf. *I*A11 and *I*A12.) That strict inequality may hold in A16 is seen by the example in Figure A17. (Cf. *I*A17.)

A17

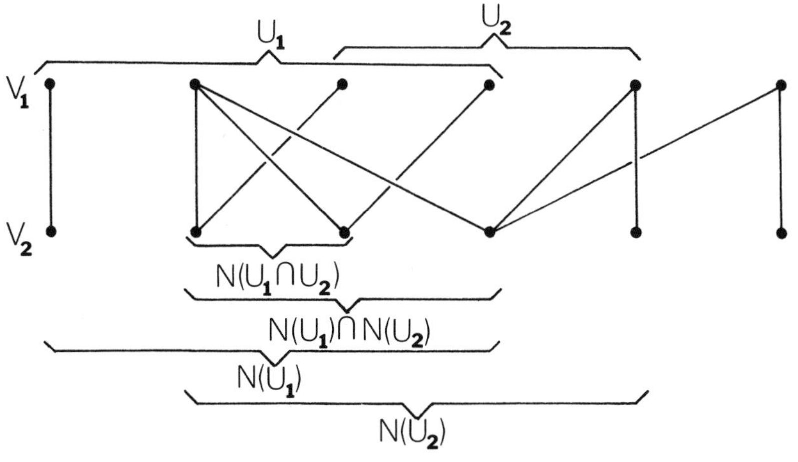

Evaluating cardinalities in A15 and A16 and then adding gives

$$\begin{aligned} |N(U_1 \cup U_2)| + |N(U_1 \cap U_2)| &\leq |N(U_1) \cup N(U_2)| + |N(U_1) \cap N(U_2)| \\ &= |N(U_1)| + |N(U_2)|. \end{aligned}$$

When this inequality is subtracted from

$$|U_1 \cup U_2| + |U_1 \cap U_2| = |U_1| + |U_2|$$

(cf. *I*C3), one obtains

A18 $$\delta(U_1) + \delta(U_2) \leq \delta(U_1 \cup U_2) + \delta(U_1 \cap U_2).$$

A19 *Exercise.*

(a) Prove that $N(U_1) + N(U_2) \subseteq N(U_1 + U_2)$. (Cf. *I*A15.)

(b) Determine what inequality, if any, relates $\delta(U_1 + U_2)$ with $\delta(U_1) + \delta(U_2)$.

A20 Proposition. *The set of critical subsets of V_1 is closed with respect to union and intersection.*

Proof. If U_1 and U_2 are critical sets, A18 gives

$$2\delta_1 \leq \delta(U_1 \cup U_2) + \delta(U_1 \cap U_2).$$

Since no set has deficiency greater than δ_1, it must hold that $\delta(U_1 \cup U_2) = \delta_1 = \delta(U_1 \cap U_2)$. □

Since the Main Matching Theorem establishes the cardinality of a largest matching as $|V_1| - \delta_1$, we know that certain overly large subsets of V_1 cannot be "matched," i.e., if fewer than δ_1 vertices of V_1 are deleted, the remaining set is still too big to admit a matching. The next proposition, on the other hand, gives affirmative information; it says that the complement in V_1 of any critical set U has a matching and that, moreover, the vertices of $N(U)$ are not required for it.

A21 Proposition. *If $U \subseteq V_1$ and U is a critical set in the bipartite graph* B, *then there exists a matching of $V_1 + U$ into $V_2 + N(U)$.*

Proof. Let U be a critical set and let $\mathrm{B}' = (\{V_1 + U, V_2 + N(U)\}, \mathscr{E}')$ be the subgraph of B induced by $(V_1 + U) \cup (V_2 + N(U))$. Let N' denote the function for B′ analogous to N for B, and let δ' denote deficiencies in B′.

If $T \subseteq V_1 + U$, then

$$\begin{aligned}
\delta'(T) &= |T| - |N'(T)| \\
&= |T| - |N(T) + (N(U) \cap N(T))| \\
&= \delta(U) + |T| - (|N(T)| - |N(U) \cap N(T)|) - \delta(U) \\
&= |U| + |T| - (|N(U)| + |N(T)| - |N(U) \cap N(T)|) - \delta(U) \\
&= |U| + |T| - |N(U \cup T)| - \delta(U) \qquad \text{(A15 and } I\text{C3)} \\
&= \delta(U \cup T) - \delta_1(\mathrm{B}) \leq 0
\end{aligned}$$

since U is critical and $U \cap T = \varnothing$. The proposition now follows from Corollary A6. □

The abundance of existence results above notwithstanding, it is worthwhile to note that in any bipartite graph one can actually construct a largest matching, merely by applying the Flow Algorithm (*IV*D3) to the network constructed in the proof of the Main Matching Theorem.

VB 1-Factors

The notions developed for bipartite graphs in the previous section admit extensions to arbitrary graphs. Analogously to the concept of a mutual matching, we define a **1-factor** of a graph $\Gamma = (V, \mathscr{E})$ to be a subset of $\mathscr{E}$ which is at once both independent and vertex-covering. Thus the existence of a 1-factor in Γ is equivalent to $\nu_0(\Gamma) = 2\alpha_1(\Gamma)$ and implies that Γ has no isolated vertices.

For any subset $U \in \mathscr{P}(V)$, we consider the components of Γ_U which contain an odd number of vertices and we let $\|U\|$ denote the number of such components. In case U is independent, then $\|U\| = |U|$, and so our definition

$$\delta(U) = \|U\| - |N(U)|$$

as the **deficiency** of U is clearly seen to include the definition of deficiency in §A as a special case. Analogously we define

$$\hat{\delta}(\Gamma) = \max\{\delta(U) \colon U \in \mathscr{P}(V)\}$$

and note that $\hat{\delta}(\Gamma) \geq 0$ since $\delta(\varnothing) = 0$.

Analogous to a critical set of §A is the following notion: U is **extremal** if $\delta(U) = \hat{\delta}(\Gamma)$. The precise relationship between these two terms will be made clear in the following proof.

B1 Proposition. *For any bipartite graph,* $\hat{\delta} = \hat{\delta}_1 + \hat{\delta}_2$.

Proof. Let $B = (\{V_1, V_2\}, \mathscr{E})$ be a bipartite graph. We first show that if $U \subseteq V_1 \cup V_2$ is a smallest extremal set in B, then U is independent. Let W be the vertex set of a component of B_U and let $W_i = W \cap V_i$ $(i = 1, 2)$. For definiteness suppose that $|W_1| \leq |W_2|$. One readily verifies that $\|U + W_1\| = \|U\| - \varepsilon + |W_2|$, where $\varepsilon = 1$ if $|W|$ is odd and $\varepsilon = 0$ if $|W|$ is even. Since $W_2 \neq \varnothing$, we have $N(U + W_1) \subseteq N(U) + W_1$. Hence

$$\begin{aligned}\delta(U + W_1) &= \|U + W_1\| - |N(U + W_1)| \\ &\geq \|U\| - |N(U)| + |W_2| - |W_1| - \varepsilon \geq \hat{\delta}(B),\end{aligned}$$

which is contrary to the choice of U unless $W_1 = \varnothing$ and U is independent. Therefore,

$$\begin{aligned}\hat{\delta}(B) &= |U \cap V_1| + |U \cap V_2| - |N(U \cap V_1)| - |N(U \cap V_2)| \\ &= \delta_1(U \cap V_1) + \delta_2(U \cap V_2) \leq \hat{\delta}_1(B) + \hat{\delta}_2(B).\end{aligned}$$

To prove the reverse inequality, let U_1 and U_2 be critical subsets of V_1 and V_2, respectively, and let $W = (U_1 \cap N(U_2)) \cup (U_2 \cap N(U_1))$. Then

$$\begin{aligned} \delta_1(B) + \delta_2(B) &= |U_1| + |U_2| - (|N(U_1)| + |N(U_2)|) \\ &= |U_1 + U_2| - (|N(U_1 + U_2)| + |W|) \\ &\le \|U_1 + U_2\| + |W| - |N(U_1 + U_2)| - |W| \\ &= \delta(U_1 + U_2) \le \delta(B). \end{aligned}$$ □

B2 Corollary. *For any bipartite graph, $\nu_0 = 2\alpha_1 + \hat{\delta}$.*

PROOF. Combine the above proposition with Corollary A5. □

Much of the work of the present section is motivated by the fact (Theorem B14 below) that the identity in B2 holds for all graphs.

Since $|U|$ is odd if and only if Γ_U has an odd number of components with odd vertex sets, one has

B3 $$|U + N(U)| - \delta(U) \equiv |U| - \|U\| \equiv 0 \pmod 2.$$

B4 Lemma. *For any extremal set U there exists an extremal set W containing U such that $W + N(W) = V$ and $\|U\| \le \|W\|$.*

PROOF. If U is extremal, let $W = V + N(U)$. Clearly $U \subseteq W$, and every component of Γ_U is a component of Γ_W. Hence $\|U\| \le \|W\|$ while $N(U) = N(W)$. Thus $\delta(W) \ge \delta(U)$, and so W is extremal. □

B5 Lemma. *For any graph, $\nu_0 \ge 2\alpha_1 + \hat{\delta}$.*

PROOF. Let U be an extremal set of $\Gamma = (V, \mathscr{E})$ and let $C_1, \ldots, C_m$ be the odd vertex sets from among the components of Γ_U. Let $\mathscr{F}$ be an independent α_1-subset of $\mathscr{E}$. Note that if each vertex in C_i is incident with some edge in $\mathscr{F}$, then at least one edge in $\mathscr{F}$ is incident with both a vertex in C_i and a vertex in $N(U)$ ($i = 1, \ldots, m$). Since $\mathscr{F}$ is independent, there cannot be more than $|N(U)|$ such edges. Hence at least $m - |N(U)|$ sets C_i contain a vertex incident with no edge in $\mathscr{F}$. In other words, at least $m - |N(U)| = \|U\| - |N(U)| = \hat{\delta}(\Gamma)$ vertices are incident with no edge in $\mathscr{F}$. Hence $2|\mathscr{F}| + \hat{\delta} \le \nu_0$. □

B6 *Example.* Consider the graph $\Gamma = (V, \mathscr{E})$ shown in the figure:

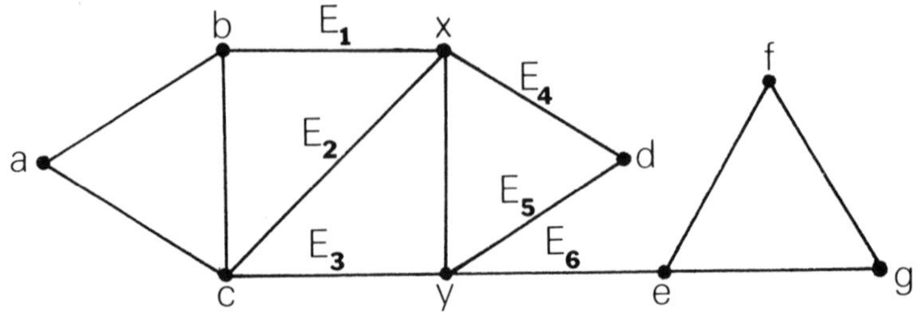

Let $U = \{a, b, c, d, e, f, g\}$, and so $N(U) = \{x, y\}$. We let $C_1 = \{a, b, c\}$, $C_2 = \{d\}$, and $C_3 = \{e, f, g\}$. Thus $\delta(U) = \|U\| - |N(U)| = 3 - 2 = 1$. By Lemma B5, $\delta \leq \nu_0 - 2\alpha_1 = 9 - 2 \cdot 4 = 1$, and so U is extremal.

An obvious necessary condition for a graph $\Gamma = (V, \mathscr{E})$ to admit a 1-factor is that $\nu_0(\Gamma)$ be even. If, however, Γ has the property that $\Gamma_{(x)}$ admits a 1-factor for all $x \in V$, then certainly $\nu_0(\Gamma)$ is odd, but Γ is as close to admitting a 1-factor as an odd graph can be. Such graphs are said to be **almost factorable**. For example, for all n, the graph K_{2n} admits a 1-factor, and so K_{2n+1} is almost factorable.

A subset $U \subseteq V$ is said to be **normal** if it is extremal, $V = U + N(U)$, and every component of Γ_U is itself almost factorable (and hence odd). Clearly the set U in Example B6 is normal. However, the graph Γ in the example is not almost factorable since $\Gamma_{(y)}$ has no 1-factor.

B7 Exercise. Prove that *a graph* $\Gamma = (V, \mathscr{E})$ *is almost factorable if and only if it is connected,* $\nu_0(\Gamma)$ *is odd, and* V *is a normal set.*

B8 *Exercise.*

(a) Let Θ be a spanning subgraph of Γ. Show that if Θ is almost factorable, then so is Γ.

(b) Find all "edge-minimal" almost factorable graphs on ≤ 5 vertices.

Corresponding to any normal subset U of V, there exists a bipartite multigraph B_U of particular interest. B_U is a subcontraction of Γ and is constructed as follows. First delete any and all edges incident only to vertices in $N(U)$. Let the vertices of B_U consist of the singletons $\{x\}$ for $x \in N(U)$ and of the vertex sets of the odd components of Γ_U. Thus B_U is a bipartite multigraph, but need not be a graph as we shall see presently. (Clearly all the results of §A, although stated only for graphs, are extendable in an obvious way to multigraphs.) The multigraph B_U corresponding to U and Γ of Example B6 has the form of Figure B9. Observe that it is not a graph.

B9

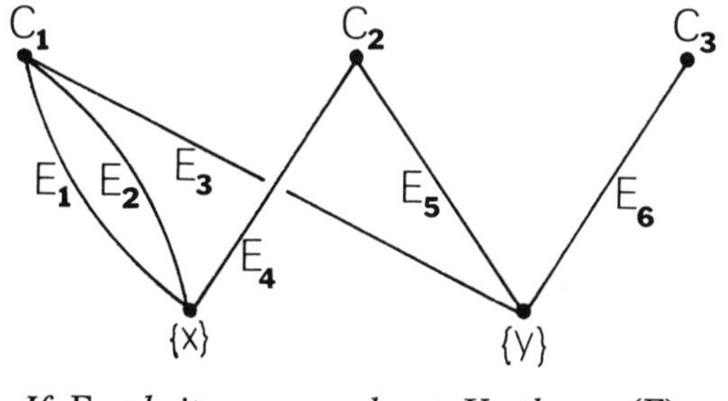

B10 Theorem. *If* Γ *admits a normal set* U*, then* $\nu_0(\Gamma) = 2\alpha_1(\Gamma) + \delta(\Gamma)$. *Moreover, the associated bipartite multigraph* B_U *admits a matching of the*

set $\{\{x\}: x \in N(U)\}$. *Finally, each largest independent set* $\mathscr{F}$ *of edges of* Γ *can be constructed as follows:*

(a) *Let* $\mathscr{F}'$ *be a matching of* $\{\{x\}: x \in N(U)\}$ *in* B_U.

(b) *Let* C *be the vertex set of a component of* Γ_U. *Regarding* C *as a vertex of* B_U,

(i) *if* C *is incident with an edge* $E \in \mathscr{F}'$, *then let* x_C *be the vertex in* C *incident with* E *in* Γ.

(ii) *if* C *is incident with no edge in* $\mathscr{F}'$, *then arbitrarily select* $x_C \in C$.

(c) *Since* U *is normal, there exists a* 1*-factor* $\mathscr{F}_C$ *in the component of* Γ_U *with vertex set* C.

(d) *Let* $\mathscr{F} = \mathscr{F}' + \sum_C \mathscr{F}_C$.

Before proving this theorem, let us first illustrate how the construction of $\mathscr{F}$ works for the graph Γ and set U of Example B6. First we must consider the graph B_U of Figure B9. Let $\mathscr{F}' = \{E_1, E_5\}$. Then $x_{C_1} = b$ and $x_{C_2} = d$. Let us arbitrarily choose $x_{C_3} = f$. Then $\mathscr{F}_{C_1} = \{\{a, c\}\}$, $\mathscr{F}_{C_2} = \varnothing$, and $\mathscr{F}_{C_3} = \{\{e, g\}\}$. Hence $\mathscr{F} = \{E_1, E_5, \{a, c\}, \{e, g\}\}$.

Proof of Theorem B10. In the graph B_U, let $V_1 = \{C: C$ is the vertex set of a component of $\Gamma_U\}$, and let V_2 denote the set of vertices complementary to V_1. Since U is extremal, $\delta_1(B_U) \geq \delta(V_1) \geq \|U\| - |N(U)| = \delta(U) = \hat{\delta}(B_U)$. By Proposition B1, $\delta_1(B_U) = \hat{\delta}(B_U)$ and $\delta_2(B_U) = 0$. We may obviously identify V_2 with $N(U)$, and by the Main Matching Theorem, there exists a matching $\mathscr{F}'$ of $N(U)$ into V_1. It is clear that steps (b), (c), and (d) can now be carried out to yield an independent edge set $\mathscr{F}$ of cardinality

$$|\mathscr{F}| = |N(U)| + \sum_C \alpha_1(\Gamma_C) = |N(U)| + \sum_C \tfrac{1}{2}(|C| - 1)$$

$$= \tfrac{1}{2}[2|N(U)| + \nu_0(\Gamma) - |N(U)| - \|U\|] = \tfrac{1}{2}(\nu_0(\Gamma) - \hat{\delta}(U)).$$

By Lemma B5 it follows that $\mathscr{F}$ is indeed a largest independent set of edges. It is now straightforward to verify that every largest independent set of edges can be constructed in exactly this way. The details are left to the reader. □

The following result appears in papers by T. Gallai [g.3], J. Edmonds [e.2], and W. Mader [m.4].

B11 Theorem. *Every graph admits a normal set of vertices.*

Proof. The method of proof is by induction on $\nu_0(\Gamma)$. The assertion is trivially valid in case $\nu_0(\Gamma) = 1$. Let $\Gamma = (V, \mathscr{E})$ be given and suppose as induction hypothesis that every proper subgraph of Γ admits a normal set. By Lemma B4 we may let U be an extremal set such that $U + N(U) = V$. Subject to this condition, we may assume that $\|U\|$ is as large as possible.

We first show that *every component of* Γ_U *has an odd number of vertices.* For let C be the vertex set of a component of Γ_U, suppose that C is even, and let $x \in C$. Since $C + \{x\}$ is odd, we have $\|U + \{x\}\| \geq \|U\| + 1$. Meanwhile, since x is incident to some other vertex in C, we have $N(U + \{x\}) \subseteq N(U) + \{x\}$. Thus $\delta(U + \{x\}) \geq \delta(U)$, and so $U + \{x\}$ is extremal. By B4 there exists an extremal set W containing $U + \{x\}$ such that $W + N(W) = V$ and $\|W\| \geq \|U + \{x\}\| > \|U\|$, contrary to the maximality of $\|U\|$.

To show that the set U is normal, it remains only to show that every component of Γ_U is almost factorable. We have two cases.

Case 1: $U = V$ *and* Γ *is connected.* In this case $\Gamma = \Gamma_V$ consists of a single odd component, and so $\hat{\delta}(\Gamma) = \|V\| - |\varnothing| = 1$. Suppose that Γ is not almost factorable; there exists a vertex $x \in V$ such that $\Gamma_{(x)}$ admits no 1-factor. By the induction hypothesis, $\Gamma_{(x)}$ admits a normal set W, and so by Theorem B10, $\hat{\delta}(\Gamma_{(x)}) = \nu_0(\Gamma_{(x)}) - 2\alpha_1(\Gamma_{(x)})$. Since this quantity is even and positive, $\hat{\delta}(\Gamma_{(x)}) \geq 2$. Let $N_{(x)}$ and $\delta_{(x)}$ denote the functions for $\Gamma_{(x)}$ corresponding to N and δ. Since W is normal, $W + N_{(x)}(W) = V + \{x\}$.

If $x \in N(W)$, then $\delta(W) = \|W\| - |N(W)| = \|W\| - (|N_{(x)}(W)| + 1) = \delta_{(x)}(W) - 1 \geq 1$. Hence W is an extremal set in Γ, and $\|W\| = \delta(W) + |N(W)| > 1 = \|V\|$, giving a contradiction. If $x \notin N(W)$, then $\|W \cup \{x\}\| = \|W\| + 1$, while $N(W + \{x\}) = N_{(x)}(W)$. Hence $\delta(W + \{x\}) = \|W\| + 1 - |N_{(x)}(W)| = \delta_{(x)}(W) + 1 \geq 3$, contrary to the fact that $\hat{\delta}(\Gamma) = 1$.

Case 2: $U \subset V$, *or* $U = V$ *and* Γ *is not connected.* Let C be the vertex set of some component of Γ_U. Then $C \subset V$ and by the induction hypothesis, Γ_C admits a normal set W. Hence $W + N_C(W) = C$, where N_C is the restriction of N to Γ_C. We define δ_C analogously. Thus

B12 $$\delta_C(W) = \|W\| - |N_C(W)| = \|W\| - (|C| - |W|) \geq 1$$

by B3 since $|C|$ is odd. Let $X = U + W$. Then

B13 (a) $$N(X) \subseteq N(U) + N_C(W)$$

(b) $$\|X\| = (\|U\| - 1) + \|W\|.$$

Thus $\delta(X) = \|X\| - |N(X)| \geq \|U\| + \|W\| - 1 - |N(U)| - |N_C(W)| = \delta(U) + \delta_C(W) - 1 \geq \delta(U)$ by B13a and B12. Since U is extremal, strict equality must hold throughout this chain and hence in B13a as well. We see that X is extremal and that $X + N(X) = V$. Therefore $\|X\| \leq \|U\|$, which by B13b implies that $\|W\| \leq 1$. Hence $W = C$ by B12, and Γ_C is almost factorable by Exercise B7. □

Combining Theorems B10 and B11 we have

B14 Theorem (C. Berge [b.3]). *For any graph,* $\nu_0 = 2\alpha_1 + \hat{\delta}$.

Since $\hat{\delta}(\Gamma) = 0$ if and only if Γ admits a 1-factor, we deduce immediately

B15 Corollary (W. T. Tutte [t.4]). *The graph* $\Gamma = (V, \mathscr{E})$ *admits a* 1*-factor if and only if* $\|U\| \leq |N(U)|$ *for all* $U \subseteq V$.

Note in particular how Tutte's result generalizes Corollary A6 from bipartite graphs to all graphs.

B16 *Exercise.* Let $\Gamma = (V, \mathscr{E})$ be a graph with $\nu_0(\Gamma)$ even. Show that Γ admits a 1-factor if and only if $\|U\| \leq |N(U)| + 1$ for all $U \subseteq V$.

B17 *Exercise* (Errera [e.7]). Let Γ be a connected trivalent graph in which all isthmuses lie on a single path. Show that Γ has a 1-factor. [*Hint*: use Exercise B16.]

B18 *Exercise* (C. Berge [b.5]). Let $\Gamma = (V, \mathscr{E})$ be h-valent and suppose that whenever $\varnothing \subset S \subset V$, then the number of edges incident with exactly one element of S is at least $h - 1$. Show that Γ either admits a 1-factor or is almost factorable.

B19 *Exercise.* Show that $\alpha_1 \geq \min\{[\nu_0/2], \breve{\rho}\}$ for any graph.

B20 *Exercise* (P. Erdős and T. Gallai [e.4]). Prove for any graph

$$\nu_1 \leq \begin{cases} \dbinom{2\alpha_1}{2} & \text{if } \nu_0 = 2\alpha_1; \\[2ex] \dbinom{2\alpha_1 + 1}{2} & \text{if } 2\alpha_1 < \nu_0 \leq \dfrac{5\alpha_1 + 2}{2}; \\[2ex] \dbinom{\alpha_1}{2} + \alpha_1(\nu_0 - \alpha_1) & \text{if } (5\alpha_1 + 2)/2 < \nu_0. \end{cases}$$

For further information on 1-factors, see Lovász [ℓ.5].

VC Coverings and Independent Sets in Graphs

Throughout this section, $\Gamma = (V, \mathscr{E})$ denotes a graph. An **externally stable set** (also called a "dominating set") is a vertex-covering subset of V. The cardinality of a smallest externally stable set in V is denoted by $\beta_{00}(\Gamma)$, or simply β_{00}, and is called the **external stability** of Γ. In a like manner, but without specific names, we define:

$\beta_{01}(\Gamma)$ = cardinality of a smallest edge-covering subset of V;
$\beta_{10}(\Gamma)$ = cardinality of a smallest vertex-covering subset of $\mathscr{E}$;
$\beta_{11}(\Gamma)$ = cardinality of a smallest edge-covering subset of $\mathscr{E}$.

Example. The graph in Figure C1 has $\nu_0 = 9$, $\nu_1 = 11$, $\alpha_0 = 5$, $\alpha_1 = 4$, $\beta_{00} = 3$, $\beta_{01} = 4$, $\beta_{10} = 5$, $\beta_{11} = 2$.

C1

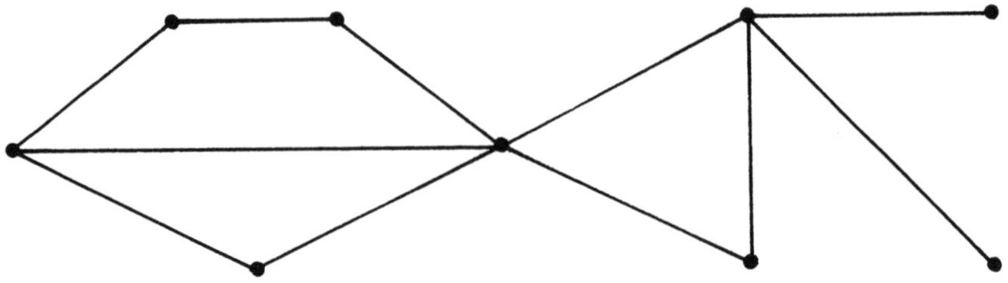

C2 *Exercise.* Compute each of the above parameters for:
(a) the complete graph K_n;
(b) the n-circuit Δ_n, $n \geq 3$;
(c) the graph obtained by removing an elementary n-cycle from K_n, $n \geq 3$.

C3 *Exercise.* Find two nonisomorphic connected graphs Γ_1 and Γ_2 for which $\alpha_i(\Gamma_1) = \alpha_i(\Gamma_2)$ and $\beta_{ij}(\Gamma_1) = \beta_{ij}(\Gamma_2)$ for all $i, j \in \{0, 1\}$.

C4 *Exercise.* Show that if Γ has no isolated vertices and if W is any minimal externally stable set, then $V + W$ contains another minimal externally stable set, and hence $\beta_{00} \leq \nu_0/2$.

C5 Exercise. *Let* $W \subseteq V$. Show that W *is an independent vertex set if and only if* $V + W$ *is an edge-covering set.*

The present section is concerned with the relationships of various graph-theoretical parameters to each other. Theorem B14 of the previous section was the first of the present sequence of identities and inequalities.

C6 Proposition. *For any graph with k isolated vertices,*
(a) $\beta_{10} + k = \alpha_1 + \delta$;
(b) $\delta \leq \alpha_0 \leq \beta_{10} + k$.

PROOF. (a) Let $\mathscr{F}$ be a vertex-covering β_{10}-set of edges. By the minimality of $\mathscr{F}$, $\Gamma_{\mathscr{F}}$ must be a forest. Since its cycle space is trivial, *III*A15b yields $\nu_0(\Gamma) = \nu_0(\Gamma_{\mathscr{F}}) = \nu_1(\Gamma_{\mathscr{F}}) + \nu_{-1}(\Gamma_{\mathscr{F}}) \leq \beta_{10}(\Gamma) + \alpha_1(\Gamma) + k$, since the selection of one edge from each component of $\Gamma_{\mathscr{F}}$ yields an independent set. Combining this with Theorem B14 yields $\alpha_1 + \delta \leq \beta_{10} + k$. To obtain the reverse inequality, let $\mathscr{F}$ denote instead an independent α_1-set of edges. By Theorem B14, there are precisely δ vertices incident with no edge in $\mathscr{F}$. If for each of the $\delta - k$ nonisolated vertices, an edge is adjoined to $\mathscr{F}$ which is incident with that vertex, the resulting set is vertex-covering and contains $\alpha_1 + \delta - k$ edges. Hence $\beta_{10} \leq \alpha_1 + \delta - k$.

(b) By Theorem B14, exactly δ vertices are incident with no edge of a largest independent edge set. Such a δ-set of vertices must be independent. Hence $\delta \leq \alpha_0$. If the isolated vertices of Γ are deleted, then each edge of a vertex-covering β_{10}-set of edges is incident with at most one vertex of any independent set of vertices. Hence $\alpha_0 - k \leq \beta_{10}$. □

If $\mathscr{A} \subseteq \mathscr{P}(U)$, let $\bar{\mathscr{A}} = \{U + T: T \in \mathscr{A}\}$. Then a set $T \subseteq U$ is a minimal element of $(\mathscr{A}, \subseteq)$ if and only if $U + T$ is maximal in $(\bar{\mathscr{A}}, \subseteq)$. It follows that for any $\mathscr{A} \subseteq \mathscr{P}(U)$,

C7 $$\min\{|T|: T \in \mathscr{A}\} + \max\{|T|: T \in \bar{\mathscr{A}}\} = |U|.$$

The first part of the next proposition is a direct application of this principle.

C8 Theorem. (T. Gallai [g.2], 1959).

(a) *For any graph,* $\alpha_0 + \beta_{01} = \nu_0$.

(b) *For any graph without isolated vertices,* $\alpha_1 + \beta_{10} = \nu_0$.

PROOF. (a) Let $\mathscr{A} = \{W \in \mathscr{P}(V): W \text{ is edge-covering}\}$. By C5, $\bar{\mathscr{A}} = \{W \in \mathscr{P}(V): W \text{ is independent}\}$. Then by C7,

$$\beta_{01} + \alpha_0 = \min\{|W|: W \in \mathscr{A}\} + \max\{|W|: W \in \bar{\mathscr{A}}\} = \nu_0.$$

(b) Eliminate δ from B14 and C6a. □

C9 Proposition. *For any graph without isolated vertices, the following inequalities hold:*

(a) $\alpha_0 \leq \beta_{10} \leq \alpha_0 + \alpha_1$;

(b) $\alpha_1 \leq \beta_{01} \leq 2\alpha_1$;

(c) $\alpha_1 \leq \nu_0 - \alpha_0 \leq 2\alpha_1$.

Furthermore, if equality holds in any one of the left-hand inequalities above, then it holds in all three of the left-hand inequalities. If equality holds in any one of the right-hand inequalities or in the inequality $\delta \leq \alpha_0$ (cf. C6b), *then it holds in all three right-hand inequalities and* $\delta = \alpha_0$.

PROOF. From the two parts of Gallai's theorem, $\beta_{10} - \alpha_0 = \beta_{01} - \alpha_1 = \nu_0 - \alpha_0 - \alpha_1$, and by Proposition C6b, $\beta_{10} - \alpha_0 \geq 0$. This establishes the three left-hand inequalities as well as the fact that if equality holds in any one of them, then it holds in all three. By Theorem B14 and Proposition C6b, $2\alpha_1 - (\nu_0 - \alpha_0) = \alpha_0 - \delta \geq 0$. This establishes (c) and that equality holds in the right-hand inequality of (c) if and only if $\delta = \alpha_0$. The rest follows by substitution from Gallai's theorem. □

Actually portions of the above proposition hold when the prohibition against isolated vertices is relaxed. These are the portions not requiring C8b. The greater generality obtained by sorting them out, however, is in our opinion not worth the effort.

The following is an equivalent formulation of the Main Matching Theorem.

C10 Theorem (D. König [k.5], 1936). *In a bipartite graph,* $\alpha_1 = \beta_{01}$.

PROOF. Let $\Gamma = (\{V_1, V_2\}), \mathscr{E})$. By Proposition C9b, $\alpha_1 \leq \beta_{01}$, and so it suffices to find an edge-covering α_1-subset of $V_1 \cup V_2$. Clearly α_1 is the cardinality of a largest matching in Γ. By the Main Matching Theorem, $\alpha_1 = |V_1| - \delta_1$. Let $U \subseteq V_1$ be a critical set. One observes that $V_1 + U + N(U)$ is an edge-covering set with cardinality

$$|V_1| - (|U| - |N(U)|) = |V_1| - \delta_1.$$ □

C11 Corollary. *In a bipartite graph without isolated vertices,* $\alpha_0 = \beta_{10}$ *and* $\alpha_0 + \alpha_1 = \nu_0$.

C12 Proposition. *In any graph,*

(a) $\beta_{00} \leq \alpha_0$;

(b) $\beta_{11} \leq \alpha_1$;

(c) $\beta_{00} \leq \alpha_1$ *if there are no isolated vertices.*

PROOF. (a) Clearly every maximal independent vertex set is externally stable.

(b) The argument is the same as in (a).

(c) Consider a smallest externally stable set V_1 of vertices of Γ and let $V_2 = V + V_1$. Let $\mathscr{F}$ be the subset of $\mathscr{E}$ consisting of edges incident with one vertex in V_1 and one vertex in V_2. Then $\mathrm{B} = (\{V_1, V_2\}, \mathscr{F})$ is a bipartite graph, and by the Main Matching Theorem,

$$\beta_{00}(\Gamma) - \delta_1(\mathrm{B}) = |V_1| - \delta_1(\mathrm{B}) = \alpha_1(\mathrm{B}) \leq \alpha_1(\Gamma).$$

It suffices to show that $\delta_1(\mathrm{B}) = 0$. Suppose that in B, $\delta(U) > 0$ for some $U \subseteq V_1$, and let $W = V_1 + U + N(U)$. Then $|W| < |V_1|$, and since Γ has no isolated vertices, W is externally stable, contrary to the definition of V_1. □

C13 *Exercise.* From C6 and C9 we deduce that for $j \neq k$, we have $\alpha_i \leq \beta_{j,k}$ in three of the four possible cases. From C12, we infer $\beta_{j,j} \leq \alpha_i$ in three of four possible cases. Determine whether the remaining inequalities, namely $\alpha_0 \leq \beta_{01}$ and $\beta_{11} \leq \alpha_0$, are true or false, and show that in any graph without isolated vertices, we have the weaker result $\beta_{11} \leq \beta_{01}$.

C14 *Exercise.* Let $\alpha(\Gamma) = \max\{|U| : U \subseteq V;\ \Gamma_U \text{ is a forest}\}$, and let $\beta(\Gamma) = \min\{|U| : U \subseteq V;\ \text{every circuit in } \Gamma \text{ meets } U\}$. Prove $\alpha(\Gamma) + \beta(\Gamma) = \nu_0(\Gamma)$.

C15 *Exercise.* State and prove a result analogous to C14 with V replaced by $\mathscr{E}$ in the definitions of $\alpha(\Gamma)$ and $\beta(\Gamma)$.

C16 *Exercise.* Show that in a graph without isolated vertices, the cardinality of a minimum subset $\mathscr{F} \subseteq \mathscr{E}$ with the property that $\mathscr{F}$ excludes at most one edge from each vertex cocycle is $\beta_{10} - \nu_1 + \nu_0$.

C17 *Exercise.* Prove:

(a) Every smallest vertex-covering edge set contains a largest independent edge set.

(b) Every largest independent edge set is contained in a smallest vertex-covering edge set.

VD Systems with Representatives

For this section $\Lambda = (V, f, E)$ will denote an arbitrary system. We define a **list of distinct representatives (LDR)** for Λ to be an injection $\lambda: E \to V$ such that $\lambda(e) \in f(e)$ for all $e \in E$. A system admitting an LDR is called a **system with distinct representatives (SDR)**. The fundamental result in the literature concerning SDR's is the following:

D1 Theorem (Philip Hall [h.4], 1935). *The system* (V, f, E) *is an SDR if and only if*

$$F \subseteq E \Rightarrow |F| \le \left|\bigcup_{e\in F} f(e)\right|.$$

The Philip Hall Theorem should readily be seen to be equivalent to Corollary A6 above. For let $\Gamma = (\{V, F\}, \mathscr{E})$ be the bipartite graph of Λ. If $F \subseteq E$, then $N(F) = \bigcup_{e\in F} f(e)$, and the equivalence is immediate. It should be remarked that Philip Hall's theorem is strictly a statement of existence. It indicates neither how an LDR can be found nor, when an LDR does exist, how many distinct LDR's Λ may have. The enumerative problem is considered below in §F. When an LDR exists, it can be found in the following way. Since an LDR $\lambda\colon E \to V$ of Λ exists if and only if the set of edges $\{\{e, \lambda(e)\}: e \in E\}$ is a matching of E in Γ, the algorithm for constructing a largest matching in Γ can thus be employed to find a "longest" LDR for Λ.

D2 Exercise. Just as the Philip Hall Theorem is the analog for systems of Corollary A6, devise and justify *system analogs for*:

(a) The Main Matching Theorem;
(b) *Corollary A7*;
(c) The König Theorem (C10).

D3 *Exercise.* Let $\mathscr{B}$ be an independent subset of the vector space $(\mathscr{P}(U), +)$. Prove that the set system $(U, \mathscr{B})$ is an SDR.

Let V be a fixed set and consider two systems $\Lambda_i = (V, f_i, E_i)$ for $i = 1, 2$. A **list of common representatives (LCR)** for the two systems is a pair (λ_1, λ_2) where λ_i is an LDR for Λ_i $(i = 1, 2)$ and $\lambda_1[E_1] = \lambda_2[E_2]$. An obvious necessary condition for the existence of a LCR is that $|E_1| = |E_2|$.

D4 Proposition. *Let the systems* $\Lambda_1 = (V, f_1, E_1)$ *and* $\Lambda_2 = (V, f_2, E_2)$ *be given with* $|E_1| = |E_2| = m$. *A necessary and sufficient condition for* Λ_1 *and* Λ_2 *to have an LCR is that for all* $D_1 \subseteq E_1$ *and* $D_2 \subseteq E_2$,

$$|D_1| + |D_2| - m \le \left|\left(\bigcup_{e\in D_1} f_1(e)\right) \cap \left(\bigcup_{e\in D_2} f_2(e)\right)\right|.$$

PROOF. *Necessity.* Suppose that (λ_1, λ_2) is an LCR for Λ_1 and Λ_2, and let $D_1 \subseteq E_1$ and $D_2 \subseteq E_2$. Clearly $|\lambda_1[D_1] \cup \lambda_2[D_2]| \le m$, while $\lambda_1[D_1] \cap \lambda_2[D_2] \subseteq (\bigcup_{e\in D_1} f(e)) \cap (\bigcup_{e\in D_2} f(e))$. Hence

$$\begin{aligned} |D_1| + |D_2| &= |\lambda_1[D_1]| + |\lambda_2[D_2]| \\ &= |\lambda_1[D_1] \cup \lambda_2[D_2]| + |\lambda_1[D_1] \cap \lambda_2[D_2]| \\ &\le m + \left|\left(\bigcup_{e\in D_1} f(e)\right) \cap \left(\bigcup_{e\in D_2} f(e)\right)\right|. \end{aligned}$$

Sufficiency. Let V' be a $|V|$-set disjoint from V and all other sets in question. There is a bijection $V \to V'$ given by $x \mapsto x'$. If $S \subseteq V$, let $S' = \{x' \in V' : x \in V\}$. We now form the bipartite graph $\mathrm{B} = (\{V \cup E_1, V' \cup E_2\}, \mathscr{E})$, where

$$\mathscr{E} = \{\{x, x'\} : x \in V\} \cup \{\{x, e\} : e \in E_2;\, x \in f_2(e)\} \cup \{\{e, x'\} : e \in E_1;\, x \in f_1(e)\}.$$

Observe that $|V \cup E_1| = |V' \cup E_2|$. See Figure D5.

D5

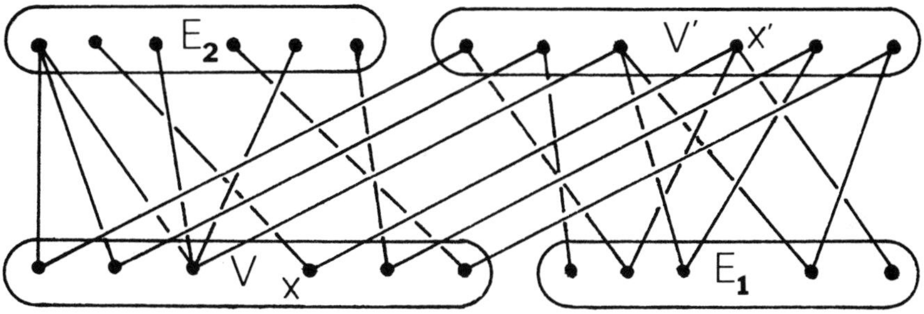

The bipartite graph of Λ_2 The bipartite graph of Λ_1 (isomorphic copy)

Let us first suppose that B admits a mutual matching. Let U denote the set of vertices in V thereby matched with elements of E_2. Thus $U \in \mathscr{P}_m(V)$. It follows that $V + U$ is matched with $(V + U)'$, and so U' is necessarily matched with E_1. For $i = 1, 2$, define $\lambda_i : E_i \to U$ so that for each $e \in E_1$, $\lambda_1(e)'$ is the vertex in U' matched with e, while for each $e \in E_2$, $\lambda_2(e)$ is the vertex in U matched with e. Clearly (λ_1, λ_2) is an LCR.

It suffices therefore to prove that B admits a mutual matching. By Corollary A6, this is equivalent to showing that $\delta(A) \leq 0$ for all $A \subseteq V \cup E_1$. We may write $A = W + D_1$, where $W \subseteq V$ and $D_1 \subseteq E_1$. Then

$$N(A) = W' \cup \left(\bigcup_{e \in D_1} f_1(e)\right)' + \{e \in E_2 : f_2(e) \cap W \neq \varnothing\}.$$

Let $D_2 = \{e \in E_2 : f_2(e) \subseteq V + W\}$. Invoking the condition, we compute

$$\begin{aligned}
\delta(A) &= |A| - |N(A)| \\
&= |D_1| + |W| - \left(|W'| + \left|\left(\bigcup_{e \in D_1} f_1(e)\right)' \cap (V + W)'\right| + (|E_2| - |D_2|)\right) \\
&= |D_1| + |D_2| - m - \left|\left(\bigcup_{e \in D_1} f_1(e)\right) \cap (V + W)\right| \\
&\leq \left|\left(\bigcup_{e \in D_1} f_1(e)\right) \cap \left(\bigcup_{e \in D_2} f_2(e)\right)\right| - \left|\left(\bigcup_{e \in D_1} f_1(e)\right) \cap (V + W)\right| \leq 0. \quad \square
\end{aligned}$$

D6 *Exercise*. Let $\Lambda_i = (V, f_i, E_i)$ be a system with $|E_i| = m$ for $i = 1, \ldots, k$. An LCR $(\lambda_1, \ldots, \lambda_k)$ for these systems is defined in the obvious way. Consider the condition

$$\sum_{i=1}^{k} |D_i| - (k-1)m \leq \left| \bigcap_{i=1}^{k} \bigcup_{e \in D_i} f_i(e) \right|, \quad \text{for all } D_i \subseteq E_i.$$

(a) Show that this condition is necessary for the existence of an LCR.

(b) Show that it is sufficient for the existence of "pairwise LCR's" (λ_i, λ_j) for Λ_i and Λ_j when $1 \leq i < j \leq k$.

(c) Give a counter-example to show that this condition is not sufficient for the existence of an LCR when $k = 3$ and $m = 2$.

D7 *Exercise* (H. J. Ryser [r.9]). Let $\mathcal{Q}_i \in \mathbb{P}_m(U)$ for $i = 1, 2$.

(a) Prove that $(U, \mathcal{Q}_1)$ and $(U, \mathcal{Q}_2)$ admit an LCR if and only if for all $\mathcal{S} \subseteq \mathcal{Q}_1 \cup \mathcal{Q}_2$,

$$\mathcal{S} \text{ covers } U \Rightarrow |\mathcal{S}| \geq m.$$

(b) Prove that if $\mathcal{Q}_1 \cup \mathcal{Q}_2 \subseteq \mathcal{P}_n(U)$ for some n, then $(U, \mathcal{Q}_1)$ and $(U, \mathcal{Q}_2)$ admit an LCR.

D8 *Exercise*. Let G be a finite group, let H be an arbitrary subgroup, and let m be the index of H in G. Prove that there exist $x_1, \ldots, x_m \in G$ such that

$$G = \bigcup_{i=1}^{m} x_i H = \bigcup_{i=1}^{m} H x_i.$$

We conclude this section with an application of the foregoing theory.

An $r \times s$ **Latin rectangle** is a rectangular array of symbols with r rows and s columns such that each symbol appears at most once in each row and each column. A **Latin square of order** n is an $n \times n$ Latin rectangle on n symbols.

D9 Proposition. *If $r < s$, an $r \times s$ Latin rectangle on s symbols can always be extended to a Latin square of order s.*

Proof. Let $V = \{1, 2, \ldots, s\}$ be the set of symbols of the given rectangle. Let $E = \{e_1, \ldots, e_s\}$ be an arbitrary s-set and define $f: E \to \mathcal{P}(V)$ by letting $f(e_i)$ be the $(s - r)$-set of symbols not appearing in the ith column. Since each symbol $j \in V$ appears precisely once in each of the r rows, it must appear in precisely r columns. Hence j is absent in precisely $s - r$ columns. By the analog for systems of Corollary A7 (see Exercise D2b), if $r < s$ then the system (V, f, E) admits an LDR λ, and an $(r + 1)$-st row $\lambda(e_1), \ldots, \lambda(e_s)$ may be adjoined to the Latin rectangle to form an $(r + 1) \times s$ Latin rectangle. If $r + 1 < s$, the above argument may be repeated until a Latin square of order s is obtained. □

VE {0, 1}-Matrices

In this section $M = [m_{ij}]$ will denote an $r \times s$ matrix where $m_{ij} = 0$ or 1 for all $(i, j) \in \{1, \ldots, r\} \times \{1, \ldots, s\}$. Such a matrix is called a {0, 1}-matrix. By a **line** of M is meant either a row or a column of M.

We shall now define a rather broad "incidence relation" among lines and positions in a {0, 1}-matrix. The reader should perceive that this definition is motivated by the definitions of incidence in multigraphs (§A) and of incidence matrix of a system (§*I*D).

We say that the ith row and jth column of M are **incident** if $m_{ij} = 1$. The ith and jth rows are **incident** if their inner product satisfies

$$\sum_{k=1}^{s} m_{ik} m_{jk} > 0.$$

(Thus $m_{ik} = 1 = m_{jk}$ for some $k \in \{1, \ldots, s\}$.) Similarly two columns are **incident** if their inner product is positive. We say that the position (i, j) is incident with the ith row and with the jth column. Finally, two positions on a common line are **incident**.

The number of lines of an $r \times s$ matrix M is denoted by $\pi(M) = r + s$.

If M is an incidence matrix of a system $\Lambda = (V, f, E)$, then the matrix $MM^* = A = [a_{ij}]$ is called a **vertex-vertex incidence matrix** (or an **adjacency matrix**) of Λ. The set $V = \{x_1, \ldots, x_r\}$ may be indexed so that for $i, j \in \{1, \ldots, r\}$,

$$a_{ij} = |\{e \in E: \{x_i, x_j\} \subseteq f(e)\}| = |f^*(x_i) \cap f^*(x_j)|.$$

E1 Exercise. Show that *a multigraph is uniquely determined (up to isomorphism) by any one of its adjacency matrices*. Show that this does not hold for systems in general.

A **block-block incidence matrix** of a system (V, f, E) with incidence matrix M is the matrix $M^*M = H = [h_{ij}]$. The set $E = \{e_1, \ldots, e_s\}$ may be indexed so that for $i, j \in \{1, \ldots, s\}$,

$$h_{ij} = |f(e_i) \cap f(e_j)| = |\{x \in V: \{e_i, e_j\} \subseteq f^*(x)\}|.$$

(The significance of these entries a_{ij} and h_{ij} will be pursued in Chapter IX.)

E2 *Exercise.* Characterize all graphs not uniquely determined (up to isomorphism) by their block-block incidence matrices.

Let $\Gamma = (V, \mathscr{E})$ be a graph with $V = \{x_1, \ldots, x_r\}$ and $\mathscr{E} = \{E_1, \ldots, E_s\}$. Let $M = [m_{ij}]$ be the incidence matrix of Γ which conforms to this indexing. Then the entry a_{ij} in the corresponding adjacency matrix $A = MM^*$ of Γ is

$$a_{ij} = \begin{cases} \rho(x_i) & \text{if } i = j; \\ 0 & \text{if } i \neq j \text{ and } \{x_i, x_j\} \notin \mathscr{E}; \\ 1 & \text{if } i \neq j \text{ and } \{x_i, x_j\} \in \mathscr{E}. \end{cases}$$

Moreover,

$$\rho(x_i) = \sum_{j \neq i} a_{ij} = \sum_{j \neq i} a_{ji}.$$

In the block-block incidence matrix of Γ, the diagonal entries are all 2 and the others are each 0 or 1. Graphs are not the only systems with the property that the nondiagonal entries in these two matrices are 0 or 1, as we see in the following exercise.

E3 *Exercise.* Prove that for any {0, 1}-matrix M the following three statements are equivalent.
  (a) The nondiagonal entries on MM^* are 0 or 1.
  (b) The nondiagonal entries of M^*M are 0 or 1.
  (c) M has no 2×2 submatrix of the form

$$\begin{bmatrix} 1 & 1 \\ 1 & 1 \end{bmatrix}.$$

E4 *Exercise.* Let A be the adjacency matrix of a graph $\Gamma = (V, \mathscr{E})$ which corresponds to the indexing $\{x_1, \ldots, x_r\} = V$. Let B be obtained from A by replacing all the diagonal entries with 0. For $n \in \mathbb{N}$, consider the nth power $B^n = [b_{ij}(n)]$ of B. Show that there exists an x_ix_j-path of length at most n if and only if $b_{ij}(n) > 0$. If in the definition of path (§*II*C) we relax the condition that two consecutive edges must be distinct, show that $b_{ij}(n)$ is the precise number of "x_ix_j-paths" of length n.

Consider the bipartite graph $\Gamma = (\{V_1, V_2\}, \mathscr{E})$ where $|V_i| = r_i$. Its (vertex-edge) incidence matrices are incidence matrices of the set system $(V_1 \cup V_2, \mathscr{E})$ and are not of great interest; they assume the form of Figure E5, where the $r_i \times |\mathscr{E}|$ submatrix M_i has exactly one 1 in each column and at most r_j $(j \neq i)$

E5

$$\begin{array}{c} \\ V_1 \\ V_2 \end{array} \begin{array}{c} \mathscr{E} \\ \left[\begin{array}{c} M_1 \\ \hdashline M_2 \end{array}\right] \end{array}$$

1's in each row. The adjacency matrices of Γ, on the other hand, have the form of Figure E6, where D_i is a diagonal matrix whose diagonal is a list of

E6

$$\begin{array}{c} \\ V_1 \\ V_2 \end{array} \begin{array}{c} \quad V_1 \qquad V_2 \\ \left[\begin{array}{c:c} D_1 & M \\ \hdashline M^* & D_2 \end{array}\right] \end{array}$$

the valences of the vertices in V_i $(i = 1, 2)$ and M is a {0, 1}-matrix of order $r_1 \times r_2$. Observe that M is an incidence matrix of a system for which Γ is the bipartite graph. In the light of Exercise E1, M alone suffices to characterize Γ. In fact, there is a natural identification of the set of all $r_1 \times r_2$ {0, 1}-matrices with the set of all subgraphs of Γ obtained from K_{r_1, r_2} by deleting edges only. Permutations of rows or of columns of M correspond to automorphisms of the system Γ leaving V_1 and V_2 fixed setwise. It is thus natural that some of the graph-theoretical parameters introduced in §A and §C yield the following interpretations, which may be considered as definitions, in the context of the {0, 1}-matrices:

$\alpha_0(M)$ = $\max(\{r_1, r_2\} \cup \{\pi(L): L \text{ is a zero-submatrix of } M\})$.

$\alpha_1(M)$ = cardinality of a largest set of positions occupied by 1's in M, no two of which are incident. (This parameter is called "term rank" in the literature, notably by H. J. Ryser [r.9].)

$\beta_{01}(M)$ = cardinality of a smallest set of lines of M such that every position occupied by a 1 is incident with a line in the set.

$\beta_{10}(M)$ = cardinality of a smallest set of positions occupied by 1's in M at least one of which lies on every line containing a 1.

The parameters β_{00} and β_{11} do not appear to have interesting matrix analogs. Clearly equivalent to the König Theorem (C10) is

E7 Theorem (Egarváry [e.3]). *In a* {0, 1}*-matrix, the cardinality of a largest set of positions occupied by* 1's *no two of which are incident equals the cardinality of a smallest set of lines such that every position occupied by a* 1 *is incident with a line in the set.*

Occasionally we may require a matrix over an arbitrary integral domain, in which case the symbol 1 in the above theorem can be replaced by the words "nonzero element."

An isolated vertex of the bipartite graph Γ corresponds to a line of 0's in each of its adjacency matrices. Thus the analog of Corollary C11 is:

E8 Proposition. *If there is a* 1 *in each line of an* $r_1 \times r_2$ {0, 1}*-matrix* M, *then the cardinality of a smallest set of positions occupied by* 1's *at least one of which lies on every line equals* $\alpha_0(M)$, *and*

$$\alpha_0(M) + \alpha_1(M) = \pi(M).$$

Applying Theorem C8 to bipartite graphs yields

E9 Proposition. *For a* {0, 1}*-matrix* M,

$$\alpha_0(M) + \beta_{01}(M) = \pi(M),$$

and if there is a 1 *in each line of* M, *then* $\alpha_1(M) + \beta_{10}(M) = \pi(M)$.

We state an analog for matrices of Corollary A6.

E10 Proposition. *Given any $r_1 \times r_2$ $\{0, 1\}$-matrix with $r_1 \leq r_2$, there exist r_1 positions occupied by* 1's, *no two of which are incident, if and only if every $n \times r_2$ submatrix has at most $r_2 - n$ columns of* 0's $(n = 1, \ldots, r_2)$.

The four parameters and four results in this section pertain to $\{0, 1\}$-matrices in general, i.e., not just to those matrices reflecting the particular kind of incidence defined for systems in §A. These results should therefore yield information about arbitrary systems associated, by some yet-to-be-defined kind of incidence, with some given $\{0, 1\}$-matrix. The reader may find it fruitful to explore various types of incidence in multigraphs or other systems, interpreting such "incidence matrices" as edge-cycle, edge-cocycle, cycle-spanning forest, vertex-circuit, etc. Perhaps the results of this section will yield some new results or at least some short proofs of old results. This is indeed a very open-ended problem.

E11 *Exercise.* Let L be a $\{0, 1\}$-matrix and let n be the largest number of 1's in any line. Prove that L is a submatrix of a $\{0, 1\}$-matrix each of whose lines contains exactly n 1's.

VF Enumerative Considerations

The results in the first four sections of this chapter have been largely concerned with questions of existence—existence of matchings, of LDR's, of sets of lines in matrices with given properties, and so forth. Here in the final section of the chapter, we consider questions of the form, "Given existence, how many are there?" The flavor is reminiscent of §*I*C, but the earlier sections of this chapter permit extension again of our numerical answers to a wide variety of superficially dissimilar models.

Let $M = [m_{ij}]$ be an $r \times s$ matrix over some integral domain. If $r \leq s$, the **permanent** of M is given by

$$\operatorname{perm}(M) = \sum_{\varphi} m_{1,\varphi(1)} \cdot \ldots \cdot m_{r,\varphi(r)}$$

where the summation is over all injections

$$\varphi: \{1, \ldots, r\} \to \{1, \ldots, s\}.$$

By Proposition *I*C13, perm(M) is the sum of $s!/(s - r)!$ products. If $r = s$, then perm(M) can be thought of as the "unsigned determinant" of M. The following proposition should be evident.

F1 Proposition. *Let $\Lambda = (V, f, E)$ be a system with $|E| \leq |V|$. Let $\Gamma = (\{E, V\}, \mathscr{E})$ be the bipartite graph of Λ and let M be an incidence matrix of Λ. Then* perm(M) *equals both the number of LDR's of Λ and the number of matchings of E in Γ.*

Before the next theorem, due to Marshall Hall, it may be well to stand off for a moment and consider the logical relationships among the results of Chapters IV, V, and VI. The Max-Flow–Min-Cut Theorem was proved in Chapter IV essentially from first principles developed in Chapters I, II, and III. From it we deduced the Main Matching Theorem, one of whose corollaries (A6) has been demonstrated (§C, §D, §E) to be equivalent to various other results, notably the Philip Hall Theorem. The Main Matching Theorem can straightforwardly be shown to imply the Max-Flow–Min-Cut Theorem. Looking ahead to Chapter VI, we anticipate proving the equivalence between the Max-Flow–Min-Cut Theorem and the Menger Theorem for separation in graphs. Marshall Hall's theorem below is totally independent of this logical chain. It will be proved from first principles; yet it yields the Philip Hall Theorem as a trivial corollary. Marshall Hall's theorem, therefore, might have been used as our logical starting point rather than Max-Flow–Min-Cut. The choice of a starting point for this presentation is very much a matter of taste.

F2 Theorem (Marshall Hall [h.1], 1948). *Let* $q \in \mathbb{N} + \{0\}$. *Let the system* $\Lambda = (V, f, E)$ *satisfy*

F3
$$D \subseteq E \Rightarrow |D| \leq \left|\bigcup_{e \in D} f(e)\right|,$$

and suppose that all blocks have size at least q. *The number of LDR's admitted by* Λ *is at least:*

(a) $q!$ *if* $q < |E|$;

(b) $q!/(q - |E|)!$ *if* $|E| \leq q$.

Proof. (Adapted from H. B. Mann and H. J. Ryser [m.6].) We proceed by induction on $|E|$. If $E = \{e\}$, then $|E| \leq q$ and Λ admits exactly $|f(e)|$ LDR's, where $|f(e)| \geq q = q!/(q-1)!$ as required. The induction hypothesis is that the theorem holds for all systems with fewer than m blocks. We assume $|E| = m$. Condition F3 can be broken into two cases.

Case 1: $\varnothing \subset D \subset E \Rightarrow |D| < |\bigcup_{e \in D} f(e)|$. Let $e_0 \in E$. Since $|f(e_0)| \geq q \geq 1$, pick $x_0 \in f(e_0)$. Let $E' = E + \{e_0\}$ and $V' = V + \{x_0\}$, and form the subsystem $\Lambda' = (V', f', E')$ where

$$f'(e) = f(e) \cap V', \quad \text{for all } e \in E'.$$

If $D \subseteq E'$, then by assumption

$$|D| \leq \left|\bigcup_{e \in D} f(e)\right| - 1 \leq \left|\bigcup_{e \in D} f'(e)\right|,$$

i.e., Λ' satisfies F3. Clearly $q < |E|$ if and only if $q - 1 < |E'|$. By the induction hypothesis, Λ' admits at least $(q-1)!$ LDR's if $q < |E|$ and at least

$$\frac{(q-1)!}{[(q-1) - |E'|]!} = \frac{(q-1)!}{(q - |E|)!}$$

LDR's if $q \geq |E|$. To each LDR λ' of Λ' corresponds the unique LDR λ of Λ given by

$$\lambda(e) = \begin{cases} \lambda'(e) & \text{if } e \in E'; \\ x_0 & \text{if } e = e_0. \end{cases}$$

As there exist at least q initial choices for x_0, Λ admits at least q times as many LDR's as Λ' does, as required.

Case 2: there exists non-empty $D \subset E$ such that $|D| = |\bigcup_{e \in D} f(e)|$. In this case, for all $e \in D$, one has $q \leq |f(e)| \leq |D| < |E|$. By the induction hypothesis, the subsystem Λ_D admits at least $q!$ LDR's. Now form another subsystem $\Theta = (W, g, E + D)$ where $W = V + \bigcup_{e \in D} f(e)$ and $g(e) = f(e) \cap W$ for all $e \in E + D$.

We show that Θ satisfies F3. Suppose that for some $A \subseteq E + D$ it held that $|A| > |\bigcup_{e \in A} g(e)|$. Since $D \cap A = \varnothing$, our various assumptions give

$$\left|\bigcup_{e \in D + A} f(e)\right| \leq \left|\bigcup_{e \in D} f(e)\right| + \left|\bigcup_{e \in A} f(e)\right|$$
$$< |D| + |A| = |D + A|,$$

contrary to the assumption that Λ satisfies F3.

Since Θ satisfies F3, it satisfies the hypothesis of the theorem with 1 in place of q. By the induction hypothesis, Θ admits at least one LDR λ_1. (Note: λ_1 could have been obtained here by the Philip Hall Theorem, but as we said in the foregoing remarks, we wished to give this result an independent proof.) Let λ_0 be an LDR for Λ_D and define $\lambda: E \to V$ by

$$\lambda(e) = \begin{cases} \lambda_0(e) & \text{if } e \in D; \\ \lambda_1(e) & \text{if } e \in E + D. \end{cases}$$

Since Λ_D and Θ have disjoint vertex sets, λ is an injection, and hence λ is an LDR of Λ. Since Λ_D admits at least $q!$ LDR's, so does Λ. □

The next three corollaries are immediate consequences of Marshall Hall's theorem and other results of this chapter. Their proofs are left to the reader. From the proof of D9 we obtain

F4 Corollary (Marshall Hall [h.3], 1945). *If $1 \leq r < s$, the number of ways to extend an $r \times s$ Latin rectangle on s symbols to an $(r + 1) \times s$ Latin rectangle is at least $(s - r)!$.*

F5 *Exercise.* If $1 \leq r < s$, determine lower bounds for the number of $r \times s$ Latin rectangles on s symbols and for the number of Latin squares of order s.

F6 Corollary. *Let $\Gamma = (\{V_1, V_2\}, \mathscr{E})$ be a bipartite graph, let $q \in \mathbb{N}$, and suppose*

$\rho(x) \geq q$ *for all* $x \in V_1$. *If* Γ *admits a matching of* V_1, *then the number of such matchings is at least*

(a) $q!$ *if* $q < |V_1|$;

(b) $q!/(q - |V_1|)!$ *if* $q \geq |V_1|$.

F7 *Exercise.* Apply Corollary F6 to complete bipartite graphs and thereby obtain "half" of Proposition *I*C13, namely that

$$|\text{inj}(Y^X)| \geq \frac{|Y|!}{(|Y| - |X|)!} \quad \text{if } |X| \leq |Y|.$$

The *i*th **row-sum** of an $r \times s$ matrix $A = [a_{ij}]$ over an integral domain is $\rho_i = \sum_{j=1}^{s} a_{ij}$, and the *j*th **column-sum** is $\sigma_j = \sum_{i=1}^{r} a_{ij}$. We say that A has constant row-sums (respectively, has constant column-sums) if $\rho_i = \rho_j$ (respectively, $\sigma_i = \sigma_j$) for all i, j.

F8 Corollary. *Let* M *be an* $r \times s$ $\{0, 1\}$*-matrix and let* $q \in \mathbb{N} + \{0\}$ *be a lower bound for the row-sums of* M. *If* $\text{perm}(M) \neq 0$, *then*

$$\text{perm}(M) \geq \begin{cases} q! & \text{if } q < r; \\ q!/(q - r)! & \text{if } q \geq r. \end{cases}$$

A **permutation matrix** P is an $r \times s$ $\{0, 1\}$-matrix such that $PP^* = I_r$, where

$$I_r = \begin{bmatrix} 1 & & & 0 \\ & \cdot & & \\ & & \cdot & \\ & & & \cdot \\ 0 & & & 1 \end{bmatrix}$$

denotes the identity matrix of order r. It is a well-known fact of linear algebra that if P is an $r \times s$ permutation matrix, then $r \leq s$. Moreover, $r = s$ if and only if all row- and column-sums are 1. Clearly $\text{perm}(P) = 1$.

F9 Proposition. *Let* A *be a nonzero* $r \times s$ *matrix whose entries are nonnegative elements of some field* $\mathbb{F} \subseteq \mathbb{R}$. *Let* $r \leq s$ *and suppose* A *has constant row-sums and constant column-sums. Then there exist* $r \times s$ *permutation matrices* $P_1, \ldots, P_n$ *and positive elements* $c_1, \ldots, c_n \in \mathbb{F}$ *such that*

$$A = \sum_{i=1}^{n} c_i P_i.$$

PROOF. Let all row-sums equal ρ and all column-sums equal σ.

If $r < s$, let J be the $(s - r) \times s$ matrix each of whose entries equals 1. Consider the augmented matrix

$$A' = \begin{bmatrix} A \\ \hdashline \frac{\rho}{s} J \end{bmatrix}$$

which also has the property that its entries are nonnegative elements of $\mathbb{F}$. Since $\sum_{i,j} a_{ij} = r\rho = s\sigma$, the column-sums of A' are all equal to $\sigma + (s - r)\rho/s = \rho$; i.e., all line sums are equal to ρ. Clearly if the proposition holds for A', then it holds for A. We may therefore assume that A is an $r \times r$ matrix with all line-sums equal to ρ.

We assert that A has an r-set of positions occupied by positive elements, no two of which are incident. Were this not so, one could invoke Theorem E7 and the remark following it to infer: some set of fewer than r lines of A, say p rows and q columns with $p + q < r$, would be incident with every position occupied by a positive entry in A. Hence

$$r\rho = \sum_{i,j} a_{ij} \leq (p + q)\rho < r\rho,$$

which is absurd.

Let P_1 be the $r \times r$ matrix with 1's in the aforementioned r-set of positions and 0's elsewhere. Then P_1 is a permutation matrix. Let c_1 be the least of the entries of A occupying one of these r positions. Hence $c_1 > 0$, and $A - c_1P_1$ is also an $r \times r$ matrix whose entries are nonnegative elements of $\mathbb{F}$ and has constant line-sums $\rho - c_1$.

The foregoing argument may be repeated with $A - c_1P_1$ in place of A. Since $A - c_1P_1$ has more 0's than A, this process must terminate in a finite number of steps. □

F10 Corollary. *Let M be a $\{0, 1\}$-matrix of order r with constant line-sums ρ. Then there exist distinct permutation matrices $P_1, \ldots, P_\rho$ such that*

$$M = \sum_{i=1}^{\rho} P_i.$$

Proof. In the proof of the proposition, $c_1 = 1$ and each successive coefficient c_j in the jth iteration of the argument is also 1. The process must terminate in exactly ρ steps. □

F11 Corollary. *Let $\Lambda = (V, f, E)$ be a system with $|E| \leq |V|$ and with blocksize k. Suppose that Λ^* has constant blocksize. Then Λ admits a k-set of LDR's $\lambda_1, \ldots, \lambda_k$ such that $\lambda_i(e) \neq \lambda_j(e)$ for all $e \in E$ and $1 \leq i < j \leq k$.*

Systems satisfying the hypothesis of this corollary are of particular interest and will be studied in some depth in Chapter IX.

CHAPTER VI

Separation and Connectivity in Multigraphs

The present chapter will deal with multigraphs, principally graphs, and with directed graphs. What will be done here is, in a sense, "dirty" graph theory, since it will lack the elegance that comes from showing that graphs inherit properties from more general and more abstract combinatorial models. The motivation will come rather from pictorial representations of multigraphs or directed graphs, and we shall rarely escape from this vein in the course of the proofs.

There are several reasons for inserting this chapter at this point.

1. The "lead off" theorem is the now classical Menger Theorem, but our proof is not classical; it will employ the Max-Flow–Min-Cut Theorem, which will be shown in turn to follow from Menger's theorem.
2. The notion of k-connectivity, introduced here, generalizes the concepts of biconnectivity and triconnectivity introduced in Chapter III.
3. Connectivity is not only of some interest itself in graph theory, but is prerequisite to the study of further graph-theoretical topics, including chromatic problems to be presented in Chapters VII and VIII.

VIA The Menger Theorem

In this section and the next, Γ will denote either the multigraph (V, f, E) or the directed graph (V, D).

Let $A, Z \subseteq V$ and $A \cap Z = \varnothing$. If Γ is (V, f, E) (respectively, (V, D)), the term **AZ-path** will denote an elementary (directed) az-path such that $a \in A$, $z \in Z$, and no other vertex of it is in $A + Z$. An **AZ-edge** is an AZ-path of length 1. Possibly A is a singleton $\{a\}$ or $Z = \{z\}$, in which case we shall suppress the braces and speak of an aZ-path, az-edge, etc. (Cf. §*II*B and §*III*A.)

A set $S \subseteq V + A + Z$ **separates** Z **from** A if $\Gamma_{(S)}$ contains no AZ-path. (If $\Gamma = (V, D)$, we understand $\Gamma_{(S)} = (V + S, D \cap [(V + S) \times (V + S)])$.) Such a set S is called a **separating set of** Z **from** A. The cardinality of a smallest separating set of Z from A, if one exists, will be denoted by $\sigma(A, Z)$. Clearly such a set exists if and only if there exists no AZ-edge. If there exists no AZ-path in Γ, then of course $\sigma(A, Z) = 0$. If $\Gamma = (V, f, E)$, then $\sigma(A, Z) = \sigma(Z, A)$.

An m-family of paths is **openly-disjoint** if whenever two members have any common vertex at all, then that vertex is for each of the paths either an initial or a terminal vertex.

A1 Theorem (K. Menger [m.9], 1927). *Given* Γ, *let* $a, z \in V$, $a \neq z$. *Suppose* Γ *contains no az-edge. If* $\sigma(a, z) \geq m$ *then* Γ *contains an openly-disjoint m-family of az-paths.*

PROOF. If the theorem is true for graphs, then it is obviously true for multigraphs. Moreover, if it is true for directed graphs, then it is true for graphs, since any graph $(V, \mathscr{E})$ can be replaced by the directed graph (V, D) where $D = \{(x, y) : \{x, y\} \in \mathscr{E}\}$. Thus $(x, y) \in D \Leftrightarrow (y, x) \in D$, and every path in $(V, \mathscr{E})$ corresponds to a directed path in (V, D). The theorem will now be proved for the case $\Gamma = (V, D)$. We assume the hypothesis of the theorem.

Let x_0 be some object not in V and let $V' = V + \{a, z, x_0\}$. Let $D' \subseteq V' \times V'$ consist of all edges of the form:

$$\begin{array}{ll} (x, y) & \text{if } (x, y) \in D \text{ and } \{x, y\} \cap \{a, z\} = \varnothing; \\ (x_0, y) & \text{if } (a, y) \in D; \\ (x, x_0) & \text{if } (x, z) \in D. \end{array}$$

Let $\Gamma' = (V', D')$. (For example, see Figure A2.)

A2

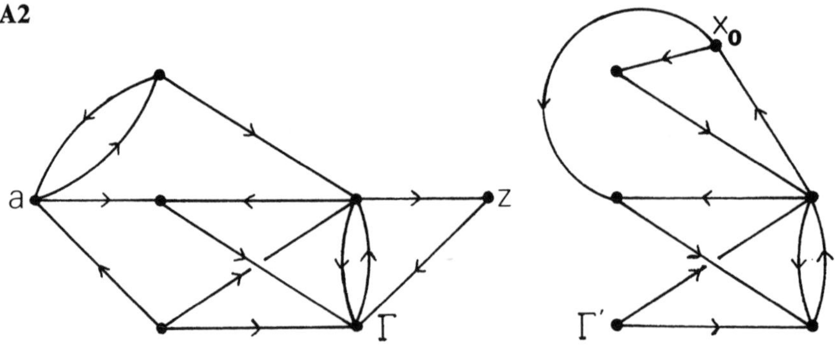

There is an obvious bijection assigning to each az-path in Γ a unique directed circuit in Γ' through x_0, obtained by "fusing" a and z. In order to apply the vertex form of the Max-Flow–Min-Cut Theorem (*IV*F2), we define a vertex capacity $j: V' \to \mathbb{N}$ by

$$j(x) = \begin{cases} 1 & \text{if } x \neq x_0; \\ |V| & \text{if } x = x_0. \end{cases}$$

Every openly-disjoint m-family of az-paths in Γ yields a feasible flow in Γ' whose value at x_0 is m. Conversely, let h be an integral feasible flow in Γ' with $h(x_0) = m$. By *IV*A17, h is the sum of elementary flows. The supports of these elementary flows correspond to directed circuits in Γ' and no two of them have any common vertex except possibly x_0. In fact, exactly m of them pass through x_0. Hence they correspond to an openly-disjoint m-family of az-paths in Γ. Thus,

A3 $\max\{m$: there exists an openly-disjoint m-family of az-paths in $\Gamma\}$
$$= \max\{h(x_0): h \text{ is a feasible flow in } \Gamma'\}.$$

We have observed that a set $S \subseteq V' + \{x_0\}$ meets every directed circuit through x_0 in Γ' if and only if S meets every az-path in Γ. For such a set, $j(S) = \sum_{x \in S} j(x) = |S|$. Thus,

A4 $\min\{j(S): S \text{ is a cut through } x_0 \text{ in } \Gamma'\}$
$$= \min\{|S|: S \text{ separates } z \text{ from } a \text{ in } \Gamma\} = \sigma(a, z).$$

The result follows from A3, *IV*F2, and A4. □

Assuming the Menger Theorem to hold, we now indicate a proof of the Max-Flow–Min-Cut Theorem based on it. By the same reasoning as in our first proof of the Max-Flow–Min-Cut Theorem (*IV*C6), we see that it suffices to prove the theorem for an arbitrary integral network (V, k). Let $W = \{(x, y) \in V \times V: x \neq y\}$ as in Chapter IV, and let $e_0 = (y_0, x_0) \in W$.

For each $e \in W$, let U_e be a $k(e)$-set and assume that $U_e \cap U_{e'} = \varnothing$ when $e \neq e'$. Let a and z be two additional objects, and let

$$U = \{a, z\} \cup \bigcup_{e \in W} U_e.$$

We form the directed graph (U, D) where

$$D = (\{a\} \times U_{e_0}) + \bigcup_{\substack{w,x,y \in V \\ (x,y) \neq e_0}} (U_{(w,x)} \times U_{(x,y)}) + \bigcup_{y \in V} (U_{(y,y_0)} \times \{z\}).$$

For an example of this construction see Figure A5.

A5

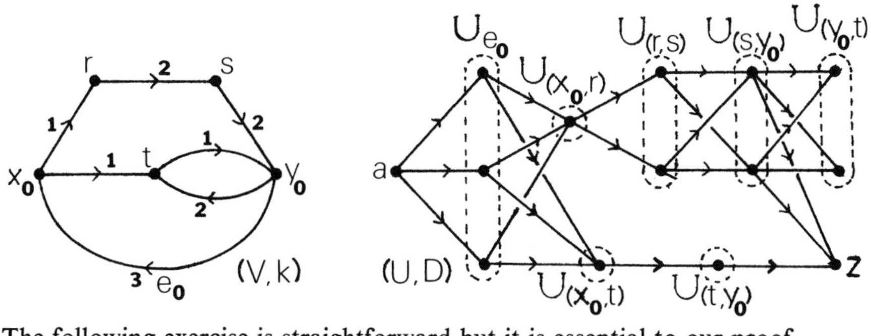

The following exercise is straightforward but it is essential to our proof.

A6 *Exercise.* In the above notation, show

(a) $\max\{h(e_0): h \text{ is a feasible flow in } (V, k)\} = \max\{m: (U, D) \text{ admits an openly-disjoint } m\text{-family of } az\text{-paths}\}$.

(b) $\sigma(a, z) = \min(\{k(e_0)\} \cup \{k(C; e_0): C \text{ is a cut through } e_0 \text{ in } (V, k)\})$.

The Max-Flow–Min-Cut Theorem follows at once from this exercise and the Menger Theorem.

VIB Generalizations of the Menger Theorem

We continue with the convention that Γ will denote either the multigraph (V, f, E) or the directed graph (V, D). When in the Menger Theorem the vertices a and z are replaced by arbitrary disjoint subsets of V, one obtains the following:

B1 Proposition. *Let Γ be given and let $A, Z \subset V$, $A \cap Z = \varnothing$. Suppose Γ contains no AZ-edge. If $\sigma(A, Z) \geq m$, then Γ contains an openly-disjoint m-family of AZ-paths.*

PROOF. From $\Gamma = (V, D)$ we form $\Theta = (U, C)$ as follows. Let $U = V + A + Z + \{a, z\}$, where a and z are new objects not in V. Let

$$\begin{aligned} C = (D \cap [(V + A + Z) \times (V + A + Z)]) \\ \cup \{(a, y): y \in V + A + Z \text{ and } (x, y) \in D \text{ for some } x \in A\} \\ \cup \{(x, z): x \in V + A + Z \text{ and } (x, y) \in D \text{ for some } y \in Z\}. \end{aligned}$$

If $\Gamma = (V, f, E)$, then Θ is the contraction of Γ given by the partition $\{A, Z\} \cup \{\{x\}: x \in V + A + Z\}$. Intuitively speaking, Θ is obtained from Γ by the coalescing of A and Z to single vertices a and z, respectively, and the elimination of all edges both of whose incident vertices belong to A or Z. Clearly the sets separating Z from A in Γ are identical to those separating z from a in Θ, and vice versa. By the Menger Theorem, Θ contains an openly-disjoint m-family of az-paths. These determine the required AZ-paths in Γ. □

In order to present a powerful generalization of Menger's theorem, due to G. A. Dirac, we shall require a broader interpretation of the notion of a separating set. Let $S \subseteq V \cup E$ (respectively, $S \subseteq V \cup D$). In the directed case we let $\Gamma_{(S)}$ acquire the meaning

$$(V + (V \cap S), [D + (D \cap S)] \cap [(V + (V \cap S)) \times (V + (V \cap S))]).$$

In both the directed and "undirected" cases, $\Gamma_{(S)}$ is obtained by the deletion of all elements of S and all edges incident with vertices in S. Now let A and Z be nonempty disjoint proper subsets of V. Then S **separates** Z **from** A if

(a) $\Gamma_{(S)}$ contains no AZ-path,
(b) S contains neither A nor Z.

B2 Theorem (Dirac [d.6], 1960). *Let* $\Gamma = (V, f, E)$ *or* (V, D). *Let* $\varnothing \subset A$, $Z \subset V$, *and* $A \cap Z = \varnothing$. *Suppose that whenever* $S \subset V \cup E$ *separates* Z *from* A, *then* $|S| \geq m$. *Then*

(**a**) Γ *contains an openly-disjoint m-family* $\mathscr{F}$ *of AZ-paths.*

(**b**) *The family* $\mathscr{F}$ *may be chosen to satisfy both of the following conditions:*

(**i**) *If* $|A| \geq m$, *then no two paths in* $\mathscr{F}$ *have a common vertex in* A. *If* $|A| < m$ *and if* $g: A \to \mathbb{N} + \{0\}$ *satisfies* $\sum_{a \in A} g(a) = m$, *then each* $a \in A$ *is the initial vertex of precisely* $g(a)$ *paths in* $\mathscr{F}$.

(**ii**) *If* $|Z| \geq m$, *then no two paths in* $\mathscr{F}$ *have a common vertex in* Z. *If* $|Z| < m$ *and if* $h: Z \to \mathbb{N} + \{0\}$ *satisfies* $\sum_{z \in Z} h(z) = m$, *then each* $z \in Z$ *is the terminal vertex of precisely* $h(z)$ *paths in* $\mathscr{F}$.

PROOF. Let A and Z be given as in the hypothesis.

(a) We form from Γ a new directed graph $\Gamma' = (V', D')$ where $V' = V \cup E$ (respectively, $V \cup D$) and D' consists of all pairs of the following forms:

$$(x, e) \quad \text{if } x \in f(e) \text{ (respectively, } e = (x, y) \in D \text{ for some } y \in V),$$

and

$$(e, y) \quad \text{if } y \in f(e) \text{ (respectively, } e = (x, y) \in D \text{ for some } x \in V).$$

(For example, see Figure B3.) Observe that Γ' contains no AZ-edge. Suppose $S \subseteq V'$ separates Z from A in Γ'. Then S surely separates Z from A in Γ. By hypothesis, $|S| \geq m$ and by Proposition B1, Γ' admits an openly-disjoint

B3

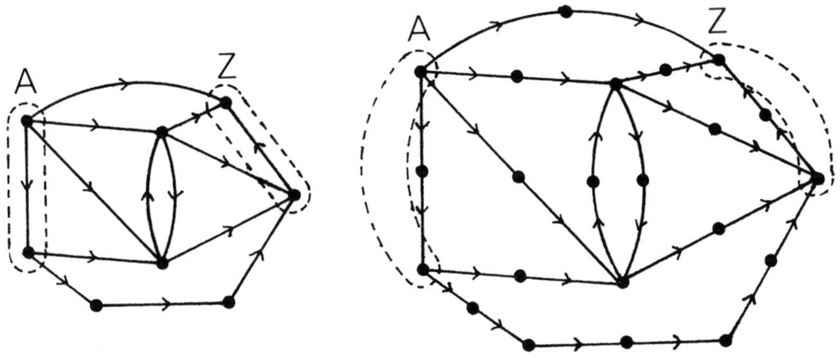

$\Gamma = (V, D)$ $\Gamma' = (V', D')$

m-family of AZ-paths. It is immediate that Γ also admits such a family. For if an AZ-path in Γ' has the form

$$a, (a, e_1), e_1, (e_1, a_1), a_1, \ldots, a_{r-1}, (a_{r-1}, e_r), e_r, (e_r, z), z,$$

then the corresponding path in Γ is simply

$$a, e_1, a_1, \ldots, a_{r-1}, e_r, z.$$

(b) (i) First suppose $|A| \geq m$. We form a new multigraph $\Gamma' = (V', f', E')$ (respectively, a new directed graph Γ') from Γ by letting $V' = V \cup \{c\}$, where c is a new vertex. We let $E' = E \cup \{\{c, a\}: a \in A\}$ and let f' be the appropriate extension of f (respectively, let $D' = D \cup \{(c, a): a \in A\}$). See Figure B4. Let S' separate Z from c in Γ'. Corresponding to S' is a set $S \subseteq V \cup E$ (respectively, $V \cup D$) given by $S = [S' \cap (V \cup E)] \cup \{a \in A: \{c, a\} \in S'\}$.

B4

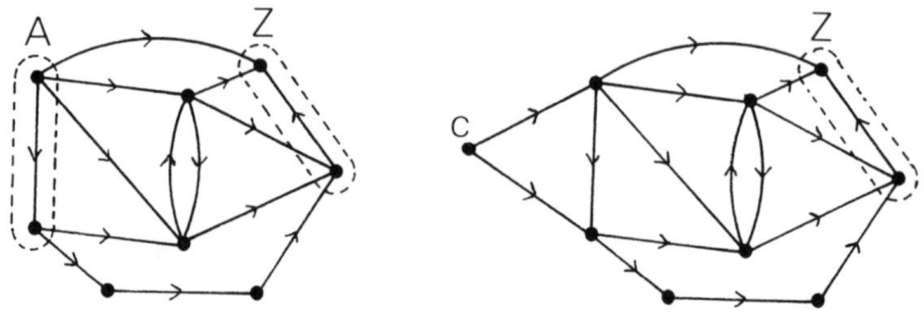

$\Gamma = (V, D)$ $\qquad$ $\Gamma' = (V', D')$

Hence $|S'| \geq |S|$, with a nearly identical argument from here on if Γ is a directed graph.

If S did not separate Z from A in Γ, there would exist an AZ-path Π in $\Gamma_{(S)}$ whose initial vertex is some $a \in A + (A \cap S)$. This would imply that neither a nor $\{c, a\}$ (respectively, (c, a)) belongs to S'. By adjoining c and $\{c, a\}$ (respectively, (c, a)) to Π, one obtains a cZ-path in $\Gamma'_{(S')}$, and so S' fails to separate Z from c in Γ' as assumed. Hence S separates Z from A in Γ. By hypothesis, $|S| \geq m$. Therefore $|S'| \geq m$.

Since every set separating Z from c in Γ' has cardinality at least m, part (a) of this proposition assures the existence of an openly-disjoint m-family of cZ-paths in Γ'. Since no two of these intersect in A, the required AZ-paths in Γ are clearly obtainable from these paths.

Now suppose $|A| < m$. Let the function g be given as in (i). We transform Γ into a directed graph in two stages. First we form $\Gamma' = (V', D')$ precisely as in the proof of part (a) above. From Γ' we obtain $\Gamma'' = (V'', D'')$ as follows. Let $A'' = \{a_j: a \in A; j = 1, \ldots, g(a)\}$ be disjoint from V' and let $V'' = (V' + A) \cup A''$; let

$$D'' = \{(x, y) \in D': x, y \in V + A\} \cup \{(a_j, y): (a, y) \in D'; j = 1, \ldots, g(a)\}$$
$$\cup \{(x, a_j): (x, a) \in D'; j = 1, \ldots, g(a)\}.$$

(See Figure B5.)

B5

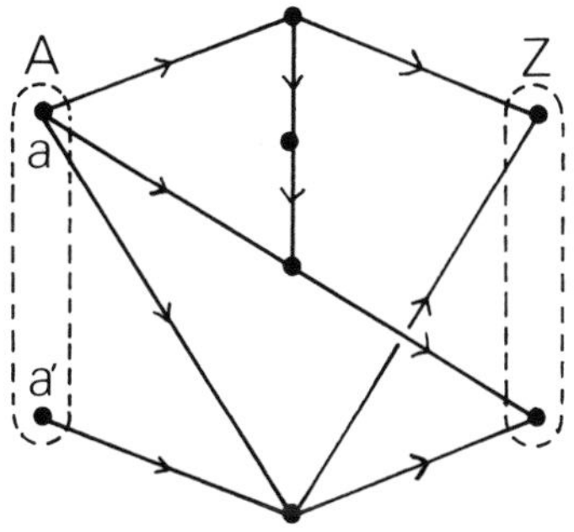

$\Gamma' = (V', D')$

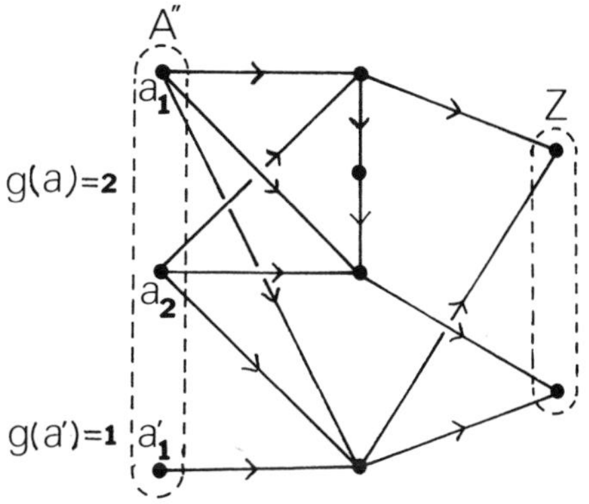

$\Gamma'' = (V'', D'')$

Note that

B6 $$|A''| = \sum_{a \in A} g(a) = m.$$

Now let $S'' \subset V''$ separate Z from A'' in Γ''. Corresponding to S'' is a set $S' \subseteq V' \cup D'$ given by

$$S' = [S'' \cap (V' \cup D')] \cup \{(a, x): a \in A; (a_j, x) \in S'' \text{ for some } j = 1, \ldots, g(a)\} \\ \cup \{a \in A: \{a_1, \ldots, a_{g(a)}\} \subseteq S''\}.$$

Hence $|S''| \geq |S'|$.

We assert that $|S''| \geq m$. Were that not so, we would have $A'' \nsubseteq S''$ by B6, and consequently $A \nsubseteq S'$. Also $|S'| < m$, implying, by the way that Γ' was constructed from Γ and by the hypothesis, that S' fails to separate Z from A in Γ'. Hence there would exist an AZ-path Π' in $\Gamma'_{(S')}$ whose initial vertex is some $a \in A + (A \cap S')$. Also Π' contains an edge (a, x) for some $x \in V'$. It follows that for some $j \in \{1, \ldots, g(a)\}$, neither a_j nor the edge (a_j, x) is an element of S''. If a and (a, x) are replaced in Π' by a_j and (a_j, x), one obtains an $A''Z$-path in $\Gamma''_{(S'')}$, and so S'' fails to separate Z from A'' as assumed.

Since every set separating Z from A'' in Γ'' has cardinality at least m, the first case in part (b)(i) assures the existence of an openly-disjoint m-family of $A''Z$-paths in Γ'' having no common vertex in A''. The reader can straightforwardly deduce how this family determines the family of AZ-paths required.

(ii) Let Γ denote the graph Γ' or Γ'' constructed in part (i) according as $|A| \geq m$ or $|A| < m$, respectively. The proof is completed by applying the appropriate construction from (i) where one interchanges the roles of Z and A' or A'', respectively. □

Suppose that the sets A and Z in the above theorem are each m-subsets of V and that any set separating Z from A is at least an m-set. Then Γ admits an m-family of *disjoint* AZ-paths. It is not hard to see that the converse is also true. Another condition equivalent to these two lies implicitly in various arguments of M. J. Piff [p.4] and is stated explicitly as follows.

B7 Proposition. *Let* $\Gamma = (V, f, E)$ *or* (V, D) *and let* $A, Z \in \mathscr{P}_m(V)$. *There exists an m-family of disjoint AZ-paths in* Γ *if and only if there exists a bijection* $b\colon V + Z \to V + A$ *such that for each* $x \in V + Z$, *either* $b(x) = x$ *or, when* $\Gamma = (V, f, E)$ *then* $\{x, b(x)\} \in f[E]$, *and when* $\Gamma = (V, D)$ *then* $(x, b(x)) \in D$.

PROOF. As in earlier proofs in this chapter, we may assume that $\Gamma = (V, D)$.

Suppose that an m-family $\mathscr{F}$ of disjoint AZ-paths has been given. Since we have not assumed that A and Z are disjoint, $\mathscr{F}$ may contain some paths of length 0. Any vertex comprising such a trivial path belongs to neither $V + Z$ nor $V + A$, and so we may assume that if $x \in V + Z$, then either x lies on no path in $\mathscr{F}$ or there exists an edge (x, x') on some path in $\mathscr{F}$. In the former case, let $b(x) = x$ and in the latter case let $b(x) = x'$. The paths are disjoint and, except for the trivial ones, they are $[A \cap (V + Z)][Z \cap (V + A)]$-paths. Hence b is a bijection.

Conversely, let the bijection b be given. Let $A = \{a_1, \ldots, a_m\}$ and let

$$V_i = \begin{cases} \{a_i\} & \text{if } a_i \in Z; \\ \{b^j(a_i) \colon j \in \mathbb{N};\ b^{j-1}(a_i) \in V + Z\} & \text{if } a_i \in V + Z. \end{cases}$$

Since $b\colon V + Z \to V + A$ is a bijection, the sets $V_1, \ldots, V_m$ are pairwise-disjoint. If $a_i \in Z$, we take a path of length 0 from a_i to itself. Otherwise, V_i has the form $\{a_i, b(a_i), b^2(a_i), \ldots, b^{j_i}(a_i)\}$, where all the $j_i + 1$ elements are distinct and only the last one listed belongs to Z. In this case, form the path $a_i, (a_i, b(a_i)), b(a_i), \ldots, b^{j_i}(a_i)$. This construction yields the required m-family of disjoint paths. □

B8 *Exercise.* Suppose that whenever S separates Z from A in $\Gamma = (V, \mathscr{E})$ and $S \subseteq \mathscr{E}$, then $|S| \geq m$. Prove that Γ admits an m-family of AZ-paths no two of which have a common edge.

VIC Connectivity

In this section, $\Gamma = (V, \mathscr{E})$ will denote a graph. If $S \subseteq V \cup \mathscr{E}$, then $\Gamma_{(S)}$ denotes $(\Gamma_{(V \cap S)})_{(\mathscr{E} \cap S)}$.

Every graph is **0-connected**. If $|V| \geq m + 1$, then Γ is m-**connected** if

$$S \subseteq V,\ |S| < m \Rightarrow \Gamma_{(S)} \text{ is connected.}$$

By convention, the complete graph K_1 is **1-connected.**

Thus Γ is connected if and only if Γ is 1-connected. If Γ is n-connected and $0 \leq m \leq n$, then Γ is m-connected.

If Γ is n-connected but not $(n + 1)$-connected, we say that Γ has **connectivity** (sometimes called "vertex-connectivity") equal n, and we write,

$$\kappa(\Gamma) = n.$$

For example, $\kappa(K_n) = n - 1$ for $n = 2, 3, \ldots$.

C1

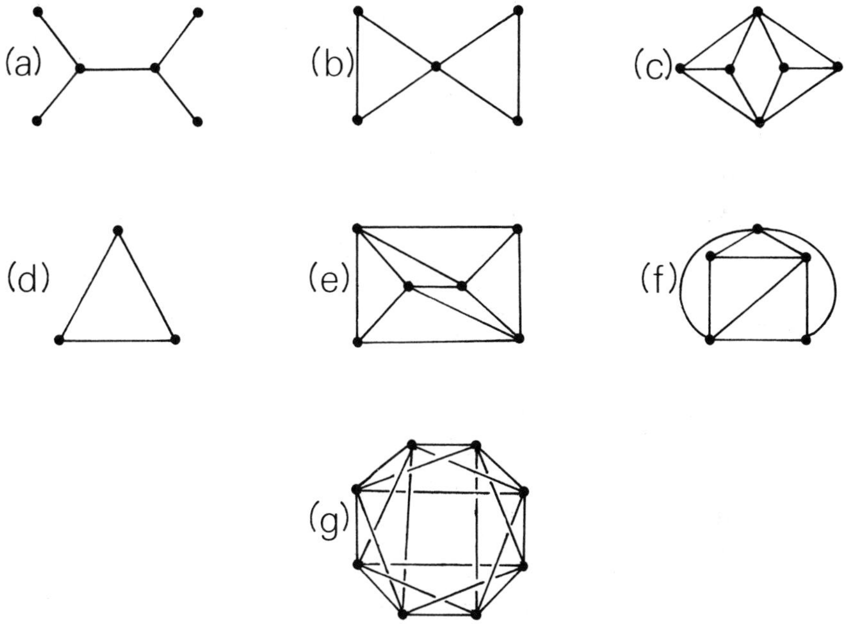

In Figure C1, graphs (a) and (b) have connectivity 1, (c) and (d) have connectivity 2, (e) and (f) have connectivity 3, and (g) has connectivity 4.

If Γ is m-connected and if a and z are nonincident vertices of Γ, then $\sigma(a, z) \geq m$. On the other hand, if Γ is not complete, then there exist a, $z \in V$ and a $\kappa(\Gamma)$-subset of V which separates z from a. Hence

C2 $$\kappa(\Gamma) = \min\{\sigma(a, z) : \{a, z\} \in \mathscr{P}_2(V) + \mathscr{E}\}$$

whenever Γ is not complete. In particular, if $\kappa(\Gamma) = 1$ and $|V| \geq 3$, then Γ has an articulation vertex (cf. Exercise *III*B11). The terms "biconnected" and "2-connected" are equivalent for graphs. Similarly, the terms "triconnected" and "3-connected" are equivalent for graphs.

If $|V| \geq 2$, then

C3 $$\kappa(\Gamma) \leq |V| - 1,$$

with equality holding if and only if Γ is complete. We observe that

C4 $$\kappa(\Gamma) \leq \check{\rho}(\Gamma).$$

C5 Proposition. *Let Γ be m-connected for some $m \geq 1$. Let $x \in V$ and $E \in \mathscr{E}$. Then the subgraphs $\Gamma_{(x)}$ and $\Gamma_{(E)}$ are each $(m-1)$-connected.*

PROOF. If $\Gamma_{(x)}$ is complete, then $\kappa(\Gamma_{(x)}) = |V| - 2 \geq \kappa(\Gamma) - 1$ by C3. If $\Gamma_{(x)}$ is not complete, we let $S \in \mathscr{P}_{m-2}(V + \{x\})$. Since $|S + \{x\}| = m - 1$, $(\Gamma_{(x)})_{(S)} = \Gamma_{(S+\{x\})}$ is connected. Hence $\Gamma_{(x)}$ is $(m-1)$-connected.

Let $E = \{x, y\}$, and suppose that $\Gamma_{(E)}$ is not $(m-1)$-connected. Since Γ is m-connected, $|V| \geq m + 1$ by C3. Hence there exists $S \in \mathscr{P}_{m-2}(V)$ such that $\Gamma_{(S \cup \{E\})}$ is not connected. Select a and z in different components of $\Gamma_{(S \cup \{E\})}$. We may choose $\{a, z\} \neq \{x, y\}$, for otherwise $V = S \cup \{x, y\}$, which implies that $|V| \leq |S| + 2 = m$. We may suppose $x \notin \{a, z\}$. Now S does not separate z from a in Γ. It follows that every az-path in Γ must include E, and hence include x. But then $S \cup \{x\}$ separates z from a in Γ while $|S \cup \{x\}| \leq m - 1$, contrary to the hypothesis that Γ is m-connected. □

C6 *Exercise.* Let $S \subseteq V \cup \mathscr{E}$. Prove that

$$\kappa(\Gamma) - \kappa(\Gamma_{(S)}) \leq |S|.$$

The basic theorem on connectivity of graphs is the following:

C7 Theorem (H. Whitney [w.8], 1932). *A necessary and sufficient condition for a graph $\Gamma = (V, \mathscr{E})$ to be m-connected is:*

C8 *For any $\{a, z\} \in \mathscr{P}_2(V)$ there exists an openly-disjoint m-family of az-paths in Γ.*

PROOF. If Γ is the complete graph K_n, then obviously both C8 holds and Γ is m-connected if and only if $m \leq n - 1$. Suppose, therefore, that Γ is not complete. Since the theorem is trivial for $m = 0$ and 1, we suppose $m \geq 2$.

Necessity. Let Γ be m-connected and let $\{a, z\} \in \mathscr{P}_2(V)$ be given. First suppose $\{a,z\} \notin \mathfrak{E}$. By C2, $\sigma(a,z) \geqslant m$. Theorem A1 assures the existence of an openly-disjoint m-family of az-paths. On the other hand, if $E = \{a,z\} \in \mathfrak{E}$, then by Proposition C5, the subgraph $\Gamma_{(E)}$ is $(m-1)$-connected. Since $\Gamma_{(E)}$ has no az-edge and since $\sigma(a, z) \geq m - 1$ in $\Gamma_{(E)}$, Theorem A1 assures the existence of an openly-disjoint $(m-1)$-family $\mathscr{F}$ of az-paths in $\Gamma_{(E)}$. Adjoining to $\mathscr{F}$ the path a, E, z gives the required m-family for Γ.

Sufficiency. Assume that Condition C8 holds for some $m \geq 2$. Let $S \subseteq V$ and suppose that $\Gamma_{(S)}$ is not connected. Select a and z in different components of $\Gamma_{(S)}$. By C8, there exists an openly-disjoint m-family $\mathscr{F}$ of az-paths in Γ. Since no path in $\mathscr{F}$ lies entirely in $\Gamma_{(S)}$, each of the m paths in $\mathscr{F}$ contains a vertex in S. Since $a, z \notin S$ and $\mathscr{F}$ is openly-disjoint, one obtains $|S| \geq m$, and Γ is m-connected. □

The following result is essentially an adaptation of Theorem B2 to the terminology of the present section.

C9 Proposition. *Let Γ be m-connected. Let $A, Z \subset V$ such that $A \cap Z = \varnothing$ and $|A|, |Z| \leq m$. Let there exist functions $g: A \to \mathbb{N} + \{0\}$ and $h: Z \to \mathbb{N} + \{0\}$ such that*

$$\sum_{a \in A} g(a) = m = \sum_{z \in Z} h(z).$$

Then Γ admits an openly-disjoint m-family $\mathscr{F}$ of AZ-paths such that for each $a \in A$ and $z \in Z$, precisely $g(a)$ of the paths in $\mathscr{F}$ contain a and precisely $h(z)$ of the paths contain z.

Observe carefully what the above proposition does *not* say. It does not imply that for some given $a_0 \in A$ and $z_0 \in Z$, the family $\mathscr{F}$ includes an a_0z_0-path. Such a condition is of a fundamentally different nature, and demands much more from a graph, as the next result indicates.

We shall say that Γ satisfies the condition O_n ($n = 1, 2, \ldots$) if $|V| \geq 2n$ and, given $\{a_1, \ldots, a_n, z_1, \ldots, z_n\} \in \mathscr{P}_{2n}(V)$, there exists an openly-disjoint n-family $\{\Pi_1, \ldots, \Pi_n\}$ of paths such that Π_i is an a_iz_i-path ($i = 1, \ldots, n$). Clearly O_n implies O_m if $n > m$.

C10 Theorem [w.3]. *If Γ satisfies O_n, then $\kappa(\Gamma) \geq 2n - 1$.*

PROOF. Suppose $\kappa(\Gamma) \leq 2n - 2$, and let $k = [\kappa(\Gamma)/2]$. V contains a separating set $S = \{a_1, \ldots, a_k, z_1, \ldots, z_m\}$ where

$$m = \begin{cases} k & \text{if } \kappa(\Gamma) \text{ is even;} \\ k + 1 & \text{if } \kappa(\Gamma) \text{ is odd.} \end{cases}$$

If $\kappa(\Gamma)$ is even, $k \leq n - 1$. Choose vertices a_{k+1} and z_{k+1} in different components of $\Gamma_{(S)}$. Clearly Γ does not satisfy O_{k+1}. Hence Γ does not satisfy O_n.

If $\kappa(\Gamma)$ is odd, $k \leq n - 2$. Choose a_{k+2} and z_{k+2} in different components of $\Gamma_{(S)}$ and let a_{k+1} be any other vertex in $V + S$. Since Γ does not satisfy O_{k+2}, it does not satisfy O_n. □

Note that $\kappa(K_{2n}) = 2n - 1$ and that K_{2n} satisfies O_n. This shows that the inequality in Theorem C10 cannot be sharpened. The reader interested in the "O_n Problem" may read [w.3], [j.1], [ℓ.1].

We now give an application of Proposition C9.

C11 Theorem (G. A. Dirac [d.7], 1960). *If $\Gamma = (V, \mathscr{E})$ is n-connected for some $n \geq 2$ and if $1 \leq m \leq n$, then for any $S \in \mathscr{P}_m(V)$, there exists an elementary circuit in Γ which contains S.*

PROOF. The proof is by induction on m. Since Γ is assumed to be a biconnected graph, $|V| \geq 3$, and the result holds for $m = 1, 2$ by Exercise *III*B13.

Suppose now that $3 \leq m \leq n$ and that the conclusion holds for $m - 1$. Let $S = \{z_1, \ldots, z_m\} \subseteq V$ be given, and let $Z = \{z_1, \ldots, z_{m-1}\}$. By the induction hypothesis there exists an elementary circuit Δ which contains Z, and we may assume that z_m is not on Δ. Let us reindex Z so that $z_1, \ldots, z_{m-1}$ appear in cyclic order as one proceeds around Δ.

Case 1: every vertex in Δ is in Z. Since Γ is $(m - 1)$-connected, we have by Proposition C9 that Γ admits an openly-disjoint $(m - 1)$-family of z_mZ-paths $\{\Pi_1, \ldots, \Pi_{m-1}\}$ such that Π_i is a z_mz_i-path for $i = 1, \ldots, m - 1$. Let Π_0 be the elementary path obtained when the edge $\{z_{m-1}, z_1\}$ is removed from Δ. An elementary circuit containing S is obtained by linking up the paths Π_1, Π_0, and Π_{m-1} in the obvious way.

Case 2: Δ contains a vertex $u \in V + S$. Since Γ is m-connected, Proposition C9 assures the existence of an openly-disjoint m-family $\{\Pi_1, \ldots, \Pi_m\}$ of $z_m(Z + \{u\})$-paths, where $\Pi_1, \ldots, \Pi_{m-1}$ are as in Case 1 and Π_m is a z_mu-path. Proceeding along Π_i from z_m, let y_i be the first vertex in Δ encountered ($i = 1, \ldots, m$). Now Δ is formed from an openly-disjoint $(m - 1)$-family of paths $\Delta_1, \ldots, \Delta_{m-1}$ linked up "end-to-end" where Δ_i is a z_iz_{i+1}-path for $i = 1, \ldots, n - 2$ and Δ_{m-1} is a $z_{m-1}z_1$-path. Since the union of the sets of vertices of these $m - 1$ paths Δ_i includes the m vertices $y_1, \ldots, y_m$, some one of these paths, say Δ_r, contains (the old Pigeon-hole Principle!) two of these vertices, say y_s and y_t. (See Figure C12.)

C12

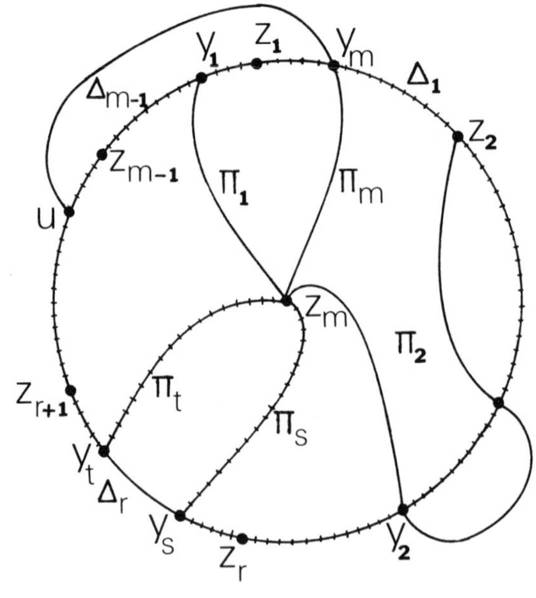

Now consider the y_sy_t-path contained in Δ which contains all the paths Δ_i for $i \neq r$, and hence all of Z. The elementary circuit obtained by properly linking up this path with Π_s and Π_t is an elementary circuit containing all of S as required. (This circuit is shown in "railroad tracks" in Figure C12.) □

Let $\zeta(\Gamma)$ denote the largest integer m for which it holds nonvacuously that if $U \in \mathscr{P}_m(V)$, then there is an elementary circuit in Γ which contains U. Theorem C11 asserts that if $\kappa(\Gamma) \geq 2$, then

C13 $$\zeta(\Gamma) \geq \kappa(\Gamma).$$

C14 *Exercise*. Show that for each $n \geq 2$, the graph Γ such that $\zeta(\Gamma) = \kappa(\Gamma) = n$ having the fewest vertices and the fewest edges is the complete bipartite graph $K_{n,n+1}$.

The above exercise shows that C13 cannot be sharpened, but the graphs $K_{n,n+1}$ serve also as prototypes for all graphs for which equality holds in C13. These have been characterized [w.6] when $\kappa(\Gamma) \geq 3$ as possessing a $\kappa(\Gamma)$-subset S of V such that

$$\nu_{-1}(\Gamma_{(S)}) \geq \kappa(\Gamma) + 1.$$

When $\kappa(\Gamma) = 2$, then Γ is one of three types. The smallest representative of each type is shown in Figure C15.

C15

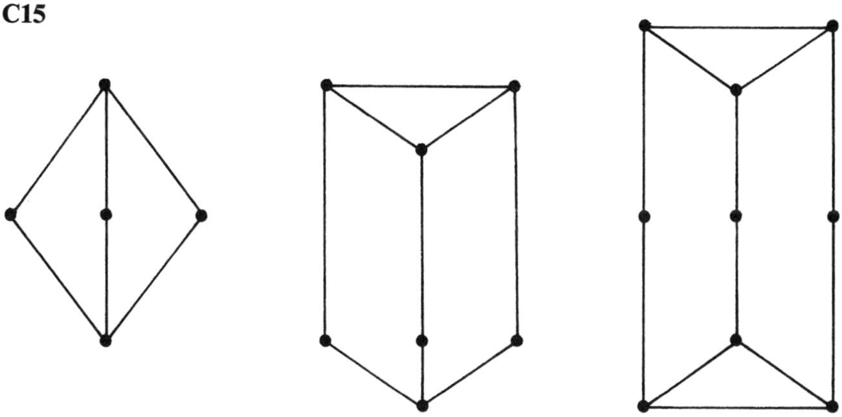

At the other extreme, a moment's reflection will yield examples of graphs Γ with $\zeta(\Gamma) - \kappa(\Gamma)$ equal to the upper bound of $\zeta(\Gamma) - 2$. If $\zeta(\Gamma) = \nu_0(\Gamma)$, then Γ is said to be **hamiltonian**, and a circuit which contains V is called a **hamilton circuit**. Conditions for a graph to be hamiltonian are generally rather complicated and lead to a whole area of graph theory not included in this text. The interested reader may consult [c.3].

Another but equivalent definition of m-connectedness was given by O. Ore [o.1]. Let the graph $\Gamma = (V, \mathscr{E})$ be given, and let $\Theta = (U, \mathscr{F})$ be a subgraph of Γ. If $x \in V$, we say x is an **interior vertex** of Θ if $\mathscr{F}$ contains the vertex-cocycle of x. We say x is an **exterior vertex** of Θ if $\mathscr{E} + \mathscr{F}$ contains the vertex-cocycle of x. Otherwise, x is an **attachment vertex** of Θ. The following exercise consists of proving the equivalence between Ore's definition and the one we have adopted.

C16 Exercise. *Let Γ be connected and not complete.*

(a) Show that Γ *is m-connected if and only if every subgraph* Θ *of* Γ *having at least one interior vertex and one exterior vertex has at least m attachment vertices.*

(b) Prove that (a) *holds when "subgraph" has been replaced by "vertex-induced subgraph."*

We shall require this exercise as we prove the following proposition, which generalizes Whitney's theorem in much the way that Dirac generalized Menger's theorem. A partial converse of Proposition C9 is hereby obtained. (The "full" converse is false. See Exercise C19 below.)

C17 Theorem [m.10]. *Let m be a positive integer, and let $\Gamma = (V, \mathscr{E})$ be a graph with $\nu_0(\Gamma) \geq m + 1$. The following 2m statements are equivalent:*

A. Γ *is m-connected.*

B_k $(k = 1, \ldots, m)$. *Given $\{a, z_1, \ldots, z_k\} \in \mathscr{P}_{k+1}(V)$, there exists an openly disjoint m-family of $a\{z_1, \ldots, z_k\}$-paths of which $m - k + 1$ are az_1-paths and one is an az_i-path for each $i = 2, \ldots, k$.*

C_k $(k = 1, \ldots, m - 1)$. *Given $S \in \mathscr{P}_k(V)$, and $\{a, z\} \in \mathscr{P}_2(V + S)$, there exists an openly-disjoint $(m - k)$-family of az-paths in $\Gamma_{(S)}$.*

Proof. Observe that if $n \geq m + 1$, then all the $2m$ statements are true for K_n. We assume, therefore, that Γ is not complete. The entanglement of the implications that we prove is indicated by the directed graph in Figure C18. (It *is* "strongly connected.")

C18

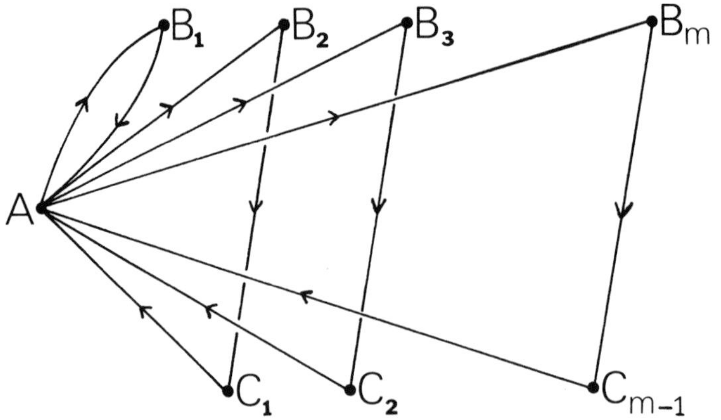

$\mathsf{A} \Rightarrow \mathsf{B}_k$ $(k = 1, \ldots, m)$. This is precisely Proposition C9 with

$$h(z_i) = \begin{cases} m - k + 1 & \text{if } i = 1; \\ 1 & \text{if } i = 2, \ldots, k. \end{cases}$$

$\mathsf{B}_1 \Rightarrow \mathsf{A}$. The condition B_1 is precisely C8. The implication follows from Whitney's theorem (C7).

$\mathsf{B}_k \Rightarrow \mathsf{C}_{k-1}$ ($k = 2, \ldots, m$). Let $S \in \mathscr{P}_{k-1}(V)$ and $\{a, z\} \in \mathscr{P}_2(V + S)$ be given. Assume B_k with $z = z_1$ and $S = \{z_1, \ldots, z_k\}$. This provides an openly-disjoint $[m - (k - 1)]$-family of az-paths which avoid S.

$\mathsf{C}_k \Rightarrow \mathsf{A}$ ($k = 1, \ldots, m - 1$). Let $U \subseteq V$, and suppose that a is an interior vertex of Γ_U and that z is an exterior vertex of Γ_U. Let S be the set of attachment vertices of Γ_U. By the definitions, every az-path contains a vertex in S.

Suppose $|S| \leq k$. By C_k, there exists an openly-disjoint $(m - k)$-family of az-paths in $\Gamma_{(S)}$, and since $m - k > 0$, this family is not empty; i.e., some az-path contains no vertex of S. Hence $|S| > k$. Let $T \in \mathscr{P}_k(S)$. Similarly, there exists an openly-disjoint $(m - k)$-family of az-paths in $\Gamma_{(T)}$. Since no two of these paths contain the same vertex in $S + T$, we have $m - k \leq |S + T| = |S| - k$. Hence Γ_U has $|S| \geq m$ attachment vertices, and Γ is m-connected by Exercise C16. □

C19 *Exercise* (A. C. Green). Construct a graph Γ such that for each $\{a, z_1, z_2, z_3\} \in \mathscr{P}_4(V)$, there is an openly-disjoint 6-family of $a\{z_1, z_2, z_3\}$-paths, two of which are az_i-paths for each $i = 1, 2, 3$, but Γ is not 6-connected.

If Γ is connected, we define a **distance function** $d\colon V \times V \to \mathbb{N}$ by letting $d(x, y)$ denote the least number of edges in an xy-path.

C20 *Exercise*. Prove that d is a metric on V; i.e., for all $x, y, z \in V$, (a) $d(x, y) = 0$ if and only if $x = y$, (b) $d(x, y) = d(y, x)$, and (c) $d(x, z) \leq d(x, y) + d(y, z)$.

If Γ is connected, we define the **diameter** of Γ by

$$\delta(\Gamma) = \max\{d(x, y) \colon (x, y) \in V \times V\}.$$

The following proposition is an elementary application of Whitney's theorem. Its proof is an exercise. Intuitively, it states that if $\nu_0(\Gamma)$ is fixed, then diameter and connectivity vary inversely; if Γ is "long" it must be "thin" and vice versa.

C21 Proposition [w.2]. **(a)** *If Γ is connected and $\nu_0(\Gamma) \geq 2$, then*

$$\nu_0 \geq \kappa(\delta - 1) + 2.$$

(b) *Given integers $n > k$, $k \geq 1$, and $d \geq 1$ satisfying $n = k(d - 1) + 2$, there exists a graph Γ with $\nu_0(\Gamma) = n$, $\kappa(\Gamma) = k$, and $\delta(\Gamma) = d$.*

VID Fragments

In §*III*B it was seen how the edges of a graph admit a decomposition into edge sets of lobes while vertices lying in more than one lobe were found to be articulation vertices. Each such vertex comprises in itself a set which separates the graph. Thus lobe "decompositions" of a graph are of interest only when the connectivity of the graph is at most 1. This section presents a method for decomposing a graph which is of interest as well for highly-connected graphs.

Since complete graphs do not admit separating sets, *we shall assume for the rest of this section that* Γ *is connected and not complete.*

Let $\mathscr{S}(\Gamma)$ denote the set of separating sets of Γ and let $\mathscr{S}_0(\Gamma)$ denote the subset of $\mathscr{S}(\Gamma)$ consisting of the smallest separating sets of Γ. Thus $|S| = \kappa(\Gamma)$ for all $S \in \mathscr{S}_0(\Gamma)$.

If $W \subseteq V$, we write $\overline{W} = V + W + N(W)$. If $W, \overline{W} \neq \varnothing$, then $N(W) \in \mathscr{S}(\Gamma)$, and so

$$\kappa(\Gamma) = \min\{|N(W)| : W, \overline{W} \neq \varnothing\}.$$

A **fragment** of Γ is a subset $W \subseteq V$ such that $W, \overline{W} \neq \varnothing$ and $|N(W)| = \kappa(\Gamma)$.

D1 Exercise.

(a) *Show that* W *is a fragment of* Γ *if and only if* $\overline{W}$ *is a fragment of* Γ. *Let* W_1 *and* W_2 *be fragments of* Γ. Show that

(b) $N(W_i) = N(\overline{W}_i)$; (c) $W_i = \overline{\overline{W}}_i$;

(d) $N(W_1 \cap W_2) \subseteq (W_1 \cap N(W_2)) + (N(W_1) \cap W_2) + (N(W_1) \cap N(W_2))$;

(e) $N(W_1 \cup W_2) = (\overline{W}_1 \cap N(W_2)) + (N(W_1) \cap \overline{W}_2) + (N(W_1) \cap N(W_2))$;

(f) $\overline{W_1 \cup W_2} = \overline{W}_1 \cap \overline{W}_2$;

(g) $\overline{W}_1 \cup \overline{W}_2 \subseteq \overline{W_1 \cap W_2}$.

Which of (b)–(g) hold if we allow W_1 and W_2 to be arbitrary subsets of V?

D2 Proposition. *Let* W_1 *and* W_2 *be fragments of the graph* Γ *and suppose that* $W_1 \cap W_2 \neq \varnothing$. *If* $|W_1| < |\overline{W}_2|$, *then both* $W_1 \cap W_2$ *and* $W_1 \cup W_2$ *are fragments.*

Proof. For $i, j \in \{1, 2\}$ and $i \neq j$, define

$$s_i = |N(W_i) \cap W_j|, \qquad t_i = |N(W_i) \cap \overline{W}_j|, \qquad q = |N(W_1) \cap N(W_2)|.$$

From these definitions, the fact that $W_1 \cap W_2$, $\overline{W_1 \cap W_2} \neq \varnothing$ (by D1g), and Exercise D1d, we have

D3 $$s_1 + q + t_1 = |N(W_1)| = \kappa \leq |N(W_1 \cap W_2)| \leq s_1 + q + s_2.$$

Hence

D4 $$t_1 \leq s_2.$$

If it were so that $\overline{W}_1 \cap \overline{W}_2 = \varnothing$, then D4 would yield

$$|W_1| = |W_1 \cap W_2| + s_2 + |W_1 \cap \overline{W}_2| > t_1 + |W_1 \cap \overline{W}_2| = |\overline{W}_2|,$$

contrary to hypothesis. Hence $\overline{W}_1 \cap \overline{W}_2 \neq \varnothing$, and so by Exercise D1(a,d) together with D4,

D5 $$|N(\overline{W}_1 \cap \overline{W}_2)| \leq t_1 + q + t_2 \leq s_2 + q + t_2 = |N(W_2)|.$$

Since W_2 is a fragment, equality must hold throughout D5, which implies that $\overline{W}_1 \cap \overline{W}_2$ is also a fragment. By Exercise D1(f,a), $W_1 \cup W_2$ is a fragment. Equality in D5 implies equality in D4 and hence in D3. Since W_1 is a fragment, so is $W_1 \cap W_2$. □

Let us define $\alpha(\Gamma) = \min\{|W| : W \text{ is a fragment of } \Gamma\}$. A fragment W of Γ is called an **atom** if $|W| = \alpha(\Gamma)$.

D6 Corollary. *If A is an atom and W is a fragment of Γ, then exactly one of the following holds: $A \subseteq W$; $A \subseteq \overline{W}$; or $A \subseteq N(W)$.*

Proof. Suppose $A \nsubseteq N(W)$. Then $A \cap P \neq \varnothing$, where $P = W$ or $P = \overline{W}$. By D2, $A \cap P$ is a fragment, but by definition of an atom, $A = A \cap P$, whence $A \subseteq P$. □

In particular,

D7 Corollary. *In any graph, distinct atoms are disjoint.*

D8 Corollary. *If A is an atom and U is a union of atoms of Γ, then exactly one of the following holds: $A \subseteq U$, $A \subseteq \overline{U}$, or $A \subseteq N(U)$.*

Proof. Suppose $A \cap N(U) \neq \varnothing$. Then there exists an atom $B \subseteq U$ such that some vertex of A is incident with some vertex of B. By D6, $A \subseteq N(B) \subseteq U + N(U)$. Since U is a union of atoms, D7 implies $A \subseteq N(U)$. □

D9 *Exercise.* Show that if the condition $|W_1| < |\overline{W}_2|$ is removed from the hypothesis of D2, then one can obtain that either the conclusion of D2 holds or that $W_1 \cap \overline{W}_2$ and $\overline{W}_1 \cap W_2$ are fragments if they are not empty.

D10 Proposition. *Let W_1 and W_2 be fragments of Γ such that $W_1 \subseteq N(W_2)$. Then $2\alpha \leq \kappa$.*

Proof. Recall the symbols q, s_i, and t_i from the proof of D2. Thus $\alpha \leq |W_1| \leq s_2$.

We assert that $\alpha \leq s_1$. For if this were not so and if $\overline{W}_1 \cap W_2 = \varnothing$, then $W_2 = N(W_1) \cap W_2$, and so $|W_2| = s_1 < \alpha$, which is impossible. On the other hand, if $\overline{W}_1 \cap W_2 \neq \varnothing$, then by D1(b,d),

$$\begin{aligned} |N(\overline{W}_1 \cap W_2)| &\leq t_2 + q + s_1 = (|N(W_2)| - s_2) + s_1 \\ &< (\kappa - \alpha) + \alpha = \kappa, \end{aligned}$$

which is also impossible.

Hence $\alpha \leq s_1$. With $\overline{W}_2$ in place of W_2, the same argument yields $\alpha \leq t_1$. Hence $2\alpha \leq s_1 + t_1 \leq |N(W_1)| = \kappa$. □

If each vertex of Γ is an element of some atom of Γ, then by D7, the atoms of Γ form a unique partition of V, called the **atomic partition**. When it exists, the atomic partition always has more than a single cell. Graphs admitting the atomic partition must therefore contain an edge meeting each of two atoms, each of which (by D6) lies in the image under N of the other. By D10 it follows that $2\alpha \leq \kappa$.

On the other hand, let $x \in A$, where A is an atom of Γ. Then x is incident with at most $\alpha - 1$ vertices in A and at most $|N(A)|$ vertices in $V + A$. Hence $\rho(x) \leq \kappa + \alpha - 1$. Since $\kappa \leq \check{\rho}$ (by C4), we have shown

D11 Proposition. *If Γ admits the atomic partition, then*

$$2\alpha \leq \rho(x) \leq \kappa + \alpha - 1, \quad \textit{for all } x \in V.$$

D12 *Exercise.* Show that the following are equivalent for any connected graph Γ:

(a) $\alpha(\Gamma) \geq 2$;
(b) $\kappa(\Gamma) < \check{\rho}(\Gamma)$;
(c) $N(\{x\}) \notin \mathscr{S}_0(\Gamma)$ for all $x \in V$.

D13 Proposition. *Let $\kappa(\Gamma)$ be represented as $\kappa = m\alpha + b$ where $m \geq 1$ and $0 \leq b < \alpha$. Then for any vertex x in an atom,*

$$\rho(x) \leq \frac{m+1}{m}\kappa - 1 - \frac{b}{m}.$$

Moreover, if Γ admits the atomic partition, then $b = 0$ and $m \geq 2$.

PROOF. If $x \in A$ for some atom A, then by D11,

$$\rho(x) \leq (\alpha - 1) + \kappa = \frac{\kappa - b}{m} - 1 + \kappa = \frac{m+1}{m}\kappa - 1 - \frac{b}{m}.$$

If Γ admits the atomic partition, then by D6, $N(A)$ is a union of atoms. Hence α divides κ, whence $b = 0$. By D11, $\kappa \geq \alpha + 1 > \alpha$. Hence $m \geq 2$. □

D14 Corollary. *Let Γ be a graph which admits the atomic partition. Then $\hat{\rho}/\kappa < \frac{3}{2}$, and the bound of $\frac{3}{2}$ is best possible but is never attained.*

PROOF. By D13, $\hat{\rho}/\kappa \leq 1 + 1/m - 1/\kappa < \frac{3}{2}$. To show that $\frac{3}{2} - \hat{\rho}/\kappa$ can be an arbitrarily small positive number, consider the graph Γ constructed as follows.

Let $V_0, V_1, \ldots, V_{p-1}$ be p disjoint n-sets where $p \geq 4$ and $n \geq 1$. Let $V = V_0 \cup \ldots \cup V_{p-1}$. Let

$$\mathscr{E} = \{\{x, y\} \in \mathscr{P}_2(V) : x \in V_i;\, y \in V_j;\, j - i \equiv 0, \pm 1 \pmod{p}\}.$$

Let $\Gamma = (V, \mathscr{E})$. Then Γ is ρ-valent with $\rho = 3n - 1$. The elements of $\mathscr{S}_0(\Gamma)$ are of the form $V_i + V_j$ where $j - i \not\equiv 0, \pm 1 \pmod{p}$. Hence $\kappa = 2n$ and the atoms are the sets V_i $(i = 0, \ldots, p - 1)$. Thus $\frac{3}{2} - \rho/\kappa = 1/2n$, and n may be chosen arbitrarily large. □

Clearly every graph which is vertex-transitive admits the atomic partition. Such a graph is, moreover, ρ-valent, and so the last two propositions and corollary apply in particular to vertex-transitive graphs.

D15 *Exercise.* Show that in a vertex-transitive graph any two atoms induce isomorphic subgraphs which are themselves vertex-transitive graphs. If you are acquainted with the rudiments of permutation groups, show further that the atomic partition of a vertex-transitive graph is a complete system of imprimitivity for $G_0(\Gamma)$.

D16 *Exercise.* Suppose that Γ admits the atomic partition and that $\kappa(\Gamma) < \check{\rho}(\Gamma)$.

(a) Determine the graph Γ such that $\nu_0(\Gamma)$ is minimal, and show that it is unique.

(b) Show that $\kappa(\Gamma)$ cannot be prime and that $\check{\rho}(\Gamma)$ cannot equal 4 or 6.

(c) Show that if $\check{\rho} = 8$, then Γ is a graph of the form in the proof of D14 with $n = 3$.

(d) Show that if $\check{\rho} = 7$, then $\kappa = 6$ and either $\alpha = 2$ or $\alpha = 3$; in the latter case the atoms induce paths of length 2. Construct examples in both cases.

D17 *Exercise.* Show that if Γ admits the atomic partition, then $\hat{\rho} - \check{\rho} < \alpha$.

The results in the section were brought together from the following papers: [j.1], [j.2], [j.3], [m.2], [m.3], and [w.4].

VIE Tutte Connectivity and Connectivity of Subspaces

Throughout this section, Γ will denote the multigraph (V, f, E). We extend to multigraphs the definition of connectivity defined for graphs in §C. Thus

$$\kappa(\Gamma) = \begin{cases} \kappa(V, f[E]) & \text{if } f[E] \subset \mathscr{P}_2(V); \\ \nu_0(\Gamma) - 2 + \min\{|f^{-1}[S]| : S \in \mathscr{P}_2(V)\} & \text{if } f[E] = \mathscr{P}_2(V). \end{cases}$$

This extension is designed to conform with Whitney's theorem (C7), for one can easily see that this theorem is now valid for multigraphs.

The symbol B_n for $n \in \mathbb{N} + \{0\}$ is the multigraph with $\nu_0(\mathrm{B}_n) = 2$ and $\nu_1(\mathrm{B}_n) = n$. Clearly, $\kappa(\mathrm{B}_n) = n$.

In this section we study a different concept of connectivity due to W. T. Tutte [t.7].

Let $\Theta = (U, g, F)$ be a submultigraph of Γ. Let $A(\Theta)$ denote the set of attachment vertices of Θ. (The extension to multigraphs of this term defined in §C is obvious: an attachment vertex of Θ is a vertex incident with an edge in F and an edge in $E + F$.) Γ is k-**separated** if there exists a proper nonempty submultigraph $\Theta = (U, g, F)$ of Γ such that

(a) $k \leq \nu_1(\Theta) \leq \nu_1(\Gamma) - k$;

(b) $|A(\Theta)| = k$.

E1 Exercise. *Assume that* $\Theta = (U, f, F)$ *satisfies* (a) *in the above definition.* Show that

(a) *Both* Γ_F *and* Γ_{E+F} *satisfy* (a) *above*;

(b) $A(\Gamma_F) = A(\Gamma_{E+F}) = A(\Theta)$.

The **Tutte connectivity** of Γ is defined by

$$\tau(\Gamma) = \min\{k\colon \Gamma \text{ is } k\text{-separated}\}$$

(where the minimum over the empty set is ∞). The multigraph Γ is m-**Tutte connected** if $m \leq \tau(\Gamma)$.

Observe that both $\tau(\Gamma)$ and $\kappa(\Gamma)$ represent cardinalities of smallest sets of attachment vertices of submultigraphs of Γ. The fine difference lies in the collection of submultigraphs considered. For $\tau(\Gamma)$, one considers submultigraphs which both include and exclude at least as many edges as they have attachment vertices. For $\kappa(\Gamma)$, except when every two vertices are incident, one considers submultigraphs having at least one interior and one exterior vertex.

E2 Exercise. Show that $\tau(\Gamma) = \infty$ *for* $\Gamma = \mathrm{K}_1, \mathrm{K}_2, \mathrm{K}_3, \mathrm{B}_2$ *and* B_3.

E3 Exercise. *Suppose* $\Gamma \neq \mathrm{K}_1, \mathrm{K}_2, \mathrm{K}_3, \mathrm{B}_2, \mathrm{B}_3$. *If* $\kappa(\Gamma)$ *or* $\tau(\Gamma)$ *is* 0 *or* 1, *then* $\kappa(\Gamma) = \tau(\Gamma)$.

E4 Proposition. *If* Γ *is a multigraph in which every two vertices are incident, then*

$$\tau(\Gamma) = \begin{cases} \infty & \textit{if } \Gamma = \mathrm{K}_1, \mathrm{K}_2, \mathrm{K}_3, \mathrm{B}_2, \mathrm{B}_3; \\ 3 & \textit{if } \Gamma = \mathrm{K}_n \textit{ for } n \geq 4; \\ 2 & \textit{otherwise.} \end{cases}$$

Proof. If Γ satisfies the hypothesis and $\nu_1(\Gamma) < 4$, then Γ is one of $\mathrm{K}_1, \mathrm{K}_2, \mathrm{K}_3, \mathrm{B}_2, \mathrm{B}_3$ and we use Exercise E2. Hence we assume $\nu_1(\Gamma) \geq 4$. Since $\kappa(\Gamma) > 1$, we have $\tau(\Gamma) > 1$ by E3. Let $\{e_1, e_2\} \in \mathscr{P}_2(E)$. Let $\Theta = \Gamma_{\{e_1, e_2\}}$. Then $f(e_1) \cup f(e_2) = A(\Theta)$, and $|A(\Theta)| \geq 2$. In fact, $|A(\Theta)| = 2$ if and only if $f(e_1) = f(e_2)$. Hence $\tau(\Gamma) = 2$ if Γ is not a graph, and $\tau(\Gamma) > 2$ if Γ is a graph. Suppose then Γ is a graph, i.e., $\Gamma = \mathrm{K}_n$ for $n \geq 4$. Then $\nu_1(\Gamma) \geq 6$. If $\Theta = \Gamma_U$ where $U \in \mathscr{P}_3(V)$, then $3 = \nu_1(\Theta) \leq \nu_1(\Gamma) - 3$, and $|A(\Theta)| = 3$. □

E5 Lemma. *If* $\Gamma \neq \mathrm{K}_1, \mathrm{K}_2, \mathrm{K}_3, \mathrm{B}_2, \mathrm{B}_3$, *then* $\tau(\Gamma) \leq \kappa(\Gamma)$.

PROOF. By E4, we may assume that some two vertices are not incident. By Exercise C16, there exists a submultigraph Θ of Γ with an interior vertex a, an exterior vertex z, and $|A(\Theta)| = \kappa(\Gamma)$. By C4, $\kappa(\Gamma) \leq \rho(a) \leq \nu_1(\Theta)$, and $\kappa(\Gamma) \leq \rho(z) \leq \nu_1(\Gamma) - \nu_1(\Theta)$. So $\kappa(\Gamma) \leq \nu_1(\Theta) \leq \nu_1(\Gamma) - \kappa(\Gamma)$. Hence $\tau(\Gamma) \leq |A(\Theta)| = \kappa(\Gamma)$. □

The **girth** of Γ is the cardinality of a smallest nonempty cycle of Γ and is denoted by $\gamma(\Gamma)$. We adopt the convention that $\gamma(\Gamma) = \infty$ if Γ is a forest. Clearly $\gamma(\Gamma) \geq 2$ for any multigraph Γ, and $\gamma(\Gamma) \geq 3$ if and only if Γ is a graph.

E6 Lemma. *If* $\Gamma \neq K_1, K_2, K_3, B_2, B_3$, *then* $\tau(\Gamma) \leq \gamma(\Gamma)$.

PROOF. If Γ contains a vertex of valence 2 or less, and if Γ is not one of the "forbidden" multigraphs, then by E5, $\tau(\Gamma) \leq \kappa(\Gamma) \leq 2 \leq \gamma(\Gamma)$. Hence assume that the smallest valence is at least 3. We may also assume $\nu_1(\Gamma) \geq 4$, since the only multigraph with smallest valence 3 and $\nu_1(\Gamma) < 4$ is B_3. By Exercise *III*A6a, Γ contains a nonempty cycle. Let $Z \in \mathscr{Z}(\Gamma)$ such that $|Z| = \gamma(\Gamma)$. The circuit Γ_Z is clearly an elementary circuit, and hence $\nu_0(\Gamma_Z) = \nu_1(\Gamma_Z) = \gamma(\Gamma)$. Since each vertex of Γ_Z has valence at least 3 in Γ, each vertex of Γ_Z is incident with an edge in $E + Z$. If $|Z| \geq 3$, these edges are obviously distinct. If $|Z| = 2$, then $|E + Z| \geq 2$. Thus $|E + Z| \geq |Z| = \gamma(\Gamma)$. It also follows that each vertex of Γ_Z is a vertex of attachment, and so $|A(\Gamma_Z)| = |Z|$. Thus Γ is $\gamma(\Gamma)$-separated. □

E7 Proposition. *If* $\Gamma \neq K_1, K_2, K_3, B_2, B_3$, *then* $\tau(\Gamma) = \min\{\kappa(\Gamma), \gamma(\Gamma)\}$.

PROOF. If $\Gamma = K_n$ or B_n for $n \geq 4$ the conclusion holds by Proposition E4. Assume then that $\Gamma \neq K_n$, $\Gamma \neq B_n$ for $n \geq 1$, and hence by E5, $\tau(\Gamma)$ is finite. Let $\Theta = (U, g, F)$ be a subgraph of Γ such that $\tau(\Gamma) \leq \nu_1(\Theta) \leq \nu_1(\Gamma) - \tau(\Gamma)$ and $|A(\Theta)| = \tau(\Gamma)$. From Exercise E1 and the definition of $\tau(\Gamma)$ we have $\tau(\Gamma) \leq \nu_1(\Gamma_{E+F}) \leq \nu_1(\Gamma) - \tau(\Gamma)$ and $|A(\Gamma_{E+F})| = \tau(\Gamma)$.

If $U = A(\Theta)$, then $\nu_0(\Theta) = \tau(\Gamma) \leq \nu_1(\Theta)$, and Θ contains an elementary cycle Z by *III*A15b. But then $\gamma(\Gamma) \leq |Z| \leq \nu_0(\Theta) = \tau(\Gamma)$, and by E6, $\gamma(\Gamma) = \tau(\Gamma)$. By a symmetric argument, $\tau(\Gamma) = \gamma(\Gamma)$ also holds if the vertex set of Γ_{E+F} equals $A(\Theta)$.

Assume then in the light of E6 that $\tau(\Gamma) < \gamma(\Gamma)$. By the argument just concluded, Θ admits both an interior and an exterior vertex. By C16, $|A(\Theta)| \geq \kappa(\Gamma)$. By E5, $\tau(\Gamma) = \kappa(\Gamma)$. □

Example. Let Γ_1 be the cube (Figure *III*F21), let Γ_2 denote the octahedron (Figure *III*F20), and let Γ_3 be the graph in Figure E8. Then we have the following values.

	κ	γ	τ
Γ_1	3	4	3
Γ_2	4	3	3
Γ_3	4	4	4

E8

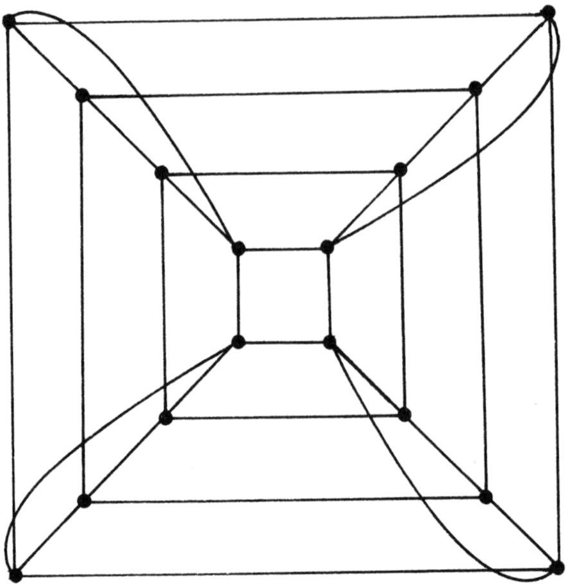

E9 Corollary. *If Γ is a graph, then Γ is 3-connected if and only if Γ is 3-Tutte connected.*

We have observed (Exercise E1) that for any submultigraph $\Theta = (U, g, F)$ of Γ, $A(\Theta) \supseteq A(\Gamma_F)$, and equality holds if $|A(\Theta)| = \tau(\Gamma)$. Hence if $\tau(\Gamma) \leq \nu_1(\Theta) \leq \nu_1(\Gamma) - \tau(\Gamma)$ and $\tau(\Gamma) = |A(\Theta)|$, then $\tau(\Gamma) \leq \nu_1(\Gamma_F) \leq \nu_1(\Gamma) - \tau(\Gamma)$ and $\tau(\Gamma) = |A(\Gamma_F)|$. We note that Θ consists only of isolated vertices if and only if $F = \varnothing$. We have shown that if Γ has no isolated vertices, then $\tau(\Gamma)$ can be determined by considering only those subgraphs of Γ which are induced by a nonempty proper subset of E. We state this formally as follows.

E10 Lemma. *Suppose Γ contains no isolated vertex. Then Γ is k-separated if and only if there exists $F \subset E$, $F \neq \varnothing$, such that*
(a) $k \leq \min\{|F|, |E + F|\}$;
(b) $k = |A(\Gamma_F)| = |A(\Gamma_{E+F})|$.

We observe that

$$|A(\Gamma_F)| = \nu_0(\Gamma) - |\{x : f^*(x) \subseteq F\}| - |\{x : f^*(x) \subseteq E + F\}|.$$

Hence by *III*A15a, we have the following result.

E11 Lemma. *If Γ is connected and $\varnothing \subset F \subset E$, then*

$$|A(\Gamma_F)| = \dim(\mathscr{Z}^\perp(\Gamma)) + 1 - \dim(\mathscr{Z}^\perp(\Gamma) \cap \mathscr{P}(F)) - \dim(\mathscr{Z}^\perp(\Gamma) \cap \mathscr{P}(E + F)).$$

Recall that Γ is biconnected if and only if $\mathscr{Z}^{\perp}(\Gamma)$ is connected. We introduce a general concept of connectivity of a subspace in such a way that for any multigraph Γ, the connectivity of $\mathscr{Z}^{\perp}(\Gamma)$ is less by 1 than $\tau(\Gamma)$. A subspace $\mathscr{A} \subseteq \mathscr{P}(E)$ is **m-separated** if there exists $\{F_1, F_2\} \in \mathbb{P}_2(E)$ such that

(a) $m < \min\{|F_1|, |F_2|\}$;
(b) $m = \dim(\mathscr{A}) - \dim(\mathscr{A} \cap \mathscr{P}(F_1)) - \dim(\mathscr{A} \cap \mathscr{P}(F_2))$.

The **connectivity of the subspace** $\mathscr{A}$ is defined analogously as $\tau(\mathscr{A}) = \min\{m: \mathscr{A}$ is m-separated$\}$.

E12 Proposition. *If Γ is a connected multigraph, then $\tau(\Gamma) = \tau(\mathscr{Z}^{\perp}(\Gamma)) + 1$.*

PROOF. Let Γ be connected. The result is trivial if $\nu_0(\Gamma) = 1$. By Lemmas E10 and E11, Γ is k-separated if and only if there exists $F \subset E$, $F \neq \varnothing$, such that

(a) $k \leq \min\{|F|, |E + F|\}$;
(b) $k = \dim(\mathscr{Z}^{\perp}(\Gamma)) + 1 - \dim(\mathscr{Z}^{\perp}(\Gamma) \cap \mathscr{P}(F))$
$\quad - \dim(\mathscr{Z}^{\perp}(\Gamma) \cap \mathscr{P}(E + F))$.

Thus Γ is k-separated if and only if there exists $\{F_1, F_2\} \in \mathbb{P}_2(E)$ such that

(c) $k - 1 < \min\{|F_1|, |F_2|\}$;
(d) $k - 1 = \dim(\mathscr{Z}^{\perp}(\Gamma)) - \dim(\mathscr{Z}^{\perp}(\Gamma) \cap \mathscr{P}(F_1)) - \dim(\mathscr{Z}^{\perp}(\Gamma) \cap \mathscr{P}(F_2))$.

Therefore, Γ is k-separated if and only if $\mathscr{Z}^{\perp}(\Gamma)$ is a $(k - 1)$-separated subspace of $\mathscr{P}(E)$. □

E13 Exercise. Prove that *a subspace $\mathscr{A} \subseteq \mathscr{P}(E)$ is a connected subspace if and only if $\tau(\mathscr{A}) \geq 1$.*

E14 Proposition. *If $\mathscr{A}$ is a subspace of $\mathscr{P}(E)$, then $\tau(\mathscr{A}) = \tau(\mathscr{A}^{\perp})$.*

PROOF. Let $\{F_1, F_2\} \in \mathbb{P}_2(E)$. From *II*A6, *II*A10, and *II*A15 we have for $\{i, j\} = \{1, 2\}$,

$$\dim(\mathscr{A} \cap \mathscr{P}(F_i)) = |F_i| - \dim(\pi_{F_i}[\mathscr{A}^{\perp}])$$

and

$$\dim(\pi_{F_i}[\mathscr{A}^{\perp}]) = \dim(\mathscr{A}^{\perp}) - \dim(\mathscr{A}^{\perp} \cap \mathscr{P}(F_j)).$$

Thus,

$$\begin{aligned}
\dim(\mathscr{A}) &- \dim(\mathscr{A} \cap \mathscr{P}(F_1)) - \dim(\mathscr{A} \cap \mathscr{P}(F_2)) \\
&= \dim(\mathscr{A}) - [|F_1| - \dim(\pi_{F_1}[\mathscr{A}^{\perp}])] - [|F_2| - \dim(\pi_{F_2}[\mathscr{A}^{\perp}])] \\
&= \dim(\mathscr{A}) - |E| + \dim(\pi_{F_1}[\mathscr{A}^{\perp}]) + \dim(\pi_{F_2}[\mathscr{A}^{\perp}]) \\
&= -\dim(\mathscr{A}^{\perp}) + [\dim(\mathscr{A}^{\perp}) - \dim(\mathscr{A}^{\perp} \cap \mathscr{P}(F_2))] \\
&\quad + [\dim(\mathscr{A}^{\perp}) - \dim(\mathscr{A}^{\perp} \cap \mathscr{P}(F_1))] \\
&= \dim(\mathscr{A}^{\perp}) - \dim(\mathscr{A}^{\perp} \cap \mathscr{P}(F_1)) - \dim(\mathscr{A}^{\perp} \cap \mathscr{P}(F_2)).
\end{aligned}$$

Therefore $\mathscr{A}$ and $\mathscr{A}^{\perp}$ are m-separated for the same values of m. □

From the definition of orthogonal multigraphs we have

E15 Corollary. *If* Γ *and* Θ *are connected multigraphs with* $\Gamma \perp \Theta$, *then* $\tau(\Gamma) = \tau(\Theta)$.

Combining this and Corollary E9, we obtain the result anticipated in Exercise *III*E20.

E16 Corollary (Whitney [w.9]). *If* $\Gamma \perp \Theta$ *and each are without isolated vertices, then* Γ *is* 3*-connected if and only if* Θ *is* 3*-connected.*

E17 *Exercise.* Prove that if Γ is a planar multigraph, then $\tau(\Gamma) \leq 3$. [*Hint*: use *III*F4.]

Using the results from §*II*A summarized in the proof of E14, one easily proves:

E18 Lemma. *Let* $\mathscr{A} \subseteq \mathscr{P}(E)$ *be a subspace and let* $\{F_1, F_2\} \in \mathbb{P}_2(E)$. *Then*

$$\begin{aligned}\dim(\mathscr{A}) - \dim(\mathscr{A} \cap \mathscr{P}(F_1)) - \dim(\mathscr{A} \cap \mathscr{P}(F_2))\\ = |F_i| - \dim(\mathscr{A} \cap \mathscr{P}(F_i)) - \dim(\mathscr{A}^{\perp} \cap \mathscr{P}(F_i))\end{aligned}$$

for $i = 1, 2$.

E19 Lemma. $\tau(\mathscr{A}) < \infty$ *if and only if* $\mathscr{A} \cup \mathscr{A}^{\perp}$ *contains a pair of nonempty disjoint elements.*

PROOF. If $\mathscr{A}$ is m-separated, there exists $\{F_1, F_2\} \in \mathbb{P}_2(E)$ such that

$$\dim(\mathscr{A}) - \dim(\mathscr{A}^{\perp} \cap \mathscr{P}(F_1)) - \dim(\mathscr{A} \cap \mathscr{P}(F_2)) < \min\{|F_1|, |F_2|\}.$$

Combining this inequality with Lemma E18, we have for $i = 1, 2$, $\dim(\mathscr{A} \cap \mathscr{P}(F_i) + \dim(\mathscr{A}^{\perp} \cap \mathscr{P}(F_i)) > 0$. Hence for $i = 1, 2$, there exists $A_i \in \mathscr{A} \cup \mathscr{A}^{\perp}$, such that $A_i \neq \varnothing$ and $A_i \subseteq F_i$. Clearly $A_1 \cap A_2 = \varnothing$.

Conversely, assume that for $i = 1, 2$, there exists $A_i \in \mathscr{A} \cup \mathscr{A}^{\perp}$ such that $A_i \neq \varnothing$ and $A_1 \cap A_2 = \varnothing$. Let $F_1 = A_1$ and $F_2 = E + F_1$. One easily verifies that $\mathscr{A}$ is m-separated for $m = \dim(\mathscr{A}) - \dim(\mathscr{A} \cap \mathscr{P}(F_1)) - \dim(\mathscr{A} \cap \mathscr{P}(F_2))$. □

E20 Proposition. $\tau(\mathscr{A}) = \infty$ *if and only if* $|E| \leq 3$ *and either* $\mathscr{A}$ *or* $\mathscr{A}^{\perp} = \{\varnothing, E\}$.

PROOF. We first dispose of the case where $\mathscr{A} = \{\varnothing, S\}$ for some $S \subseteq E$, and so $\mathscr{A}^{\perp} = \mathscr{E}(S) \oplus \mathscr{P}(E + S)$. If $\varnothing \subset S \subset E$, or if $S = \varnothing$ and $|E| > 1$, then $\mathscr{A}^{\perp}$ is not connected, and $\tau(\mathscr{A}) = \tau(\mathscr{A}^{\perp}) = 0$ by E14 and E13. If $S = E$ and $|E| > 3$, then $\tau(\mathscr{A}) < \infty$ by E19. Conversely, suppose $\mathscr{A} = \{\varnothing, E\}$ and $|E| \leq 3$. Since $\mathscr{A}^{\perp} = \mathscr{E}(E)$, Lemma E19 implies that $\tau(\mathscr{A}) = \infty$.

We may now assume that $\dim(\mathscr{A}) \geq 2$, and by symmetry, $\dim(\mathscr{A}) \leq \dim(\mathscr{A}^{\perp})$. We assume $\tau(\mathscr{A}) = \infty$. Let F be an elementary set in $\mathscr{A}$. Then $\dim(\mathscr{A} \cap \mathscr{P}(F)) = 1$. It follows that $\pi_F[\mathscr{A}^{\perp}]$, the orthogonal complement of

$\mathscr{A} \cap \mathscr{P}(F)$ in $\mathscr{P}(F)$, is $\mathscr{E}(F)$. On the other hand, $\dim(\mathscr{A}^\perp) = \dim(\pi_F[\mathscr{A}^\perp]) + \dim(\mathscr{A}^\perp \cap \mathscr{P}(E + F))$. But by Lemma E19, $\dim(\mathscr{A}^\perp \cap \mathscr{P}(E + F)) = 0$. Hence $\dim(\mathscr{A}^\perp) = \dim(\pi_F[\mathscr{A}^\perp]) = \dim(\mathscr{E}(F)) = |F| - 1$ by *II*A1. We conclude that every elementary set $F \in \mathscr{A}$ has cardinality $\dim(\mathscr{A}^\perp) + 1$, and by symmetry, every elementary set in $\mathscr{A}^\perp$ has cardinality $\dim(\mathscr{A}) + 1$.

Now let F be an elementary set in $\mathscr{A}$. Since $\dim(\mathscr{A}^\perp) \geq 2$, we have $|F| \geq 3$, and we may choose $\{x_0, x_1, x_2\} \in \mathscr{P}_3(F)$. We may also choose $H_i \in \mathscr{A}^\perp$ such that $\pi_F[H_i] = H_i \cap F = \{x_0, x_i\}$ for $i = 1, 2$. Since $|H_i| \geq \dim(\mathscr{A}) + 1$, we have $|F \cup H_i| = |F| + |H_i| - 2 \geq \dim(\mathscr{A}^\perp) + \dim(\mathscr{A}) = |E|$. So $F \cup H_i = E$. Hence $\{x_1, x_2\} = H_1 + H_2 \in \mathscr{A}^\perp$, and so $2 = |\{x_1, x_2\}| \geq \dim(\mathscr{A}) + 1$, contrary to our assumption. □

If $\mathscr{A} = \mathscr{Z}(\Gamma)$ then $\gamma(\Gamma) = \min\{|F| : F \in \mathscr{A} + \{\varnothing\}\}$. Hence it is natural to define the **girth** of a subspace $\mathscr{A}$ as

$$\gamma(\mathscr{A}) = \min\{|F| : F \in \mathscr{A} + \{\varnothing\}\}.$$

E21 *Exercise.* Prove that $\tau(\mathscr{A}) \leq \min\{\gamma(\mathscr{A}), \gamma(\mathscr{A}^\perp)\}$.

CHAPTER VII

Chromatic Theory of Graphs

VIIA Basic Concepts and Critical Graphs

Throughout this section Γ will denote the graph $(V, \mathscr{E})$. A **vertex m-coloring** of Γ is a surjection h of V onto an m-set C, subject to the condition: *If $\{x_1, x_2\} \in \mathscr{P}_2(V)$ and $h(x_1) = h(x_2)$, then x_1 and x_2 are not incident.* The elements of C are called **colors** and the sets $h^{-1}[j]$ for $j \in C$ are called **color classes**. If $x \in V$ and $h(x) = j$, one also says, "x has been colored j." We say that Γ is **vertex m-colorable** if Γ admits a vertex m-coloring. The **vertex chromatic number** of Γ is

$$\chi_0(\Gamma) = \min\{m\colon \Gamma \text{ has a vertex } m\text{-coloring}\}.$$

One says that Γ is **vertex m-chromatic** if $\chi_0(\Gamma) = m$. Clearly Γ has a vertex m-coloring if and only if $\chi_0(\Gamma) \leq m \leq \nu_0(\Gamma)$, and equality holds throughout if and only if Γ is a complete graph.

In dual fashion we define an **edge m-coloring** of Γ to be a surjection h of $\mathscr{E}$ onto an m-set C subject to the condition: *If $\{E_1, E_2\} \in \mathscr{P}_2(\Gamma)$ and $h(E_1) = h(E_2)$, then E_1 and E_2 are not incident*, or equivalently, that $E_1 \cap E_2 = \varnothing$. The terms **color** and **color class** are defined analogously for edge colorings. The **edge chromatic number** of Γ is defined by

$$\chi_1(\Gamma) = \min\{m\colon \Gamma \text{ has an edge } m\text{-coloring}\}.$$

Γ has an edge m-coloring if and only if $\chi_1(\Gamma) \leq m \leq \nu_1(\Gamma)$. Equality holds throughout if and only if Γ is either K_3 or a tree of diameter 2.

The following assertions are in the nature of observations and are easily verified. Here $i = 0$ or 1.

A1 *If Θ is a subgraph of Γ, then $\chi_i(\Theta) \leq \chi_i(\Gamma)$.*

A2 $$\chi_i(\Gamma) = \max\{\chi_i(\Theta)\colon \Theta \textit{ is a component of } \Gamma\}.$$

A3
$$\chi_i(\Delta_n) = \begin{cases} 2 & \textit{if } n \textit{ is even;} \\ 3 & \textit{if } n \textit{ is odd.} \end{cases}$$

A4 *A color class, be it of vertices or of edges, is always an independent set.*

A5
$$\alpha_i(\Gamma)\chi_i(\Gamma) \geq \nu_i(\Gamma).$$

A6
$$\chi_0(\Gamma) \leq 2 \textit{ if and only if } \Gamma \textit{ is bipartite.}$$

A7 Exercise. Prove: *if* $S \subseteq V$, *then* $0 \leq \chi_0(\Gamma) - \chi_0(\Gamma_{(S)}) \leq |S|$; *if* $\mathscr{S} \subseteq \mathscr{E}$ *then* $0 \leq \chi_1(\Gamma) - \chi_1(\Gamma_{(\mathscr{S})}) \leq |\mathscr{S}|$.

For $m \geq 2$, we say that Γ is m-**critical** if $\chi_0(\Gamma) = m$ but $\chi_0(\Theta) < m$ for all proper subgraphs Θ of Γ. We call Γ a **critical graph** if Γ is m-critical for some $m \geq 2$. The complete graph K_m is clearly m-critical, and K_2 is the only 2-critical graph.

A8 *Exercise.* Show that the odd circuits are the only 3-critical graphs.

At this point we begin a succession of results relating the chromatic number of a graph to the graph structure. The first of these is due to G. A. Dirac [d.3].

A9 Lemma. *For any graph* Γ, *either* $K_{\chi_0(\Gamma)}$ *is a subgraph of* Γ *or* $\chi_0(\Gamma) \leq \nu_0(\Gamma) - 2$.

PROOF. Let $\nu_0(\Gamma) = n$ and suppose K_{n-1} is not a subgraph of Γ. There exists $\{x, y\} \in \mathscr{P}_2(V) + \mathscr{E}$. Let $U = V + \{x, y\}$. If Γ_U is not complete, then $\chi_0(\Gamma_U) \leq n - 3$, and any vertex $(n - 3)$-coloring of Γ_U can be extended to a vertex $(n - 2)$-coloring of Γ. Assume, therefore, that $\Gamma_U = K_{n-2}$. Since K_{n-1} is not a subgraph of Γ, there exist $v, w \in U$ such that $\{x, v\} \notin \mathscr{E}$ and $\{y, w\} \notin \mathscr{E}$. Hence any vertex $(n - 2)$-coloring h of Γ_U can be extended to a vertex $(n - 2)$-coloring of Γ by assigning $h(x) = h(v)$ and $h(y) = h(w)$. □

A10 *Exercise.* Show that for any graph Γ, either $K_{\chi_0(\Gamma)}$ is a subgraph of Γ or $\chi_0(\Gamma) \leq \nu_0(\Gamma) - \alpha_0(\Gamma)$, thus proving Lemma A9.

A11 Lemma. *Let* Γ *be a vertex m-chromatic. Let* $x_0 \in V$ *and suppose* h_0 *is a vertex m-coloring of* Γ *with respect to which* $\{x_0\}$ *is a color class. Then*

(a) x_0 *is incident with at least one vertex of each color except* $h_0(x_0)$. *Hence* $\rho(x_0) \geq m - 1$.

(b) *If* $\rho(x_0) = m - 1$ *and if* $\{x_0, x_1\} \in \mathscr{E}$, *then* x_1 *is incident with at least one vertex of every color except* $h_0(x_1)$. *In this case* $\rho(x_1) \geq m - 1$. *Moreover, if y and z are incident with* x_0, *then y and z belong to the same component of* $\Gamma_{h_0^{-1}[h_0[\{y,z\}]]}$.

(c) *If* $\rho(x_1) = m - 1$ *and if* $\{x_1, x_2\} \in \mathscr{E}$, *then* x_2 *is incident with at least one vertex of every color except* $h_0(x_2)$ *and perhaps* $h_0(x_0)$. *In this case* $\rho(x_2) \geq m - 1$.

PROOF. (a) For definiteness, suppose $h_0: V \to \{1, \ldots, m\}$ and that $h_0(x_0) = m$. If x_0 were incident with no vertex of color, say $j \in \{1, \ldots, m-1\}$, then one could define an $(m-1)$-coloring h of Γ by changing the color of x_0 from m to j, contrary to assumption. Hence $\rho(x_0) \geq m - 1$.

(b) If $\rho(x_0) = m - 1$, then clearly x_0 is incident with exactly one vertex of each of the colors $1, \ldots, m - 1$. Let x_1 be incident with x_0. For definiteness, suppose that $h_0(x_1) = 1$. If x_1 is incident with no vertex of some color $j \in \{2, \ldots, m-1\}$, then one could define a vertex $(m-1)$-coloring h of Γ by

$$h(x) = \begin{cases} h_0(x) & \text{if } x \in V + \{x_0, x_1\}; \\ 1 & \text{if } x = x_0; \\ j & \text{if } x = x_1, \end{cases}$$

contrary to assumption. Hence $\rho(x_1) \geq m - 1$.

Let y and z be given as in the statement. Let $(W, \mathscr{F})$ be the component of $\Gamma_{h_0^{-1}[h_0[\{y,z\}]]}$ containing y, but suppose $z \notin W$. Define a vertex $(m-1)$-coloring h of Γ by

$$h(x) = \begin{cases} h_0(x) & \text{if } x \notin W + \{x_0\}; \\ h_0(z) & \text{if } x \in W \cap h_0^{-1}[h_0(y)]; \\ h_0(y) & \text{if } x \in W \cap h_0^{-1}[h_0(z)] + \{x_0\}, \end{cases}$$

contrary to assumption.

(c) If $\rho(x_1) = m - 1$, then x_1 is clearly incident with exactly one vertex of each color $2, \ldots, m$. By an argument like that in the proof of (b), one deduces that every vertex x_2 incident with x_1 has valence at least $m - 2$, since x_2 must be incident with a vertex of each color except the colors $h_0(x_1)$ and possibly $h_0(x_0)$. If $\rho(x_2) = m - 2$, then x_2 is incident with exactly one vertex of each of the other $m - 2$ colors. Supposing for definiteness that $h_0(x_2) = 2$, one easily verifies that the following is an $(m-1)$-coloring of Γ:

$$h(x) = \begin{cases} h_0(x) & \text{if } x \in V + \{x_0, x_1, x_2\}; \\ 1 & \text{if } x = x_0 \text{ or } x_2; \\ 2 & \text{if } x = x_1, \end{cases}$$

contrary to assumption. □

The above arguments do not apply as well to a vertex x_3 incident with x_2 if one merely assumes that $\rho(x_2) = m - 1$. This is because unlike x_0 and x_1, the vertex x_2 need not be incident with representatives of $m - 1$ different color classes.

Components of the subgraphs induced by the union of two color-classes are referred to in the literature as **Kempe chains**. Observe that a new m-coloring can always be obtained when the two colors used in a Kempe chain are interchanged on that Kempe chain alone.

If Γ is m-critical, then given any vertex $x \in V$, there exists an m-coloring of Γ with respect to which $\{x\}$ is a color class. Hence $\rho(x) \geq m - 1$ for all $x \in V$. We have proved

A12 Corollary. *For a critical graph,* $\chi_0 \leq \check{\rho} + 1$.

The next corollary will be useful in the following section.

A13 Corollary. *Let m be a positive integer. If* $\check{\rho}(\Theta) \leq m - 1$ *for some* $\chi_0(\Gamma)$*-critical subgraph* Θ *of* Γ, *then* $\chi_0(\Gamma) \leq m$.

PROOF. Applying A12 to the subgraph Θ, we obtain $\chi_0(\Gamma) = \chi_0(\Theta) \leq \check{\rho}(\Theta) + 1 \leq m$. □

From A12 and *III*A1 we readily deduce

A14 Corollary. *For a critical graph,* $2\nu_1 \geq (\chi_0 - 1)\nu_0$.

We shall presently prove two inequalities stronger than Corollary A14 and shall obtain further results about valences of critical graphs. In order to do this, we shall require some information relating chromatic properties and connectedness.

A15 Proposition. *Let* Γ *be connected, let* $S \subseteq V$ *be a separating set of* Γ, *let* $W_1, \ldots, W_k$ *be the vertex sets of the components of* $\Gamma_{(S)}$, *and let* $\Theta_i = \Gamma_{W_i + S}$. *Suppose that for each* $i = 1, \ldots, k$, *there exists a* $\chi_0(\Theta_i)$*-coloring of* Θ_i *which is injective when restricted to S. Then*

$$\chi_0(\Gamma) = \max\{\chi_0(\Theta_i): i = 1, \ldots, k\}.$$

PROOF. Let $m = \max\{\chi_0(\Theta_i): i = 1, \ldots, k\}$. Let $S = \{x_1, \ldots, x_{|S|}\}$. For each $i = 1, \ldots, k$, let $h_i: W_i + S \to \{1, \ldots, \chi_0(\Theta_i)\}$ be a $\chi_0(\Theta_i)$-coloring. Without loss of generality we may impose that $h_i(x_j) = j$ for $j = 1, \ldots, |S|$ and $i = 1, \ldots, k$. Define $h: V \to \{1, \ldots, m\}$ by $h(x) = h_i(x)$ if $x \in W_i + S$. Then h is clearly an m-coloring, and by A1, $\chi_0(\Gamma) = m$. □

A16 Corollary. *Let* Γ *be critical, and let* $S \subseteq V$ *be a separating set. Then* Γ_S *is not complete.*

In particular, if $|S| = 1$, this corollary yields

A17 Corollary. *If* Γ *is critical, then either* $\Gamma = \mathrm{K}_2$ *or* Γ *is biconnected.*

A18 Corollary. $\chi_0(\Gamma) = \max\{\chi_0(\Theta): \Theta \text{ is a lobe of } \Gamma\}$.

A19 *Exercise.* Show that A17 cannot be sharpened; show *a fortiori* that for each $m \geq 2$, there exists an m-critical graph which is not triconnected.

We are now prepared to state a major result in chromatic graph theory, which will be seen to be considerably stronger than Corollary A14. It is due to G. A. Dirac [d.4]. The proof given here is due to H. V. Kronk and J. Mitchem [k.6]. It is much shorter than the original one.

A20 Theorem. *Every m-critical graph* ($m \geq 4$) *which is not complete satisfies*

$$2\nu_1 \geq (m - 1)\nu_0 + m - 3.$$

PROOF. Let an integer $m \geq 4$ be given. If the assertion is false, there exists an m-critical graph $\Gamma = (V, \mathscr{E})$ which is not K_m, and

A21 $$2\nu_1 \leq (m - 1)\nu_0 + m - 4.$$

Since Γ is m-critical, A21 together with Corollary A14 imply that the average valence ρ of Γ satisfies

$$m - 1 \leq \rho \leq m - 1 + (m - 4)/\nu_0.$$

Hence if we define $U = \{x \in V: \rho(x) \geq m\}$, then

A22 $$|U| \leq m - 4 < \nu_0.$$

By Corollary A12, $\rho(x) = m - 1$ for all $x \in V + U$. Arbitrarily choose $x_0 \in V + U$. Since $\Gamma_{(x_0)}$ is $(m - 1)$-colorable, Γ admits an m-coloring $h: V \to \{1, \ldots, m\}$ such that $h^{-1}[m] = \{x_0\}$. Lemma A11(a,b) applies. Let $x_1, \ldots, x_{m-1}$ denote all the vertices incident with x_0, where $h(x_i) = i$ for $i = 1, \ldots, m - 1$, and let $\Theta_{ij} = \Theta_{ji} = (W_{ij}, \mathscr{F}_{ij})$ denote the Kempe chain between x_i and x_j for $1 \leq i < j \leq m - 1$.

Since Γ is m-critical and not complete, Γ does not contain K_m as a subgraph. Hence there exist vertices x_a and x_b incident with x_0 but not incident with each other.

Let $n = m - 1 - |h[U]|$. From A22 one deduces

A23 $$3 \leq n \leq m - 1.$$

Assume for definiteness that $\{1, \ldots, n\}$ is the complement in $\{1, \ldots, m - 1\}$ of $h[U]$, and so $x_1, \ldots, x_n \in V + U$. By Lemma A11b, x_i is incident with exactly one vertex of each color except i, for $i = 1, \ldots, n$.

We next show:

A24 $$1 \leq i < j \leq n \Rightarrow \Theta_{ij} \text{ is an elementary } x_i x_j\text{-path.}$$

Observe that $W_{ij} \subseteq V + U$. Clearly within Θ_{ij} both x_i and x_j have valence 1. Proceeding from x_i along any $x_i x_j$-path in Θ_{ij}, let y be the first vertex encountered whose valence in Θ_{ij} exceeds 2. Then y is incident with at least three vertices of color j or i depending upon whether $h(y) = i$ or j, respectively. Since $\rho(y) = m - 1$, y is incident with no vertex of some color

$$r \in \{1, \ldots, m - 1\} + \{i, j\}.$$

If we define

A25 $$h'(x) = \begin{cases} h(x) & \text{if } x \in V + \{y\}, \\ r & \text{if } x = y, \end{cases}$$

then h' is an m-coloring of Γ with respect to which $\{x_0\}$ is a color class, and yet Θ_{ij} (with respect to h') contains no $x_i x_j$-path, contrary to Lemma A11b. This proves A24.

We assert

A26 $$\{i, j, k\} \in \mathscr{P}_3(\{1, \ldots, n\}) \Rightarrow W_{ij} \cap W_{ik} = \{x_i\}.$$

If $y \in W_{ij} \cap W_{ik}$, then $h(y) = i$. If $y \neq x_i$, then y is incident (by A24) with two vertices colored j and two vertices colored k. Since $\rho(y) = m - 1$, y is incident with no vertex of some color $r \in \{1, \ldots, m - 1\} + \{i, j, k\}$. Defining h' as in A25, we obtain an m-coloring with respect to which Lemma A11b is again contradicted, whence A26 holds.

We next demonstrate

A27 $$1 \leq i < j \leq n \Rightarrow \{x_i, x_j\} \in \mathscr{E}.$$

For definiteness, suppose $\{x_1, x_2\} \notin \mathscr{E}$ and let $z \in W_{1,2}$ be incident with x_1. Since $\Theta_{1,2}$ is an x_1x_2-path by A24, and since $n \geq 3$ by A23, the following function h' is an m-coloring of Γ with respect to which $\{x_0\}$ is a color class:

$$h'(x) = \begin{cases} h(x) & \text{if } x \in V + W_{1,3}; \\ 1 & \text{if } x \in W_{1,3} \cap h^{-1}[3]; \\ 3 & \text{if } x \in W_{1,3} \cap h^{-1}[1]. \end{cases}$$

In effect h' reverses the colors 1 and 3 in $\Theta_{1,3}$. However, with respect to h', the vertex z lies in $W_{1,2} \cap W_{2,3}$, contrary to A26, whence A27 holds.

We define a subset P of the set $\{1, \ldots, m - 1\}$ of colors:

$$P = \{i\colon 1 \leq i \leq m - 1;\ x_a \text{ and } x_b \text{ are each incident with exactly one vertex of color } i\}.$$

Clearly $a, b \notin P$. We shall prove

A28 $$P \cap \{1, 2, \ldots, n\} \neq \varnothing.$$

If this were false, one would have $P \subseteq h[U]$. For each of the $m - (|P| + 3)$ colors not in $P + \{a, b, m\}$, either x_a or x_b is incident with at least two vertices of that color. Hence

$$\rho(x_a) + \rho(x_b) \geq 2(m - 1) + m - |P| - 3.$$

We also have $\rho(x) \geq m - 1$ for all $x \in V$ by Corollary A12. In addition, there would be at least one vertex in $U + \{x_a, x_b\}$ (that is, with valence greater than $m - 1$) of each of the colors in P. Summing all the valences would yield

$$\begin{aligned} 2\nu_1 &= \sum_{x \in V} \rho(x) \geq [2(m - 1) + m - |P| - 3] + (\nu_0 - 2)(m - 1) + |P| \\ &= (m - 1)\nu_0 + m - 3, \end{aligned}$$

contrary to A21. This proves A28. Without loss of generality, one may assume that $1 \in P$.

We assert

A29 Either $\{x_1, x_a\} \notin \mathscr{E}$ or $\{x_1, x_b\} \notin \mathscr{E}$.

If both 2-sets were edges of Γ, then since $1 \in P$, x_1 would be the unique vertex in $h^{-1}[1]$ incident with x_a as well as the unique vertex in $h^{-1}[1]$ incident with x_b. If we define

$$h_1(x) = \begin{cases} h(x) & \text{if } x \in V + \{x_1, x_a\}; \\ a & \text{if } x = x_1; \\ 1 & \text{if } x = x_a, \end{cases}$$

then x_b is incident with no vertex in $h_1{}^{-1}[1]$, contrary to Lemma A11b. For definiteness, we may assume that $\{x_1, x_b\} \notin \mathscr{E}$.

For each $i \in \{2, \ldots, n\}$, consider the Kempe chain Θ_{ib}. By the argument used to prove A24, if Θ_{ib} is not an elementary $x_i x_b$-path, then it contains a vertex $x \in U \cap h^{-1}[b]$. If x lies in Θ_{jb} for k different colors $j \in \{2, \ldots, n\}$, then $\rho(x) \geq m - 1 + k$.

Suppose for the moment that

$$\sum_{x \in U \cap h^{-1}[b]} \rho(x) \geq (m-1)|U \cap h^{-1}[b]| + n - 1.$$

Since $|U| - |U \cap h^{-1}[b]| = |U + (U \cap h^{-1}[b])| \geq |h[U]| - 1$, we obtain

$$\begin{aligned} 2\nu_1 &= \sum_{x \in U \cap h^{-1}[b]} \rho(x) + \sum_{\substack{x \in U \\ h(x) \neq b}} \rho(x) + \sum_{x \in V + U} \rho(x) \\ &\geq (m-1)|U \cap h^{-1}[b]| + n - 1 + m|U + (U \cap h^{-1}[b])| \\ &\quad + (m-1)(\nu_0 - |U|) \\ &= (m-1)\nu_0 + |U| - |U \cap h^{-1}[b]| + n - 1 \\ &\geq (m-1)\nu_0 + (|h[U]| - 1) + n - 1 \\ &\geq (m-1)\nu_0 + m - 3, \end{aligned}$$

contrary to A21. Hence

A30 $$\sum_{x \in U \cap h^{-1}[b]} \rho(x) \leq (m-1)|U \cap h^{-1}[b]| + n - 2.$$

This inequality proves that Θ_{ib} is an elementary $x_i x_b$-path for at least one color $i \in \{2, \ldots, n\}$, and we suppose now that Θ_{ib} is an elementary $x_i x_b$-path for all $i = 2, \ldots, p$.

Let $w \in W_{1,b}$ be incident with x_1. Then $h(w) = b$ and by assumption, $w \neq x_b$. We shall show that

A31 $$w \notin \bigcap_{i=2}^{p} W_{i,b}.$$

For if A31 were false, we would have $\rho(w) \geq (m-1) + (p-1)$. For each $i \in \{2, \ldots, n\}$ such that Θ_{ib} is not an elementary path, there is a vertex in $U \cap h^{-1}[b] \cap W_{i,b}$. The sum of the valences of such vertices is at least $(m-1) + (n-p)$. Hence

$$\sum_{x \in U \cap h^{-1}[b]} \rho(x) \geq (m-1)|U \cap h^{-1}[b]| + (p-1) + (n-p),$$

contrary to A30. This proves A31. For definiteness, suppose $w \notin W_{2,b}$.

We define

$$h_2(x) = \begin{cases} h(x) & \text{if } x \in V + W_{2,b}; \\ b & \text{if } x \in W_{2,b} \cap h^{-1}[2]; \\ 2 & \text{if } x \in W_{2,b} \cap h^{-1}[b]. \end{cases}$$

Then h_2 is an m-coloring of Γ, since $\Theta_{2,b}$ is a Kempe chain. Since $w \notin W_{2,b}$, $h_2(w) = b = h_2(x_2)$. By A27, $\{x_1, x_2\} \in \mathscr{E}$. Hence, x_1 is incident with two vertices in $h_2{}^{-1}[b]$, contrary to Lemma A11b. □

Our first corollary is a now classical result in chromatic graph theory known as "Brooks' theorem."

A32 Corollary (R. L. Brooks, [b.18]). *Let $m \geq 4$ be an integer. If no component of Γ is the complete graph K_m, and if $\rho(x) \leq m - 1$ for all $x \in V$, then $\chi_0(\Gamma) \leq m - 1$.*

PROOF. Let $m \geq 4$ be given and let Θ be a graph satisfying the hypothesis. Let $\Gamma = (V, \mathscr{E})$ be a $\chi_0(\Theta)$-critical subgraph of Θ. Then Γ satisfies the hypothesis, too. By A12, $\chi_0 - 1 \leq \rho(x) \leq m - 1$ for all $x \in V$. If $\chi_0 = m$, then Γ is an m-critical, $(m - 1)$-valent graph different from K_m. By *III*A1, $2\nu_1 = (m - 1)\nu_0$. But by Theorem A20

$$2\nu_1 \geq (m - 1)\nu_0 + m - 3 > (m - 1)\nu_0$$

since $m \geq 4$. Hence $\chi_0 \leq m - 1$. □

A33 *Exercise.* Show that Brooks' theorem is equivalent to the following: Every m-critical graph $(m \geq 4)$ which is not complete satisfies:

$$2\nu_1 \geq (m - 1)\nu_0 + 1.$$

Kronk's and Mitchem's proof of Dirac's theorem (A20) reduces to a proof of Brooks' theorem as formulated in A33 when the set U in their proof is empty and so $n = m - 1$. The ultimate contradiction is then attained with the proof of A27. (See L. S. Mel'nikov and V. G. Vizing, [m.8].)

An immediate consequence of Brooks' theorem is

A34 Corollary. *If Γ is connected but not complete and if $\hat{\rho}(\Gamma) \geq 3$, then*

$$\chi_0(\Gamma) \leq \hat{\rho}(\Gamma).$$

If $\hat{\rho}(\Gamma) = 2$, this last inequality may fail; the odd circuits are counterexamples. However, if $\hat{\rho}(\Gamma) \geq 3$, then the difference between the two sides in the inequality can be made arbitrarily large, as the next exercise indicates.

A35 *Exercise* (Dirac [d.4]). Let r be an integer, $r \geq 3$. Construct a 4-critical graph $\Gamma = (V, \mathscr{E})$ such that $\rho(x_0) = r$ for some $x_0 \in V$ and $\rho(x) = 3$ for all $x \in V + \{x_0\}$.

A36 *Exercise.* Determine whether the following statement is true or false: K_4 is the only critical 3-valent graph.

The best analog to A34 for edge colorings is due to V. G. Vizing. We state it without proof.

A37 Theorem (V. G. Vizing, [v.1]). *For any multigraph* $\Theta = (V, f, E)$ *with selection* s,

$$\chi_1(\Theta) \leq \hat{\rho}(\Theta) + \max\{s(U): U \subseteq V\}.$$

To see that Vizing's theorem cannot be strengthened, consider the multigraph obtained by replicating exactly r times each edge of the complete graph K_3. Then since any two edges are incident, $\chi_1 = \nu_1 = 3r = 2r + r$, but $\rho = \hat{\rho} = 2r$.

A38 *Exercise.* Let the multigraph Θ be obtained from Δ_{2k+1} $(k \geq 2)$ by replicating each edge exactly r times. Determine $\chi_1(\Theta)$.

Clearly the inequality

A39
$$\hat{\rho}(\Theta) \leq \chi_1(\Theta)$$

gives a lower bound for the edge chromatic number of any multigraph Θ. It, too, cannot be sharpened, for let Θ be obtained by replicating exactly r times the edges of any even circuit. Then $\hat{\rho} = \rho = 2r = \chi_1$. From A37 we deduce

A40 Corollary. *For any graph,* $\hat{\rho} \leq \chi_1 \leq \hat{\rho} + 1$.

Each bound in A40 can be attained; it suffices to consider even or odd circuits, respectively.

A41 *Exercise.* Prove that if Θ is a bipartite multigraph, then $\chi_1(\Theta) = \hat{\rho}(\Theta)$. [*Hint*: use a result from §*V*C.]

Particular attention has been paid for about a century to the edge chromatic number of **(3-valent)** graphs. In 1880, P. G. Tait [t.1] conjectured that if Γ is trivalent and has no isthmus, then $\chi_1(\Gamma) = 3$. Consequently the name **Tait coloring** has come to mean an edge 3-coloring of a trivalent graph. In 1898, J. Petersen [p.3] disproved Tait's conjecture. His counter-example, shown in two isomorphic representations in Figure A42, is known as the **Petersen graph**. This graph recurs in numerous far-flung contexts. For example, it is the smallest trivalent graph of girth 5. (See also §*IX*E below.)

A42

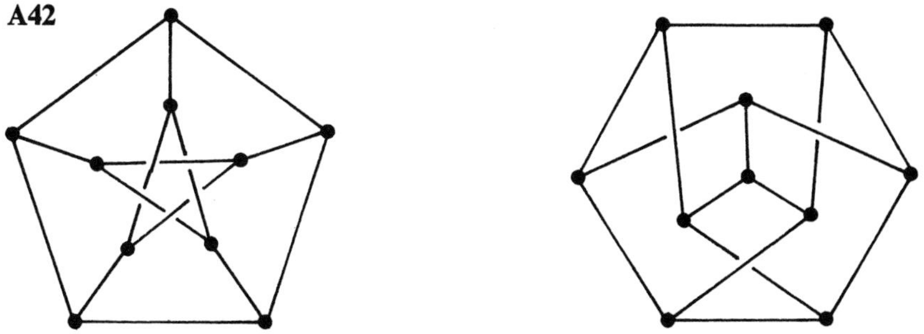

A43 *Exercise.* Show that if a graph admits a Tait coloring, then it is biconnected.

A44 *Exercise.* Let Γ be a trivalent graph. Prove that if Γ has a Hamilton circuit, then Γ has a Tait coloring, but that the converse is false.

A45 Exercise. Show that *the Petersen graph is the unique smallest biconnected, trivalent graph with edge chromatic number* 4. What if the condition of biconnectedness were relaxed?

A46 Exercise (J. Petersen-B. Descartes). *Let* $h: \mathscr{E} \to \{1, 2, 3\}$ *be a Tait coloring of the trivalent graph* Γ, *and let* $\mathscr{F}$ *be any minimal nonempty cocycle of* Γ. Prove that *the three integers* $|\mathscr{F} \cap h^{-1}[i]|$ *for* $i = 1, 2, 3$ *are pairwise congruent modulo* 2.

A47 *Exercise.* Let $\Gamma = (V, \mathscr{E})$.

(a) (Ore, [o.2, p. 122].) Prove that $\chi_0(\Gamma) \leq 4$ if and only if there exists a set $S \subseteq V$ such that Γ_S and Γ_{V+S} are bipartite.

(b) (Woodall, [w.13].) Prove that $\chi_0(\Gamma) \leq 4$ if and only if there exists a set $\mathscr{F} \subseteq \mathscr{E}$, such that $\Gamma_{(\mathscr{F})}$ and $\Gamma_{(\mathscr{E}+\mathscr{F})}$ are each bipartite.

Further Reference

Blanche Descartes, [d.1].

VIIB Chromatic Theory of Planar Graphs

The most noteworthy achievement in combinatorics in 1976 was the decision, in the affirmative, of the so-called

B1 Four-Color Conjecture. $\chi_0(\Gamma) \leq 4$ *for every planar graph* Γ.

For over one hundred years many mathematicians had labored on this problem. An entire lengthy book by Oystein Ore [o.2] is devoted to its history and such progress as had been made toward its solution, equivalent formulations, and generalizations. Since much of this theory is of interest in its own right, we present some of these results in the first part of this section.

The proof of B1 announced by K. Appel and W. Haken [a.1] requires some thousands of hours of computer time and consequently is not within the scope of this book. Nonetheless, a general description of their approach together with an example or two are very much in the spirit of this chapter and will be presented in the second part of this section. Since we do not include any formal proof of the Four-Color "Theorem," it is reasonable to treat it as only a conjecture while presenting some of the more classical results. The form of many of these is in effect that B1 is equivalent to some other condition. We *now* know, however, that the other condition always holds, too.

Originally, the Four-Color Problem was posed concerning the coloring of the regions of a planar imbedding of a graph, subject to the condition that two regions incident with a common edge be assigned different colors. As such the problem was motivated by the question of coloring the countries of an idealized map so that countries with a common boundary receive different colors. We say "idealized" since on the one hand we require that all countries be connected (e.g., Pakistan is acceptable only since 1972), while on the other hand no rules apply for noncontiguous regions (e.g., the Atlantic Ocean need not be the same color as Lake Ontario, but may be the same color as Hungary). Of course, the face-coloring problem is entirely "orthogonal" to the vertex problem which we have already stated and to which we now turn our attention.

B2 Lemma (The Six-Color Theorem). *If* Γ *is planar, then* $\chi_0(\Gamma) \leq 6$.

PROOF. Let Γ be planar, and let Θ be a $\chi_0(\Gamma)$-critical subgraph of Γ. By *III*E15, Θ is planar, and so by *III*F12, $\check{\rho}(\Theta) \leq 5$. The result follows by A13. □

B3 Theorem (The Five-Color Theorem). *If* Γ *is planar, then* $\chi_0(\Gamma) \leq 5$.

PROOF. If $\chi_0(\Gamma) > 5$, then $\chi_0(\Gamma) = 6$ by B2, and we select a 6-critical subgraph $\Theta = (V, \mathscr{E})$ of Γ. By *III*E15, Θ is planar, and by *III*F12, Θ admits a vertex x_0 of valence at most 5. By A12, $\rho(x_0) = 5$ (in Θ). Since Θ is 6-critical, it admits a 6-coloring $h_0\colon V \to \{0, 1, \ldots, 5\}$ with respect to which $\{x_0\} = h_0^{-1}[0]$.

Let $x_1, \ldots, x_5$ be the vertices incident with x_0 where the indices are such that the edges $\{x_0, x_i\}$ and $\{x_0, x_{i+1}\}$ are incident with a common face Z_i of a planar imbedding of Θ ($i = 1, \ldots, 5$; $x_6 = x_1$). (See Figure B4.) By Lemma A11a, we may assume without loss of generality that $h_0(x_i) = i$ for $i = 1, \ldots, 5$. By virtue of Lemma A11b, we may let Ω_1 be an elementary x_1x_3-path lying in $\Theta_{h_0^{-1}[\{1,3\}]}$ and let Ω_2 be an elementary x_2x_4-path lying in $\Theta_{h_0^{-1}[\{2,4\}]}$. Then the circuit consisting of x_3, $\{x_3, x_0\}$, x_0, $\{x_0, x_1\}$ followed by Ω_1 and the circuit consisting of x_4, $\{x_4, x_0\}$, x_0, $\{x_0, x_2\}$ followed by Ω_2 cross at the vertex x_0 and have no other common vertex. This is impossible by *III*E22. □

B4

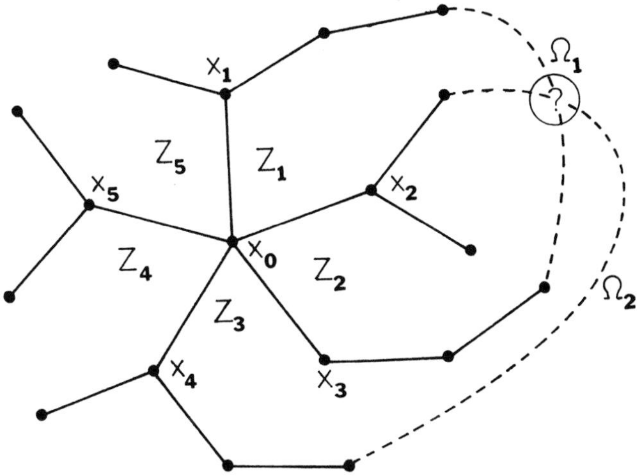

The technique used in the above proof can be adapted to prove the following.

B5 Proposition. *Let Γ be planar and let x_0 be a vertex of Γ such that $\rho(x_0) \leq 4$. If $\chi_0(\Gamma_{(x_0)}) \leq 4$, then $\chi_0(\Gamma) \leq 4$.*

Proof. Let $x_1, \ldots, x_m$ be the vertices incident with x_0 and let $h\colon V + \{x_0\} \to \{1, 2, 3, 4\}$ be a $\chi_0(\Gamma_{(x_0)})$-coloring of $\Gamma_{(x_0)}$. If $|h[V + \{x_0\}]| < 4$, then it is clear that h can be extended to a vertex n-coloring of Γ with $n \leq 4$. Hence it may be assumed that $m = 4$ and that h is a 4-coloring of $\Gamma_{(x_0)}$ such that $h(x_i) = i$ $(i = 1, 2, 3, 4)$. If $\chi_0(\Gamma) > 4$, then there exists a 5-coloring $h_0\colon V \to \{0, 1, \ldots, 4\}$ of Γ such that

$$h_0(x) = \begin{cases} 0 & \text{if } x = x_0; \\ h(x) & \text{if } x \in V + \{x_0\}. \end{cases}$$

Without loss of generality, assume that there are faces Z_i incident with edges $\{x_0, x_i\}$ and $\{x_0, x_{i+1}\}$ $(i = 1, \ldots, 4;\ x_5 = x_1)$. We again invoke Lemma A11b and continue exactly as in the proof of Theorem B3. □

If a planar graph Γ were a counter-example to the Four-Color Conjecture, then by B5, any subgraph of Γ obtained by deleting vertices of valence 4 or less would also be a counter-example. (Cf. B9 below.)

The method of proof of B5 is inadequate if the hypothesis is changed to allow $\rho(x_0) = 5$. Let us consider why. In any vertex 4-coloring of $\Gamma_{(x_0)}$, the five vertices incident with x_0 would represent at most four colors. When we then extend a 4-coloring of $\Gamma_{(x_0)}$ to a 5-coloring of Γ by coloring the vertex x_0 a new color, the hypothesis of Lemma A11b is not fulfilled, and we are not assured of the existence of the necessary Kempe chains.

B6 Corollary. *If* Γ *is planar and* $\nu_0(\Gamma) \le 11$, *then* $\chi_0(\Gamma) \le 4$.

PROOF. By *III*F12 and *III*F13, $\check{\rho}(\Theta) \le 4$ for any subgraph Θ of Γ. Let $S_0 = V$. Inductively, for $i \ge 0$, there exists a vertex $x_i \in S_i = S_{i-1} + \{x_{i-1}\}$ whose valence in Γ_{S_i} is at most 4. Clearly $\chi_0(\Gamma_{S_k}) \le 4$ for some k. By B5, if $\chi_0(\Gamma_{S_i}) \le 4$, then $\chi_0(\Gamma_{S_{i-1}}) \le 4$ and the result follows by induction. □

B7 Corollary. *If* Γ *is planar and* $\gamma(\Gamma) \ge 4$, *then* $\chi_0(\Gamma) \le 4$.

PROOF. Without loss of generality one may assume that Γ is connected. Let Θ be any connected subgraph of Γ with $\nu_1(\Theta) > 0$. Since $\gamma(\Theta) \ge 4$, $\rho^{\perp}(\Theta) \ge 4$, and so by *III*F4,

$$\frac{1}{\rho(\Theta)} \ge \frac{1}{4} + \frac{1}{\nu_1(\Theta)} > \frac{1}{4},$$

whence $\check{\rho}(\Theta) \le \rho(\Theta) < 4$. Let $V = S_0, S_1, \ldots, S_k$ be constructed inductively as in the previous corollary, where $|S_k| < 12$. The result follows from B6 by k iterations of B5. □

B8 *Exercise*. Prove: if Γ is planar and $\gamma(\Gamma) \ge 6$, then $\chi_0(\Gamma) \le 3$.

Actually a much weaker hypothesis than $\gamma(\Gamma) \ge 6$ is sufficient for $\chi_0(\Gamma) \le 3$. H. Grötzsch [g.7] proved in 1958 that $\gamma(\Gamma) \ge 4$ is sufficient. (See especially O. Ore, [o.2, Chapter 13].)

For the remainder of this section we consider some of the many equivalent formulations of the Four-Color Conjecture (B1). The first three of these are in the form of the comment following the proof of B5, namely: the Four-Color Conjecture is true in general if it is true for a certain subclass of planar graphs.

B9 Proposition. *The Four-Color Conjecture is true if and only if all planar graphs* Γ *with* $\check{\rho}(\Gamma) = 5$ *are vertex 4-colorable.*

PROOF. Suppose that

B10 $$\check{\rho}(\Gamma) = 5 \Rightarrow \chi_0(\Gamma) \le 4$$

is true for all planar graphs Γ. Let Θ be a planar graph with the least possible number of vertices such that $\check{\rho}(\Theta) \le 4$ but $\chi_0(\Theta) = 5$. Let x_0 be a vertex of Θ such that $\rho(x_0) \le 4$. If $\check{\rho}(\Theta_{(x_0)}) = 5$ then $\chi_0(\Theta_{(x_0)}) \le 4$ by B10. If $\check{\rho}(\Theta_{(x_0)}) \le 4$, then $\chi_0(\Theta_{(x_0)}) \le 4$ by the assumption of minimality on $\nu_0(\Theta)$. Either way, we deduce from B5 that $\chi_0(\Theta) = 4$, giving a contradiction.

The converse is obvious. □

A planar imbedding of a graph is called a **triangular imbedding** if $|Z| = 3$ for every region Z of the imbedding. I. Fáry [f.1] proved in 1948 that if a graph is planar, then it can be realized in the plane in such a way that every edge is a segment of a straight line. Consequently, if a graph admits a triangular imbedding, then it admits the intuitively appealing realization wherein every region is a geometric triangle.

B11 *Exercise.* Prove that every planar graph without isthmus which admits a triangular imbedding is biconnected, and with the exception of K_3, is triconnected.

B12 Proposition. *The Four-Color Conjecture is true if and only if all planar graphs which admit a triangular imbedding are vertex* 4-*colorable.*

PROOF. Suppose that $\chi_0(\Theta) \leq 4$ for every planar graph Θ which admits a triangular imbedding. Were there to exist a planar graph with vertex chromatic number 5, consider a 5-critical subgraph $\Gamma = (V, \mathscr{E})$ of it, and let $\{Z_1, \ldots, Z_m\}$ be a planar imbedding of Γ.

By A17, Γ is biconnected. Hence by *III*B3 and *III*E9 each region Z_i is an elementary cycle, and so by *III*A9, Z_i induces an elementary circuit

$$x_{i,0}, E_{i,1}, x_{i,1}, E_{i,2}, \ldots, E_{i,q_i}, x_{i,q_i} = x_{i,0}$$

in Γ for each $i = 1, \ldots, m$. Let $y_1, \ldots, y_m$ be m distinct objects not in V and define

$$U = V \cup \{y_1, \ldots, y_m\},$$

$$\mathscr{F} = \mathscr{E} \cup \{\{x_{i,j}, y_i\} : j = 1, \ldots, q_i;\ i = 1, \ldots, m\},$$

and $$\Omega = (U, \mathscr{F}).$$

Since $\Gamma = \Omega_V$, it suffices to prove that Ω admits a triangular imbedding.

We define

$$Z_{i,j} = \{E_{i,j+1}, \{x_{i,j+1}, y_i\}, \{y_i, x_{i,j}\}\}, \quad \text{for } j = 0, \ldots, q_i - 1;\ i = 1, \ldots, m.$$

B13

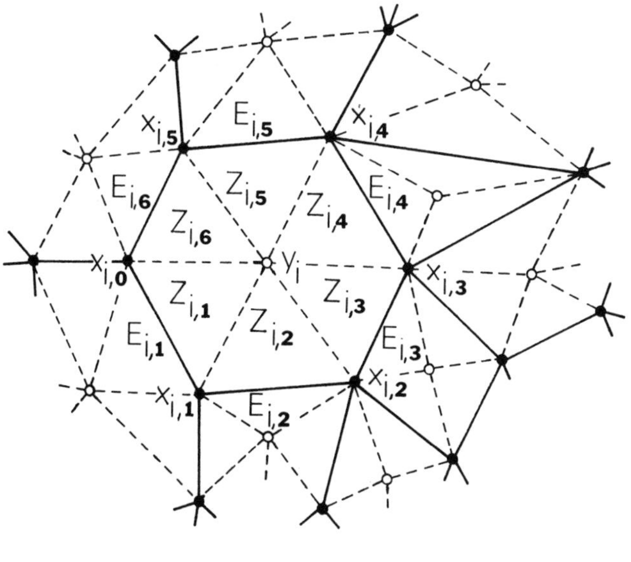

(See Figure B13.) It is then straightforward to verify that the collection

$$\{Z_{i,j}: j = 0, \ldots, q_i - 1; i = 1, \ldots, m\}$$

is a triangular imbedding of Θ.

Once again, the converse is obvious. □

B14 Proposition. *Let Γ be planar with $v_0(\Gamma) \geq 4$. A necessary and sufficient condition for $\chi_0(\Gamma) \leq 4$ is that there exist three pairwise-disjoint subsets $\mathscr{E}_1, \mathscr{E}_2, \mathscr{E}_3 \subseteq \mathscr{E}$ such that $\mathscr{E} = \mathscr{E}_1 + \mathscr{E}_2 + \mathscr{E}_3$, and if Z is any region of a planar imbedding of Γ, then $|Z \cap \mathscr{E}_1| \equiv |Z \cap \mathscr{E}_2| \equiv |Z \cap \mathscr{E}_3|$ (mod 2).*

PROOF. *Necessity.* Suppose $\chi_0(\Gamma) \leq 4$ and let

$$h_0: V \to \mathbb{K} \times \mathbb{K}$$

be a vertex 4-coloring of Γ. We represent the four elements of $\mathbb{K} \times \mathbb{K}$ in the natural way: (0, 0), (0, 1), (1, 0), and (1, 1). Based on h_0, we define a function $h_1: \mathscr{E} \to \{(0, 1), (1, 0), (1, 1)\}$ given by

$$h_1(\{x, y\}) = h_0(x) + h_0(y) \quad \text{for each } \{x, y\} \in \mathscr{E}.$$

(In general h_1 will *not* be an edge coloring.) We define

B15 $\quad \mathscr{E}_1 = h_1^{-1}[(0, 1)], \quad \mathscr{E}_2 = h_1^{-1}[(1, 0)], \quad$ and $\quad \mathscr{E}_3 = h_1^{-1}[(1, 1)].$

Now let Z be a region of some imbedding of Γ. By *II*C1, Z is a sum of pairwise-disjoint elementary cycles; let Z' be one of these. By *III*A9, Z' determines an elementary circuit $x_0, \{x_0, x_1\}, x_1, \{x_1, x_2\}, \ldots, \{x_{k-1}, x_k\}, x_k = x_0$.

In the vector space $\mathbb{K} \times \mathbb{K}$,

$$\begin{aligned} |Z' \cap \mathscr{E}_1|(0, 1) + |Z' \cap \mathscr{E}_2|(1, 0) + |Z' \cap \mathscr{E}_3|(1, 1) &= \sum_{i=0}^{k-1} h_1(\{x_i, x_{i+1}\}) \\ &= 2 \sum_{i=0}^{k-1} h_0(x_i) = (0, 0). \end{aligned}$$

Hence $|Z' \cap \mathscr{E}_1| + |Z' \cap \mathscr{E}_3| \equiv 0$ (mod 2) and $|Z' \cap \mathscr{E}_2| + |Z' \cap \mathscr{E}_3| \equiv 0$ (mod 2). By addition, these same congruences hold for Z.

Sufficiency. Let the sets $\mathscr{E}_1, \mathscr{E}_2, \mathscr{E}_3$ be given and let them determine a function $h_1: \mathscr{E} \to \{(0, 1), (1, 0), (1, 1)\}$ consistent with B15. We prove that h_1 determines a vertex coloring $h_0: V \to \mathbb{K} \times \mathbb{K}$.

Arbitrarily choose $x_0 \in V$ and define $h_0(x_0) = (0, 0)$. If $y \in V$ is in the same component of Γ as x_0, let $x_0, \{x_0, x_1\}, x_1, \{x_1, x_2\}, \ldots, \{x_{k-1}, x_k\}, x_k = y$ be an x_0y-path in Γ. Define

$$h_0(y) = \sum_{i=0}^{k-1} h_1(\{x_i, x_{i+1}\}).$$

We assert that h_0 is well-defined, for the sum of the edge sets of any two x_0y-paths is a cycle, which in turn is a sum of regions. It follows from our assumption that the sum of the values of h_1 on any region is 0. This process may be carried out for each component of Γ. It is immediate that h_0 is a vertex coloring. □

The condition of the above proposition implies that if a region Z is a 3-cycle, then Z contains exactly one edge in each set $\mathscr{E}_i$. Thus $\mathscr{E}_i \neq \varnothing$ and $\{\mathscr{E}_1, \mathscr{E}_2, \mathscr{E}_3\} \in \mathbb{P}_3(\mathscr{E})$. In particular, when all the regions are 3-cycles, we have the following.

B16 Corollary. *Let Γ be a planar graph without isthmus having a triangular imbedding $\{Z_1, \ldots, Z_m\}$. A necessary and sufficient condition for $\chi_0(\Gamma) \leq 4$ is that the graph Θ orthogonal to Γ with respect to $\{Z_1, \ldots, Z_m\}$ admit a Tait coloring.*

Proof. Since by definition, $\{Z_1, \ldots, Z_m\}$ is the set of vertex cocycles of Θ, we have that Θ is trivalent. The sets $\mathscr{E}_1, \mathscr{E}_2, \mathscr{E}_3$ of the proposition correspond to the color classes of a Tait coloring of Θ (cf. Exercise A46), whence the result follows. □

B17 Corollary (P. G. Tait [t.1], 1880). *The Four-Color Conjecture is equivalent to the following conjecture; every planar trivalent graph without isthmus admits a Tait coloring.*

Proof. By the definition of a graph orthogonal with respect to an imbedding, we have that each trivalent graph without isthmus which admits a planar imbedding is orthogonal, with respect to that imbedding, to a graph without isthmus admitting a triangular imbedding, and conversely. By Proposition B12, the Four-Color Conjecture is equivalent to the proposition that $\chi_0(\Gamma) \leq 4$ for every graph Γ admitting a triangular imbedding. The corollary now follows from Corollary B16. □

As though it was not already obvious from Figure A42, let us give a quick but rigorous proof that the Petersen graph Π is nonplanar. With $\nu_0 = 10$ and $\nu_1 = 15$, we deduce from the Euler Formula *III*F2b that if Π were indeed planar, then $\nu_0(\Gamma) = 7$ where $\Gamma \perp \Pi$. By B6, $\chi_0(\Gamma) \leq 4$. Since Π has no isthmus, it follows from B16 that Π admits a Tait coloring, contrary to Exercise A45.

B18 *Exercise.* For each integer $n \geq 6$ construct a planar graph Γ such that $\nu_0(\Gamma) = n$, $\chi_0(\Gamma) = 4$, and Γ does not contain K_4 as a subgraph.

B19 *Exercise.* Show that if Γ is planar, $\Theta \perp \Gamma$, and Θ admits a Hamilton circuit, then $\chi_0(\Gamma) \leq 4$. Show also that the converse is false.

Two other conjectures equivalent to B1 come readily from Exercise A47. Many further equivalent formulations are presented in T. L. Saaty [s.1].

The effect of many of the foregoing results is to tell us that if we are to search for a counter-example Γ to the Four-Color Conjecture, then we lose no generality in making a number of graph-theoretical assumptions about Γ. Not only do we have that $\nu_0(\Gamma) \geq 12$ and $\chi_0(\Gamma) = 5$, but we may assume that Γ belongs to a class which we shall call G consisting of all planar, 5-chromatic graphs Γ on m vertices having a triangular imbedding, where all planar graphs Γ with $\nu_0(\Gamma) < m$ satisfy $\chi_0(\Gamma) \leq 4$. The approach of Appel and Haken, like that of their predecessors, was to show, naturally, that G is empty.

Let $\Gamma = (V, \mathscr{E})$ and let $S \subseteq V$. We say a pair (Γ, S) is a **configuration** if Γ_S is an elementary circuit and $\Gamma_{(S)}$ is connected. (Each graph Γ_i in Figure B20 becomes a configuration when paired with the set of vertices S_i indicated in the figure as bounding its exterior region.) A configuration (Γ, S) is said to be **immersed** in the graph Θ if Γ is a subgraph of Θ and $\Gamma_{(S)}$ is a component of $\Theta_{(S)}$. A configuration (Γ, S) is said to be **reducible** if any planar graph Θ with $\nu_0(\Theta) = m$ and in which (Γ, S) can be immersed satisfies $\chi_0(\Theta) \leq 4$. Clearly a reducible configuration cannot be immersed in any graph in G. Finally, a collection H of configurations is **unavoidable** if each graph in G has a configuration from H immersed in it.

Thus if a configuration is reducible and belongs to an unavoidable collection H, its removal from H will result in a smaller unavoidable collection. Hence G is empty if and only if there exists a nonempty unavoidable collection of reducible configurations.

B20

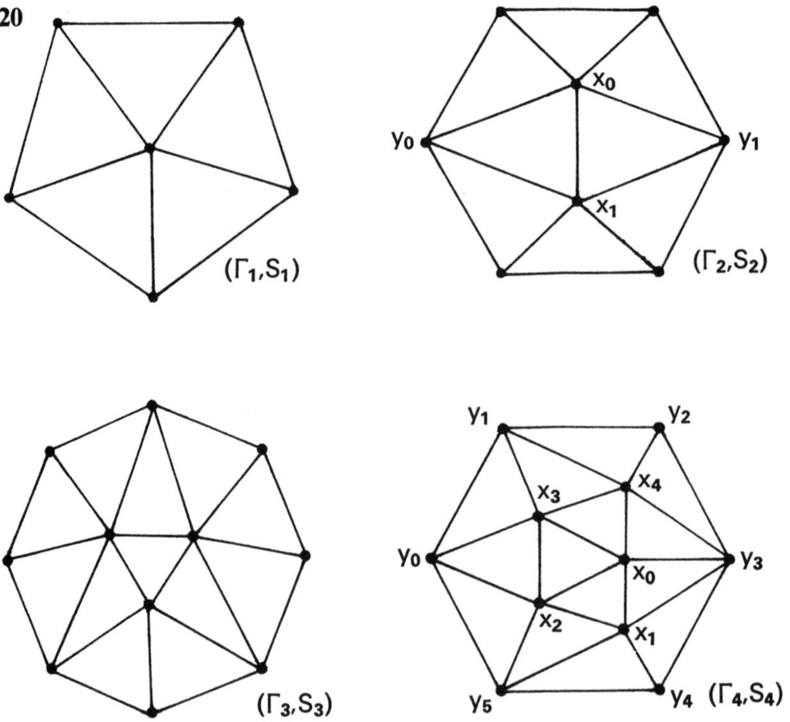

Kempe [k.2] had quite correctly shown that the collection consisting of the single configuration (Γ_1, S_1) in Figure B20 is unavoidable. (His error lay in his "proof" that (Γ_1, S_1) is reducible.) The Appel–Haken approach is similar in that it consists of constructing an unavoidable collection of reducible configurations. Their collection, however, has about 2000 of them. (Apparently it was easier to construct this collection than to prove directly that (Γ_1, S_1) is reducible!) What we shall do here is to present another unavoidable collection, this one consisting of just two configurations, and to demonstrate that a third configuration is reducible. Our purpose is to demonstrate some of the Appel–Haken techniques and thereby to give some of the flavor of their work. To do this, we require three lemmas which have been standard tools of the trade.

B21 Lemma. *If Z is a 3-cycle of a graph $\Gamma \in \mathsf{G}$, then Z is a region for every planar imbedding of Γ.*

PROOF. Consider a planar imbedding of Γ which does not include Z as a region. By Exercise *III*E24, the set S of vertices incident with the edges in Z form a separating set of Γ. Since Γ_S is complete, the restriction to S of any coloring of a subgraph of Γ containing S is an injection. By Proposition A15 and the definition of G, $\chi_0(\Gamma) \leq 4$, contrary to assumption. □

B22 Lemma. *Let Γ belong to G, and let x be a vertex of Γ. Then $\Gamma_{N(x)}$ is an elementary circuit.*

PROOF. Since Γ contains a 5-critical subgraph on the same vertex set, it is biconnected. By Lemma *III*E21, the edges in the vertex cocycle of x may be denoted by $E_0, E_1, \ldots, E_{\rho(x)-1}$ so that for some triangular imbedding, E_i and E_{i+1} lie on a common region (where subscripts are read modulo $\rho(x)$). Let $E_i = \{x, y_i\}$ for $i = 0, 1, \ldots, \rho(x) - 1$. Since every region is a 3-cycle, $\{\{y_0, y_1\}, \{y_1, y_2\}, \ldots, \{y_{\rho(x)-1}, y_0\}\}$ is a cycle of Γ.

If $\{y_i, y_j\}$ is an edge of Γ, then since the three edges $E_i, \{y_i, y_j\}, E_j$ determine a region by Lemma B21 and since every edge of a biconnected graph belongs to exactly two regions, y_j must be $y_{i\pm 1}$ (subscripts read modulo $\rho(x)$). □

B23 Lemma. *Every graph in G admits either* (i) *a pair of incident* 5*-valent vertices or* (ii) *a pair of incident* 6*-valent vertices both of which are incident with a common* 5*-valent vertex.*

PROOF. Let $\Gamma = (V, \mathscr{E})$ be a graph in G which satisfies neither condition, and define $f\colon V \to \mathbb{Z}$ by $f(x) = 6 - \rho(x)$ for all $x \in V$. Since Γ admits a triangular imbedding, $\nu_2 = 2\nu_1/3$, which when substituted into the Euler Formula *III*F2b, yields $\nu_0 - (\nu_1/3) = 2$. Thus

$$\sum_{x \in V} f(x) = 6\nu_0 - \sum_{x \in V} \rho(x) = 6\nu_0 - 2\nu_1 = 12.$$

For $x \in V$, let $U(x)$ denote the set of 5-valent vertices in $N(x)$, and let $U = \bigcup_{x \in V} U(x)$. For each $u \in U$, let $n(u)$ denote the number of vertices in $N(u)$ whose valence is at least 7. Since condition (i) fails, there are $(5 - n(u))$ 6-valent vertices in $N(u)$ for each $u \in U$. No two of these 6-valent vertices are incident, since condition (ii) fails. By Lemma B22, each such vertex u is incident with at most two 6-valent vertices. Thus $3 \le n(u) \le 5$ for all $u \in U$.

Let us now define $g\colon V \to \mathbb{Q}$ by

$$g(x) = \begin{cases} -1 & \text{if } \rho(x) = 5, \\ 0 & \text{if } \rho(x) = 6, \\ \sum_{u \in U(x)} 1/n(u) & \text{if } \rho(x) \ge 7. \end{cases}$$

Appel and Haken refer to g as a "discharging function"; one imagines each vertex of Γ as possessing a certain amount of "charge" and that the function g removes one unit of charge from each 5-valent vertex u and redistributes it by transferring $1/n(u)$ units to each of the $n(u)$ vertices incident with u having valence at least 7. Thus there is no net change in "charge" of the whole graph. In other words, $\sum_{x \in V} g(x) = 0$. Now let us define $h\colon V \to \mathbb{Q}$ by $h = f + g$, whence it follows that $\sum_{x \in V} h(x) = 12$. We shall obtain a contradiction by showing that $h(x) \le 0$ for all $x \in V$.

One verifies easily that $h(x) = 0$ if $\rho(x) = 5$ or 6. Let $x \in V$ and suppose $\rho(x) \ge 7$. Since no two 5-valent vertices are incident while $\Gamma_{N(x)}$ is an elementary circuit, it follows that $|U(x)| \le \rho(x)/2$. Since $n(u) \ge 3$ for all $u \in U(x)$, we have

$$g(x) = \sum_{u \in U(x)} \frac{1}{n(u)} \le \frac{[\rho(x)/2]}{3} \le \frac{\rho(x)}{6}.$$

If $\rho(x) = 7$, then $h(x) \le -1 + 1 = 0$, while if $\rho(x) \ge 8$, then $h(x) \le 6 - (5\rho(x)/6) < 0$. □

B24 Proposition. *The two configurations* (Γ_2, S_2) *and* (Γ_3, S_3) *of Figure* B20 *form an unavoidable collection.*

PROOF. Let Θ belong to **G**. Referring to Lemma B23, we assert that if Θ satisfies (i), then (Γ_2, S_2) can be immersed in Θ, while if Θ satisfies (ii), then (Γ_3, S_3) can be immersed in Θ. We demonstrate the first assertion, leaving the second to be completed by the reader.

Let x_0 and x_1 be a pair of incident 5-valent vertices. Then $\Theta_{N(x_0)}$ and $\Theta_{N(x_1)}$ are both 5-circuits. Since Θ admits a triangular imbedding, there exist vertices y_0 and y_1 such that $\{\{x_0, x_1\}, \{x_1, y_0\}, \{y_0, x_0\}\}$ and $\{\{x_0, x_1\}, \{x_1, y_1\}, \{y_1, x_0\}\}$ are regions; this follows from Lemma *III*E21, B21, and B22. This yields the graph Γ_2 as a subgraph of Θ. Furthermore, the vertices in Γ_2 account for all vertices incident with x_0 and x_1. Hence, $\Gamma_{(S_2)} = \Gamma_{\{x_0, x_1\}}$ is a component of $\Theta_{(S_2)}$. □

B25 Proposition. *The configuration* (Γ_4, S_4) *of Figure* B20 *is reducible.*

PROOF. Let Θ be a planar graph with $\nu_0(\Theta) = m$, and suppose that the configuration (Γ_4, S_4) is immersed in Θ. We must show that $\chi_0(\Theta) \le 4$. We suppose that $\chi_0(\Theta) > 4$. By Lemma B21, every 3-cycle of Θ is a region of any planar imbedding of Θ. The 3-cycles of Γ_4 are also 3-cycles of Θ; their sum is the cycle $\{\{y_0, y_1\}, \{y_1, y_2\}, \ldots, \{y_5, y_0\}\}$. (The set of vertices on the corresponding circuit is S_4.) It follows that $\Theta_{(\{x_0,\ldots,x_4\})}$ has a planar imbedding in which this cycle is a region. Let Θ' be formed from $\Theta_{(\{x_0,\ldots,x_4\})}$, by adjoining any of the edges $\{y_0, y_2\}$, $\{y_0, y_3\}$, $\{y_0, y_4\}$ not already in Θ. Clearly Θ' is planar. Since $\nu_0(\Theta') < \nu_0(\Theta) = m$, there exists a vertex k-coloring h of Θ' for some $k \leq 4$. Clearly h is also a k-coloring of $\Theta_{(\{x_0,\ldots,x_4\})}$. Using colors from the set $\{0, 1, 2, 3\}$, we list in Table B26 the six possible restrictions of h to S_4 subject to permutations of the colors or the permutation on S_4 given by $(y_1, y_5)(y_2, y_4)$, which is extendable to a vertex-automorphism of Γ_4. In all but the first case, h is easily extended to a 4-coloring of Θ, as indicated in the table, yielding a contradiction. Hence, the restriction to S_4 of every vertex

B26

	y_0	y_1	y_2	y_3	y_4	y_5	x_0	x_1	x_2	x_3	x_4
Case 1	0	1	2	1	2	1					
Case 2	0	1	2	1	2	3	2	0	1	3	0
Case 3	0	1	2	1	3	1	0	2	3	2	3
Case 4	0	1	2	1	3	2	3	0	1	2	0
Case 5	0	1	2	3	1	2	1	0	3	2	0
Case 6	0	1	2	3	2	1	1	0	2	3	0

k-coloring of Θ' must yield the values of Case 1 of Table B26, subject to the above permutation. If there exists a y_2y_4-Kempe chain in Θ with colors 0 and 2, then there can exist no y_1y_3-Kempe chain with colors 1 and 3, the argument being identical to that of the Five-Color Theorem. In this case, the color of y_3 may be changed to 3 without affecting any other vertices of S_4, and we return to Case 6 of Table B26. Since there is no such y_2y_4-Kempe chain, there exists either no y_0y_2-Kempe chain or no y_0y_4-Kempe chain. If there exists no y_0y_4-Kempe chain, then y_4 may be recolored with 0. Note that in Θ, $\{y_0, y_4\}$ need not be an edge, and so we must consider a vertex coloring $h|_{S_4}$ not listed in the table. However, the assignments $h(x_0) = 0$, $h(x_1) = 2$, $h(x_2) = 3$, $h(x_3) = 2$, and $h(x_4) = 3$ extend h to a vertex 4-coloring of Θ. If there exists no y_0y_2-Kempe chain, the argument is symmetric. □

B27 *Exercise.* (a) Using B25 and arguments similar to those in the proof of B23, show that the m defined above is at least 16, i.e., show that if Γ is planar and $\nu_0(\Gamma) < 16$, then $\chi_0(\Gamma) \leq 4$.

(b) By showing that the only triangulated planar graph on 16 vertices with twelve 5-valent vertices and four 6-valent vertices is 4-chromatic, show that $m \geq 17$.

VIIC The Imbedding Index

In Chapter III, we gave a purely combinatorial definition of a planar imbedding of a graph. The fundamental observation leading to this definition is that the edge sets of the boundaries of the regions of a topological imbedding of a graph Γ include every edge twice and span the cycle space of Γ. In §*III*E, we showed that any set $\mathscr{S}$ of cycles which

(a) spans $\mathscr{Z}(\Gamma)$, and
(b) includes each edge, except isthmuses, exactly twice

may be taken as the set of boundaries of the regions of a topological imbedding of Γ—with suitable adjustments for isthmuses.

It is natural to attempt to extend this combinatorial approach to imbeddings in other surfaces. (A **surface** is a compact, connected topological space in which every point has a neighborhood homeomorphic to an open disk in the Euclidean plane. It is also called a "compact 2-manifold.") On surfaces other than the sphere, there exist circuits which are not homotopically trivial. On the other hand, condition (a) above means, in the light of §*III*E, that all circuits are homotopically trivial. The natural extension therefore, is simply to relax condition (a). We do this in the definition of a "cycle cover" below. The difficulty with this approach is that, while one always obtains an imbedding in a finite 2-dimensional cell complex, this complex need not be a surface. In particular, the vertices of the graph may appear as singular points on this complex.

A purely combinatorial definition of an imbedding in an orientable surface has been given by Edmonds [e.1] and will be discussed in §*E*, where the topological approach of §*III*E will be extended to arbitrary surfaces. For the present, we confine ourselves to a combinatorial substitute for the genus of a topological imbedding, called the "imbedding index." The main result of this section is a general theorem which has Heawood's inequality and Ringel's analog for the Euler characteristic as special cases.

A **cycle cover** of a graph Γ is a sequence $Z_1, \ldots, Z_m \in \mathscr{Z}(\Gamma) + \{\varnothing\}$ such that each edge of Γ belongs to Z_i for at most two indices i. Observe that if the cycle $Z_1 + \cdots + Z_m$ is not empty and is appended to a cycle cover $Z_1, \cdots, Z_m$, then the new sequence is also a cycle cover. We define $\nu_2(Z_1, \ldots, Z_m) = 1 + \dim(\langle\{Z_1, \ldots, Z_m\}\rangle)$ and

$$\iota(Z_1, \ldots, Z_m) = \dim(\mathscr{Z}(\Gamma)/\langle\{Z_1, \ldots, Z_m\}\rangle).$$

By *III*A15b,

C1 $\quad \iota(Z_1, \ldots, Z_m) = 1 + \nu_{-1}(\Gamma) - \nu_0(\Gamma) + \nu_1(\Gamma) - \nu_2(Z_1, \ldots, Z_m).$

We let $\nu_2(\Gamma)$ and $\iota(\Gamma)$ denote respectively the maximum of $\nu_2(Z_1, \ldots, Z_m)$ and the minimum of $\iota(Z_1, \ldots, Z_m)$, taken over all cycle covers $Z_1, \ldots, Z_m$ of Γ. The parameter $\iota(\Gamma)$ is called the **imbedding index** of Γ, and hence is given by

C2 $$\iota(\Gamma) = 1 + \nu_{-1}(\Gamma) - \nu_0(\Gamma) + \nu_1(\Gamma) - \nu_2(\Gamma).$$

C3 Exercise. Prove that $\iota(\Gamma) \geq 0$, *with equality holding if and only if* Γ *is planar.*

C4 Proposition. *If* Θ *is a subgraph of* Γ, *then* $\iota(\Theta) \leq \iota(\Gamma)$.

Proof. Any subgraph of Γ may be obtained by first deleting some edges of Γ and then deleting some isolated vertices. Since isolated vertices affect neither the dimension of a cycle space nor the size of a cycle cover, we need show only that $\iota(\Gamma_{(E)}) \leq \iota(\Gamma)$ for any edge E of Γ.

By *III*C1a, $\dim(\mathscr{Z}(\Gamma_{(E)})) = \dim(\mathscr{Z}(\Gamma)) - 1$ unless E is an isthmus, in which case $\dim(\mathscr{Z}(\Gamma_{(E)})) = \dim(\mathscr{Z}(\Gamma))$. Turning to cycle covers, we note that every cycle cover of $\Gamma_{(E)}$ is a cycle cover of Γ, and if E is an isthmus, the converse also holds. Hence $\nu_2(\Gamma) \geq \nu_2(\Gamma_{(E)})$, with equality holding if E is an isthmus. Hence $\iota(\Gamma_{(E)}) = \iota(\Gamma)$ for E an isthmus. If E is not an isthmus, then because of C2, it suffices to show that $\nu_2(\Gamma) \leq \nu_2(\Gamma_{(E)}) + 1$.

Suppose $\nu_2(\Gamma) = \nu_2(Z_1, \ldots, Z_m)$. If $E \notin \bigcup_{i=1}^m Z_i$, then $Z_1, \ldots, Z_m$ is also a cycle cover of $\Gamma_{(E)}$, and $\nu_2(\Gamma) = \nu_2(\Gamma_{(E)})$. If $E \in \bigcup_{i=1}^m Z_i$, then by including $Z_1 + \ldots + Z_m$ if necessary, we may assume that E belongs to exactly two of the cycles, say Z_{m-1} and Z_m. It follows that $Z_1, \ldots, Z_{m-1}, (Z_{m-1} + Z_m)$ is a cycle cover of $\Gamma_{(E)}$, and therefore $\nu_2(\Gamma) - 1 \leq \nu_2(\Gamma_{(E)})$. □

Define the function G by $G(j) = 2j/(j-2)$ for each integer $j \geq 3$, and define $G(\infty) = \lim_{j\to\infty} G(j) = 2$. If $k \in \mathbb{N}$, we define the integer-valued function

$$H(j,k) = \begin{cases} \left[\dfrac{G(j) + 1 + \sqrt{G^2(j) - 6G(j) + 1 + 4G(j)k}}{2}\right] & \text{for } k \geq 2; \\ \{G(j)\} & \text{for } k = 0, 1, \end{cases}$$

where for $x \in \mathbb{R}$, $\{x\}$ denotes the least integer such that $x \leq \{x\}$. The reader already familiar with Heawood's inequality will recognize that $H(3, 2k)$ is the Heawood number of the orientable surface of genus k. We will use the function $H(j, k)$ to derive a generalization of the Heawood inequality. In order to do this we must prove some preliminary results.

One obtains by direct calculation

C5 $$H(j,2) = [G(j) + 1] \geq \{G(j)\} = H(j,1) = H(j,0).$$

C6 Lemma. *$H(j, k)$ is nonincreasing in the first variable and nondecreasing in the second variable.*

Proof. Since $G(j)$ is nonincreasing in j, $H(j, k)$ is clearly nonincreasing in j for $k = 0, 1, 2$ (by C5). On the other hand, $H(j, k)$ is nondecreasing in $G(j)$ for $k \geq 2$, and therefore nonincreasing in j. Since $G(j) > 0$, $H(j, k)$ is strictly increasing in k for $k \geq 2$. The rest follows from C5. □

C7 Lemma. *For any graph Γ, $\rho(\Gamma) < H(\gamma(\Gamma), \iota(\Gamma))$.*

Proof. Let $\nu_2(\Gamma) = \nu_2(Z_1, \ldots, Z_m)$. Adjoining $Z_1 + \ldots + Z_m$ to this cover if necessary, we may assume that $\{Z_1, \ldots, Z_m\}$ is dependent. Thus $m \geq \nu_2(\Gamma)$, and so by C2 (with the argument Γ suppressed), $\iota \geq \nu_1 - \nu_0 + 2 - m$. From the definitions of cycle cover and girth, $m\gamma \leq 2\nu_1$, from which we deduce

C8 $$\iota - 2 \geq \frac{\gamma - 2}{\gamma}\nu_1 - \nu_0.$$

Recall from the definition of average valence that

C9 $$\rho\nu_0 = 2\nu_1.$$

Case 1: $\iota \leq 2$. Using C9 to eliminate ν_0 from C8 we obtain

$$\frac{1}{\rho} \geq \frac{1}{G(\gamma)} + \frac{2 - \iota}{2\nu_1}.$$

If $\iota = 0$ or 1, $\rho < G(\gamma) \leq \{G(\gamma)\} = H(\gamma, \iota)$. If $\iota = 2$, then $\rho = G(\gamma) < [G(\gamma) + 1] = H(\gamma, 2)$ by C5.

Case 2: $\iota > 2$. Using C9 to eliminate ν_1 from C8, we obtain

C10 $$\rho \leq G(\gamma)\left(1 + \frac{\iota - 2}{\nu_0}\right).$$

We define the function $f(x) = G(\gamma)(1 + ((\iota - 2)/x))$ for all real $x > 0$. Since f is a decreasing function, there exists a least $H_0 \in \mathbb{N}$ such that $f(H_0 + 1) < H_0$. We assert that $\rho < H_0$; for if $\nu_0 \leq H_0$, then $\rho < \nu_0 \leq H_0$; while if $\nu_0 > H_0$, then $\nu_0 \geq H_0 + 1$, and by C10, $\rho \leq f(\nu_0) \leq f(H_0 + 1) < H_0$. It remains only to show that $H_0 = H(\gamma, \iota)$.

By definition, H_0 is the smallest positive integer satisfying

$$H_0 > G(\gamma) + \frac{G(\gamma)(\iota - 2)}{H_0 + 1},$$

or equivalently, $H_0^2 - (G(\gamma) - 1)H_0 - G(\gamma)(\iota - 1) > 0$. That is to say that H_0 is the least integer greater than the larger root of the quadratic equation

$$x^2 - (G(\gamma) - 1)x - G(\gamma)(\iota - 1) = 0.$$

In other words, $H_0 = [x_0 + 1]$ where, by the quadratic formula,

$$x_0 = \frac{G(\gamma) - 1 + \sqrt{(G(\gamma) - 1)^2 + 4G(\gamma)(\iota - 1)}}{2}.$$

The lemma follows directly by algebraic manipulation. □

C11 Proposition. *For any graph* Γ, $\chi_0(\Gamma) \leq H(\gamma(\Gamma), \iota(\Gamma))$.

Proof. Let Θ be a $\chi_0(\Gamma)$-critical subgraph of Γ. By C7, $\check{\rho}(\Theta) \leq \rho(\Theta) < H(\gamma(\Theta), \iota(\Theta))$. Clearly $\gamma(\Theta) \geq \gamma(\Gamma)$, while by C4, $\iota(\Theta) \leq \iota(\Gamma)$. Hence by C6 $H(\gamma(\Theta), \iota(\Theta)) \leq H(\gamma(\Gamma), \iota(\Gamma))$. The proposition follows from A13. □

Combining this proposition with Lemma C6 yields

C12 Corollary. *Given j an integer ≥ 3 or $j = \infty$ and given $k \in \mathbb{N}$, let Γ be any graph satisfying $\gamma(\Gamma) \geq j$ and $\iota(\Gamma) \geq k$. Then $\chi_0(\Gamma) \leq H(j, k)$.*

Note that for $k \geq 1$, $H(3, k) = [(7 + \sqrt{1 + 24k})/2]$. Hence by C12 and Exercise C3, we have

C13 Corollary. *If* Γ *is a nonplanar graph, then*

$$\chi_0(\Gamma) \leq \left[\frac{7 + \sqrt{1 + 24\iota(\Gamma)}}{2}\right].$$

Observe that if C13 were also valid for planar graphs, it would imply the Four-Color Conjecture. However, since $H(3, 0) = \{G(3)\} = 6$, we obtain no new information in the planar case.

From *III*A15b and C6 we obtain:

C14 Corollary. *If* Γ *is a connected nonplanar graph and if* $Z_1, \ldots, Z_m$ *is any cycle cover of* Γ, *then*

$$\chi_0 \leq \left[\frac{7 + \sqrt{49 - 24(\nu_2(Z_1, \ldots, Z_m) - \nu_1 + \nu_0)}}{2}\right].$$

C15 Corollary. *Let* Γ *be any graph with* $\iota(\Gamma) < 2$. *If* $\gamma(\Gamma) \geq 4$, *then* $\chi_0(\Gamma) \leq 4$; *if* $\gamma(\Gamma) \geq 6$, *then* $\chi_0(\Gamma) \leq 3$. (Cf. B7 and B8.)

Quite possibly the reader will immediately recognize that C14 differs from Heawood's formula [h.7] only insofar as $\nu_2(Z_1, \ldots, Z_m) - \nu_1 + \nu_0$ stands in place of the Euler characteristic. The relationship between these two parameters is considered in the next section.

VIID The Euler Characteristic and Genus of a Graph

Prerequisite to an understanding of this and the next section are some basic point-set topology and an acquaintance with the classification of surfaces. (This material is not presumed for the other sections of this text.) In particular, it will be assumed that the reader is acquainted with the Euler characteristic $\varepsilon(S)$ of a surface S. If S is an orientable surface, it is convenient to work with its **genus** $\eta(S)$ given by

D1 $$\eta(S) = 1 - \tfrac{1}{2}\varepsilon(S).$$

If S denotes the 2-dimensional sphere, for example, then $\varepsilon(S) = 2$ and $\eta(S) = 0$. While not absolutely essential, some familiarity with the fundamental homotopy group would also be helpful. We recommend as a topological reference, W. S. Massey [m.7].

Throughout this section, $\Gamma = (V, \mathscr{E})$ will denote a graph. As suggested in §*III*E, Γ may be identified with a 1-manifold, called a **topological realization** of Γ. A **topological imbedding** of Γ in a surface S is a homeomorphism from a topological realization of Γ into S. A **region** of a topological imbedding of

Γ is a connected component of the complement in S of the image of the topological imbedding of Γ. If every such region is homeomorphic to an open disk, the topological imbedding is called a **2-cell imbedding**. By abuse of terminology, the vertices and edges of Γ are identified with the images under a topological imbedding of their corresponding subspaces in a topological realization of Γ. If R is a region, the set of edges, each of which belongs to the boundary of both R and some region other than R, corresponds to a cycle of Γ, called the **cycle** of R.

We begin by indicating rather intuitively how an arbitrary graph Γ always has a topological imbedding in *some* orientable surface. Consider a topological realization of Γ in 3-space and "thicken" it in the sense that edges become rods. Clearly Γ can be topologically imbedded on the surface of such a structure.

The **genus** $\eta(\Gamma)$ of the graph Γ is the genus of the orientable surface S of least genus such that Γ can be topologically imbedded in S. An imbedding of Γ in an orientable surface of genus $\eta(\Gamma)$ is called a **minimal imbedding.** The **Euler characteristic** $\varepsilon(\Gamma)$ of Γ is the Euler characteristic of the surface S of greatest Euler characteristic such that Γ can be topologically imbedded in S. An imbedding of Γ in a surface of Euler characteristic $\varepsilon(\Gamma)$ is called a **simplest imbedding**. Clearly $\varepsilon(\Theta) \geq \varepsilon(\Gamma)$ and $\eta(\Theta) \leq \eta(\Gamma)$ for any subgraph Θ of Γ.

We next indicate how *a minimal imbedding of a connected graph is always a 2-cell imbedding*. If there were a region R of the imbedding which were not a 2-cell, one could imbed a closed loop (homeomorphic image of a 1-sphere) which is homotopically nontrivial in S so that it lies entirely within R and hence is disjoint from the realization of Γ. Now cut S along this loop and cap off both ends with disks. We have produced a surface T with $\eta(T) < \eta(S) = \eta(\Gamma)$, and yet Γ is topologically imbedded in T, which is absurd. A similar argument clearly can be used to show that *any simplest imbedding of* Γ *must also be a* 2-*cell imbedding*. (See J. W. T. Youngs [y.1] for a rigorous treatment of these notions.) If a graph has a 2-cell imbedding, then the graph must be connected.

Let $R_1, \ldots, R_m$ be the regions of a topological imbedding of Γ in the surface S. Let Z_i be the cycle of R_i, and let $\bar{R}_i$ be the closure in S of R_i. If $M \subseteq \{1, \ldots, m\}$, then the boundary of $\bigcup_{i \in M} \bar{R}_i$ corresponds to the cycle $\sum_{i \in M} Z_i$. Since $\bigcup_{i \in M} \bar{R}_i$ has a nonempty boundary if and only if $\varnothing \subset M \subset \{1, \ldots, m\}$, it follows that $\sum_{i=1}^{m} Z_i = \varnothing$ is the only relation holding among the list of cycles $Z_1, \ldots, Z_m$. We have just shown that $\nu_2(Z_1, \ldots, Z_m) = m$ *when* $Z_1, \ldots, Z_m$ *are the cycles of the regions of a topological imbedding.*

We presume that the reader is acquainted with the fact that if a graph Γ admits a 2-cell imbedding on a surface S with precisely ν_2 regions, then

D2 $$\varepsilon(S) = \nu_2 - \nu_1(\Gamma) + \nu_0(\Gamma).$$

Equation D2 is known as the Euler Formula, and *III*F2b is but the planar case.

D3 Proposition. *For any graph* Γ, $2 - \varepsilon(\Gamma) \leq 2\eta(\Gamma)$, *and if* Γ *is connected, then* $\iota(\Gamma) \leq 2 - \varepsilon(\Gamma)$.

PROOF. In determining $\varepsilon(\Gamma)$ one is not restricted to consideration of orientable manifolds. Hence the first inequality follows from D1.

Now suppose that Γ is connected. Let $Z_1, \ldots, Z_m$ be the cycles of the regions of some simplest—and hence 2-cell—imbedding of Γ. Since $\nu_{-1}(\Gamma) = 1$, C1 and D2 yield

$$\iota(\Gamma) \leq \iota(Z_1, \ldots, Z_m) = 2 - \nu_0(\Gamma) + \nu_1(\Gamma) - m = 2 - \varepsilon(\Gamma). \qquad \square$$

D4 Corollary. *For any nonplanar graph* Γ,

$$\chi_0(\Gamma) \leq \left[\frac{7 + \sqrt{49 - 24\varepsilon(\Gamma)}}{2}\right] \leq \left[\frac{7 + \sqrt{1 + 48\eta(\Gamma)}}{2}\right].$$

PROOF. The right-hand inequality follows from D3 and C6. By A2, Γ has a component Θ such that $\chi_0(\Theta) = \chi_0(\Gamma)$. If Θ is planar, then $\chi_0(\Gamma) = \chi_0(\Theta) \leq 5 < 6 \leq H(3, 2 - \varepsilon(\Gamma))$, since $\varepsilon(\Gamma) \leq 1$. If Θ is nonplanar, then by C13, C6, D3, and the fact that $\varepsilon(\Theta) \geq \varepsilon(\Gamma)$, we have $\chi_0(\Gamma) = \chi_0(\Theta) \leq H(3, \iota(\Theta)) \leq H(3, 2 - \varepsilon(\Theta)) \leq H(3, 2 - \varepsilon(\Gamma))$. This proves the left-hand inequality. $\square$

D5 *Exercise.* Prove that the connectedness hypothesis may be dropped in Proposition D3.

Equality need not hold among the three terms ι, $2 - \varepsilon$, and 2η of Proposition D3. For example, K_5 can be topologically imbedded in both the projective plane and the torus but (see *III*F9) not on the sphere. Hence $2 - \varepsilon(K_5) = 1$ while $2\eta(K_5) = 2$. More strongly yet, Auslander, Brown, and Youngs [a.2] have produced a sequence of graphs Γ_n ($n = 1, 2, \ldots$) which can be imbedded in the projective plane, yielding $2 - \varepsilon(\Gamma_n) = 1$, but with genus $\eta(\Gamma_n) = n$.

That $\iota(\Gamma) < 2 - \varepsilon(\Gamma)$ can hold is shown by the next exercise.

D6 *Exercise.* Let $V = \{x_{i,j}: i = 1, 2, 3, 4; j = 1, 2\}$ and let

$$\mathscr{E} = \{\{x_{i,j}, x_{p,q}\}: \text{either } j = q \text{ and } i \neq p \text{ or } i = p \text{ and } j \neq q\}.$$

Let $\Gamma = (V, \mathscr{E})$. Show that $\iota(\Gamma) = 1$ while $\varepsilon(\Gamma) = 0$.

For any surface S we define

$$\chi_0(S) = \max\{\chi_0(\Gamma): \Gamma \text{ is topologically imbeddable in } S\}.$$

It is presumed that the reader knows that if S_1 and S_2 are orientable surfaces with $\eta(S_1) \leq \eta(S_2)$, then any graph topologically imbeddable in S_1 is also topologically imbeddable in S_2. It follows immediately that

D7 $$\eta(S_1) \leq \eta(S_2) \Rightarrow \chi_0(S_1) \leq \chi_0(S_2).$$

D8 Theorem (P. J. Heawood [h.7], 1890). *If S is an orientable surface with $\eta(S) \geq 1$, then*

$$\chi_0(S) \leq \left[\frac{7 + \sqrt{1 + 48\eta(S)}}{2}\right].$$

PROOF. By the definition of $\chi_0(S)$, one may select a graph Γ which is topologically imbeddable in S and such that $\chi_0(\Gamma) = \chi_0(S)$. Clearly $\eta(\Gamma) \leq \eta(S)$. The result now follows from D4. □

In his 1890 paper, Heawood asserted (but did not in general prove) that, moreover,

D9 $$\chi_0(S) = \left[\frac{7 + \sqrt{1 + 48\eta(S)}}{2}\right] \quad \text{for } \eta(S) \geq 1,$$

i.e., that equality holds in D8. This equality D9 came to be known as the "Heawood conjecture." The first major advance in proving it was initiated by Gerhard Ringel in the 1950's when he divided the problem into twelve cases according to the congruence class modulo 12 of $\chi_0(S)$. A few of these cases were easily disposed of, but the last three cases (the congruence classes 2, 8, and 11) were finally resolved in 1967–68 through the joint work of Ringel and J. W. T. Youngs. Somewhat surprisingly, the corresponding result for nonorientable surfaces had been completed much more easily by Ringel [r.4] in 1959. Combining Ringel's result with D1 and D9 yields

D10 Theorem. *If S is any surface with $\varepsilon(S) \leq 1$ (i.e., except for the 2-sphere), then*

$$\chi_0(S) = \left[\frac{7 + \sqrt{49 - 24\varepsilon(S)}}{2}\right].$$

Combining the Four-Color Theorem, with Theorem D10 yields

D11 Theorem. *If S is a surface, then*

$$\chi_0(S) \leq \left[\frac{7 + \sqrt{49 - 24\varepsilon(S)}}{2}\right].$$

For a self-contained and comprehensive proof of D9 and D10, the reader should consult Ringel [r.5].

D12 Proposition. *For all integers $n \geq 3$,*

$$\eta(K_k) \geq \left\{\frac{(n-3)(n-4)}{12}\right\}.$$

PROOF. For some minimal imbedding of K_n, let n_i denote the number of regions which are incident with exactly i edges ($i \geq 3$). Hence the total number of regions is

$$\nu_2 = \sum_{i=3}^{\infty} n_i,$$

whence

$$2\nu_1(\mathrm{K}_n) = \sum_{i=3}^{\infty} in_i = 3\nu_2 + \sum_{i=3}^{\infty} (i-3)n_i.$$

Suppressing the argument K_n, we remark that $\nu_0 = n$ and $\nu_1 = n(n-1)/2$. Combining these equalities with D1 and D2, one straightforwardly obtains

$$\begin{aligned}\eta(\mathrm{K}_n) &= 1 - \frac{\nu_0}{2} + \frac{\nu_1}{2} - \frac{1}{2}\left(\frac{2}{3}\nu_1 - \frac{1}{3}\sum_{i=3}^{\infty}(i-3)n_i\right) \\ &= 1 - \frac{n}{2} + \frac{n(n-1)}{12} + \frac{1}{6}\sum_{i=3}^{\infty}(i-3)n_i \\ &\geq \frac{(n-3)(n-4)}{12}.\end{aligned}$$

The proposition now follows since $\eta(\mathrm{K}_n)$ is an integer. □

The Ringel-Youngs proof of Heawood's conjecture (D9) was accomplished by proving that, in fact,

D13 $$\eta(\mathrm{K}_n) = \left\{\frac{(n-3)(n-4)}{12}\right\} \quad \text{for all } n \geq 3.$$

Let us see how D9 may be derived from D13.

We define the two functions

$$g(x) = \left\{\frac{(x-3)(x-4)}{12}\right\} \quad \text{and} \quad h(y) = \left[\frac{7 + \sqrt{1 + 48y}}{2}\right].$$

Clearly for all x, $g(x)$ is the least integer satisfying

D14 $$(x-3)(x-4) \leq 12g(x).$$

On the other hand, applying the quadratic formula to compute the zeros of the polynomial in t, $(t-3)(t-4) - 12y$, yields that $h(y)$ is the largest integer such that

D15 $$(h(y) - 3)(h(y) - 4) \leq 12y.$$

Now let S be any orientable surface with $\eta(S) \geq 1$, and let T be an orientable surface such that $\eta(T) = g(h(\eta(S)))$. It is, of course, necessary to note that $g(h(y)) \geq 1$ if $y \geq 1$, and so T exists and $\eta(T) \geq 1$. Substituting $n = h(\eta(S))$ into D13 yields $\eta(\mathrm{K}_n) = \eta(T)$, and so

D16 $$h(\eta(S)) = n = \chi_0(\mathrm{K}_n) \leq \chi_0(T).$$

Letting $y = \eta(S)$ in D15 gives

$$(h(\eta(S)) - 3)(h(\eta(S)) - 4) \leq 12\eta(S).$$

But letting $x = h(\eta(S))$ in D14 gives that $g(h(\eta(S)))$ is the least integer satisfying

$$(h(\eta(S)) - 3)(h(\eta(S)) - 4) \leq 12g(h(\eta(S))).$$

Hence $\eta(T) = g(h(\eta(S))) \leq \eta(S)$. By D7 and D16, $h(\eta(S)) \leq \chi_0(S)$, i.e.,

$$\chi_0(S) \geq \left\lfloor \frac{7 + \sqrt{1 + 48\eta(S)}}{2} \right\rfloor,$$

which is just the reverse inequality of D8, whence D9 follows.

We next present a result due to Dirac, first published in 1952 [d.3]. The short proof below [d.5] appeared in 1957, well before D10 was proved, which would have made the proof even a little shorter. We nonetheless use the 1957 proof to underline that D10 is not essential. Taking this naïve approach, we define a function

$$h(S) = [\tfrac{1}{2}(7 + \sqrt{49 - 24\varepsilon(S)})]$$

for any surface S, ignoring temporarily that $h(S) = \chi_0(S)$, but bearing in mind from D4 that, if any nonplanar graph Γ is topologically imbeddable in S, then $\chi_0(\Gamma) \leq h(S)$.

D17 Theorem (G. A. Dirac). *Let S be a surface such that either $\varepsilon(S) = 0$ or $\varepsilon(S) \leq -2$. Suppose that the graph Γ can be topologically imbedded in S and that $\chi_0(\Gamma) = h(S)$. Then Γ contains a complete subgraph on $\chi_0(\Gamma)$ vertices.*

Proof. Let Γ be topologically imbedded in S and let Θ be a $\chi_0(\Gamma)$-critical subgraph of Γ. Thus Θ is topologically imbedded in S, and it suffices to prove the theorem for Θ. Let ν_2 denote the number of regions of such an imbedding, and suppose that $\Theta \neq K_{h(S)}$. We suppress the argument Θ for the remainder of the proof.

Since each region is incident with at least three edges of Θ while each edge is incident with at most 2 regions, we have $\nu_2 \leq 2\nu_1/3$. We substitute this into D2 and, since Θ is critical but not complete, we also invoke Theorem A20 to obtain

$$(\chi_0 - 1)\nu_0 + \chi_0 - 3 \leq 2\nu_1 \leq 6\nu_0 - 6\varepsilon(S),$$

whence

D18 $$(\chi_0 - 7)\nu_0 + \chi_0 - 3 + 6\varepsilon(S) \leq 0.$$

Let us at this point dispose of two special cases. If $\varepsilon(S) = 0$, then $\chi_0 = h(S) = 7$, but this yields an absurdity when substituted into D18. If $\varepsilon(S) = -2$, then $\chi_0 = h(S) = 8$, which yields $\nu_0 \leq 7$ when substituted into D18. But $\chi_0 \leq \nu_0$ always.

Recalling that K_{χ_0} is not a subgraph of Θ, we combine A9 with D18 to get

$$\chi_0^2 - 4\chi_0 - 17 + 6\varepsilon(S) \leq 0.$$

By the quadratic formula,

D19 $$\chi_0 \leq 2 + [\sqrt{21 - 6\varepsilon(S)}].$$

Now suppose that $\varepsilon(S) \leq -10$. One verifies by elementary operations that this assumption implies

$$\sqrt{21 - 6\varepsilon(S)} \leq \tfrac{1}{2}(1 + \sqrt{49 - 24\varepsilon(S)}),$$

and so by D19,

$$\chi_0 \leq 2 + [\tfrac{1}{2}(1 + \sqrt{49 - 24\varepsilon(S)})] = h(S) - 1,$$

contrary to hypothesis. One can dispose of the remaining cases $-9 \leq \varepsilon(S) \leq -3$ one by one in a similar fashion. □

D20 *Exercise*. Verify D17 for $-9 \leq \varepsilon(S) \leq -3$, but show that it fails when $\varepsilon(S) = 2$.

Feigning naïveté no longer, we may combine D10 with D17, to obtain

D21 Corollary. *Let S be a surface such that either $\varepsilon(S) = 0$ or $\varepsilon(S) \leq -2$. The only $\chi_0(S)$-critical graph topologically imbeddable in S is complete.*

We use the Heawood Theorem once again to show that a graph Γ with $\varepsilon(\Gamma) \leq -2$ can always be colored in such a way that some one vertex itself comprises a color class.

D22 Proposition (N. Sider [s.6], 1971). *Let S be a surface such that $\varepsilon(S) \leq -2$. Suppose that the graph Γ can be topologically imbedded in S. Then there exists a vertex x of Γ such that $\chi_0(\Gamma_{(x)}) < \chi_0(S)$.*

Proof. We shall assume that $\chi_0(\Gamma) = \chi_0(S)$ and that $\nu_0(\Gamma) \geq 2\chi_0(\Gamma)$; otherwise the result is immediate.

By D10, $\chi_0(\Gamma) > \frac{1}{2}(5 + \sqrt{49 - 24\varepsilon(S)})$, whence

D23
$$(\chi_0(\Gamma))^2 - 5\chi_0(\Gamma) - 6 > -6\varepsilon(S).$$

If $\varepsilon(S) \leq -3$, then $\chi_0(\Gamma) \geq 9$, which implies

$$(\chi_0(\Gamma))^2 - 9\chi_0(\Gamma) + 6 > 0.$$

Adding this to D23 and dividing by $2\chi_0(\Gamma)$ yields

D24
$$\chi_0(\Gamma) - 1 > 6\left(1 - \frac{\varepsilon(\Gamma)}{2\chi_0(\Gamma)}\right).$$

If $\varepsilon(S) = -2$, then $\chi_0(S) = 8$, and one verifies directly that D24 still holds.

We next prove that for any subgraph Θ of Γ

D25
$$6\left(1 - \frac{\varepsilon(\Theta)}{\nu_0(\Theta)}\right) \geq \rho(\Theta).$$

Consider a simplest imbedding of Θ. If Θ is not connected, we adjoin additional edges to Θ in order to obtain a connected graph Θ' imbeddable on the same surface. Since $\varepsilon(\Theta) = \varepsilon(\Theta')$, $\nu_0(\Theta) = \nu_0(\Theta')$, and $\rho(\Theta) \leq \rho(\Theta')$, we may assume that Θ is connected and that this topological imbedding is a 2-cell imbedding. Let ν_2 denote the number of regions of this topological imbedding. Again as in the proof of D17, $\nu_2 \leq 2\nu_1/3$. By definition, $\rho(\Theta)\nu_0(\Theta) = 2\nu_1(\Theta) \leq 2\nu_1$. These two inequalities together with D2 yield D25.

Suppose now that Θ is any subgraph of Γ such that $\nu_0(\Theta) \geq 2\chi_0(\Gamma)$. Since $\varepsilon(\Theta) \geq \varepsilon(\Gamma)$,

$$-\frac{\varepsilon(\Gamma)}{2\chi_0(\Gamma)} \geq -\frac{\varepsilon(\Theta)}{\nu_0(\Theta)},$$

and so by D24 and D25,

$$\nu_0(\Theta) \geq 2\chi_0(\Gamma) \Rightarrow \chi_0(\Gamma) - 2 \geq \rho(\Theta).$$

Thus whenever $\nu_0(\Theta) \geq 2\chi_0(\Gamma)$, Θ admits a vertex incident with at most $\chi_0(\Gamma) - 2$ other vertices in Θ. Specifically, if $p = \nu_0(\Gamma) - 2\chi_0(\Gamma) + 1$, then Γ admits a sequence $x_1, \ldots, x_p \in V$ such that the valence of x_i in $\Gamma_{V+\{x_{i+1},\ldots,x_p\}}$ is at most $\chi_0(\Gamma) - 2$, for $i = 1, \ldots, p-1$.

Let $h_0 \colon V + \{x_1, \ldots, x_p\} \to \{1, \ldots, \chi_0(\Gamma)\}$ be a vertex coloring of $\Gamma_{V+\{x_1,\ldots,x_p\}}$. Since $|V + \{x_1, \ldots, x_p\}| = 2\chi_0(\Gamma) - 1$, there exists a vertex $x \in V + \{x_1, \ldots, x_p\}$ and a color $j \in \{1, \ldots, \chi_0(\Gamma)\}$ such that $\{x\} = h_0^{-1}[j]$.

We proceed by induction to extend h_0 to a vertex coloring h of Γ such that the property $\{x\} = h^{-1}[j]$ holds. It will then be evident that $\chi_0(\Gamma_{(x)}) < \chi_0(\Gamma)$ as required. Assume that h_0 has been extended to a vertex coloring h_{i-1} of $\Gamma_{V+\{x_i,\ldots x_p\}}$ such that $\{x\} = h_{i-1}^{-1}[j]$, where $i \in \{1, \ldots, p\}$. Because the valence of x_i in $\Gamma_{V+\{x_{i+1},\ldots,x_p\}}$ is at most $\chi_0(\Gamma) - 2$, at least two color classes contain no vertex incident with x_i. Therefore, it is possible to extend h_{i-1} to a vertex coloring h_i of $\Gamma_{V+\{x_{i+1},\ldots,x_p\}}$ so that $h_i(x_i) \neq j$. In particular, it still holds that $\{x\} = h_i^{-1}[j]$, which completes the induction. □

Using D25 and arguments similar to those in the above proof, Sider was able further to prove:

D26 *Exercise*. Given any graph $\Gamma = (V, \mathscr{E})$, there exists a subset $U \subseteq V$ such that $|U| \leq 6|\varepsilon(\Gamma)|$ and $\chi_0(\Gamma_{V+U}) \leq 7$.

Further Reference

J. Battle, F. Harary, Y. Kodama and J. W. T. Youngs [b.1].

VIIE The Edmonds Imbedding Technique

In this section we describe a procedure for producing all the 2-cell imbeddings in orientable surfaces of an arbitrary graph. Since this method was first announced by Jack Edmonds [e.1] it has come to be known as the "Edmonds imbedding technique."

Let $\Gamma = (V, \mathscr{E})$ denote an arbitrary connected graph. For each vertex $x \in V$, let x^* denote the vertex-cocycle of x and let π_x denote some given permutation on x^*. We turn our attention to alternating "cyclic" lists of vertices and edges of the form

E1 $$\mathsf{D} = (x_0, E_0, x_1, E_1, \ldots, E_{k-1}, x_k = x_0)$$

where all subscripts are to be understood modulo k, and $E_i = \{x_i, x_{i+1}\}$ for all $i \in \mathbb{Z}_k$. (The cyclic list D need not denote a true circuit, since condition (a) in the definition of an *st*-path (§*II*C) is not required to be satisfied.) By a "cyclic" list, we mean that D is identified with

$$(x_j, E_j, x_{j+1}, E_{j+1}, \ldots, E_{k-1}, x_0, E_0, \ldots, E_{j-1}, x_j)$$

for any $j = 1, \ldots, k - 1$. Such a cyclic list will be called **admissible** (with respect to $\{\pi_x : x \in V\}$) if

$$E_i = \pi_{x_i}(E_{i-1}), \quad \text{for all } i \in \mathbb{Z}_k,$$

and its **length** will be said to be k. For the present, let $\mathscr{A}$ denote the set of all admissible cyclic lists.

Let $E = \{x, y\} \in \mathscr{E}$. Given the sequence x, E, y, it should be clear how to extend it to an admissible sequence. One begins with $x, E, y, \pi_y(E), \pi_y(E) + \{y\}$, and continues until the vertex x appears again immediately following the edge $\pi_x{}^{-1}(E)$. We have shown that the sequence x, E, y thus determines a unique element of $\mathscr{A}$. Moreover, each edge $E = \{x, y\}$ yields two sequences x, E, y and y, E, x, each of which determines an admissible cyclic list. These two cyclic lists need not be distinct; they never are if E is an isthmus, as the reader can readily verify. We conclude that the sequences x, E, y, and y, E, x each appear exactly once in the totality of cyclic lists in $\mathscr{A}$.

For each cyclic list $\mathsf{D}_j \in \mathscr{A}$ of length k_j, let D_j be a copy of the unit disk whose boundary is a topological imbedding of the k_j "circuit" D_j. A cell complex can then be constructed when identically labeled vertices and edges on the boundaries of the various disks, or 2-cells, are identified. Intuitively speaking, two 2-cells are "sewn together" along the length of a commonly labeled edge so that the endpoints also match up, or in the case that x, E, y and y, E, x, for example, appear in the same cyclic list D_j, then the two portions of the boundary of D_j are "sewn together." Clearly the cell complex so constructed is locally a surface at every point with the possible exception of those points corresponding to vertices of Γ. Moreover, if $x \in V$, then the cell complex will be locally a surface at the point corresponding to x if and only if the permutation π_x consists of a single cycle of length $\rho(x)$. In this case, we have constructed a 2-cell imbedding of Γ in some surface.

E2 *Exercise*. Continuing the notation of the foregoing description, show that each admissible cyclic list contains no sequence of the form E, x, E if and only if none of the permutations π_x has a fixed point.

We assert that the above-constructed 2-cell imbedding is always in an orientable surface. An orientation for each 2-cell D_j is induced by the corresponding cyclic list D_j. Then, as we have mentioned, each edge appears twice, once with each possible orientation, as is required for orientability.

On the other hand, suppose now that a graph $\Gamma = (V, \mathscr{E})$ has a 2-cell imbedding in some orientable surface S, and let one of the two possible orientations ("clockwise" or "counterclockwise") be chosen. This orientation induces in a natural way a cyclic permutation π_x of x^* at each vertex $x \in V$; to wit, for each $E \in x^*$, $\pi_x(E)$ is the "next" edge in x^* after E in the sense of the given orientation. The set $\{\pi_x : x \in V\}$ of permutations determined in this way determines in turn the original 2-cell imbedding with which we started. Thus every 2-cell imbedding of Γ in an orientable surface can be obtained from a set $\{\pi_x \in \Pi(x^*) : x \in V\}$ of cyclic permutations.

To illustrate the above abstract description, we shall apply the Edmonds imbedding technique to construct a 2-cell imbedding in the torus of the graph Γ represented by Figure E3. It should be emphasized that we do not know beforehand that the surface obtained will be the torus; we only know that it will be orientable.

E3

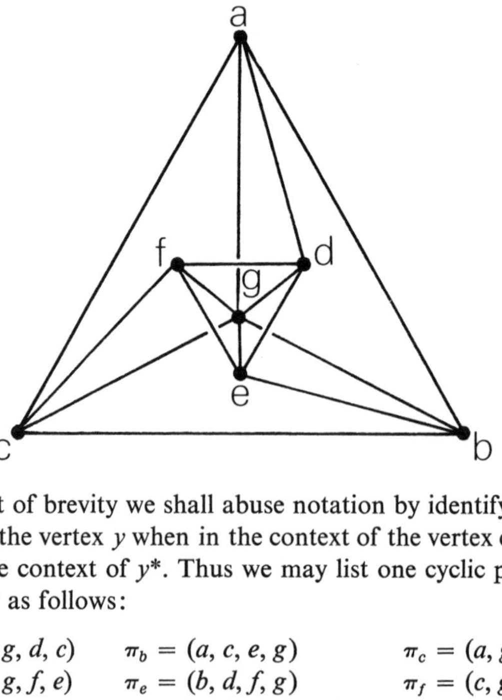

In the interest of brevity we shall abuse notation by identifying each edge $E = \{x, y\}$ with the vertex y when in the context of the vertex cocycle x^* and by x when in the context of y^*. Thus we may list one cyclic permutation at each vertex of Γ as follows:

E4
$$\pi_a = (b, g, d, c) \qquad \pi_b = (a, c, e, g) \qquad \pi_c = (a, g, f, b)$$
$$\pi_d = (a, g, f, e) \qquad \pi_e = (b, d, f, g) \qquad \pi_f = (c, g, e, d)$$
$$\pi_g = (a, b, e, f, c, d).$$

The following admissible cyclic lists are obtained (the symbols for the edges will be suppressed):

E5
$$\mathsf{D}_1 = (a, b, c, a) \qquad \mathsf{D}_2 = (a, c, g, d, f, c, b, e, d, a)$$
$$\mathsf{D}_3 = (a, d, g, a) \qquad \mathsf{D}_4 = (a, g, b, a)$$
$$\mathsf{D}_5 = (b, g, e, b) \qquad \mathsf{D}_6 = (c, f, g, c)$$
$$\mathsf{D}_7 = (d, e, f, d) \qquad \mathsf{D}_8 = (e, g, f, e).$$

(Although it is not particularly instructive, the reader may wish to sing these cyclic lists, or he may see if some labeling and 2-cell imbedding corresponds to his favorite symphonic theme.) These admissible cyclic lists yield in turn the 2-cell imbedding on the torus of Figure E6. Comparing Figure E6 with the set of permutations E4, we see that the permutations arise from this 2-cell imbedding when a clockwise orientation at each point on the surface is taken. (On the other hand, the cyclic lists E5 induce a counterclockwise orientation on each 2-cell.)

E6

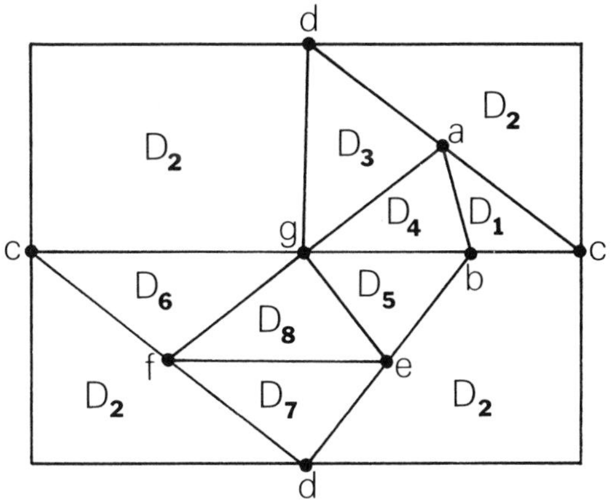

E7 *Exercise.*

(a) In the above example, there are clearly $(\rho(a) - 1)! = 6$ possible choices for π_a if π_a is to be cyclic. Keeping $\pi_b, \ldots, \pi_g$ as in E4, find the other five 2-cell imbeddings of Γ. Are they all toroidal?

(b) Give an upper bound for the number of distinct 2-cell imbeddings in orientable surfaces for an arbitrary graph.

E8 *Exercise.* Determine the genus of the Petersen graph.

E9 *Exercise.* Let the vertices of the complete graph K_n be denoted by the elements of the additive group $\mathbb{Z}_n$, and let $\pi_0 = (x_1, \ldots, x_{n-1})$ be a cyclic permutation of $\{1, \ldots, n - 1\}$, where again notation is abused as in E4. A 2-cell imbedding of K_n in an orientable surface is then defined by assigning

$$\pi_i = (x_1 + i, \ldots, x_{n-1} + i) \quad \text{for } i = 1, \ldots, n - 1.$$

(a) Prove that the cyclic permutation $(x_1, \ldots, x_{n-1})$ determines a triangular imbedding of K_n if and only if

$$x_i = x_{i-1} + x_{j-1} \quad \text{where } x_j = -x_{i-1}$$

holds for all $i = 1, \ldots, n - 1$.

(b) Verify that (1, 3, 2, 6, 4, 5) yields a triangular imbedding of K_7 on the torus, and construct it.

E10 *Exercise*. Prove:

(a) If K_n admits a triangular imbedding on some surface, then $n \equiv 0, 3, 4$, or 7 (modulo 12).

(b) Under the assumption that $n \equiv 0, 3, 4$, or 7 (modulo 12), K_n has a triangular imbedding if and only if

$$\eta(K_n) = \left\{\frac{(n-3)(n-4)}{12}\right\}.$$

(c) Use the sequence (1, 11, 14, 13, 15, 3, 8, 9, 7, 4, 17, 10, 18, 5, 16, 12, 2, 6) to prove that $\eta(K_{19}) = 20$.

(d) Prove that $\eta(K_{12}) = 6$.

Further References

E. A. Nordhaus, [n.1], and A. T. White, [w.7].

CHAPTER VIII

Two Famous Problems

In the first section of this chapter two new graph-theoretical parameters are introduced, and their relationship to the vertex independence number α_0 and the vertex chromatic number χ_0 are discussed.

In the second section we give a formal proof of a classical combinatorial theorem known as Ramsey's Theorem and include two applications of Ramsey's Theorem which are not graph-theoretical.

Various graph-theoretical applications of Ramsey's Theorem are the subject of §*C*.

All four graph-theoretical parameters play a role in the final sections of this chapter as we give a proof of the "weak Berge conjecture" and discuss the "strong Berge conjecture."

VIIIA Cliques and Scatterings

Let $\Gamma = (V, \mathscr{E})$ be a graph. The **complement** of Γ, denoted by Γ', is the graph $\Gamma' = (V, \mathscr{P}_2(V) + \mathscr{E})$. Clearly $(\Gamma')' = \Gamma$, and $(\Gamma')_S = (\Gamma_S)'$ for all $S \subseteq V$.

A1 *Exercise.* For any graph Γ, prove that either Γ or Γ' has diameter at most 3. (If Γ is not connected, its diameter is understood to be ∞.)

A subset $S \subseteq V$ will be called a **complete set** if $\mathscr{P}_2(S) \subseteq \mathscr{E}$; i.e., Γ_S is a complete subgraph. A largest complete set in Γ is called a **clique**, and $\omega(\Gamma)$ will be used to denote its cardinality. A largest independent subset of V will be called a **scattering**. Its cardinality, of course, is $\alpha_0(\Gamma)$. A subset $S \subseteq V$ is independent in Γ if and only if S is a complete set in Γ'. Thus

A2 $$\alpha_0(\Gamma) = \omega(\Gamma').$$

The least integer m such that there exists a surjection $h: V \to \{1, 2, \ldots, m\}$ such that $h^{-1}[i]$ is a complete set for each $i = 1, \ldots, m$ is denoted by $\theta(\Gamma)$. Since h is such a surjection if and only if it is a vertex m-coloring of Γ', we have

A3 $$\theta(\Gamma) = \chi_0(\Gamma').$$

In a vertex coloring of Γ, no two vertices of the same complete set receive the same color. Hence

A4 $$\omega(\Gamma) \leq \chi_0(\Gamma),$$

which together with A2 and A3 yields

A5 $$\alpha_0(\Gamma) \leq \theta(\Gamma).$$

Applying complements to *VII*A5 with $i = 0$ one obtains

A6 $$\nu_0(\Gamma) \leq \omega(\Gamma)\theta(\Gamma).$$

Ramsey's Theorem is one of the two main results in this chapter. A special case of it yields an upper bound for ν_0 in terms of the parameters α_0 and ω. The second main result, conjectured by C. Berge and proved by L. Lovász, states that $\omega(\Gamma_S) = \chi_0(\Gamma_S)$ for all $S \subseteq V$ if and only if $\alpha_0(\Gamma_S) = \theta(\Gamma_S)$ for all $S \subseteq V$.

VIIIB Ramsey's Theorem

Let integers $q_1, \ldots, q_m, s$ be given such that

B1 $$1 \leq s \leq q_i \quad (i = 1, \ldots, m)$$

where $m \geq 1$. Let V be any set, and let H be the set of all functions h of the form

$$h: \mathscr{P}_s(V) \to \{1, 2, \ldots, m\}.$$

The **Ramsey number** $n(q_1, \ldots, q_m: s)$, when it exists, is defined by

B2 $$n(q_1, \ldots, q_m: s) = \min\{|V|: (\forall h \in H)(\exists i \in \{1, \ldots, m\})(\exists U \in \mathscr{P}_{q_i}(V))[\mathscr{P}_s(U) \subseteq h^{-1}[i]]\}.$$

Otherwise, we write $n(q_1, \ldots, q_m: s) = \infty$. An equivalent formulation in words would be: $n(q_1, \ldots, q_m: s)$ is greater by 1 than the cardinality of a largest set V, if such a set exists, for which there exists an "ordered partition" of its s-subsets into at most m cells such that no q_i-subset of V has all of *its* s-subsets belonging to the ith cell.

The essence of Ramsey's Theorem is that all Ramsey numbers are finite. That is not to say that explicit formulas exist for calculating them. On the contrary, relatively few actual Ramsey numbers are known. (The known Ramsey numbers when $s = 2$ are presented in the next section.) There is some literature, however, concerning asymptotic approximations and upper and lower bounds for certain classes of Ramsey numbers. Prior to the main proof, we obtain some essential preliminary results.

First, it is immediate that if $n(q_1, \ldots, q_m: s)$ exists, then any set W such that $|W| \geq n(q_1, \ldots, q_m: s)$ also satisfies the condition stated in the right-hand member of B2. Also, if $q_i \leq \bar{q}_i$ for $i = 1, \ldots, m$, then $n(q_1, \ldots, q_m: s) \leq n(\bar{q}_1, \ldots, \bar{q}_m: s)$.

Next we note that if $m = 1$, and if $1 \leq s \leq q$, then

B3 $$n(q: s) = q. \text{ then}$$

The next result is sometimes called the "pigeon-hole principle" or the "box principle."

B4 Lemma. $$n(q_1, \ldots, q_m: 1) = \left(\sum_{i=1}^{m} q_i\right) - m + 1.$$

Proof. One may regard any $h \in H$ as a function from V into $\{1, \ldots, m\}$. It is clear that $|V| < n(q_1, \ldots, q_m: 1)$ if and only if for all $h \in H$, $|h^{-1}[i]| \leq q_i - 1$ for each $i = 1, \ldots, m$; that is, if and only if $|V| \leq (\sum_{i=1}^{m} q_i) - m$. □

B5 Exercise. Show that

(a) *If* $m \geq 2$, *and if* B1 *is satisfied, then*

$$n(q_1, \ldots, q_{m-1}: s) = n(q_1, \ldots, q_{m-1}, s: s).$$

In particular, if $m = 2$, then $n(q, s: s) = q$.

(b) *If* $\pi \in \Pi(\{1, \ldots, m\})$, *then*

$$n(q_1, \ldots, q_m: s) = n(q_{\pi(1)}, \ldots, q_{\pi(m)}: s).$$

We are now ready to state and prove a major combinatorial theorem.

B6 Theorem (F. P. Ramsey [r.2], 1930). *If* $1 \leq s \leq q_i$ *for all* $i = 1, \ldots, m$, *then the Ramsey number* $n(q_1, \ldots, q_m: s)$ *is finite.*

Proof. We proceed by induction on the variable m. The step $m = 1$ is precisely B3.

Suppose now that $m = 2$. This is the most complex step, for here we proceed by a double induction on the two variables $q_1 + q_2$ and s. Using B4 and B5 for the initial stages, our induction hypothesis is the following: for some integers q_1, q_2, and s with $q_1, q_2 > s \geq 2$, the Ramsey number $n(k_1, k_2: t)$ is finite whenever $t < s$ and whenever both $t = s$ and $k_1 + k_2 < q_1 + q_2$. The induction hypothesis implies in particular the finiteness of the following three Ramsey numbers:

$$p_1 = n(q_1 - 1, q_2: s), \quad p_2 = n(q_1, q_2 - 1: s), \text{ and } p_0 = n(p_1, p_2: s - 1).$$

We shall prove the finiteness of $n(q_1, q_2: s)$.

Suppose that V is a $(p_0 + 1)$-set and let $h: \mathscr{P}_s(V) \to \{1, 2\}$. Let $x \in V$ and define $g: \mathscr{P}_{s-1}(V + \{x\}) \to \{1, 2\}$ by

$$g(S) = h(S + \{x\}) \quad \text{for all } S \in \mathscr{P}_{s-1}(V + \{x\}).$$

Since $V + \{x\}$ is a p_0-set, the definition B2 implies existence for some $i \in \{1, 2\}$ of a set $U_i \in \mathscr{P}_{p_i}(V + \{x\})$ such that $\mathscr{P}_{s-1}(U_i) \subseteq g^{-1}[i]$. Without loss of generality we may assume the existence of the p_1-subset U_1. By the definition of p_1 and consideration of the restriction of h to $\mathscr{P}_s(U_1)$, we infer the existence of either a set $T_1 \in \mathscr{P}_{q_1-1}(U_1)$ such that $\mathscr{P}_s(T_1) \subseteq h^{-1}[1]$ or a set $T_2 \in \mathscr{P}_{q_2}(U_2)$ with $\mathscr{P}_s(T_2) \subseteq h^{-1}[2]$.

If the set T_1 exists, then $T_1 + \{x\} \in \mathscr{P}_{q_1}(U_1 + \{x\})$, and $\mathscr{P}_s(T_1 + \{x\}) \subseteq h^{-1}[1]$. Thus we have the existence of either a subset $T_1 + \{x\} \in \mathscr{P}_{q_1}(V)$ such that $\mathscr{P}_s(T_1 + \{x\}) \subseteq h^{-1}[1]$ or a subset $T_2 \in \mathscr{P}_{q_2}(V)$ such that $\mathscr{P}_s(T_2) \subseteq h^{-1}[2]$. We have in fact shown that

B7 $$n(q_1, q_2 \colon s) \le p_0 + 1,$$

and so $n(q_1, q_2 \colon s)$ is finite.

Suppose now that $m \ge 3$ and that the Ramsey number $n(q_1, \ldots, q_k \colon s)$ is finite whenever B1 holds and $k < m$. We shall prove the finiteness of $n(q_1, \ldots, q_m \colon s)$. From the case $m = 2$ above, we have the finiteness of the Ramsey number $q = n(q_{m-1}, q_m \colon s)$, and by our assumption we have the finiteness of the Ramsey number $p = n(q_1, \ldots, q_{m-2}, q \colon s)$.

Let V be a p-set and let $h \colon \mathscr{P}_s(V) \to \{1, \ldots, m\}$. By the definition of p, either for some $i \in \{1, \ldots, m - 2\}$ there exists a subset $T_i \in \mathscr{P}_{q_i}(V)$ such that $\mathscr{P}_s(T_i) \subseteq h^{-1}[i]$ or there exists a subset $U \in \mathscr{P}_q(V)$ such that $\mathscr{P}_s(U) \subseteq h^{-1}[\{m - 1, m\}]$. In the latter case, for some $i = m - 1$ or m, there exists a subset $T_i \in \mathscr{P}_{q_i}(U)$ such that $\mathscr{P}_s(T_i) \subseteq h^{-1}[i]$. We have shown that $n(q_1, \ldots, q_m \colon s) \le p$. □

The following application of the Ramsey theorem to a problem of plane geometry is due to Erdős and Szekeres [e.6]. The proof requires two preliminary results offered as exercises.

B8 Exercise. *If five points in the plane have no three points collinear, then four of the points are the vertices of a convex quadrilateral.*

B9 Exercise. *If m points in the plane have no three points collinear and if all the quadrilaterals formed by 4-subsets of these m points are convex, then the m points are the vertices of a convex m-gon.*

B10 Proposition. *Let the function f be defined by $f(3) = 3$ and $f(m) = n(5, m \colon 4)$ for $m \ge 4$. If $m \ge 3$ and $p \ge f(m)$, then any p-set of points in the plane, no three of which are collinear, contains an m-subset which is the set of vertices of a convex polygon.*

Proof. We may assume that $m \ge 4$. Let $p \ge n(5, m \colon 4)$ and let V be a p-set of points in the plane such that no three are collinear. Define $h \colon \mathscr{P}_4(V) \to \{1, 2\}$ by $h(U) = 2$ if U determines a convex quadrilateral and $h(U) = 1$ otherwise. By Exercise B8, there exists no 5-subset of V none of whose 4-subsets determines a convex quadrilateral. Since $p \ge n(5, m \colon 4)$, there exists a subset

$U \in \mathscr{P}_m(V)$ each of whose 4-subsets determines a convex quadrilateral. By Exercise B9, U determines a convex m-gon. $\square$

It is believed that the above function f is not best possible.

The next result is due to I. Schur [s.4]:

B11 Proposition. *If $\mathscr{Q} \in \mathbb{P}(\mathbb{N} + \{0\})$ and $|\mathscr{Q}| < \infty$, then there exists $Q \in \mathscr{Q}$ and $x, y \in Q$ such that $x + y \in Q$.*

Schur's result follows as an immediate corollary of the next proposition.

B12 Proposition. *Given m and $k \geq n(q_1, \ldots, q_m: 2) - 1$ where $q_1 = \ldots = q_m = 3$, then for any $\mathscr{Q} \in \mathbb{P}_m(\{1, 2, \ldots, k\})$ there exists $Q \in \mathscr{Q}$ and $x, y \in Q$ such that $x + y \in Q$.*

Proof. Denote $\{1, 2, \ldots, k + 1\}$ by V and let $\mathscr{Q} \in \mathbb{P}_m(\{1, 2, \ldots, k\})$. Define $f: \mathscr{P}_2(V) \to \{1, \ldots, k\}$ by $f(\{a, b\}) = |a - b|$. Since f is clearly surjective, there exists $\mathscr{R} \in \mathbb{P}_m(\mathscr{P}_2(V))$ whose cells are $f^{-1}[Q]$, $Q \in \mathscr{Q}$. Since $k + 1 \geq n(q_1, \ldots, q_m: 2)$, there exists a cell $Q \in \mathscr{Q}$ and a subset $\{a, b, c\} \in \mathscr{P}_3(V)$ such that $\{a, b\}, \{a, c\}, \{b, c\} \in f^{-1}[Q]$. We may assume $a < b < c$, and we then have $x = b - a$, $y = c - b$, and $z = c - a$, all belonging to Q. Thus $x, y \in Q$ and $x + y = z \in Q$. $\square$

The reader interested in the history of this problem should see the expository paper by L. Mirsky [m.12].

It is natural to define for each m the number $S(m)$ as the smallest integer such that any m-partition of $\{1, 2, \ldots, S(m)\}$ contains a cell which in turn contains two (not necessarily distinct) integers and their sum. We have shown that $S(m) \leq n(q_1, \ldots, q_m: 2) - 1$, where $q_1 = \ldots = q_m = 3$. The Ramsey numbers $n(3, 3, \ldots, 3: 2)$ will be encountered in a different context in the next section.

B13 *Exercise.* (a) Show that $S(1) = 2$ and that $S(2) = 5$. (b) Show that $S(m) \geq 2^m$. (c) Find $S(3)$.

VIIIC The Ramsey Theorem for Graphs

For brevity in this section, let us write $n(q_1, \ldots, q_m)$ for the Ramsey number $n(q_1, \ldots, q_m: 2)$. Thus we are going to restrict ourselves to the case where $s = 2$. In all other respects we continue the notation of the previous section. Now the tools of graph theory are at our disposal, for with $s = 2$, the functions $h: \mathscr{P}_2(V) \to \{1, \ldots, m\}$ correspond to ordered m-partitions of the edges of the complete graph $K_{|V|}$. The number $n(q_1, \ldots, q_m)$ may be interpreted as the number of vertices in a smallest complete graph such that for any assignment h of its edges to "colors" $1, \ldots, m$, the subgraph induced by those edges of some "color" i will contain K_{q_i} as a subgraph.

When moreover $m = 2$, the graph-theoretical problem takes on a new appearance, for one may consider h as an assignment of the 2-subsets of V to either a graph $\Gamma = (V, \mathscr{E})$ or its complement Γ'. Since a complete set in Γ' is an independent set in Γ, the number $n(q_1, q_2)$ has the following interpretation:

C1 $$n(q_1, q_2) = 1 + \max\{\nu_0(\Gamma) \colon \omega(\Gamma) < q_1 \text{ and } \alpha_0(\Gamma) < q_2\},$$

if such a maximum exists. Were C1 taken as a definition, it would make sense in spite of B1 to allow q_1 or q_2 to equal 1. For the sake of ease in proving the next proposition it is convenient to do so, and we have immediately that

C2 $$n(1, q_2) = n(q_1, 1) = 1 \quad \text{for all } q_1, q_2 \geq 1.$$

The "Ramsey Theorem for Graphs" which follows is certainly a special case of B6. Despite its redundance, its graph-theoretical proof may add insight to the proof of the general theorem.

C3 Theorem. *For all* $q_1, q_2 \geq 1$, $n(q_1, q_2)$ *exists. For all* $q_1, q_2 \geq 2$,

$$n(q_1, q_2) \leq n(q_1 - 1, q_2) + n(q_1, q_2 - 1).$$

Proof. We proceed by induction on the variable $k = q_1 + q_2$. By C2 and B5a, we know that the theorem holds for $k = 2, 3, 4$. As induction hypothesis, we choose $k \geq 5$ and suppose that the theorem holds whenever $q_1 + q_2 < k$.

Now suppose $q_1 + q_2 = k$ and let $\Gamma = (V, \mathscr{E})$ be any graph such that $\omega(\Gamma) < q_1$ and $\alpha_0(\Gamma) < q_2$. Let $x \in V$, let $S = N(x)$, and let $T = V + S + \{x\}$. We assert that $\omega(\Gamma_S) < \omega(\Gamma)$, for if U is a complete set in Γ_S, then $U + \{x\}$ is a complete set in Γ. Thus $\omega(\Gamma_S) < q_1 - 1$ and $\alpha_0(\Gamma_S) \leq \alpha_0(\Gamma) < q_2$. It follows that

C4 $$\nu_0(\Gamma_S) \leq n(q_1 - 1, q_2) - 1.$$

On the other hand, if U is independent in Γ_T, then $U + \{x\}$ is independent in Γ. So $\alpha_0(\Gamma_T) < q_2 - 1$, while $\omega(\Gamma_T) \leq \omega(\Gamma) < q_1$, and so

C5 $$\nu_0(\Gamma_T) \leq n(q_1, q_2 - 1) - 1.$$

We point out that the Ramsey numbers $n(q_1 - 1, q_2)$ and $n(q_1, q_2 - 1)$ both exist by the induction hypothesis.

Adding C4 and C5, we get

$$\begin{aligned} \nu_0(\Gamma) &= \nu_0(\Gamma_S) + \nu_0(\Gamma_T) + 1 \\ &\leq n(q_1 - 1, q_2) + n(q_1, q_2 - 1) - 1. \end{aligned}$$

Since Γ is an arbitrary graph satisfying $\omega(\Gamma) < q_1$ and $\alpha_0(\Gamma) < q_2$, the theorem follows from C1. □

C6 *Exercise.* Show that the inequality stated in Proposition C3 follows from B7 and B4.

C7 *Exercise.* Prove the following inequalities:

(a) $$n(q_1, q_2) \leq \binom{q_1 + q_2 - 2}{q_1 - 1};$$

(b) $$n(q_1, \ldots, q_m) \leq \sum_{i=1}^{m} n(q_1 - \delta_{1i}, q_2 - \delta_{2i}, \ldots, q_m - \delta_{mi}),$$

where the **Kronecker delta** $\delta_{ji} = 0$ if $j \neq i$ and 1 if $j = i$;

(c) $$n(q_1, \ldots, q_m) \leq \frac{\left[\sum_{i=1}^{m} (q_i - 1)\right]!}{\prod_{i=1}^{m} [(q_i - 1)!]}.$$

We introduce some working terminology for the present section. A graph Γ will be called a (q_1, q_2)-**graph** if $\omega(\Gamma) < q_1$ and $\alpha_0(\Gamma) < q_2$. A (q_1, q_2)-graph will be called d-**deficient**, or will be said to have **deficiency** d, if $d = n(q_1, q_2) - \nu_0(\Gamma) - 1$. Clearly $d \geq 0$.

C8 Exercise. **(a)** Show that $n(3, 3) = 6$.

(b) Show that *the 5-circuit is the only* 0-*deficient* (3, 3)-*graph.*

(c) Determine all 1-deficient (3, 3)-graphs, and show that *of these only the path of length* 3 *has exactly three edges.*

C9 Lemma. *Let $\Gamma = (V, \mathscr{E})$ be a d-deficient (q_1, q_2)-graph and let $x \in V$. Let $S = N(x)$ and let $T = V + S + \{x\}$. Then*

(a) *Γ_S is a $(q_1 - 1, q_2)$-graph and Γ_T is a $(q_1, q_2 - 1)$-graph.*

(b) *If Γ_S is a d_1-deficient $(q_1 - 1, q_2)$-graph and Γ_T is a d_2-deficient $(q_1, q_2 - 1)$-graph, then*

$$d_1 + d_2 = d + n(q_1 - 1, q_2) + n(q_1, q_2 - 1) - n(q_1, q_2).$$

(c) $n(q_1 - 1, q_2) - 1 \geq \rho(x) \geq n(q_1, q_2) - n(q_1, q_2 - 1) - d - 1.$

Proof. (a) is embodied in the proof of C3.

To prove (b) we merely add up the three equations:

$$d_1 = n(q_1 - 1, q_2) - |S| - 1$$

$$d_2 = n(q_1, q_2 - 1) - |T| - 1$$

$$0 = d - n(q_1, q_2) + \nu_0(\Gamma) + 1.$$

(c) follows from the equation $\rho(x) = |S| = n(q_1 - 1, q_2) - 1 - d_1$ and the inequality $0 \leq d_1 \leq d + n(q_1 - 1, q_2) + n(q_1, q_2 - 1) - n(q_1, q_2)$. □

C10 Proposition. $n(3, 4) = 9$.

Proof. By B5a, C8a, and C3, $n(3, 4) \leq n(2, 4) + n(3, 3) = 4 + 6 = 10$. Supposing $n(3, 4) = 10$, let Γ be a (3, 4)-graph with $\nu_0(\Gamma) = 9$; i.e., suppose Γ is

0-deficient. By Lemma C9c, Γ is 3-valent, which is impossible since $\nu_0(\Gamma)$ is odd. Hence $n(3, 4) \le 9$. But equality holds since the graphs in Figure C11 are (3, 4)-graphs with eight vertices. □

C11

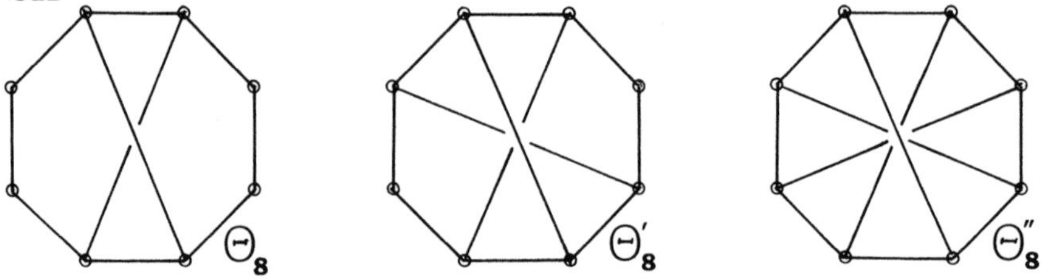

C12 Proposition. *The only 0-deficient* (3, 4)*-graphs are* Θ_8, Θ_8', *and* Θ_8'' *represented in Figure* C11.

Proof. Let $\Gamma = (V, \mathscr{E})$ be a 0-deficient (3, 4)-graph; thus $\nu_0(\Gamma) = 8$. Let $x \in V$. By Lemma C9c, $3 \ge \rho(x) \ge 2$.

Case 1: *there exists a vertex of valence* 2. Let $\rho(x) = 2$ and let y and z be incident with x. If $T = V + \{x, y, z\}$, then by Lemma C9a, Γ_T is a 0-deficient (3, 3)-graph, which by Exercise C8b must be a 5-circuit. Thus Γ must contain the subgraph shown in Figure C13, together with a set $\mathscr{F}$ of edges such that

C13

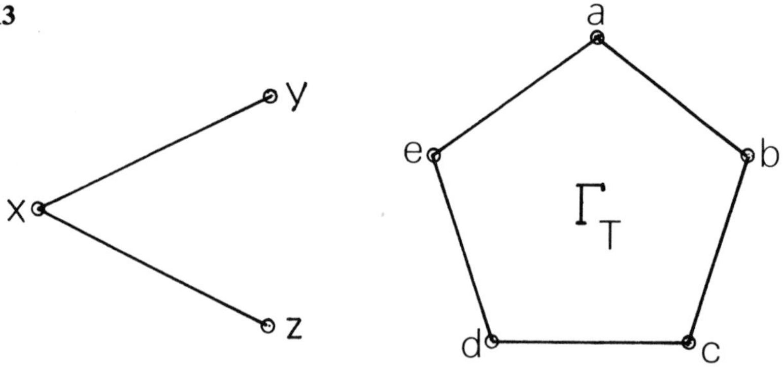

for each $F \in \mathscr{F}$, one vertex of F is in $\{y, z\}$ and the other is in $\{a, b, c, d, e\}$. Since by C9c each vertex has valence 2 or 3, y and z must each be incident with either one or two edges in $\mathscr{F}$, and each of a, b, c, d, and e can be incident with at most one edge in $\mathscr{F}$. We conclude that one, two, or three of the vertices a, b, c, d, and e are incident with no edge in $\mathscr{F}$. Without loss of generality we may assume that a is incident with neither y nor z. If c (or d) were incident with neither y nor z, then $\{a, c, y, z\}$ (or $\{a, d, y, z\}$) would be an independent 4-set—which is impossible. Similarly it is not possible that both b and e are incident with neither y nor z. Hence we may assume that a alone or a and b are the only vertices of Γ_T which are incident with neither y nor z.

We may further assume without loss of generality that z has valence 3. Since Γ contains no complete 3-set, z is incident with b and d or with c and e. For definiteness say z is incident with c and e. Finally, y is incident with d and possibly with b. Without the edge $\{y, b\}$ we have Θ_8 and with $\{y, b\}$ we have Θ_8'.

Case 2: all vertices are 3-valent. Let $w \in V$ and suppose w is incident with x, y, and z. Let $T = V + \{w, x, y, z\}$. Since $\{x, y, z\}$ must be an independent set, the number of edges in Γ_T is $|\mathscr{E}| - (\rho(x) + \rho(y) + \rho(z)) = 12 - 9 = 3$. By Exercise C8c, Γ_T is a path of length 3. Thus Γ must contain the subgraph in Figure C14, together with six edges each having one vertex in T and the

C14

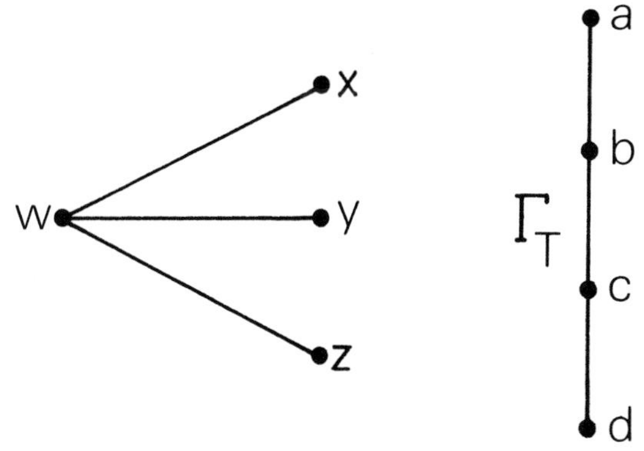

other in $\{x, y, z\}$. Since Γ contains no complete 3-set, the two vertices of Γ_T joined to each vertex in $\{x, y, z\}$ must be one of the three pairs $\{a, c\}$, $\{b, d\}$, and $\{a, d\}$. Moreover, each pair must be used exactly once in order to preserve isovalence. For example, the six edges in question could be: $\{x, a\}$, $\{x, c\}$, $\{y, b\}$, $\{y, d\}$, $\{z, a\}$, $\{z, d\}$, but any of the 3! possibilities yields the graph Θ_8''. □

C15 Exercise. *Consider the graph* $\Theta_{13} = (\mathbb{Z}_{13}, \mathscr{F})$ where $\mathscr{F} = \{\{x, y\}: x - y = 1 \text{ or } 5\}$. Show that Θ_{13} *is the only* (3, 5)-*graph with* 13 *vertices.*

C16 Proposition. $n(3, 5) = 14$.

Proof. By B5a, C10, and C3, $n(3, 5) \leq 14$. Equality follows from Exercise C15. □

C17 Exercise. Show that *the graph* $\Theta_{17} = (\mathbb{Z}_{17}, \mathscr{F})$ *where* $\mathscr{F} = \{\{x, y\}: x - y = 1, 2, 4, \text{ or } 8\}$ *is a* (4, 4)-*graph.*

C18 Proposition. $n(4, 4) = 18$.

Proof. The proposition follows from C10, B5b, C3, and C17. □

The only other known Ramsey numbers of the form $n(q_1, q_2)$ with $q_1 \leq q_2$ are $n(3, 6) = 18$ (see Exercise C20 below) and $n(3, 7) = 23$. Some estimates for other numbers are readily obtainable by C3, C9c, and various *ad hoc* techniques. Table C19 summarizes the known results for $q_1 \leq q_2$ and is footnoted by bibliographical references.

C19

$q_1 \backslash q_2$	1	2	3	4	5	6	7	8	9	
1	1	1	1	1	1	1	1	1	1	...
2		2	3	4	5	6	7	8	9	...
3			6‡	9‡	14‡	18†*	23*	27–30	36–37	
4				18‡	≤30*					

‡ R. E. Greenwood and A. M. Gleason [g.6].
† J. G. Kalbfleisch [k.1].
* J. E. Graver and J. Yackel [g.5].

Graver and Yackel showed further that for $q_1 \geq 3$, there exists a constant C such that

$$n(q_1, q_2) \leq C{q_2}^{(q_1-1)}\left(\frac{\log\log q_2}{\log q_2}\right).$$

C20 *Exercise.* Show that $n(3, 6) = 18$ by carrying out the following steps.

Step 1. Show that $n(3, 6) \leq 19$.

Step 2. Show that if $\Gamma = (V, \mathscr{E})$ is a (3, 6)-graph on 18 vertices, then $\rho(x) = 4$ or 5 for all $x \in V$. Show that if $\rho(x) = 4$ and $T = V + N(x) + \{x\}$, then Γ_T is Θ_{13}. Observe from this that each vertex of valence 4 must be incident with at least three other vertices of valence 4 and not incident with at most one vertex of valence 4. Then show that Γ is 5-valent.

Step 3. Show that Γ_T is a (3, 5)-graph with $\nu_0(\Gamma_T) = 12$ and $\nu_1(\Gamma_T) = 20$.

Step 4. Show that any (3, 5)-graph on 12 vertices with 20 edges has exactly 4 vertices of valence 4, 8 vertices of valence 3, 4 edges between vertices of valence 4, 8 edges between vertices of valence 3, and 8 edges between vertices of valence 3 and vertices of valence 4.

Step 5. Count the 5-circuits in Γ through x (each such 5-circuit has exactly one edge in Γ_T), and observe that this number is not divisible by 5. From this conclude that a (3, 6)-graph on 18 vertices cannot exist.

Step 6. Verify that $\Lambda' = (V', \mathscr{E}')$ as described below is a (3, 6)-graph on 17 vertices. Let $V' = \{0, 1, \ldots, 12, 0', 1', 8', 9'\}$, $\mathscr{E}' = \{\{a, b\}: a - b \equiv 1$ or $5 \pmod{13}\} \cup \{(a, b'\}: a - b \equiv 1$ or $5 \pmod{13}\} \cup \{\{0, 9'\}, \{0', 9\}, \{1, 8'\}, \{1', 8\}\}$. Note $\Lambda_{\{0,1,\ldots,12\}} = \Theta_{13}$. □

We conclude this section with a look at Ramsey numbers of the form $n(3, \ldots, 3)$, which we have already encountered in Proposition B12. Let us write $n(3^{(m)})$ for $n(q_1, \ldots, q_m)$, where $q_1 = \ldots = q_m = 3$. Intuitively one may think of finding a smallest complete graph such that every edge m-coloring admits a monochromatic triangle.

C21 Proposition. $n(3^{(m+1)}) \leq (m + 1)[n(3^{(m)}) - 1] + 2$, *for all* $m \geq 1$.

Proof. We proceed by induction on m. If $m = 1$, then in fact equality holds by C8a and B3.

As induction hypothesis, suppose that the given inequality holds for some $m \geq 1$. Let $p = (m + 1)[n(3^{(m)}) - 1] + 2$, and let V be a p-set. Arbitrarily choose $h: \mathscr{P}_2(V) \to \{1, \ldots, m + 1\}$.

Let $x \in V$, and define for each $i = 1, \ldots, m + 1$,

$$Q_i = \{y \in V + \{x\}: h(\{x, y\}) = i\}.$$

Noting that $(p - 1)/(m + 1) = n(3^{(m)}) - 1 + 1/(m + 1)$, we infer that $|Q_i| \geq n(3^{(m)})$ for some $i = 1, \ldots, m + 1$.

If $h(\{y, z\}) = i$ for some $y, z \in Q_i$, then $U = \{x, y, z\}$ is a 3-subset such that $\mathscr{P}_2(U) \subseteq h^{-1}[i]$. Otherwise, consider the restriction h_1 of h, namely $h_1: \mathscr{P}_2(Q_i) \to \{1, \ldots, m + 1\} + \{i\}$. But since $|Q_i| \geq n(3^{(m)})$, there exists a subset $T \in \mathscr{P}_3(Q_i)$ such that $\mathscr{P}_2(T) \subseteq h^{-1}[j]$ for some $j \in \{1, \ldots, m + 1\} + \{i\}$. Hence $n(3^{(m+1)}) \leq p$. □

Greenwood and Gleason [g.6] have determined the only other known Ramsey number as follows.

C22 Proposition. $n(3^{(3)}) = 17$.

Proof. By C8a and C21 with $m = 2$, we obtain $n(3^{(3)}) \leq 17$.

Let $V = \{0, x, x^2, \ldots, x^{15} = 1\} = GF(2^4)$, the field with 16 elements, where x satisfies $x^4 + x + 1 = 0$. We use the fact that the cubic residues of $GF(2^4)$ are the set

$$R_0 = \{x^3, x^3 + x^2, x^3 + x, x^3 + x^2 + x + 1, 1\}.$$

Thus $|R_0| = 5$ and, moreover, R_0 is a subgroup of the multiplicative group G of $GF(2^4)$. Hence $G/R_0 \cong \mathbb{Z}_3$. Let $\{R_0, R_1, R_2\}$ be the corresponding coset decomposition of G, and define $h: \mathscr{P}_2(V) \to \{0, 1, 2\}$ by

$$h(\{a, b\}) = i \Leftrightarrow a + b \in R_i, \quad i = 0, 1, 2.$$

Suppose $\mathscr{P}_2(U) \subseteq h^{-1}[i]$ for some i and some $U \in \mathscr{P}_3(V)$. Without loss of generality, say $U = \{0, 1, a\}$, and so $i = 0$. Thus $a, 1 - a \in R_0$. One sees by inspection of R_0 that this cannot happen. Hence $n(3^{(3)}) > 16$. □

It follows from Proposition C21 that $n(3^{(4)}) \leq 66$. It has been shown by Jon Folkman [f.2] that $n(3^{(4)}) \leq 65$. Greenwood and Gleason have shown that $n(3^{(4)}) > 41$.

C23 *Exercise.* Prove the following inequalities: (a) $n(3^{(m)}) \leq 3(m!)$. (b) $n(3^{(m)}) \leq m!\,e + 1$.

VIIID Perfect Graphs

A graph $\Gamma = (V, \mathscr{E})$ is said to be **color-perfect** if $\omega(\Gamma_S) = \chi_0(\Gamma_S)$ for all $S \in \mathscr{P}(V)$. By A2 and A3, Γ' is color-perfect if and only if $\alpha_0(\Gamma_S) = \theta(\Gamma_S)$ for all $S \in \mathscr{P}(V)$. A graph Γ is said to be **perfect** if both Γ and Γ' are color-perfect In 1961, C. Berge first published [b.4] the following conjecture:

D1 The Berge Conjecture. *Every color-perfect graph is perfect.*

In its complementary form, D1 asserts:

D2 *A graph is color-perfect if and only if its complement is color perfect.*

It is immediate that if $\Gamma = (V, \mathscr{E})$ is color-perfect (respectively: perfect), then so is Γ_S for all $S \in \mathscr{P}(V)$.

D3 Proposition. *All bipartite graphs are perfect.*

Proof. Let Γ be bipartite and let S be a set of vertices of Γ. If S is independent, then $\omega(\Gamma_S) = 1 = \chi_0(\Gamma_S)$, and $\alpha_0(\Gamma_S) = |S| = \theta(\Gamma_S)$. If S is not independent, then since Γ_S is also bipartite, $\omega(\Gamma_S) = 2 = \chi_0(\Gamma_S)$. (Cf. *VII*A6.) Since suppressing the isolated vertices of Γ_S would decrease $\alpha_0(\Gamma_S)$ and $\theta(\Gamma_S)$ each by the exact number of such vertices, we may assume that Γ_S has no isolated vertices. Hence $\theta(\Gamma_S)$ is precisely $\beta_{10}(\Gamma_S)$, and by *V*C11, $\beta_{10}(\Gamma_S) = \alpha_0(\Gamma_S)$.

A graph Γ will be defined to be **crucial** if Γ is not color-perfect, but $\Gamma_{(x)}$ is color-perfect for every vertex x of Γ.

D4 Corollary. *Odd circuits of length at least* 5 *and their complements are crucial.*

Proof. Let $\Delta = (V, \mathscr{E})$ be a $(2n + 1)$-circuit for $n \geq 2$. Thus

$$\omega(\Delta) = 2; \ \chi_0(\Delta) = 3; \ \alpha_0(\Delta) = n; \ \theta(\Delta) = n + 1.$$

Neither Δ nor Δ' is color-perfect. On the other hand, for any $x \in V$, the subgraph $\Delta_{(x)}$ is bipartite. By the proposition, both $\Delta_{(x)}$ and $\Delta'_{(x)}$ are color-perfect. □

The notion of a crucial graph is somewhat analogous to the notion of a critical graph with respect to vertex-colorings.

D5 *Exercise.* Prove that every crucial graph Γ satisfies $\chi_0(\Gamma) = \chi_0(\Gamma_{(x)}) + 1$ for every vertex x of Γ.

Still another formulation of the Berge Conjecture is that crucial graphs are closed under complementation. The only crucial graphs known are those described in Corollary D4. Since the latter class of graphs is closed with respect to complementation, the following sharper conjecture is indicated.

D6 Strong Berge Conjecture. *The only crucial graphs are the odd circuits of length at least* 5 *and their complements.*

In the remainder of this section we prove the Berge Conjecture (D1) and we obtain some further properties of crucial graphs. Most of these results are due to L. Lovász [ℓ.3] and [ℓ.4], but their presentation here has been considerably transformed. Those lemmas below which we have ascribed to Lovász constitute various fragments of his original proof of the conjecture.

Let $\Gamma = (V, \mathscr{E})$ be any graph, let W be a set, and let $g\colon W \to V$ be a function. We define the **Lovász graph** Λ determined by the triple (Γ, W, g) to be the graph $\Lambda = (W, \mathscr{F})$ where

$$\mathscr{F} = \{F \in \mathscr{P}_2(W)\colon g[F] \in \mathscr{E} \cup \mathscr{P}_1(V)\}.$$

We will later require the following immediate consequence of this definition.

D7 Lemma. *If* Λ *is the Lovász graph determined by* (Γ, W, g) *and if* $T \subseteq W$, *then* Λ_T *is the Lovász graph determined by both* $(\Gamma, T, g_{|T})$ *and* $(\Gamma_{g[T]}, T, g_{|T})$.

We make some elementary observations concerning Λ. Suppose $U \subseteq W$. The set U is a complete set in Λ if and only if $g[U]$ is a complete set in Γ. Moreover, if S is a complete set in Γ, then $g^{-1}[S]$ is a complete $(\sum_{x \in S} |g^{-1}[x]|)$-set in Λ. Hence $\omega(\Lambda) \geq \omega(\Gamma_{g[W]})$. If U is independent in Λ, then $g[U]$ is independent in Γ; in that case $g_{|U}$ is injective. Hence $\alpha_0(\Lambda) \leq \alpha_0(\Gamma)$.

D8 Lemma. *Let* $\Gamma = (V, \mathscr{E})$ *be a color-perfect graph. Then the Lovász graph determined by* (Γ, W, g) *for any set* W *and any* $g \in V^W$ *is also color-perfect.*

Proof. Let $\Lambda = (W, \mathscr{F})$ be the Lovász graph in question. The proof is by induction on $|W|$. If $|W| = 1$, then Λ is clearly perfect.

Suppose $|W| > 1$, and as induction hypothesis, suppose that Λ_U is color-perfect whenever $U \subset W$.

The lemma follows if g is injective, for then Λ would be isomorphic to $\Gamma_{g[W]}$. Suppose therefore that $g(u_0) = g(u_1)$ for some two distinct vertices $u_0, u_1 \in W$. Note that by definition of Λ, $\{u, u_1\} \in \mathscr{F}$ if and only if either $g(u) = g(u_0)$ or $\{u, u_0\} \in \mathscr{F}$. Let $U = W + \{u_1\}$. By the induction hypothesis, Λ_U is color-perfect.

Case 1: $u_0 \in K$ *for some clique* K *of* Λ_U. Since $K + \{u_1\}$ is thus a clique in Λ, it follows that $\omega(\Lambda) = \omega(\Lambda_U) + 1$. On the other hand, any $\chi_0(\Lambda_U)$-coloring of Λ_U can be extended to a coloring of Λ by assigning u_1 to a "new" color. Hence $\chi_0(\Lambda) \leq \chi_0(\Lambda_U) + 1 = \omega(\Lambda_U) + 1 = \omega(\Gamma)$. Combining this inequality with A4, we have $\chi_0(\Lambda) = \omega(\Lambda)$.

Case 2: u_0 *belongs to no clique of* Λ_U. Let $h\colon U \to \{1, \ldots, \chi_0(\Lambda_U)\}$ be a vertex coloring of Λ_U, and suppose $h(u_0) = 1$. Since $\omega(\Lambda_U) = \chi_0(\Lambda_U)$, each clique K of Λ_U must meet each color class $h^{-1}[i]$; i.e., $h_{|K}$ is always a bijection onto the set of colors. Since u_0 belongs to no clique of Λ_U, the set $T = h^{-1}[1] + \{u_0\}$ also meets each clique of Λ_U. Hence $\omega(\Lambda_{U+T}) = \omega(\Lambda_U) - 1$, and so $\chi_0(\Lambda_{U+T}) = \chi_0(\Lambda_U) - 1$. Let $h_1: U + T \to \{2, \ldots, \chi_0(\Lambda_U)\}$ be a vertex coloring of Λ_{U+T}. Since $T + \{u_1\}$ is independent, h_1 can be extended to a vertex $\chi_0(\Lambda_U)$-coloring h_1' of Λ by assigning $h_1'(u) = 1$ for all $u \in T + \{u_1\}$. Thus $\chi_0(\Lambda) = \chi_0(\Lambda_U) = \omega(\Lambda_U) \leq \omega(\Lambda)$, which together with A4 yields $\chi_0(\Lambda) = \omega(\Lambda)$. □

An independent set of vertices of a graph which meets every clique is called a **spread**. A complete set which meets every scattering is called a **clump**. Clearly if S is a spread in a graph Γ, then S is a clump in Γ'. Moreover, if S is a spread in Γ and C is a clump in Γ, then $\omega(\Gamma_{(S)}) = \omega(\Gamma) - 1$ and $\alpha_0(\Gamma_{(C)}) = \alpha_0(\Gamma) - 1$.

D9 Lemma (L. Lovász). *If a graph* $\Gamma = (V, \mathscr{E})$ *is color-perfect and if* $\varnothing \subset U \subseteq V$, *then* Γ_U *contains a clump.*

Proof. The proof is by induction on $|V|$. The result is trivial if $|V| = 1$. Let $|V| > 1$, and suppose that Γ is color-perfect. As induction hypothesis, assume that whenever $\varnothing \subset U \subset V$, then (since Γ_U is color-perfect) the subgraph Γ_U contains a clump. Suppose that Γ itself, however, contains no clump.

Let $\mathscr{C}$ be the collection of complete sets in Γ. For each complete set $C \in \mathscr{C}$, there exists a scattering S_C such that $S_C \cap C = \varnothing$. For each $C \in \mathscr{C}$, let W_C be an $\alpha_0(\Gamma)$-set, let $g_C: W_C \to S_C$ be a bijection, and assume that the sets $\{W_C: C \in \mathscr{C}\}$ are pairwise disjoint. Let $W = \bigcup_{C \in \mathscr{C}} W_C$ and define $g: W \to V$ by $g(w) = g_C(w)$ if $w \in W_C$. Let Λ be the Lovász graph determined by (Γ, W, g).

Let K be a complete set in Λ. Since $g[K]$ is complete in Γ, $|g[K] \cap S| \leq 1$ for any scattering S in Γ. Hence $|K \cap W_C| \leq 1$ for any complete set $C \in \mathscr{C}$ (cf. *I*A14). In particular, however, since $g[K] \cap S_{g[K]} = \varnothing$, we have $K \cap W_{g[K]} = \varnothing$. Hence $|K| \leq |\mathscr{C}| - 1$, and we conclude that $\omega(\Lambda) < |\mathscr{C}|$.

By Lemma D8, $\chi_0(\Lambda) = \omega(\Lambda)$, and so $\chi_0(\Lambda)\alpha_0(\Lambda) < |\mathscr{C}|\alpha_0(\Gamma) = |W| = \nu_0(\Lambda)$, contrary to *VII*A5. □

D10 Lemma (L. Lovász). *Let* $\Gamma = (V, \mathscr{E})$. *If* Γ_U *contains a spread whenever* $\varnothing \subset U \subseteq V$, *then* Γ *is color-perfect.*

Proof. The proof again proceeds by induction on $|V|$ and is trivial for $|V| = 1$. Now suppose $|V| > 1$ and suppose that Γ_U contains a spread whenever $\varnothing \subset U \subseteq V$. Applying the induction hypothesis to Γ_U for $\varnothing \subset U \subset V$, we have that Γ_U is color perfect.

Let S be a spread in Γ. Since $\Gamma_{(S)}$ is color-perfect, $\chi_0(\Gamma_{(S)}) = \omega(\Gamma_{(S)}) = \omega(\Gamma) - 1$. Clearly any vertex $(\omega(\Gamma) - 1)$-coloring of $\Gamma_{(S)}$ can be extended to a vertex $\omega(\Gamma)$-coloring of Γ by the assignment of every vertex in S to the one "new" color. Hence $\chi_0(\Gamma) \leq \omega(\Gamma)$, which together with A4 yields that Γ is color-perfect. □

These two lemmas by Lovász allow us to prove the Berge Conjecture D2, which we present more comprehensively as follows:

D11 Theorem. *Let $\Gamma = (V, \mathscr{E})$ be a graph. The following four statements are equivalent:*

(a) *Γ is color-perfect;*
(b) *Γ_U contains a clump whenever $\varnothing \subset U \subseteq V$;*
(c) *Γ' is color-perfect;*
(d) *Γ_U contains a spread whenever $\varnothing \subset U \subseteq V$.*

Proof. (a) $\Rightarrow$ (b) by Lemma D9.
(b) $\Rightarrow$ (c) by Lemma D10 applied to Γ'.
(c) $\Rightarrow$ (d) by Lemma D9 applied to Γ'.
(d) $\Rightarrow$ (a) by Lemma D10. □

A characterization of crucial graphs is an immediate consequence of this theorem.

D12 Corollary. *A graph $\Gamma = (V, \mathscr{E})$ is crucial if and only if Γ contains no spread and no clump, but whenever $\varnothing \subset U \subset V$, then Γ_U contains both a spread and a clump.*

In order to give a further characterization of crucial graphs, we require the following lemma.

D13 Lemma (Lovász). *Let $\Gamma = (V, \mathscr{E})$. Suppose that Γ_U is perfect whenever $U \subset V$ and that $\nu_0(\Gamma) \leq \alpha_0(\Gamma)\omega(\Gamma)$. Then for any Lovász graph Λ determined by Γ, $\nu_0(\Lambda) \leq \alpha_0(\Lambda)\omega(\Lambda)$.*

Proof. Let Λ be determined by (Γ, W, g). We proceed by induction on $|W|$. When $|W| = 1$, the desired inequality reads $1 \leq 1$. Assume as induction hypothesis that $\nu_0(\Lambda_T) \leq \alpha_0(\Lambda_T)\omega(\Lambda_T)$ whenever $T \subset W$.

If g is not surjective, then $\Gamma_{g[W]}$ is perfect by the hypothesis of this lemma. Since Λ is also determined by $(\Gamma_{g[W]}, W, g)$, Lemma D8 yields that Λ is color-perfect, and the desired inequality follows from *VII*A5.

We assume, therefore, that g is a surjection. It may be presumed that g is not an injection, for otherwise Γ and Λ would be isomorphic graphs. Therefore, there exists $x \in V$ such that $g^{-1}[x] = C$ where $|C| > 1$. As mentioned earlier, C must be complete in Λ. Fix $u \in C$, and observe that since the function $g_{|W+\{u\}}$ is also a surjection (onto V), $\alpha_0(\Lambda_{(u)}) = \alpha_0(\Gamma) = \alpha_0(\Lambda)$. (See the remark preceding Lemma D8.) This together with the induction hypothesis yields

$$\nu_0(\Lambda) - 1 = \nu_0(\Lambda_{(u)}) \leq \alpha_0(\Lambda_{(u)})\omega(\Lambda_{(u)}) \leq \alpha_0(\Lambda)\omega(\Lambda).$$

Thus if the desired inequality fails for Λ, then it fails only by 1; i.e., we may suppose

D14 $$\nu_0(\Lambda) = \alpha_0(\Lambda)\omega(\Lambda) + 1.$$

Since $\Lambda_{(C)}$ is the Lovász graph determined by $(\Gamma_{(x)}, W + C, g_{|W+C})$ and since $\Gamma_{(x)}$ is perfect, Lemma D8 yields that $\Lambda_{(C)}$ is color-perfect. Since $\chi_0(\Lambda_{(C)}) = \omega(\Lambda_{(C)}) \leq \omega(\Lambda)$, there exists a vertex coloring $h: W + C \to \{1, \ldots, \omega(\Lambda)\}$ of $\Lambda_{(C)}$. By D14,

$$\alpha_0(\Lambda)\omega(\Lambda) = \nu_0(\Lambda_{(C)}) + |C| - 1 = \sum_{i=1}^{\omega(\Lambda)} |h^{-1}[i]| + |C| - 1.$$

Since no color class $h^{-1}[i]$ contains more than $\alpha_0(\Lambda)$ vertices, at most $|C| - 1$ color classes fail to be scatterings. Let T denote the union of exactly $\omega(\Lambda) - |C| + 1$ color classes in $W + C$ with respect to h which *are* scatterings. Thus

$$|T + \{u\}| = \alpha_0(\Lambda)(\omega(\Lambda) - |C| + 1) + 1.$$

But by the induction hypothesis, $\nu_0(\Lambda_{T+\{u\}}) \leq \alpha_0(\Lambda_{T+\{u\}})\omega(\Lambda_{T+\{u\}}) \leq \alpha_0(\Lambda)\omega(\Lambda_{T+\{u\}})$, whence

D15 $$\omega(\Lambda) - |C| + 1 < \omega(\Lambda_{T+\{u\}}).$$

Recall that by the definition of T, $\omega(\Lambda_T) \leq \omega(\Lambda) - |C| + 1$. Combining this with D15 yields $\omega(\Lambda_{T+\{u\}}) > \omega(\Lambda_T)$. Therefore $u \in C'$ for some clique C' of $\Lambda_{T+\{u\}}$. Since $C' \cup C$ is a complete set, $\omega(\Lambda) \geq |C' \cup C| = |C'| + |C| - 1 = \omega(\Lambda_{T+\{u\}}) + |C| - 1$, contrary to D15. □

D16 Theorem (Lovász). *A graph* $\Gamma = (V, \mathscr{E})$ *is perfect if and only if* $\nu_0(\Gamma_S) \leq \alpha_0(\Gamma_S)\omega(\Gamma_S)$ *for all* $S \in \mathscr{P}(V)$.

Proof. Suppose that Γ is perfect, and let $S \in \mathscr{P}(V)$. Since Γ_S is then perfect, too, we have by *VII*A5, $\nu_0(\Gamma_S) \leq \alpha_0(\Gamma_S)\chi_0(\Gamma_S) = \alpha_0(\Gamma_S)\omega(\Gamma_S)$.

We prove the converse by induction on $|V|$. The assertion being trivial for $|V| = 1$, we assume that Γ_S is perfect whenever $\varnothing \subset S \subset V$ and that $\nu_0(\Gamma) \leq \alpha_0(\Gamma)\omega(\Gamma)$.

Suppose that Γ is not perfect. By Theorem D11, Γ contains no clump. We now repeat precisely the construction in the proof of Lemma D9. Thus all the symbols $\mathscr{C}$, S_C, W_C, g_C, W, g, and Λ will reassume their previous meanings, and by the same argument as in the proof of D9, we obtain:

D17 $$\omega(\Lambda) < |\mathscr{C}|.$$

However, by Lemma D13 together with the means by which Λ is constructed, we have

$$\alpha_0(\Lambda)\omega(\Lambda) \geq |W| = |\mathscr{C}|\alpha_0(\Gamma) \geq |\mathscr{C}|\alpha_0(\Lambda),$$

and so $|\mathscr{C}| \leq \omega(\Lambda)$, contrary to D17. □

D18 Corollary. *If* Γ *is crucial, then*

$$\nu_0(\Gamma) = \alpha_0(\Gamma)\omega(\Gamma) + 1.$$

PROOF. If Γ is crucial, then certainly $\nu_0(\Gamma) > 1$. Thus if x is a vertex of Γ, then $\Gamma_{(x)}$ is perfect while Γ is not perfect. By the theorem

$$\begin{aligned}\nu_0(\Gamma) = \nu_0(\Gamma_{(x)}) + 1 &\leq \alpha_0(\Gamma_{(x)})\omega(\Gamma_{(x)}) + 1\\ &\leq \alpha_0(\Gamma)\omega(\Gamma) + 1\\ &\leq \nu_0(\Gamma).\end{aligned}$$ □

It is easy to see that for any crucial graph Γ, $\alpha_0(\Gamma) \geq 2$ and $\omega(\Gamma) \geq 2$.

D19 *Exercise*. Let Γ be crucial. Prove that if $\omega(\Gamma) = 2$, then Γ is an odd circuit of length at least 5, and if $\alpha_0(\Gamma) = 2$, then Γ' is an odd circuit of length at least 5.

D20 *Exercise*. A graph $\Gamma = (V, \mathscr{E})$ is a **comparability graph** if there exists a transitive, antisymmetric relation R on V such that $\{x, y\} \in \mathscr{E}$ if and only if either $(x, y) \in R$ or $(y, x) \in R$. Using Dilworth's theorem *IV*G7, show that every comparability graph is color-perfect, and hence perfect.

CHAPTER IX

Designs

Most of the combinatorial objects studied in this book have been defined as systems satisfying certain conditions. The term "design" is reserved for a system $\Lambda = (V, f, E)$ with selection s such that $\bar{s}$ and $\overline{s^*}$ are constant on some of the collections $\mathscr{P}_t(V)$ $(t = 1, \ldots, |V| - 1)$ and $\mathscr{P}_a(E)$ $(a = 1, \ldots, |E| - 1)$ respectively. The reader may wish before proceeding to review the properties of the functions $\bar{s}$ and $\overline{s^*}$ in §*I*E.

IXA Parameters of Designs

Throughout this chapter $\Lambda = (V, f, E)$ will denote a system such that $V \neq \varnothing$ and $E \neq \varnothing$, and s will denote the selection of Λ. We also fix the letters $v = |V|$ and $b = |E|$.

Let $\{t_1, \ldots, t_p\}$ and $\{a_1, \ldots, a_q\}$ be sets of integers, at least one of which is nonempty, and suppose that $0 < t_1 < \ldots < t_p < v$ and $0 < a_1 < \ldots < a_q < b$. We say that the system Λ is a $(t_1, \ldots, t_p; a_1, \ldots, a_q)$-**design** if $\bar{s}$ is a positive constant on $\mathscr{P}_{t_i}(V)$ for $i = 1, \ldots, p$, and $\overline{s^*}$ is a positive constant on $\mathscr{P}_{a_i}(E)$ for $i = 1, \ldots, q$. For each such design we define functions $\lambda: \{0, t_1, \ldots, t_p\} \to \mathbb{N} + \{0\}$ and $\lambda^*: \{0, a_1, \ldots, a_q\} \to \mathbb{N} + \{0\}$ by

$$\lambda(0) = \bar{s}(\varnothing) = b; \ \lambda^* = \overline{s^*}(\varnothing) = v;$$

$$\lambda(t_i) = \bar{s}(T) \quad \text{for all } T \in \mathscr{P}_{t_i}(V),\ i = 1, \ldots, p; \quad \text{and}$$

$$\lambda^*(a_i) = \overline{s^*}(A) \quad \text{for all } A \in \mathscr{P}_{a_i}(E),\ i = 1, \ldots, q.$$

The functions λ and λ^* are called the **design parameters** of Λ. When $q = 0$, we write that Λ is a $(t_1, \ldots, t_p;)$-design, and when $p = 0$, we write that Λ is a $(; a_1, \ldots, a_q)$-design. Such designs are called **one-sided** and will be considered at the end of the next section. If $pq > 0$, then Λ is **two-sided**. Observe that Λ is a $(t_1, \ldots, t_p; a_1, \ldots, a_q)$-design if and only if Λ^* is a $(a_1, \ldots, a_q; t_1, \ldots, t_p)$-design.

Suppose that Λ is a $(t_1, \ldots, t_p; a_1, \ldots, a_q)$-design and let $0 < t_1' < \ldots < t_m' < v$ and $0 < a_1' < \ldots < a_n' < b$. Consider the following two statements:

A1 $\{t_1', \ldots, t_m'\} \subseteq \{t_1, \ldots, t_p\}$ *and* $\{a_1', \ldots, a_n') \subseteq \{a_1, \ldots, a_q\}$.

A2 Λ *is a* $(t_1', \ldots, t_m'; a_1', \ldots, a_n')$*-design.*

Clearly A1 implies A2. Whenever for a design Λ, A2 implies A1, we say that $(t_1, \ldots, t_p; a_1, \ldots, a_q)$ is the **design-type** of Λ.

The design Λ is said to be **degenerate** if either $\bar{s}(V) > 0$ or $\overline{s^*}(E) > 0$. By *I*E8 this means that either some vertex lies in every block or some block contains every vertex. Clearly Λ is degenerate if and only if Λ^* is degenerate. Degenerate designs are pursued in the next section.

We shall say that Λ is a **complete design** (respectively, **transposed-complete design**) if for some positive integer $k < v$ (respectively, $k < b$), Λ (respectively, Λ^*) is isomorphic to the set system $(V, \mathscr{P}_k(V))$. (Observe that if $k = 2$, then Λ is a complete graph on v vertices. Thus complete designs are generalizations of complete graphs.) If $\Lambda = (V, \mathscr{P}_k(V))$, then clearly $\overline{s^*}(A) = k$ for $A \in \mathscr{P}_1(E)$. It follows from this and Exercise *I*E11 that Λ is a $(1, 2, \ldots, k; 1)$-design with parameters

A3
$$\lambda(t) = \binom{v - t}{k - t} \quad \textit{for } t = 0, 1, \ldots, k;$$
$$\lambda^*(0) = b; \qquad \lambda^*(1) = k.$$

Furthermore, Λ is not a $(t;)$-design for $t > k$. If $k < v - 1$ and $a > 1$, it can be readily verified that Λ is not a $(; a)$-design. If $k < v - 1$, then $(1, 2, \ldots, k; 1)$ is the design type of Λ. However, when $k = v - 1$, then Λ is isomorphic to Λ^*, and both designs have design-type

$$(1, \ldots, v - 1; 1, \ldots, v - 1).$$

A design isomorphic to $(V, \mathscr{P}_{v-1}(V))$ is called a **trivial design**.

Since Λ is a $(t_1, \ldots, t_p; a_1, \ldots, a_q)$-design if and only if Λ^* is a $(a_1, \ldots, a_q; t_1, \ldots, t_p)$-design, there corresponds to any assertion regarding the properties of one of the design parameters of Λ an obvious assertion regarding the properties of the other design parameter. In the arguments and assertions which follow, we will not always include both a statement and its "transpose," but we shall always feel free to use one if the other has been presented. (Cf. Proposition *I*D4.)

Observe that a $(; 1)$-design has (constant) blocksize $\lambda^*(1)$. Following established convention, we shall (in this chapter) reserve the letter k for this quantity. Every incidence matrix of a $(; 1)$-design Λ has constant column sum k. Dually, a $(1;)$-design Λ has the property that every vertex is contained in exactly $\lambda(1)$ blocks. One traditionally denotes $\lambda(1)$ by r, called the **replication number** of Λ, and it is the constant row sum of any incidence matrix for a (1;)-design.

Our first substantial result shows that for a design Λ with blocksize k, the value of $\bar{s}$ on $\mathscr{P}_t(V)$ where $t \le k$ determines the value of $\bar{s}$ on all subsets of V with cardinality less than t.

A4 Lemma. *Let Λ be a (; 1)-design. If $S \subseteq V$ and $t \in \mathbb{N}$ satisfy $|S| \le t \le k$, then*

$$\bar{s}(S) = \frac{1}{\binom{k-|S|}{t-|S|}} \sum_{T \in \mathscr{P}_t(V)} [S, T]\bar{s}(T).$$

PROOF. The argument consists of the following chain of equalities, which rely only on the definition of $\bar{s}$ and the fact that $s(U) = 0$ for $|U| \ne k$.

$$\begin{aligned}
\bar{s}(S) &= \sum_{U \in \mathscr{P}_k(V)} [S, U]s(U) \\
&= \frac{1}{\binom{k-|S|}{t-|S|}} \sum_{U \in \mathscr{P}_k(V)} |\{T \in \mathscr{P}_t(V) : S \subseteq T \subseteq U\}| s(U) \\
&= \frac{1}{\binom{k-|S|}{t-|S|}} \sum_{U \in \mathscr{P}_k(V)} \sum_{T \in \mathscr{P}_t(V)} [S, T][T, U]s(U) \\
&= \frac{1}{\binom{k-|S|}{t-|S|}} \sum_{T \in \mathscr{P}_t(V)} [S, T] \sum_{U \in \mathscr{P}_k(V)} [T, U]s(U) \\
&= \frac{1}{\binom{k-|S|}{t-|S|}} \sum_{T \in \mathscr{P}_t(V)} [S, T]\bar{s}(T).
\end{aligned}$$

□

A5 Proposition. *Let Λ be a $(t; 1)$-design where $t \le k$. Then Λ is a $(1, \ldots, t; 1)$-design, and for each $i = 0, 1, \ldots, t$,*

$$\lambda(i) = \frac{\binom{v-i}{t-i}}{\binom{k-i}{t-i}} \lambda(t).$$

PROOF. Let $i \in \{0, 1, \ldots, t\}$ be given and let $S \in \mathscr{P}_i(V)$. By Lemma A4 and the assumption that $\bar{s}$ is constant on $\mathscr{P}_t(V)$,

$$\bar{s}(S) = \frac{1}{\binom{k-i}{t-i}} \sum_{T \in \mathscr{P}_t(V)} [S, T]\lambda(t)$$

$$= \frac{1}{\binom{k-i}{t-i}} \binom{v-i}{t-i} \lambda(t).$$

Thus $\bar{s}(S)$ is determined only by $|S| = i$; that is, Λ is an $(i;)$-design. □

A nondegenerate $(t; 1)$-design is called a **t-design**. It follows that a t-design is a t'-design for each $t' = 1, \ldots, t$. If Λ is a t-design, then each t-subset of V is contained in exactly $\lambda(t)$ blocks. Since $\lambda(t) > 0$ by definition, and since the blocksize of Λ is assumed to be k, we have

A6 *If Λ is a t-design, then $t \leq k$.*

A7 *Exercise.* Characterize all k-designs.

The following corollary to Proposition A5 is immediate.

A8 Corollary. *If Λ is a $(t; a)$-design where either $t = 1$ or $a = 1$, then Λ has design-type $(1, \ldots, t'; 1, \ldots, a')$ for some $t' \geq t$ and some $a' \geq a$.*

A9 Exercise. Show that *if Λ is a t-design, then*

$$\lambda(i+1) = \frac{k-i}{v-i} \lambda(i) \quad \text{for all } i = 0, 1, \ldots, t-1,$$

and hence λ is a decreasing function.

A10 Exercise. Show that *if Λ is a t-design, then*

$$\lambda(i) = \frac{\binom{k}{i}}{\binom{v}{i}} b \quad \text{for } i = 0, 1, \ldots, t.$$

Another term for a nondegenerate 1-design in the literature is **tactical configuration**. A ρ-valent graph is a tactical configuration with $r = \rho$ and $k = 2$. "Partially-balanced incomplete block designs" and "partial geometries," the topics of §E and §F, respectively, are also tactical configurations. Traditionally, however, the class of 2-designs has received much more attention. A 2-design which is not complete is called a **balanced incomplete block design**, or **BIB-design**. Such designs will be studied in §C.

Having already obtained some information about $(; 1)$-designs we now consider $(; 2)$-designs. If Λ is a $(; 2)$-design, the intersection of the images under f of any two blocks has precisely $\lambda^*(2)$ elements. Suppose $f(e_1) = f(e_2)$ for some distinct $e_1, e_2 \in E$. Then $|f(e_1)| = \lambda^*(2)$ and $f(e_1) \cap f(e) = f(e_1)$ for all $e \in E$. Hence $\bigcap_{e \in E} f(e) = f(e_1) \neq \varnothing$ (since $\lambda^*(2) > 0$). We have proved:

A11 Proposition. *If Λ is a $(; 2)$-design, and if $\overline{s^*}(E) = 0$, then Λ is a set system.*

The next result is sometimes known as "Fisher's inequality" [r.9].

A12 Theorem. *If* Λ *is a* $(;2)$*-design and if* $\overline{s^*}(E) = 0$, *then* $b \leq v$.

Proof (D. R. Woodall [w.12], 1970). We may clearly assume that $b > 1$. Let M be an incidence matrix for Λ. Thus M is a $v \times b$ matrix, whose jth column sum we denote by σ_j. Since the usual inner product of any two distinct columns, regarded as vectors, is exactly $\lambda^*(2)$, we have $\sigma_j \geq \lambda^*(2)$ for $j = 1, \ldots, b$. Equality can hold only if M has two identical columns. This is impossible since by A11, Λ is a set system. Hence,

A13 $$\sigma_j - \lambda^*(2) > 0, \quad j = 1, \ldots, b.$$

If we suppose $v < b$, then we can extend M to a $b \times b$ matrix N by adjoining $b - v$ rows of 0's at the bottom of M. Since $\det(N) = 0$, we have $\det(N^*N) = 0$. This latter matrix is of the form

$$N^*N = \begin{bmatrix} \sigma_1 & & & \lambda^*(2) \\ & \sigma_2 & & \\ & & \ddots & \\ \lambda^*(2) & & & \sigma_b \end{bmatrix}$$

Its main diagonal is $(\sigma_1, \ldots, \sigma_b)$ while all other entries are $\lambda^*(2)$. Next we evaluate $\det(N^*N)$ again, this time by still more elementary methods.

First subtract the first column from all of the other columns. Then, for each $i = 2, \ldots, b$, add to the first row the ith row multiplied by

$$(\sigma_1 - \lambda^*(2))/(\sigma_i - \lambda^*(2)).$$

Since only 0's remain above the main diagonal, the product of the entries on the main diagonal gives $\det(N^*N)$. But each of these entries except the first is of the form $\sigma_i - \lambda^*(2)$, while the first term is

$$\sigma_1 + (\sigma_1 - \lambda^*(2))\lambda^*(2) \sum_{i=2}^{b} (\sigma_i - \lambda^*(2))^{-1}.$$

By A13 every term is positive. Hence $\det(N^*N) > 0$, giving a contradiction. □

The above proposition was proved independently by H. J. Ryser [r.10]. We indicate his proof. Since $N^*N = M^*M$, one may prove $\det(M^*M) > 0$ by the same arguments as before. Then $b = \operatorname{rank}(M^*M) \leq \operatorname{rank}(M) \leq v$, as required.

By duality we immediately obtain:

A14 Corollary. *A nondegenerate* $(2;2)$*-design satisfies the condition* $b = v$.

A15 Proposition. *If* $\Lambda = (V, \mathscr{E})$ *is a nondegenerate* $(2;2)$*-design, then*

$$\lambda(2) = \lambda^*(2).$$

PROOF. Consider the set of ordered pairs:

$$S = \{(T, \{E_1, E_2\}) : T \in \mathscr{P}_2(V); \{E_1, E_2\} \in \mathscr{P}_2(\mathscr{E}); T \subseteq E_1 \cap E_2\}.$$

Let $\lambda(2) = m$ and $\lambda^*(2) = n$. For each $T \in \mathscr{P}_2(V)$ there are exactly m blocks containing T, and hence $(T, \{E_1, E_2\}) \in S$ for exactly $\binom{m}{2}$ 2-sets $\{E_1, E_2\} \in \mathscr{P}_2(\mathscr{E})$. Thus $|S| = \binom{v}{2}\binom{m}{2}$. By a symmetric argument one verifies that $|S| = \binom{b}{2}\binom{n}{2}$. By Corollary A14, $m = n$. □

A t-design with $t \geq 2$ which is also a (; 2)-design is by Proposition A5 a (1, 2; 1, 2)-design. Such a design is called a **symmetric block design** and will receive attention in §C when we consider BIB-designs. A **symmetric λ-linked** design is a nondegenerate, nontrivial (2; 2)-design which is not a t-design for any t.

A16 *Exercise.* Show that the system $(V, \mathscr{E})$ where $v \geq 4$, $x_0 \in V$, and

$$\mathscr{E} = \{\{x_0, x\} : x \in V + \{x_0\}\} \cup \{V + \{x_0\}\}$$

is a symmetric λ-linked design.

IXB Design-Types

We begin the classification of designs by first obtaining information about the design parameters of degenerate designs.

B1 Proposition. *Let Λ be a $(t;)$-design, and suppose that $\overline{s^*}(E) > 0$. Then for $S \in \mathscr{P}(V)$,*

$$s(S) = \begin{cases} \lambda(t) & \text{if } |S| = v; \\ 0 & \text{if } t \leq |S| \leq v - 1. \end{cases}$$

PROOF. By hypothesis, we may select $x_0 \in \bigcap_{e \in E} f(e)$. Let $T \in \mathscr{P}_{t-1}(V + \{x_0\})$. Since x_0 belongs to every block and Λ is assumed by hypothesis to be a $(t;)$-design, $\bar{s}(T) = \bar{s}(T + \{x_0\}) = \lambda(t)$. Now let $x_1 \in V + T$. Since $T + \{x_1\} \in \mathscr{P}_t(V)$, we have $\bar{s}(T + \{x_1\}) = \lambda(t) = \bar{s}(T)$. Since every block which contains $T + \{x_1\}$ also contains T, we conclude that T and $T + \{x_1\}$ are contained in the very same blocks. However, x_1 was chosen arbitrarily in $V + T$, which implies that any block containing T must contain all of V. Thus $s(V) = \bar{s}(V) = \bar{s}(T) = \lambda(t)$.

Recall that T was chosen arbitrarily in $\mathscr{P}_{t-1}(V + \{x_0\})$. Now let $S \in \mathscr{P}(V)$ where $t \leq |S| \leq v - 1$. Clearly S contains some set $T \in \mathscr{P}_{t-1}(V + \{x_0\})$. Hence if S is the image of a block, it contains V, which is impossible. We conclude that $s(S) = 0$. □

B2 Corollary. *If Λ is a $(t;)$-design and $\overline{s^*}(E) > 0$, then Λ is a $(t, t + 1, \ldots, v - 1;)$-design.*

By definition, a design may be degenerate for either of two reasons: either $\overline{s^*(E)} > 0$ or $\bar{s}(V) > 0$. We have just proved, allowing duality to play its role, that if a $(t; a)$-design is degenerate for one reason, then it is degenerate for the other reason as well. More precisely,

B3 Corollary. *If Λ is a degenerate $(t; a)$-design, then $\bar{s}(V) = \lambda(t)$ and $\overline{s^*(E)} = \lambda^*(a)$.*

B4 *Exercise.* Let $\Lambda = (V, f, E)$ be a $(t;)$-design. Prove:

(a) If $\lambda(t) > \bar{s}(V) > 0$, then the subsystem $(V, f_{|F}, F)$ where $F = \{e \in E: f(e) \neq V\}$ is a nondegenerate $(t;)$-design.

(b) If Λ is not degenerate, then Λ is a subsystem of some degenerate $(t;)$-design with vertex set V.

(c) Let W and F be any disjoint sets such that $|W| > t$ and $|F| > a$. Then there exists a function $f: F \to \mathscr{P}(W)$ such that (W, f, F) is a degenerate $(t; a)$-design.

B5 *Exercise.* Determine the design-types impossible for degenerate two-sided designs.

The following is a generalization of Theorem A12.

B6 Theorem. *Let Λ be a $(; a)$-design no vertex of which is incident with every block.*

(a) If $a \geq 2$, then $v \geq b$.

(b) If $a \geq 3$, then $v = b$ if and only if Λ is a trivial design.

Proof. The proof is by induction on a. Theorem A12 serves as the first step, with $a = 2$. For our induction hypothesis, we assume that the theorem is valid for all $(; a)$-designs where a is some fixed integer at least 2 and no vertex is incident with every block.

Suppose that Λ is a $(; a + 1)$-design such that $\overline{s^*(E)} = 0$. Select $e_0 \in E$, and consider the subsystem $\Lambda_0 = (f(e_0), f_0, E + \{e_0\})$, where $f_0(e) = f(e_0) \cap f(e)$ for all $e \in E + \{e_0\}$. Since any a-set of blocks of Λ_0 becomes an $(a + 1)$-set of blocks of Λ when e_0 is adjoined to it, Λ_0 is a $(; a)$-design. Were some vertex of Λ_0 incident with every block of Λ_0, then that vertex would be incident with every block of Λ, contrary to assumption. Applying, as we now may, the induction hypothesis to Λ_0, we obtain:

B7
$$b - 1 = |E + \{e_0\}| \leq |f(e_0)| \leq v.$$

Thus either $b - 1 = v$ or $b \leq v$. In the former case, equality holds across B7, independently of the arbitrary choice of the block e_0. Thus $f(e) = V$ for all $e \in E$, and $\overline{s^*(E)} = v > 0$. Hence $b \leq v$, proving (a).

If $b = v$, then B7 implies that all blocks have size either v or $v - 1$. Let M be an incidence matrix for Λ. Each of the b columns of M has at most one 0, but since no vertex is incident with every block, each of the v rows contains at least one 0. If m is the number of 0's in M, then $v \leq m \leq b$. By our assumption that $b = v$, there must be exactly one 0 in each row and each column. But then M is an incidence matrix of the set system $(V, \mathscr{P}_{v-1}(V))$. We have already noted (A3) that $b = v$ for trivial designs. □

The property of being a $(t; a)$-design when both $t \geq 2$ and $a \geq 2$ and one of these values is >2 is very restrictive, as we shall presently see.

B8 Corollary. *Let Λ be a $(t; a)$-design with $t \geq 2$ and $a \geq 2$. If $ta > 4$, then Λ is degenerate or trivial.*

PROOF. Suppose that Λ is nondegenerate. From Theorem B6a and its transpose, we infer that $b = v$. If $ta > 4$, we apply B6b or its transpose to infer that Λ is trivial. □

Up to this point no effort has been made to identify those design-types which are possible. Using Proposition A5 and Corollary B8 we can completely identify the design-types of all two-sided designs.

B9 Theorem. *Every nondegenerate 2-sided design has one of the following design-types:*

(a) $(1, \ldots, t; 1)$, $t \geq 1$, (*nontrivial t-design*)
(b) $(1; 1, \ldots, a)$, $a \geq 1$, (*transpose of* (**a**))
(c) $(1, 2; 1, 2)$, (*nontrivial symmetric block design*)
(d) $(2; 2)$, (*symmetric λ-linked design*)
(e) $(1, \ldots, v-1; 1, \ldots, v-1)$, (*trivial design*).

B10 Exercise. Prove Theorem B9.

We shall presently see that no restrictions such as in Theorem B9 can be put on the design-types of one-sided designs.

B11 Exercise. *Let $v \geq 3$ and let $V = \{x_1, \ldots, x_v\}$.* Show that $(1;)$ *is the design-type of the set systems $(V, \mathscr{E})$ and $(V, \mathscr{E}')$ with*

$$\mathscr{E} = \{S_1, \ldots, S_{v-1}, T_1, \ldots, T_{v-1}\}$$

where $S_i = \{x_1, \ldots, x_i\}$ and $T_i = V + S_i$ for $i = 1, \ldots, v-1$, and with $\mathscr{E}' = \mathscr{E} + \{V\}$.

B12 Proposition (J. K. Doyle and C. J. Leska [d.8]). *Let integers $t_1, \ldots, t_p, v$ be given with $0 < t_1 < \ldots < t_p < v$ and $v \geq 3$. Then there exists a degenerate design and a nondegenerate design on v vertices with design-type $(t_1, \ldots, t_p;)$.*

PROOF. In light of B11 we may assume $t_p > 1$. We construct the required designs inductively. Let V be a v-set. The constructions for the degenerate and nondegenerate cases differ only in the initial stages. In order to construct a nondegenerate design, let E_0 denote a $\binom{v}{t_p}$-set and let $f_0\colon E_0 \to \mathscr{P}_{t_p}(V)$ be a bijection. In order to construct a degenerate design in the case $t_p = v - 1$, let E_0 denote a v-set and let $f_0(e) = V$ for all $e \in E_0$. In order to construct a degenerate design when $t_p < v - 1$, let E_0 denote a $(\binom{v}{t_p} - \binom{v-1}{t_p} + 2)$-set and let $f_0\colon E_0 \to (\mathscr{P}_{t_p}(V) + \mathscr{P}_{t_p}(V')) \cup \{V, V'\}$ be a bijection where V' is a $(v - 1)$-subset of V.

In all cases, let $\Lambda_0 = (V, f_0, E_0)$. We observe that Λ_0 is a $(t_p;)$-design but is not a $(t;)$-design for $t_p < t < v$. (We emphasize that this property will be unaffected by the adjoining to E_0 of blocks of size less than t_p.) This has been the initial step, $j = 0$, for a proof by induction. Now let $0 \leq j \leq t_p - 1$, and let the following be our induction hypothesis: *There exists a system* $\Lambda_j = (V, f_j, E_j)$ *such that for all* $t \geq t_p - j$, Λ_j *is a* $(t;)$-*design if and only if* $t \in \{t_1, \ldots, t_p\}$. We shall adjoin blocks to E_j and extend f_j appropriately in order to form a design Λ_{j+1} which satisfies the condition of the induction hypothesis for all $t \geq t_p - (j + 1)$.

Case 1: $t_p - (j + 1) \in \{t_1, \ldots, t_{p-1}\}$. Let us say $t_p - (j + 1) = t_i$. If $\bar{s}$ is constant on $\mathscr{P}_{t_i}(V)$, we are done. Suppose, therefore, that $\bar{s}$ is not constant on $\mathscr{P}_{t_i}(V)$. Let $m = \max\{\bar{s}(S)\colon S \in \mathscr{P}_{t_i}(V)\}$, and let

$$n = \sum_{S \in \mathscr{P}t_i(V)} (m - \bar{s}(S)).$$

Let F be an n-set disjoint from E_j and define $E_{j+1} = E_j \cup F$. Define $f_{j+1}\colon E_{j+1} \to \mathscr{P}(V)$ so that $f_{j+1|E_j} = f_j$ and $f_{j+1}[F] \subseteq \mathscr{P}_{t_i}(V)$ with the property that for each set $S \in \mathscr{P}_t(V)$, $|f_{j_i+1}^{-1}[S]| = m - \bar{s}(S)$. This is clearly possible. The system $\Lambda_{j+1} = (V, f_{j+1}, E_{j+1})$ has the desired property.

Case 2: $t_p - (j + 1) \notin \{t_1, \ldots, t_p\}$. For brevity let $u = t_p - (j + 1)$. If $\bar{s}$ is not constant on $\mathscr{P}_u(V)$, we are done. If $\bar{s}$ is constant on $\mathscr{P}_u(V)$, then Λ_j is a $(u;)$-design, and we proceed to destroy this property as follows. Let $U \in \mathscr{P}_u(V)$ and let e_{j+1} be some object not in E_j. Define $E_{j+1} = E_j \cup \{e_{j+1}\}$, and define $f_{j+1}\colon E_{j+1} \to \mathscr{P}(V)$ by

$$f_{j+1}(e) = \begin{cases} f_j(e) & \text{if } e \in E_j; \\ U & \text{if } e = e_{j+1}. \end{cases}$$

Then $\Lambda_{j+1} = (V, f_{j+1}, E_{j+1})$ has the desired property.

By continuing this construction, we obtain a design $\Lambda = (V, f, E)$ which is a $(t;)$-design if and only if $t \in \{t_1, \ldots, t_p\}$. It remains to show that Λ is not a $(; a)$-design for all $a \geq 1$. We first show that if Λ is a $(; 1)$-design, then Λ can be replaced by another design Λ' which is also a $(t;)$-design if and only if $t \in \{t_1, \ldots, t_p\}$, but which does not have constant blocksize. If Λ is a $(; 1)$-design, then Λ is a $(1, 2, \ldots, t_p;)$-design by Proposition A5, in which case E is the set E_0 postulated at the beginning of this proof. Let F be a v-set disjoint from E and let $E' = E \cup F$. Define $f'\colon E' \to \mathscr{P}(V)$ so that $f'_{|E} = f$ and $f'_{|F}\colon F \to \mathscr{P}_1(V)$ is a bijection. Since $t_p > 1$, $\Lambda' = (V, f', E')$ has the same properties as Λ but does not have constant blocksize.

We may thus assume that Λ is a $(;a)$-design with $a \geq 2$. By B6, $v \geq b$. But $v \leq |E_0| \leq b$, with $|E_0| = b$ only if $t_i = i$ for all $i = 1, \ldots, p$, while $v = |E_0|$ only if $t_p = v - 1$. But if both equalities hold simultaneously, then $b \geq 2v$ by our construction, giving a contradiction. Hence Λ has design-type $(t_1, \ldots, t_p;)$. □

The degenerate design constructed in the above theorem has the property that $\bar{s}(V) > 0$. Due to Corollary B2, however, it generally will not have the property that $\overline{s^*}(E) > 0$.

A careful reading of the above proof also reveals that the design Λ constructed therein will be a set system when the design is nondegenerate and the integers $t_1, \ldots, t_p$ are consecutive. Otherwise Λ will in general not be a set system. Actually, set systems of type $(t_1, \ldots, t_p;)$ are also constructible under a slightly more general hypothesis than this.

B13 Proposition. *Let integers $t_1, \ldots, t_p$, v be given such that $0 < t_1 < \ldots < t_p < v$ and $v \geq 3$ and such that $t_{i+1} = t_i + 1$ holds for all indices $i \in \{2, \ldots, p-1\}$. Then there exists a set system with v vertices and design-type $(t_1, \ldots, t_p;)$.*

Proof. If $t_1, \ldots, t_p$ are consecutive, we refer to B12. Otherwise, the integers $t_1, \ldots, t_p$ can be represented as

$$t_1, t_2, t_2 + 1, \ldots, t_2 + (p-2) = t_p, \quad \text{where } t_1 < t_2 - 1.$$

Let V be a v-set. Let $S \in \mathscr{P}_{t_2 - 1}(V)$ and let $R \in \mathscr{P}_{t_1 - 1}(S)$. Define

$$\mathscr{E} = \mathscr{P}_{t_p}(V) + [\mathscr{P}_{t_1}(V) + \mathscr{P}_{t_1}(S)] + \{R, S\},$$

and let $\Lambda = (V, \mathscr{E})$.

It is immediate that for $2 \leq i \leq p$, $\lambda(t_i) = \binom{v-t_i}{t_p-t_i}$, since the only blocks containing a t_i-set have size t_p. We assert that $\lambda(t_1) = \binom{v-t_1}{t_p-t_1} + 1$, since in addition to the blocks of size t_p, every t_1-set is contained in one additional block of size $t_2 - 1$ or t_1, depending upon whether or not it is a subset of S. Hence Λ is a $(t_1, \ldots, t_p;)$-design.

Clearly Λ is no $(t;)$-design for $t > t_p$. Suppose $t_1 < t < t_2$. If $T \in \mathscr{P}_t(S)$, then $\bar{s}(T) = \binom{v-t}{t_p-t} + 1$, but if $T \in \mathscr{P}_t(V) + \mathscr{P}_t(S)$, then $\bar{s}(T) = \binom{v-t}{t_p-t}$. Finally suppose $t < t_1$. For $T \in \mathscr{P}_t(S)$, let $\tau(T)$ denote the number of blocks of size t_1 containing T. We readily compute

$$\tau(T) = \binom{v-t}{t_1-t} - \binom{(t_2-1)-t}{t_1-t}.$$

If $T \in \mathscr{P}_t(R)$, then $\bar{s}(T) = \binom{v-t}{t_p-t} + \tau(T) + 2$, while if $T \in \mathscr{P}_t(S) + \mathscr{P}_t(R)$, then

$$\bar{s}(T) = \binom{v-t}{t_p-t} + \tau(T) + 1.$$

Hence Λ is not a $(t;)$-design for all $t \notin \{t_1, \ldots, t_p\}$.

One shows that Λ is no $(;\, a)$-design exactly as in the proof of Proposition B12. Hence Λ has design-type $(t_1, \ldots, t_p;)$. $\square$

B14 *Exercise.* Let t_1, t_2, t_3, v be integers such that $0 < t_1 < t_2 < t_3 < v$, and $t_3 - t_1 - 1 < 2t_2$. Construct a set system on v vertices with design-type $(t_1, t_2, t_3;)$.

B15 *Exercise.* Determine the design-types and design parameters for each of the following systems introduced in proofs in §*I*E.

(a) (Y, Φ, Y^X) in Proposition *I*E14.
(b) $(\Pi(B), f, B)$ in Proposition *I*E16.
(c) (V, f, B) in Theorem *I*E21.

IXC *t*-Designs

Of all designs, it is unquestionably the t-designs that have been subjected to the greatest amount of study. There are applications of the theory of t-designs to algebra, geometry, number theory, and statistics.

Unless a t-design is a symmetric block design, the cardinality of the intersection of the images of any two blocks is not constant. Nonetheless, t-designs all have the following property as a consequence of the Principle of Inclusion-Exclusion (*I*E13):

C1 Exercise. Show that *if* Λ *is a t-design with design parameter* λ, *if* $0 \le j \le i \le t$, *and if* $S \in \mathscr{P}_i(V)$, *then the number of blocks whose images meet S in a j-set is*

$$\sum_{h=0}^{i-j} \frac{(-1)^h i!\, \lambda(h+j)}{h!\, j!\, (i-h-j)!}.$$

All t-designs readily generate other t-designs, as the next three propositions indicate.

C2 Proposition. *Let* $\Lambda = (V, f, E)$ *be a t-design. Define* $\hat{f}(e) = V + f(e)$ *for all* $e \in E$. *Then* $\hat{\Lambda} = (V, \hat{f}, E)$ *is a* $\hat{t}$*-design where* $\hat{t} = \min\{t, v - k\}$.

Proof. Clearly $\hat{\Lambda}$ has blocksize $v - k$. Hence $\hat{\Lambda}$ is a $(;\,1)$-design. Let d be its selection. Since $\hat{t} \le t$, Exercise C1 with $i = \hat{t}$ and $j = 0$ implies that $\bar{d}$ is constant on $\mathscr{P}_{\hat{t}}(V)$. Since $\hat{t} \le v - k$, $\bar{d}$ is not zero on $\mathscr{P}_{\hat{t}}(V)$. Hence $\hat{\Lambda}$ is a $(\hat{t};\,1)$-design. Since the blocksize of $\hat{\Lambda}$ is less than v, $\bar{d}(V) = 0$. By applying Corollary B3 to $\hat{\Lambda}$, we conclude that $\hat{\Lambda}$ is nondegenerate. $\square$

$\hat{\Lambda}$ is called the complementary design of Λ. Clearly $\hat{\hat{\Lambda}} = \Lambda$

C3 Proposition. *If the set system* $(V, \mathscr{E})$ *is a noncomplete t-design with blocksize k, then* $\bar{\Lambda} = (V, \mathscr{P}_k(V) + \mathscr{E})$ *is a t-design.*

PROOF. Clearly $\overline{\Lambda}$ has blocksize k. Let $\bar{\lambda}$ denote the appropriate design parameter of $\overline{\Lambda}$. One readily verifies that $\bar{\lambda}(t) = \binom{v-t}{k-t} - \lambda(t)$. That $\overline{\Lambda}$ is nondegenerate follows as in the previous proposition. $\square$

C4 Proposition. *Let* $\Lambda_i = (V, f_i, E_i)$, $i = 1, 2$, *be t-designs with the same blocksize k and* $E_1 \cap E_2 = \varnothing$. *Then* $\Lambda_0 = (V, f, E_1 \cup E_2)$ *where* $f(e) = f_i(e)$ *for* $e \in E_i$ *is a t-design with blocksize k. Furthermore,* $\lambda_0 = \lambda_1 + \lambda_2$, *where* λ_i *is the corresponding design-parameter of* Λ_i $(i = 0, 1, 2)$.

PROOF. Clearly Λ_0 has blocksize k. Also if s_1, s_2, and s_0 are the selections of Λ_1, Λ_2, and Λ_0, respectively, and if $S \subseteq V$, then $\bar{s}_0(S) = \bar{s}_1(S) + \bar{s}_2(S)$. $\square$

We call Λ_0 as in the above proposition the **sum** of Λ_1 and Λ_2 and denote it by $\Lambda_1 + \Lambda_2$. This leads at once to the concept of the nth **multiple** $n\Lambda$ of a design Λ, obtained by replicating each block n times. If Λ is not the sum of two other designs we say Λ is **indecomposable**. Otherwise we say Λ is **decomposable.**

C5 *Exercise.* Let $\mathsf{D}_{t,k}(V)$ denote the collection of all isomorphism classes of t-designs with vertex set V and blocksize k. (Recall that each isomorphism class is uniquely determined by its selection.) Define the sum of two isomorphism classes to be the isomorphism class of the sum of any representatives of the two classes. Define the nth multiple of a class analogously. Show that:

(a) $\mathsf{D}_{t,k}(V)$ is closed under addition and scalar multiplication by a positive integer and these operations are well-defined;

(b) addition is commutative and associative;

(c) scalar multiplication distributes over addition;

(d) every design in $\mathsf{D}_{t,k}$ is a sum of indecomposable designs in $\mathsf{D}_{t,k}$.

The three propositions C2, C3, and C4 facilitate in classification of t-designs, since our investigations can be narrowed as follows. By C2, we may restrict ourselves to the case $k \leq v/2$. By C4 we may restrict ourselves to a search for indecomposable designs. Finally in the case of designs which are set systems, we may by C3 restrict ourselves to the case $b \leq \frac{1}{2}\binom{v}{k}$. These restrictions are substantial. In particular, Graver and Jurkat [g.4] have shown that for any t and k, the number of *indecomposable* designs in $\mathsf{D}_{t,k}$ is finite. (Clearly $\mathsf{D}_{t,k}$ is always infinite for $t \leq k$.)

Very little is known about t-designs for large values of t. In fact, the complete designs are the only known t-designs for $t > 5$ which are set systems. On the other hand, infinite classes of indecomposable noncomplete 3-, 4-, and 5-designs are known (see Alltop [a.0]).

C6 *Exercise.* The following is an incidence matrix for a t-design. Determine the design-type and the design parameters of this design. Check your results against the values predicted in Proposition A5.

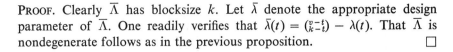

$$\begin{bmatrix} 1&1&1&1&1&1&0&0&0&0&0&0\\ 1&1&1&0&0&0&1&1&1&0&0&0\\ 1&0&0&1&1&0&1&0&0&1&1&0\\ 1&0&0&0&0&1&0&1&1&1&1&0\\ 0&1&0&1&0&0&1&1&0&1&0&1\\ 0&1&0&0&1&1&0&0&1&1&0&1\\ 0&0&1&1&0&1&0&1&0&0&1&1\\ 0&0&1&0&1&0&1&0&1&0&1&1 \end{bmatrix}$$

From A9 with $i = 0$, we obtain the identity

C7 $$vr = bk$$

for all t-designs.

Of greatest and broadest interest among all designs are balanced incomplete block-designs. They are the objects of our attention for the rest of this section. Acceding to existing convention, *we shall abbreviate* $\lambda(2)$ *by* λ. This will cause no confusion since we shall not be interested in the value of $\lambda(i)$ for $i > 2$ even if it should happen in some instances to be well-defined. H. J. Ryser [r.9] refers to a BIB-design as a "(b, v, r, k, λ)-configuration." But care must be taken, for while these five values may in a sense *name* a design, they need not *characterize* it. For example there are two nonisomorphic (26, 13, 6, 3, 1)-configurations.

The 5 values b, v, r, k, λ are by no means independent. In addition to C7, we have

C8 $$(v - 1)\lambda = (k - 1)r,$$

which follows at once from A9 with $i = 1$.

Both C7 and C8 may also be obtained by the following simple counting arguments. Counting the 1's in an incidence matrix M for Λ by rows yields vr. Counting by columns yields bk. To obtain the second identity, fix a row R_0 and count the number of pairs (R, C) consisting of one row $R \neq R_0$ and one column C such that C is incident with both R and R_0. We may choose R in $v - 1$ ways and the pair may be completed in λ ways. On the other hand, we may choose C in r ways and the pair may then be completed in $k - 1$ ways.

It follows from C7 and C8 that given any three of the values b, v, r, k, λ, one can easily compute the other two. However, one cannot begin to construct a BIB-design by arbitrarily choosing three parameter values, since C7 and C8 need not always yield positive integral values for the other two.

By A6, $k \geq 2$ for every BIB-design. However, if Λ is a 2-design with blocksize 2, then since every 2-subset of V must be contained in the same number of blocks, Λ is isomorphic to $n(V, \mathscr{P}_2(V))$ for some positive integer n. Hence $k \geq 3$ for every BIB-design.

A BIB-design with $k = 3$ is called a **triple system**. When, moreover, $\lambda = 1$, we have a **Steiner triple system**. Equation C8 becomes

$$v - 1 = 2r,$$

and so v must be odd. Solving for r and substituting into C7 gives

$$b = \frac{v(v-1)}{6}.$$

Since b must be an integer, we have proved:

C9 Proposition. *A necessary condition for the existence of a Steiner triple system with v vertices is that*

$$v \equiv 1 \text{ or } 3 \text{ (modulo 6).}$$

When $v > 6$, this condition is also sufficient, as was first shown by Kirkman [k.4] as early as 1847.

C10 *Exercise.* Give the analogous necessary condition for the existence of a triple system with $\lambda = 2$.

The condition to be determined in Exercise C10 is again sufficient for existence. Thus for all 5-tuples $(b, v, r, 3, \lambda)$ of positive integers satisfying C7 and C8 with $\lambda < 3$, there exists a $(b, v, r, 3, \lambda)$-configuration. The same is true for $k = 4$ (cf. Hanani [h.5]). The corresponding result for $k = 5$ is false. No (21, 15, 7, 5, 2)-configuration exists.

Since t-designs are by definition nondegenerate, a Steiner triple system Λ cannot have 1 or 3 vertices. Hence by C9, $v \geq 7$. The incidence matrix shown in Figure C11 is for the Steiner triple system whose parameter values are $b = v = 7$, $r = k = 3$, $\lambda = 1$. This particular configuration will keep reappearing under various disguises. For example, it is also a symmetric block design, the projective plane of order 2, an Hadamard design, and a matroid of some interest.

C11
$$\begin{bmatrix} 1 & 1 & 1 & 0 & 0 & 0 & 0 \\ 1 & 0 & 0 & 1 & 1 & 0 & 0 \\ 1 & 0 & 0 & 0 & 0 & 1 & 1 \\ 0 & 1 & 0 & 1 & 0 & 1 & 0 \\ 0 & 1 & 0 & 0 & 1 & 0 & 1 \\ 0 & 0 & 1 & 1 & 0 & 0 & 1 \\ 0 & 0 & 1 & 0 & 1 & 1 & 0 \end{bmatrix}$$

C12 Exercise. Show that *there is, up to isomorphism, only one Steiner triple system on 7 vertices.*

C13 *Exercise.* Find the "smallest" nontrivial triple system with $\lambda = 2$.

C14 Proposition. *If Λ is a symmetric block design, then*
(**a**) $k = r$;
(**b**) $(v - 1)\lambda = k(k - 1)$;
(**c**) $\lambda^*(2) = \lambda$.

PROOF. (a) and (b) are immediate consequences of A14, C7, and C8. (c) was proved in A15. □

This proposition shows why symmetric block designs are also called "(v, k, λ)-configurations." Its converse is false, unlike the situation with Steiner triple systems. The mere existence of positive integers satisfying the above three equations does not imply the existence of a symmetric block design with the given numbers as parameters. For example there is no symmetric block design with $b = v = 22$, $r = k = 7$, and $\lambda^*(2) = \lambda = 2$. [See Exercise C23 below.]

C15 Exercise. *Let M be an incidence matrix of a design Λ.*

(**a**) *Let J denote a matrix (of appropriate dimensions) wherein every entry is* 1. *Let $n_1, n_2 \in \mathbb{N}$ such that $0 < n_2 < n_1 < v$.* Prove:

$$M^*M = (n_1 - n_2)I + n_2J$$

if and only if Λ is a $(; 1, 2)$*-design with $k = n_1$ and $\lambda^*(2) = n_2$.*

(**b**) *If Λ is a* $(; 1, 2)$*-design, then*

$$\det(M^*M) = [k + (b - 1)\lambda^*(2)](k - \lambda^*(2))^{b-1}.$$

C16 Theorem (H. J. Ryser [r.8], 1950). *Let Λ be a* $(; 1, 2)$*-design for which $b = v$ and $\lambda^*(2) < k < v$. If M is an incidence matrix of Λ, then $MM^* = M^*M$.*

PROOF. Let us abbreviate $\lambda^*(2)$ by λ^*. By hypothesis, M is a square matrix. By Exercise C15a,

C17
$$M^*M = (k - \lambda^*)I + \lambda^*J.$$

By Exercise C15b and the hypothesis,

$$\det(M^*)\det(M) = \det(M^*M) \neq 0.$$

Hence M is nonsingular. From the fact that M has constant column sum k, we have

$$kJ = JM,$$

whence

C18
$$JM^{-1} = \frac{1}{k}J.$$

Since J is a $v \times v$ matrix, $J^2 = vJ$, and so by C17 followed by C18,

$$\begin{aligned} JM^* &= JM^*MM^{-1} \\ &= J[(k - \lambda^*)I - \lambda^*J]M^{-1} \\ &= (k - \lambda^*)JM^{-1} - \lambda^*vJM^{-1} \\ &= \frac{k - \lambda^* + \lambda^*v}{k} J. \end{aligned}$$

Taking the transpose gives

C19
$$MJ = \frac{k - \lambda^* + \lambda^*v}{k} J.$$

Equation C19 implies that the matrix M has constant row sums $(k - \lambda^* + \lambda^*v)/k$, and hence Λ is a $(1; 1, 2)$-design. Since $b = v$, equation C7 gives

$$k = \frac{k - \lambda^* + \lambda^*v}{k},$$

whence C19 becomes

C20
$$MJ = kJ.$$

Hence by C17, C18, and C20,

$$\begin{aligned} MM^* &= MM^*MM^{-1} \\ &= M[(k - \lambda^*)I + \lambda^*J]M^{-1} \\ &= (k - \lambda^*)I + \lambda^*J \\ &= M^*M. \end{aligned}$$

□

C21 Corollary (Ryser, *ibid*). *Let Λ be a $(; 1, 2)$-design for which $b = v$ and $\lambda^*(2) < k < v$. Then Λ is a symmetric block design.*

PROOF. Let M be an incidence matrix of Λ. By Exercise C15a,

$$M^*M = (k - \lambda^*(2))I + \lambda^*(2)J.$$

By the theorem, this quantity is equal to MM^*. Again by the exercise, M^* is an incidence matrix of a $(; 1, 2)$-design. Hence M is an incidence matrix of a $(1, 2;)$-design; i.e., Λ is a $(1, 2; 1, 2)$-design. Since $k = r < b = v$, Λ is nondegenerate. □

C22 *Exercise.* Prove that if a set system Λ is a t-design with $t + k \geq v$, then Λ is a complete design.

C23 *Exercise.* Prove that if Λ is a symmetric block design and if v is even, then $k - \lambda$ is a square. (Chowla and Ryser [c.2] and Shrikhande [s.5].)

C24 *Exercise.* Let Λ be a symmetric block design and let $e_0 \in E$. For $i = 1, 2$, determine the design-types and parameters of the designs $\Lambda_i = (V_i, f_i, E + \{e_0\})$ formed from Λ as follows:

$$V_1 = V + f(e_0), \qquad V_2 = f(e_0),$$
$$f_i(e) = f(e) \cap V_i \quad \text{for } i = 1, 2;\ e \in E + \{e_0\}.$$

The designs Λ_1 and Λ_2 constructed from Λ in Exercise C24 are called, respectively, the **residual design** and the **derived design** of Λ with respect to e_0. It should be pointed out that different choices of e_0 may result in nonisomorphic residual designs or nonisomorphic derived designs. (See K. N. Bhattacharya [b.7] and W. S. Connor and M. Hall [c.4].)

C25 *Exercise* (M. Hall, Jr. [h.2]). Determine whether Λ is isomorphic to Λ^*, where Λ has the following as an incidence matrix.

$$\begin{bmatrix}
1&1&1&1&1&1&1&0&0&0&0&0&0&0&0\\
1&1&1&0&0&0&0&1&1&1&1&0&0&0&0\\
1&1&1&0&0&0&0&0&0&0&0&1&1&1&1\\
1&0&0&1&1&0&0&1&1&0&0&1&1&0&0\\
1&0&0&1&1&0&0&0&0&1&1&0&0&1&1\\
1&0&0&0&0&1&1&1&0&1&0&1&0&1&0\\
1&0&0&0&0&1&1&0&1&0&1&0&1&0&1\\
0&1&0&1&0&1&0&1&1&0&0&0&0&1&1\\
0&1&0&1&0&1&0&0&0&1&1&1&1&0&0\\
0&1&0&0&1&0&1&1&0&0&1&0&1&1&0\\
0&1&0&0&1&0&1&0&1&1&0&1&0&0&1\\
0&0&1&1&0&0&1&1&0&1&0&0&1&0&1\\
0&0&1&1&0&0&1&0&1&0&1&1&0&1&0\\
0&0&1&0&1&1&0&1&0&0&1&1&0&0&1\\
0&0&1&0&1&1&0&0&1&1&0&0&1&1&0
\end{bmatrix}$$

We conclude this section with a brief discussion of a class of symmetric block designs known as "Hadamard designs," so called because the existence of an Hadamard design with $m - 1$ vertices is equivalent to the existence of an $m \times m$ "Hadamard matrix."

An $m \times m$ matrix H over $\mathbb{Z}$ whose entries are ± 1 is called an **Hadamard matrix** if

C26 $$MM^* = mI,$$

where I is the $m \times m$ identity matrix.

If H is an $m \times m$ Hadamard matrix, then any matrix H' obtainable from H by a permutation of rows, a permutation of columns, or multiplication of a line by -1 is said to be **equivalent** to H. Clearly H' is also an $m \times m$ Hadamard matrix, and this relation of "equivalence" is indeed an equivalence relation on the set of all $m \times m$ Hadamard matrices. An Hadamard matrix is said to be **normalized** if all the entries in the first row and the first column are $+1$. Every Hadamard matrix is clearly equivalent to a normalized Hadamard matrix. In Figure C27 are shown some normalized Hadamard matrices.

C27 $$[1], \quad \begin{bmatrix} 1 & 1 \\ 1 & -1 \end{bmatrix}, \quad \begin{bmatrix} 1 & 1 & 1 & 1 \\ 1 & 1 & -1 & -1 \\ 1 & -1 & 1 & -1 \\ 1 & -1 & -1 & 1 \end{bmatrix}$$

From the definition of an $m \times m$ Hadamard matrix, it follows that the inner product of two distinct rows is always 0. Hence each row other than the first row of a normalized Hadamard matrix must have as many entries equal $+1$ as -1. It follows that if $m > 1$, then m is even. In fact a stronger condition holds.

C28 Proposition. *If H is an $m \times m$ Hadamard matrix, then $m = 1$, or $m = 2$, or $m \equiv 0$ (modulo 4).*

Proof. By the above remarks and C27, we may assume that H is normalized and that $m \geq 4$. In the $3 \times m$ submatrix formed by the first three rows of H, exactly four types of columns are possible, namely

$$\begin{bmatrix} 1 \\ 1 \\ 1 \end{bmatrix}, \quad \begin{bmatrix} 1 \\ 1 \\ -1 \end{bmatrix}, \quad \begin{bmatrix} 1 \\ -1 \\ 1 \end{bmatrix}, \quad \text{and} \quad \begin{bmatrix} 1 \\ -1 \\ -1 \end{bmatrix},$$

which we presume to occur c_1, c_2, c_3, and c_4 times, respectively. Clearly

C29 $$c_1 + c_2 + c_3 + c_4 = m.$$

From the inner products of rows 1 and 2, rows 1 and 3, and rows 2 and 3, respectively, we obtain

$$\begin{aligned} c_1 + c_2 - c_3 - c_4 &= 0 \\ c_1 - c_2 + c_3 - c_4 &= 0 \\ c_1 - c_2 - c_3 + c_4 &= 0, \end{aligned}$$

which together with C29 yield the unique solution $c_i = m/4$, $i = 1, \ldots, 4$. Since each c_i is integral, $m \equiv 0$ (modulo 4). □

It is not at present known whether the condition $m \equiv 0$ (modulo 4) is also sufficient for the existence of an $m \times m$ Hadamard matrix.

The precise relationship between Hadamard matrices and symmetric block designs is given by the next result.

C30 Proposition. *For each integer $q \geq 2$, the equivalence classes of $4q \times 4q$ Hadamard matrices are in one-to-one correspondence with the symmetric block design with parameters $v = 4q - 1$, $k = 2q - 1$, $\lambda = q - 1$.*

Proof. Let M be an incidence matrix of a symmetric block design with parameters $v = 4q - 1$, $k = 2q - 1$, and $\lambda = q - 1$. This is equivalent by C15a, C16, and C14 to

C31 $$MM^* = qI + (q-1)J.$$

Let $H_1 = 2M - J$ where, as before, J denotes the matrix whose every entry is 1. Then H_1 is a $(4q-1) \times (4q-1)$ matrix whose entries are ± 1 and whose line sums are -1. Beginning with C31 we derive the following sequence of equations, which are equivalent under our assumptions on M or on H_1:

$$4MM^* - (4q-4)J = 4qI$$
$$4MM^* - 2(2q-1)J - 2(2q-1)J + (4q-1)J = 4qI - J$$
$$4MM^* - 2JM - 2JM^* + J^2 = 4qI - J$$
$$(H_1 + J)(H_1{}^* + J) - J(H_1 + J) - J(H_1{}^* + J) + J^2 = 4qI - J$$

C32 $$H_1H_1{}^* = 4qI - J.$$

Let H be the $4q \times 4q$ matrix such that every entry in the first row and first column is 1, and when these two lines are suppressed, the remaining matrix is H_1. One easily verifies that H is a $4q \times 4q$ matrix satisfying C26 if and only if H_1 is a $(4q-1) \times (4q-1)$ matrix whose entries are ± 1, whose line sums are -1, and which satisfies C32. Finally, we observe that permuting the rows or the columns of M has an identical effect on H and vice versa, while the multiplication of any line of M by -1 yields a matrix without constant row sums while the same operation on H yields an equivalent Hadamard matrix. □

The above symmetric block designs as well as those obtainable from them by the procedure of Proposition C2 are known as **Hadamard designs**.

If $M = [m_{ij}]$ and N are any two matrices, the **Kronecker product** of M by N, denoted by $M \times N$, is defined to be the matrix

$$M \times N = \begin{bmatrix} m_{11}N & m_{12}N & \dots & m_{1q}N \\ m_{21}N & m_{22}N & \dots & m_{2q}N \\ \vdots & \vdots & & \vdots \\ m_{p1}N & m_{p2}N & \dots & m_{pq}N \end{bmatrix}.$$

C33 *Exercise.* (a) Show that if M and N are both $\{0, 1\}$-matrices, then both $M \times N$ and $N \times M$ are incidence matrices of the same system.

(b) Show that if M_i is an incidence matrix of a $(1; 1)$-design Λ_i $(i = 1, 2)$, then $M_1 \times M_2$ is also an incidence matrix of some $(1; 1)$-design; determine its parameters in terms of those of Λ_1 and Λ_2. Prove that one cannot replace $(1; 1)$ with BIB.

Observe that for matrices M and N,

C34 $$(M \times N)^* = M^* \times N^*.$$

C35 Exercise. Prove that *for matrices M_1, M_2, N_1, N_2, of suitable dimensions,*

$$(M_1 \times N_1)(M_2 \times N_2) = (M_1M_2) \times (N_1N_2).$$

C36 Proposition. *The Kronecker product of two Hadamard matrices is an Hadamard matrix.*

PROOF. Let H_i be an $m_i \times m_i$ Hadamard matrix ($i = 1, 2$). Then

$$\begin{aligned}(H_1 \times H_2)(H_1 \times H_2)^* &= (H_1 \times H_2)(H_1^* \times H_2^*) && \text{by C34}\\ &= H_1H_1^* \times H_2H_2^* && \text{by C35}\\ &= m_1 I_{m_1 \times m_1} \times m_2 I_{m_2 \times m_2} && \text{by C26}\\ &= m_1 m_2 I_{m_1m_2 \times m_1m_2}. && \square\end{aligned}$$

C37 Corollary. *For all $n \in \mathbb{N}$, there exists a $2^n \times 2^n$ Hadamard matrix.*

Consequently, there exists an Hadamard design with $(4 \cdot 2^n) - 1$ vertices for all $n \geq 1$. In the next exercise we revisit the smallest Hadamard design:

C38 *Exercise.* Using the Kronecker product, construct an 8×8 Hadamard matrix. Show that the corresponding Hadamard design has C11 as an incidence matrix. Conclude that all 8×8 Hadamard matrices are equivalent.

IXD Finite Projective Planes

A **finite projective plane** is a symmetric block design Λ with $v \geq 4$ and $\lambda = 1$. It is traditional in this context to substitute the terms **point** for "vertex" and **line** for "block." From Proposition C14b we obtain

$$v = k^2 - k + 1.$$

The number $n = k - 1$ is called the **order** of Λ. Thus we have

D1 $$b = v = n^2 + n + 1; \qquad r = k = n + 1,$$

for a finite projective plane of order n. It follows from D1 and the condition $v \geq 4$, that the order of a finite projective plane is at least 2 and that the blocksize is at least 3.

D2 *Exercise.* Prove that all finite projective planes are nontrivial designs.

As previously remarked, the Steiner triple system given by the matrix C11 is a finite projective plane of order 2. Furthermore by Exercise C12 it is the only finite projective plane of order 2.

The next result shows the equivalence between the definition of a finite projective plane given here and the more classical definition. It is convenient to say that a set of points is **collinear** if it is contained in some line.

D3 Proposition. *A set system $\Lambda = (V, \mathscr{E})$ is a finite projective plane if and only if the following three conditions hold:*

(a) *Every pair of points is contained in exactly one line.*
(b) *Every pair of lines intersects in exactly one point.*
(c) *There exists a 4-subset of V no 3-subset of which is collinear.*

PROOF. It is immediate that (a) and (b) are equivalent to saying that Λ is a (2; 2)-design with $\lambda(2) = \lambda^*(2) = 1$.

Suppose first that Λ is a finite projective plane, in which case (a) and (b) hold, as we have noted. Since $r = k \geq 3$ we may choose a point x_0 and two lines L_1 and L_2 containing it. Now choose x_1 and x_3 on L_1 distinct from x_0, and choose x_2 and x_4 on L_2 distinct from x_0. One easily verifies that $\{x_1, \ldots, x_4\}$ satisfies condition (c). (See Figure D4.)

D4

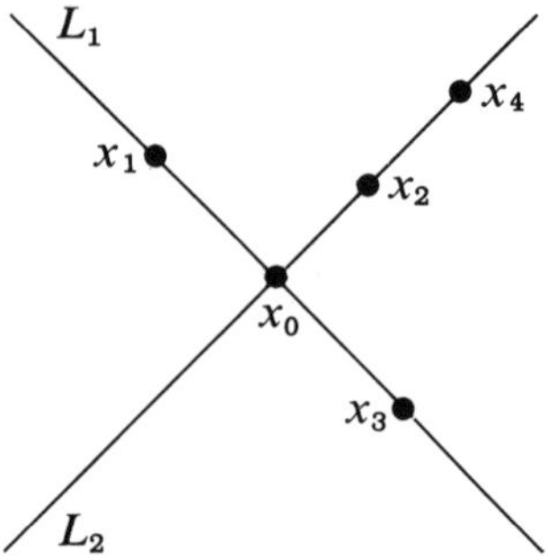

Conversely, suppose Λ satisfies (a), (b), and (c). Then Λ is a (2; 2)-design with $v \geq 4$, and it suffices, due to A5, to prove that Λ is a (; 1)-design. We do this by arbitrarily choosing two lines and establishing a bijection between them.

We assert that given any two lines L and L', there exists a point $x_0 \notin L \cup L'$. For let $\{x_1, x_2, x_3, x_4\}$ be a set postulated by (c). If $\{x_1, \ldots, x_4\} \nsubseteq L \cup L'$, then clearly one may choose $x_0 \in \{x_1, \ldots, x_4\}$. If $\{x_1, \ldots, x_4\} \subseteq L \cup L'$, we may assume that $x_1, x_2 \in L + (L \cap L')$ and $x_3, x_4 \in L' + (L \cap L')$. Let L_1 be the line through x_1 and x_3 and let L_2 be the line through x_2 and x_4 as required by (a). Let $x_0 \in L_1 \cap L_2$. By (b), $x_0 \notin L \cup L'$.

Now define the function $\varphi: L \to L'$ as follows. If $x \in L$, let L_x denote the unique line (by virtue of (b)) which contains x and x_0, and let $\varphi(x)$ be the unique point (by virtue of (a)) which is contained in $L_x \cap L'$. (See Figure D5.) By (b), φ is an injection, since exactly one line contains both x_0 and $\varphi(x)$. Thus $|L| \leq |L'|$, and by symmetry equality must hold. □

D5

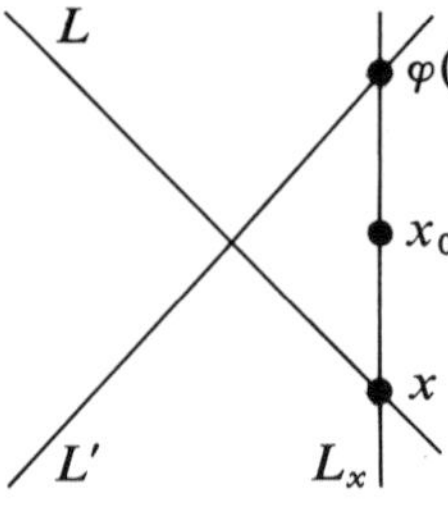

D6 *Exercise.* Prove that Proposition D3 also holds when condition (c) is replaced by

(c′) *There exists a* 4-*subset of lines, every* 3-*subset of which has an empty intersection.*

D7 *Exercise*. Use Proposition D3 to give a constructive proof that there exists only one projective plane of order 2.

D8 *Exercise*. Let Λ be a nondegenerate design. Prove that Λ satisfies conditions (a) and (b) but not (c) of Proposition D3 if and only if Λ is either $(V, \mathscr{P}_{v-1}(V))$ for $v = 3$ or Λ is the symmetric λ-linked design described in Exercise A16.

We now turn to the question of existence of finite projective planes. The following exercise gives a construction for a projective plane of order n whenever there exists a finite field of order n; i.e., whenever $n = p^m$ where p is prime and m is a positive integer.

D9 *Exercise*. Let V be the 3-dimensional vector space over the finite field $\mathbb{F}$ where $|\mathbb{F}| = n$. Let V be the set of 1-dimensional subspaces of V and let E be the set of 2-dimensional subspaces of V. Let $f\colon E \to \mathscr{P}(V)$ be defined for each $e \in E$ by $f(e) = \{x \in V\colon x \text{ is a subspace of } e\}$. Prove that (V, f, E) is a finite projective plane of order n.

The finite projective plane of order 2 can be constructed from the field $\mathbb{K}$ in the manner of Exercise D9. Although not all finite projective planes are constructible in this way (see M. Hall, Jr. [h.3, p. 175]), all known finite projective planes do have a prime power order. Exercise D9 shows that there are finite projective planes of order n with $n = 2, 3, 4, 5, 7, 8, 9$. It has been proved [t.2] that there exists no finite projective plane of order 6. The least order for which the existence of a finite projective plane is still open is 10. The existence question for finite projective planes may be studied by means of Latin squares, which were introduced in §*V*D3.

Let $n \geq 2$ and let $\mathbb{L}_n$ be the set of all Latin squares of order n with entries from $\{1, 2, \ldots, n\}$. Let $S_1, S_2 \in \mathbb{L}_n$ and let $a_{ij}{}^h$ denote the entry in S_h lying in the ith row and the jth column ($h = 1, 2$; $i, j = 1, \ldots, n$). We say $\{S_1, S_2\}$ is **orthogonal** if the n^2 ordered pairs $(a_{ij}{}^1, a_{ij}{}^2)$ for $i, j \in \{1, \ldots, n\}$ are all distinct. A subset of $\mathbb{L}_n$ is said to be **orthogonal** if it is not empty and all of its 2-subsets are orthogonal. In particular, all 1-subsets of $\mathbb{L}_n$ are orthogonal.

By Proposition *V*D9, $\mathbb{L}_n \neq \varnothing$ for all n. Thus $\mathbb{L}_n$ always contains an orthogonal t-set for some $t \geq 1$. The next exercise gives an upper bound for t, but not always a least upper bound.

D10 Exercise. Prove: *if $\mathbb{L}_n$ admits an orthogonal t-set, then $t \leq n - 1$.*

An orthogonal $(n - 1)$-subset of $\mathbb{L}_n$ is called a **complete orthogonal set.** Such a set exists trivially when $n = 2$ and one can be easily constructed when $n = 3$. A complete orthogonal set for $\mathbb{L}_4$ is shown in D11.

D11
$$\begin{bmatrix} 1 & 2 & 3 & 4 \\ 2 & 1 & 4 & 3 \\ 3 & 4 & 1 & 2 \\ 4 & 3 & 2 & 1 \end{bmatrix} \quad \begin{bmatrix} 1 & 2 & 3 & 4 \\ 3 & 4 & 1 & 2 \\ 4 & 3 & 2 & 1 \\ 2 & 1 & 4 & 3 \end{bmatrix} \quad \begin{bmatrix} 1 & 2 & 3 & 4 \\ 4 & 3 & 2 & 1 \\ 2 & 1 & 4 & 3 \\ 3 & 4 & 1 & 2 \end{bmatrix}.$$

Being unable to construct an orthogonal 2-subset of $\mathbb{L}_6$, Euler conjectured in 1779 that $\mathbb{L}_n$ admitted no orthogonal 2-subset when $n \equiv 2$ (modulo 4). This conjecture was confirmed for the special case $n = 6$ by C. Tarry [t.2] in 1901, but R. C. Bose and S. S. Shirkhande [b.17], E. T. Parker [p.1], and all three jointly [b.15], proved in 1959–1960 that Euler's conjecture is false for all other values of n in this congruence class. When $n \not\equiv 2$ (modulo 4), an orthogonal 2-subset always exists, as H. B. Mann [m.5] showed in 1942. (A concise development of this theory is to be found in H. J. Ryser [r.9, Chapter 7].) We now prove a powerful result relating complete orthogonal sets of Latin squares to finite projective planes.

D12 Theorem (R. C. Bose [b.10], 1939). *Let $n \geq 2$ be an integer. There exists a finite projective plane of order n if and only if $\mathbb{L}_n$ admits a complete orthogonal set.*

Proof. Let Λ be a finite projective plane of order n. Let

$$L_\infty = \{x_0, x_1, \ldots, x_{n-1}, x_\infty\}$$

be a line of Λ. Let $R_1, \ldots, R_n$ denote the remaining n lines of Λ through x_0, and let $C_1, \ldots, C_n$ denote the remaining n lines through x_∞. Let x_{ij} be the unique point on $R_i \cap C_j$, by virtue of D3b. By D3a, the set

$$\{x_{ij}: i, j \in \{1, 2, \ldots, n\}\}$$

consists of n^2 distinct points. These are all of the vertices in $V + L_\infty$. We will associate the point x_{ij} with the (i, j)-position in an $n \times n$ array. (See Figure D13.)

D13

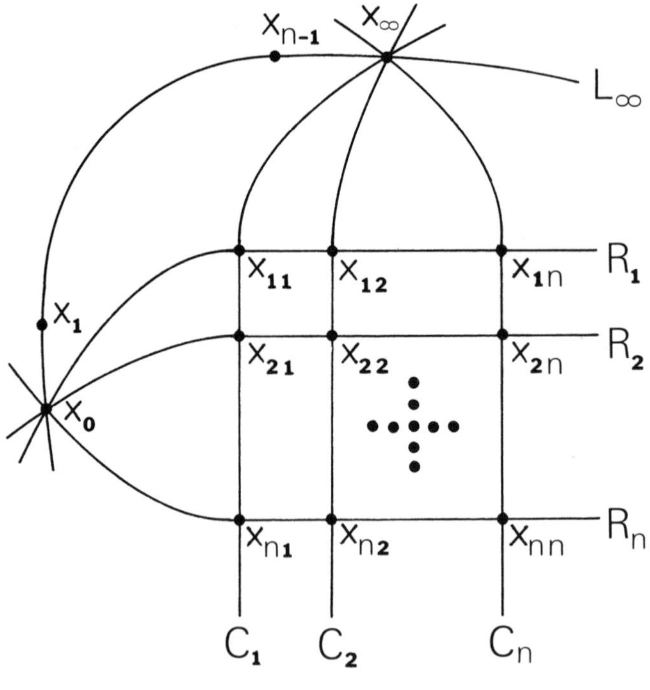

With each point x_i on L_∞ other than x_0 and x_∞ we associate a Latin square $S_h \in \mathbb{L}_n$ as follows. Label the remaining n lines through x_h (other than L_∞) by $L_{h1}, \ldots, L_{hn}$. Denoting the ijth entry of S_h by $a_{ij}{}^h$, we define $a_{ij}{}^h = q$ if $x_{ij} \in L_{hq}$. Since L_{hq} intersects each R_i and each C_j exactly once, the entry q occurs exactly once in each row and each column of S_h. Thus S_h is a Latin square for $h = 1, \ldots, n - 1$.

To see that $\{S_h : h = 1, \ldots, n - 1\}$ is an orthogonal set, observe that $(a_{ij}{}^h, a_{ij}{}^k) = (p, q)$ if and only if x_{ij} is the point of intersection of L_{hp} and L_{kq}.

Conversely, starting with a set of $n^2 + n + 1$ points and $2n + 1$ lines as in Figure D13 and a complete orthogonal set $\{S_1, \ldots, S_{n-1}\}$ of Latin squares from $\mathbb{L}_n$, we reverse the above construction, using S_h to define the remaining n lines through x_h for each $h = 1, \ldots. n - 1$. The details are left to the reader. □

IXE Partially-Balanced Incomplete Block Designs

The object of study in this section is a class of tactical configurations (nondegenerate (1; 1)-designs). This class will be defined by imposing some additional structure, but in general not enough structure, to bring the design-type to anything above (1; 1).

Let $\lambda_1 > \lambda_2 > \ldots > \lambda_m \geq 0$ be integers. For each $i = 1, \ldots, m$ and each $x \in V$, we define the function $n_i : V \to \mathbb{N}$ by

$$n_i(x) = |\{y \in V + \{x\} : \bar{s}(\{x, y\}) = \lambda_i\}|.$$

A set system $\Lambda = (V, \mathscr{E})$ is a **partially-balanced incomplete block design with** m **classes** or more briefly, an m**-class PBIB-design** if it is a tactical configuration with $\bar{s}[\mathscr{P}_2(V)] = \{\lambda_1, \ldots, \lambda_m\}$ and such that each of the functions $n_1, \ldots, n_m$ is constant on V. Henceforth we write n_i for $n_i(x)$.

For the rest of this section $\Lambda = (V, \mathscr{E})$ will be presumed to be an m-class PBIB-design. Note that if $m = 1$ and $\lambda_1 > 0$, then Λ is actually a 2-design.

If $\{x, y\} \in \mathscr{P}_2(V)$ and $\bar{s}(\{x, y\}) = \lambda_i$, we say that x and y are ***i*th-associates.** Thus each vertex in Λ has n_i ith associates for $i = 1, \ldots, m$. It is convenient to make the convention that each vertex be a 0th-associate of itself, and so $n_0 = 1$. It is immediate that

E1
$$v = \sum_{i=0}^{m} n_i.$$

If x and y are ith associates, then exactly λ_i blocks contain $\{x, y\}$, and so $\lambda_0 = r$. Since Λ is nondegenerate, $\lambda_0 > \lambda_1$.

Let M be an incidence matrix of Λ. Then MM^* is a $v \times v$ matrix with the entry λ_h in the ith row, jth column if the elements of V associated with the ith and jth rows of M are hth-associates. We observe that the sum of the terms in any row (and by symmetry in any column) of MM^* is

$$\sum_{i=0}^{m} \lambda_i n_i = \sum_{y \in V} \bar{s}(\{x, y\}),$$

where x is some fixed element of V. Subtracting λ_0 from each side yields

$$\sum_{i=1}^{m} \lambda_i n_i = \sum_{y \in V+\{x\}} \bar{s}(\{x, y\}),$$

and by the definition of $\bar{s}$ we obtain

$$\begin{aligned}\sum_{i=1}^{m} \lambda_i n_i &= \sum_{T \in \mathscr{P}(V)} s(T) \sum_{y \in V+\{x\}} [\{x, y\}, T] \\ &= \sum_{x \in T \in \mathscr{P}(V)} s(T)(|T| - 1).\end{aligned}$$

Since Λ has blocksize k and replication number r, this becomes

E2
$$\sum_{i=1}^{m} \lambda_i n_i = r(k - 1),$$

the analog for PBIB-designs of C8.

An example of a 2-class PBIB-design is any ρ-valent graph which is not complete. Two vertices are first associates if they are incident and second associates otherwise. The parameters are as follows: $v = \nu_0$, $b = \nu_1$, $r = \rho$, $k = 2$, $\lambda_1 = 1$, $n_1 = \rho$, $\lambda_2 = 0$, and $n_2 = \nu_0 - \rho - 1$.

The cube (Figure *III*F21) provides an example of a 3-class PBIB-design. Let V be the set of the eight vertices of the cube; the blocks are the 4-sets of vertices which lie on a common face. Two vertices will be first associates if they are the end-points of an edge. They will be second associates if they lie on a common face but not on a common edge. Otherwise they are third associates. We have $v = 8$, $b = 6$, $r = 3$, $k = 4$, $\lambda_1 = 2$, $n_1 = 3$, $\lambda_2 = 1$, $n_2 = 3$, $\lambda_3 = 0$, and $n_3 = 1$. Observe that in this example two vertices are ith-associated whenever the distance between them is exactly i. This notion of distance may be used in some instances to derive a PBIB-design from a graph which has the property that the number n_i of vertices at a distance i from any vertex is independent of the choice of vertex.

E3 *Exercise.* Taking as blocks the sets of vertices incident with the regions, show that the other Platonic solids (§*III*F) yield PBIB-designs. Determine their parameters.

At this point we introduce a new combinatorial object called an "association scheme." Association schemes may appear at first to be quite unrelated to PBIB-designs. However, neither of these two concepts, for reasons presently to become obvious, appears in the existing literature without some allusion to the other. An m-**class association scheme** is an ordered pair $(V, \mathscr{Q})$, where $\mathscr{Q} \in \mathbb{P}_m(\mathscr{P}_2(V))$ subject to the conditions E4 and E5 below. Let $\mathscr{Q} = \{Q_1, \ldots, Q_m\}$.

E4 *For each $i \in \{1, \ldots, m\}$, there exists a positive integer n_i such that for every $x \in V$,*

$$|\{y \in V + \{x\}: \{x, y\} \in Q_i\}| = n_i.$$

E5 *For $i, j, k \in \{1, 2, \ldots, m\}$, there exists a nonnegative integer p^i_{jk} such that for any $\{x, y\} \in Q_i$,*

$$|\{z \in V\colon \{x, z\} \in Q_j \text{ and } \{y, z\} \in Q_k\}| = p^i_{jk}.$$

If $\{x, y\} \in Q_i$, we say that x and y are ***i*th-associates**. Condition E4 states that for each i every vertex has the same number of ith associates. Observe that the ith associates in a PBIB-design also satisfy this condition as restated. With this terminology, let us restate E5 as it was first formulated by Bose and Nair [b.14] in 1939, except that in their agronomical context they wrote "variety" for vertex.

E6 "Given two vertices which are ith-associates, the number of vertices which are common to the jth-associates of one and the kth associates of the other is independent of the pair of ith associates with which we start. This number is denoted by p^i_{jk}."

It is immediate that for $i, j\ k \in \{1, 2, \ldots, m\}$,

E7
$$p^i_{jk} = p^i_{kj}.$$

The numbers n_i and p^i_{jk} $(i, j, k \in \{1, 2 \ldots, m\})$ are the **parameters** of the association scheme. It is convenient to display the m^3 parameters p^i_{jk} in $m \times m$ matrices $P_1, \ldots, P_m$ wherein the entry in the jth row, kth column of P_i is p^i_{jk}.

Let us return briefly to the example of the cube. Using the notion of ith associate described just before E3, we see that we have a 3-class association scheme with the following parameters: $n_1 = n_2 = 3$, $n_3 = 1$ and

E8
$$P_1 = \begin{bmatrix} 0 & 2 & 0 \\ 2 & 0 & 1 \\ 0 & 1 & 0 \end{bmatrix}, \qquad P_2 = \begin{bmatrix} 2 & 0 & 1 \\ 0 & 2 & 0 \\ 1 & 0 & 0 \end{bmatrix}, \qquad P_3 = \begin{bmatrix} 0 & 3 & 0 \\ 3 & 0 & 0 \\ 0 & 0 & 0 \end{bmatrix}.$$

This example illustrates the following principle:

E9 Exercise. Show that *if Λ is an m-class PBIB-design whose* ith *associates satisfy condition* E6 *for* $i = 1, \ldots, m$, *then there exists a unique m-class association scheme on the same vertex set with the same pairs of* ith *associates.*

Insofar as we have not imposed condition E6 as part of the definition of PBIB-design, we have departed from the literature. Exercise E9 shows that without this departure, every m-class PBIB-design would determine an m-class association scheme. This is no loss, however, since it is the reverse problem that is of greater interest, namely, when and how can a PBIB-design be constructed from an association scheme. Involved first of all is the selection of a set of blocks $\mathscr{E} \subseteq \mathscr{P}_k(V)$ for some appropriate k such that each element of V belongs to the same number r of elements of $\mathscr{E}$. Secondly there is the determination of *distinct* integers $\lambda_1, \ldots, \lambda_m \in \mathbb{N}$ such that each pair of ith associates is contained in exactly λ_i elements of $\mathscr{E}$ $(i = 1, \ldots, m)$. Frustrated by the requirement that $\lambda_1, \ldots, \lambda_m$ be distinct, Bose and Shimamoto [b.16] simply dropped it, thereby generalizing the definition of PBIB-design. We will not need to do so here since we will not be probing so deeply into the same problems.

We listed above $m + m^3$ parameters for an m-class association scheme. As with PBIB-designs, the parameters are not all independent. Equation E7 has already reduced this number by $\binom{m-1}{2}$. Let us now obtain some further relations for the parameters of an m-class association scheme and simultaneously some further relations for m-class PBIB-designs satisfying E6.

If $(V, \mathscr{Q})$ is an m-class association scheme on v vertices, then the parameters $n_1, \ldots, n_m$, as defined in E4, satisfy E1. Next we obtain:

E10 Proposition. *If $(V, \mathscr{Q})$ is an m-class association scheme, then for $i, j \in \{1, \ldots, m\}$,*

$$\sum_{k=1}^{m} p_{jk}^{i} = \begin{cases} n_j - 1 & \text{if } j = i; \\ n_j & \text{if } j \neq i. \end{cases}$$

Proof. Let $\{x, y\} \in Q_i$. Of the $(n_i - 1)$ ith associates of x other than y, precisely p_{ik}^{i} of them are kth associates of y, for each $k = 1, \ldots, m$. If $j \neq i$, then of the n_j jth associates of x, precisely p_{jk}^{i} are kth associates of y, for each $k = 1, \ldots, m$. □

E11 Proposition. *If $(V, \mathscr{Q})$ is an m-class association scheme, then for $i, j, k \in \{1, 2, \ldots, m\}$,*

$$n_i p_{jk}^{i} = n_j p_{ki}^{j} = n_k p_{ij}^{k}.$$

Proof. Let $x \in V$ and let $N_h = \{y \in V: \{x, y\} \in Q_h\}$. So $|N_h| = n_h$ for $h = 1, \ldots, m$. Now let i, j, k be given and let

$$S = \{(y, z): \{y, z\} \in Q_k; y \in N_i; z \in N_j\}.$$

For each $y \in N_i$, $(y, z) \in S$ for precisely p_{jk}^{i} vertices z. Hence $|S| = n_i p_{jk}^{i}$. Symmetrically, for each $z \in N_j$, $(y, z) \in S$ for precisely p_{ik}^{j} vertices y, giving $|S| = n_j p_{ik}^{j}$. This together with E7 establishes the first equality. The second follows similarly. □

E12 *Exercise.* Let $\Lambda = (V, \mathscr{E})$ be a PBIB-design satisfying E6. Show that the complementary design $\hat{\Lambda}$ is also a PBIB-design satisfying E6, and find its parameters.

E13 *Exercise.* Is the design given in C6 a PBIB-design? If so does it satisfy E6?

E14 *Exercise.* Let $(V, \mathscr{Q})$ be an m-class association scheme. Let $V = \{x_1, \ldots, x_v\}$ and let $\mathscr{Q} = \{Q_1 \ldots, Q_m\}$. For $h = 0, 1, \ldots, m$, we define the symmetric $v \times v$ matrices $B_h = [b_{ij}^{h}]$ as follows:

$$B_0 = I; \quad \text{and for } h = 1, \ldots, m,$$

$$b_{ij}^{h} = \begin{cases} 1 & \text{if } i \neq j \text{ and } \{x_i, x_j\} \in Q_h; \\ 0 & \text{otherwise.} \end{cases}$$

Verify the following assertions:

(a) $\sum_{h=0}^{m} B_h = J$.

(b) $B_s B_t = \sum_{h=0}^{m} p_{st}^h B_h$, where $p_{st}^0 = \delta_{st} n_s$, for $st > 0$.

(c) $\{B_0, B_1, \ldots, B_m\}$ is a basis for an $(m + 1)$-dimensional commutative algebra over $\mathbb{Q}$. (W. A. Thompson, Jr. [t.3] and R. C. Bose and D. M. Mesner [b.13]). In the latter paper it is further shown that if $B_0 = I, B_1, \ldots, B_m$ are any symmetric matrices satisfying (a) and (b) for some coefficients p_{st}^h, then there exists an m-class association scheme with parameters p_{st}^h yielding the matrices $B_0, B_1, \ldots, B_m$ in the manner described.)

In general, m-class association schemes have been studied in depth only for small values of m, in particular when $m = 2$. For this case, the definition of an association scheme is stronger than it needs to be, as we now prove.

E15 Theorem (Bose and Clatworthy [b.12]). *Let V be a set and let $\mathcal{Q} = \{Q_1, Q_2\} \in \mathbb{P}_2(\mathscr{P}_2(V))$. Suppose that* E4 *is satisfied for $m = 2$. For $i, j, k \in \{1, 2\}$, define the eight functions*

$$p_{jk}^i : \{(x, y) \in V \times V : \{x, y\} \in Q_i\} \to \mathbb{N}$$

by

$$p_{jk}^i(x, y) = |\{z \in V : \{x, z\} \in Q_j; \{y, z\} \in Q_k\}|.$$

For $i \in \{1, 2\}$, if p_{11}^i is a constant function, then so are p_{12}^i, p_{21}^i and p_{22}^i, and, moreover, E7 *holds.*

PROOF. Suppose $i = 1$. Let $\{x, y\} \in Q_1$. By the same arguments used to prove E10 for $m = 2$, we obtain

E16 $$p_{11}^1(x, y) + p_{12}^1(x, y) = n_1 - 1,$$

E17 $$p_{11}^1(x, y) + p_{21}^1(x, y) = n_1 - 1, \text{ and}$$

E18 $$p_{21}^1(x, y) + p_{22}^1(x, y) = n_2.$$

If we assume that $p_{11}^1(x, y)$ has constant value (which we denote by p_{11}^1), then E16 and E17 give

E19 $$p_{12}^1(x, y) = n_1 - 1 - p_{11}^1 = p_{21}^1(x, y).$$

From E18 and E19, we obtain $p_{22}^1(x, y) = n_2 - (n_1 - 1 - p_{11}^1)$, as required.

When $i = 2$, the argument is analogous, the details being left to the reader. □

E20 Corollary. *Let V be a set and let $\mathcal{Q} = \{Q_1, Q_2\} \in \mathbb{P}_2(\mathscr{P}_2(V))$. Suppose that* E4 *is satisfied for $m = 2$ and that for $i = 1, 2$ there exists $p_{11}^i \in \mathbb{N}$ such that for all $\{x, y\} \in Q_i$,*

$$p_{11}^i = |\{z \in V : \{x, z\}, \{y, z\} \in Q_1\}|.$$

Then $(V, \mathcal{Q})$ is a 2-class association scheme.

As the next exercise indicates, the hypothesis of Corollary E20 can be weakened even further.

E21 *Exercise.* Let V, $\mathcal{Q}$, and the functions p^i_{ijk} be as given in the statement of Theorem E15. Show that the conclusion of Corollary E20 also holds if instead of requiring the constancy specifically of p^1_{11} and p^2_{11}, we merely require the constancy of some one function p^1_{jk} and some one function p^2_{ih}.

Bose and Clatworthy [b.12] show further that to assume the constancy of some arbitrary pair of the eight functions p^i_{jk} is not sufficient for a 2-class association scheme. Their counter-example, being a PBIB-design (in our sense, not theirs) of blocksize 2, can be represented by the graph in Figure E22, where the blocks coincide with the edges. The PBIB-design parameters are $v = 7$, $b = 14$, $r = 4$, $k = 2$, $\lambda_1 = 1$, $n_1 = 4$, $\lambda_2 = 0$, $n_2 = 2$. The four

E22

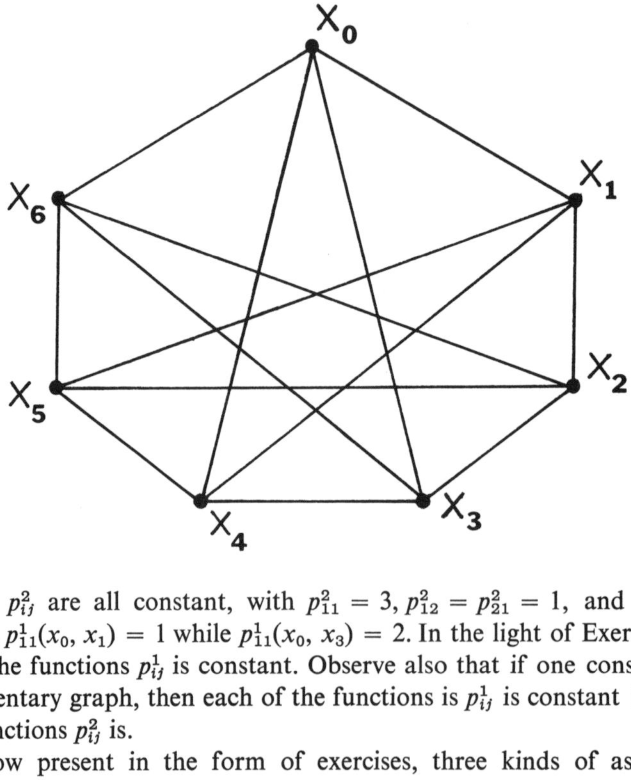

functions p^2_{ij} are all constant, with $p^2_{11} = 3$, $p^2_{12} = p^2_{21} = 1$, and $p^2_{22} = 0$. However, $p^1_{11}(x_0, x_1) = 1$ while $p^1_{11}(x_0, x_3) = 2$. In the light of Exercise E21, none of the functions p^1_{ij} is constant. Observe also that if one considers the complementary graph, then each of the functions is p^1_{ij} is constant but none of the functions p^2_{ij} is.

We now present in the form of exercises, three kinds of association schemes from which PBIB-designs satisfying E6 may be obtained. Each exercise consists of obtaining the parameters both of the association scheme and the design.

E23 *Exercise.* Let $\Lambda = (V, \mathscr{E})$ be a BIB-design with $\lambda = 1$. We form a 2-class association scheme with $\mathscr{E}$ as vertex set. We shall say that distinct elements $E_1, E_2 \in \mathscr{E}$ are first associates if $E_1 \cap E_2 \neq \varnothing$ and second associates if $E_1 \cap E_2 = \varnothing$. Then Λ^* is a 2-class PBIB-design satisfying E6 and based upon this scheme (S. Shrikhande [s.5]).

E24 *Exercise.* Let $m, n \in \mathbb{N}$ and $2 < m < n$. Let $v = mn$, and let V be a v-set whose elements are displayed in an $m \times n$ array. Two elements of V will be first associates if they lie in the same row of this array, second associates if they lie in the same column, and third associates otherwise. This yields a 3-class association scheme. A PBIB-design satisfying E6 can be formed letting each union of one row and one column of the array be a block (Bose and Nair [b. 14]).

E25 *Exercise.* Let $N = \{1, 2, \ldots, n\}$ and let $\{S_1, \ldots, S_t\} \in \mathbb{L}_n$ be orthogonal. Write $S_h = [a_{ij}^h]$. Let $V = N \times N$. We shall say that two distinct elements (i, j) and (i', j') of V are first associates if any one of the following three conditions holds:

(a) $i = i'$;
(b) $j = j'$;
(c) for some $h \in \{1, \ldots, t\}$, $a_{ij}^h = a_{i'j'}^h$.

Otherwise (i, j) and (i', j') are second associates. From this 2-class association scheme, we form a 2-class PBIB-design satisfying E6 in the following way. Let $(i, j) \in V$. The following sets are the blocks incident with (i, j):

(d) $\{(i, j'): j' = 1, \ldots, n\}$;
(e) $\{(i', j): i' = 1, \ldots, n\}$;
(f) $\{(i', j'): a_{i'j'}^h = a_{ij}^h\}$, $h = 1, \ldots, t$.

E26 *Exercise.* Prove that if $\{S_1, \ldots, S_t\}$ in Exercise E25 is a complete orthogonal set, then the above PBIB design is in fact the residual design of the finite projective plane determined by $S_1, \ldots, S_t$ (cf. D12) with respect to L_∞.

Another model of a 2-class association scheme $(V, \mathcal{Q})$ where $\mathcal{Q} = \{Q_1, Q_2\}$ is the graph $\Gamma = (V, Q_1)$ in which a pair of vertices comprise an edge if and only if the vertices are first associates. Such a graph is called a **strongly isovalent graph** ("strongly regular graph" by Bose *et al.*). It is clearly n_1-valent with $\nu_0(\Gamma) = n_1 + n_2 + 1$ (cf. E1) and $\nu_1(\Gamma) = \frac{1}{2}n_1(n_1 + n_2 + 1)$ by *III*A1.

E27 Proposition. *Let Γ be a strongly isovalent graph.*

(a) *If Γ is connected, then its diameter is at most* 2.

(b) *If Γ is not connected, then each component is a complete graph on $p_{11}^1 + 2$ vertices.*

PROOF. (a) If the distance between some two vertices of Γ is precisely 2, then $p_{11}^2 > 0$. Hence no two vertices can be at distance more than 2, since there would then exist second associates having no common first associate.

(b) If Γ is not connected, then consideration of two vertices in distinct components yields $p_{11}^2 = 0$. In this case each component has diameter at most 1. Actually it must be exactly 1 for if some component had no edges, then Γ being isovalent would have no edges, and Q_1 would be empty, contrary to the definition of a partition. The rest of the assertion is now immediate. □

From the graph-theoretical standpoint, disconnected strongly isovalent graphs are not very interesting. They arise from what Bose called "group divisible designs," which are constructed as follows. Let V be a bk-set and consider a b-partition of V, $b \geq 2$, wherein each cell is a k-subset. Let two vertices be first associates if they are elements of the same cell and second associates otherwise. We then have a 2-class association scheme with parameters

$$P_1 = \begin{bmatrix} k-2 & 0 \\ 0 & (b-1)k \end{bmatrix}, \qquad P_2 = \begin{bmatrix} 0 & k-1 \\ k-1 & (b-2)k \end{bmatrix}.$$

As a PBIB-design, the parameters are $v = bk, b, r = 1, k, \lambda_1 = 1, n_1 = k - 1, \lambda_2 = 0$, and $n_2 = (b-1)k$.

Suppose that in a set of people, each individual has the same number of acquaintances and every two people have exactly one mutual acquaintance. How many people are in the set? The answer is the so-called "Friendship Theorem" credited to Erdős, Rényi, and Sós [e.5]. Our results on strongly isovalent graphs yield a short proof as follows.

E28 Proposition. *The only strongly isovalent graph with* $p_{11}^1 = p_{11}^2 = 1$ *is* K_3.

Proof. By E11 we have

E29 $$n_1 p_{21}^1 = n_2 p_{11}^2 \quad \text{and} \quad n_2 p_{12}^2 = n_1 p_{21}^1.$$

By E10,

$$p_{11}^1 + p_{12}^1 = n_1 - 1 \quad \text{and} \quad p_{11}^2 + p_{12}^2 = n_1,$$

which yield $p_{12}^1 = n_1 - 2$ and $p_{12}^2 = n_1 - 1$. Substituting these values into E29 with $n_2 = \nu_0 - n_1 - 1$, we obtain

E30 $$\nu_0 = n_1^2 - n_1 + 1 \quad \text{and} \quad \nu_0 = (2n_1^2 - 2n_1 - 1)/(n_1 - 1).$$

Eliminating ν_0 from this pair of equations yields $n_1(n_1 - 2)^2 = 0$. Under our assumptions, $n_1 = 2$ is the only solution, which by E30 gives $\nu_0 = 3$. □

E31 *Exercise.* Determine a class of graphs other than K_3 in which every two vertices belong to a unique 3-circuit.

To conclude this section, let us consider a small but interesting class of connected strongly isovalent graphs. Suppose that $\Gamma = (V, \mathscr{E})$ is ρ-valent with diameter δ. Let $x_0 \in V$ and define

$$V_i = \{y \in V : d(x_0, y) = i\}, \quad i = 0, 1, \ldots, \delta.$$

Thus $V_0 = \{x_0\}$ and

E32 $$|V_i| \leq \rho(\rho - 1)^{i-1}, \quad i = 1, \ldots, \delta.$$

Summing over i, we obtain

E33 $$\nu_0(\Gamma) \leq 1 + \rho \sum_{i=1}^{\delta} (\rho - 1)^{i-1}.$$

If equality holds in E33, then Γ is called a **Moore graph of type** (ρ, δ). It is immediate that equality in E33 implies equality in E32 for all $i = 1, \ldots, \delta$. In this case it follows inductively for $i = 1, \ldots, \delta - 1$ that each vertex in V_i is incident with exactly one vertex in V_{i-1} and $\rho - 1$ vertices in V_{i+1}. Vertices in V_δ are incident with one vertex in $V_{\delta-1}$ and with $\rho - 1$ other vertices in V_δ. Since $\nu_0(\Gamma)$ and ρ are independent of x_0, equality in E33 imposes, moreover, that the structure just described obtains independently of the initially chosen vertex x_0 with respect to which the sets V_i are defined. It follows that *Moore graphs have girth precisely* $2\delta + 1$.

E34 Proposition. *The Moore graphs of type* $(\rho, 2)$ *are strongly isovalent.*

PROOF. Let Γ be a Moore graph of type $(\rho, 2)$. Thus $\rho = n_1$ and by Theorem E15 it suffices to prove that p_{11}^1 and p_{11}^2 are constant. Let $x_0 \in V$ and let V_0, V_1, and V_2 be as above. Let $y \in V_1$ and $z \in V_2$. Due to the structure of a Moore graph, $p_{11}^1(x_0, y) = 0$ and $p_{11}^2(x_0, z) = 1$. The result follows since x_0 was chosen arbitrarily, y was an arbitrary first associate of x_0, and z was an arbitrary second associate of x_0. □

E35 *Exercise.* Show that all Moore graphs are geodetic.

Hoffman and Singleton [h.8] have shown by powerful algebraic methods that if $\delta \leq 3$, then the only Moore graphs of type (ρ, δ) which exist are of types (2, 2), (3, 2), (7, 2), and (2, 3), with the existence of a Moore graph of type (57, 2) remaining undecided. The graph Δ_5 is of type (2, 2), Δ_7 is of type (2, 3), and the Petersen graph (Figure *VII*A42) is of type (3, 2). These graphs are the unique representatives of their respective types.

IXF Partial Geometries

In the preceding section one obtains the definition of a PBIB-design from that of a BIB-design by replacing the condition that $\bar{s}$ be constant on $\mathscr{P}_2(V)$ with a somewhat weaker condition. By similarly relaxing the requirement that every two points be on a line we may obtain "partial geometries" from finite projective planes. The terms "line" and "point" will again be used for "block" and "vertex," respectively.

A **partial geometry** is a tactical configuration $\Lambda = (V, f, E)$ such that

F1 $$\bar{s}(S) \leq 1 \quad \textit{for all } S \in \mathscr{P}_2(V),$$

and

F2 *there exists a positive integer t such that given any* $e_0 \in E$ *and* $x_0 \in V + f(e_0)$, $|\{e \in E\colon x_0 \in f(e); f(e) \cap f(e_0) \neq \varnothing\}| = t$.

It is useful to note that the transpose of F1 is also valid for a partial geometry, for if the intersection of some two lines contained a 2-set S, then $\bar{s}(S) \geq 2$. Hence

F3 $$\overline{s^*}(A) \leq 1 \quad \text{for all } A \in \mathscr{P}_2(E).$$

It follows at once that Λ is a set system. We write $\Lambda = (V, \mathscr{L})$ and rewrite F2 as follows:

F4 *There exists a positive integer t such that given any line L_0 and any point x_0 not on L_0, there exist exactly t lines through x_0 which meet L_0.*

The symbols b, v, r, and k retain their conventional meanings. The symbol t will retain its meaning from F2.

F5 Proposition. *If Λ is a partial geometry with parameters b, v, r, k, and t, then*

(a) $bk = vr$;
(b) $t \leq r$;
(c) $t \leq k$;
(d) $(v - k)t = k(k - 1)(r - 1)$;
(e) $(b - r)t = r(k - 1)(r - 1)$.

PROOF. Equation (a) is precisely C7 and (b) is immediate. To prove (c), let $L_0 \in \mathscr{L}$ and $x_0 \in V + L_0$. By F2, x_0 lies on at least t lines. Hence, $t \leq r$. By F3, no two of these lines can intersect in any point other than x_0. Hence they meet L_0 in t distinct points, and so $t \leq k$.

To prove (d), we fix $L_0 \in \mathscr{L}$ and enumerate the set

$$\{(x, L): x \in L \cap (V + L_0); L \cap L_0 \neq \varnothing\}$$

in two ways. First we may choose $x \in V + L_0$ in $(v - k)$ ways and then choose L, by virtue of F2, in t ways. Secondly, we choose the number of lines $L \neq L_0$ which meet L_0 in $k(r - 1)$ ways and then choose x on each such line in $(k - 1)$ ways. Equation (e) is easily obtained from (d) by multiplying by r/k and substituting $b = vr/k$ from (a). □

By equations (d) and (e) above, it is clear that once r, k, and t are known, v and b can be computed. Following Bose [b.11], we shall use the triple (r, k, t) to indicate the parameters of a partial geometry.

F6 *Exercise.* Let Λ be a partial geometry with parameters (r, k, k). Is Λ or Λ^* a BIB-design? Is Λ ever a finite projective plane? If so what is its order?

By definition, the transpose of a tactical configuration is a tactical configuration. We have observed that for a partial geometry Λ, the dual conditions F1 and F3 both hold. We also observe that the set of equalities and inequalities in F5 are "self dual"; i.e., if b and v as well as r and k are interchanged, the set of equalities and inequalities remains unchanged. It is not surprising then that the dual of condition F4 obtained by interchanging the notions point and line is also valid. To see this, define the "dual line" $x^* = \{L \in \mathscr{L}: x \in L\}$

for each $x \in V$. Let $L_0 \in \mathscr{L}$ and let x_0^* be a dual line such that $L_0 \notin x_0^*$. Since $x_0 \notin L_0$, there exists a t-set $\mathscr{T}$ of lines through x_0 which meet L_0. For each $L \in \mathscr{T}$, let $x_L \in L \cap L_0$. By F3, these points x_L are all distinct. It is clear then that each of the t dual lines in $\{x_L^*: L \in \mathscr{T}\}$ contains L_0 and meets x_0^*. Also no other dual lines through L_0 meet x_0^*. We have proved:

F7 Proposition. *If Λ is a partial geometry with parameters (r, k, t), then Λ^* is a partial geometry with parameters (k, r, t).*

F8 Proposition. *A partial geometry $\Lambda = (V, \mathscr{L})$ is a two-class PBIB-design satisfying* E6 *with the additional parameters $\lambda_1 = 1, n_1 = r(k-1)$, $\lambda_2 = 0$, $n_2 = (k-t)(k-1)(r-1)/t$, $p_{11}^1 = (t-1)(r-1) + k - 2$, and $p_{11}^2 = rt$.*

Proof. It is immediate from F1 that $\lambda_1 = 1$ and $\lambda_2 = 0$. Each point $x \in V$ is incident with exactly r lines, each of which contains exactly $k - 1$ points other than x. Thus each point has precisely $n_1 = r(k-1)$ first associates. Substituting this value and the value of v from F5d into $n_2 = v - n_1 - 1$ (cf. E1), we obtain $n_2 = (k-t)(k-1)(r-1)/t$.

If x and y are first associates, then the unique line containing them contains exactly $k - 2$ common first associates. Each of the $r - 1$ other lines through x contains exactly $t - 1$ first associates of y other than x. Hence $p_{11}^1 = k - 2 + (r-1)(t-1)$.

If x and y are second associates, that is to say, x and y are not collinear, then each of the r lines through x contains exactly t first associates of y. Hence $p_{11}^2 = rt$. By Corollary E20, the proof is complete. □

We state the following result without proof.

F9 Lemma (R. C. Bose and D. M. Mesner [b.13]. *If M is an incidence matrix for a connected 2-class PBIB-design satisfying* E6, *then MM^* has exactly three characteristic roots with multiplicities* 1, α, *and* β, *where*

$$\alpha = \frac{n_1 + n_2}{2} - m \quad \textit{and} \quad \beta = \frac{n_1 + n_2}{2} + m$$

for

$$m = \frac{n_1 - n_2 + (p_{12}^2 - p_{12}^1)(n_1 + n_2)}{2\sqrt{(p_{12}^2 - p_{12}^1)^2 + 2(p_{12}^2 + p_{12}^1) + 1}}.$$

We make two observations regarding this lemma. First, the values α and β are independent of λ_1, λ_2, and r. Hence when a 2-class PBIB-design Λ with incidence matrix M is derived from a 2-class association scheme $(V, \mathscr{Q})$, then α and β depend only upon the parameters of $(V, \mathscr{Q})$ and are the same for all designs derivable from $(V, \mathscr{Q})$. Secondly, a necessary condition for $(V, \mathscr{Q})$ to yield a 2-class PBIB-design at all is that α and β be positive integers.

F10 Exercise. *If* Λ *is a partial geometry with parameters* (r, k, t) *then*

$$\alpha = \frac{rk(r-1)(k-1)}{t(k+r-t-1)}.$$

This exercise along with the fact that α must be a positive integer and Proposition F4(d,e) imply

F11 Theorem (R. C. Bose [b.11]). *A necessary condition for the existence of a partial geometry with parameters* (r, k, t) *is that*

$$\frac{k(r-1)(k-1)}{t}, \quad \frac{r(r-1)(k-1)}{t}, \quad \text{and} \quad \frac{rk(r-1)(k-1)}{t(k+r-t-1)}$$

be positive integers.

This condition is not sufficient. If $(r, k, t) = (3, 11, 1)$, then $v = 231$, $b = 63$, and $\alpha = 55$. However, no partial geometry with these parameters exists (see R. C. Bose and W. H. Clatworthy [b.12]).

CHAPTER X

Matroid Theory

In his paper [w.11] entitled "On the abstract properties of linear dependence" published in 1935, H. Whitney defined systems called "matroids" and endowed them with certain properties abstracted from coordinatized vector spaces. The next notable contribution to "matroid theory" did not make its appearance until 1958–59 with three papers by W. T. Tutte; his work was revised and reappeared [t.6] in 1965.

Most of the main results of this chapter were first formulated in the above-mentioned papers and in the publication *Combinatorial Geometries* by H. Crapo and G.-C. Rota [c.5].

XA Exchange Systems

Throughout this section, $\Lambda = (V, \mathscr{A})$ will denote a set system. For $i = 1, 2$, we shall say that Λ is an **exchange system of type** i, if for every triple (A_1, A_2, x_1) such that $A_1, A_2 \in \mathscr{A}$, $|A_1| = |A_2|$, and $x_1 \in A_1 + (A_1 \cap A_2)$ all hold, there exists $x_2 \in A_2 + (A_1 \cap A_2)$ such that $A_i + \{x_1, x_2\} \in \mathscr{A}$. In other words, in an exchange system of type 1, one obtains a new block by finding some vertex x_2 to replace x_1 in A_1, while in an exchange system of type 2, a new block is obtained by letting the given vertex x_1 replace some vertex x_2 in A_2.

Given Λ, one defines the set system $\hat{\Lambda} = (V, \{V + A : A \in \mathscr{A}\})$; cf. *IXC*2. Clearly the mapping $\Lambda \mapsto \hat{\Lambda}$ takes $\hat{\Lambda}$ onto Λ for every set system Λ. It can be straightforwardly shown:

A1 Exercise. *Λ is an exchange system of type* 1 *if and only if $\hat{\Lambda}$ is an exchange system of type* 2.

A2 Proposition. *A set system is an exchange system of type* 1 *if and only if it is an exchange system of type* 2.

PROOF. Suppose that Λ is an exchange system of type 1. Let $A_1, A_2 \in \mathscr{A}$ such that $|A_1| = |A_2| = k$ and let $x_1 \in A_1 + (A_1 \cap A_2)$. Let

$$\mathscr{F} = \{A \in \mathscr{A} \cap \mathscr{P}_k(V) : x_1 \in A; A \cap A_2 \supseteq A_1 \cap A_2\}.$$

Then $\mathscr{F} \neq \varnothing$ since $A_1 \in \mathscr{F}$. Select $A_0 \in \mathscr{F}$ so that $|A_0 \cap A_2|$ is as large as possible. Since $x_1 \in A_0$ but $x_1 \notin A_2$, while $|A_0| = |A_2|$, there exists $y \in A_0 + (A_0 \cap A_2)$. Suppose $y \neq x_1$ and consider the triple (A_0, A_2, y). Since Λ is an exchange system of type 1, there exists $z \in A_2 + (A_0 \cap A_2)$ such that $A_3 = A_0 + \{y, z\} \in \mathscr{A}$. Now $|A_3| = k$ and $x_1 \in A_3$. Moreover, $A_3 \cap A_2 = (A_0 \cap A_2) + \{z\} \supset A_0 \cap A_2$. Hence $A_3 \in \mathscr{F}$, but the maximality of $|A_0 \cap A_2|$ has been contradicted. Therefore y must be x_1; that is, $A_0 + (A_0 \cap A_2) = \{x_1\}$. Since $|A_0| = |A_2|$, there exists $x_2 \in V$ such that $\{x_2\} = A_2 + (A_0 \cap A_2)$, whence $A_2 + \{x_2\} = A_0 \cap A_2$. Hence $A_2 + \{x_1, x_2\} = (A_0 \cap A_2) + \{x_1\} = A_0 \in \mathscr{A}$.

Conversely, if Λ is an exchange system of type 2, then by Exercise A1 followed by the first part of this proof, $\hat{\Lambda}$ is also an exchange system of type 2. Another application of A1 completes the proof. □

It now makes sense to define a set system to be an **exchange system** if it is an exchange system of either type and hence of both types, 1 and 2.

A3 Corollary. *Λ is an exchange system if and only if $\hat{\Lambda}$ is.*

A4 Exercise. Show that *$(V, \mathscr{A})$ is an exchange system if and only if for all k, $(V, \mathscr{A} \cap \mathscr{P}_k(V))$ is an exchange system.*

A5 *Example.* Let Γ be a multigraph with edge set E, and let $\mathscr{F}$ be the collection of edge sets of the spanning forests of Γ. We assert that $(E, \mathscr{F})$ is an exchange system. For let $F_1, F_2 \in \mathscr{F}$, and let $e \in F_1 + (F_1 \cap F_2)$. We assume also that $|F_1| = |F_2|$, although this is automatic by *III*C11. Applying the Exchange Property of *III*C16 to F_2 and e, we infer that $(E, \mathscr{F})$ is an exchange system (of type 2). The solution to *III*C19 for spanning coforests now follows from this example and Corollary A3.

A6 *Example.* The complete design $(V, \mathscr{P}_k(V))$ is an exchange system for $0 \leq k \leq |V|$. If $0 \leq j < k \leq |V|$, then $(V, \mathscr{P}_j(V) + \mathscr{P}_k(V))$ is also an exchange system.

A7 *Example.* Let V be a finite spanning set of the vector space $\mathscr{V}$ over a field $\mathbb{F}$. Let

$$\mathscr{I} = \{J \in \mathscr{P}(V) : J \text{ is independent}\},$$
$$\mathscr{S} = \{S \in \mathscr{P}(V) : S \text{ spans } \mathscr{V}\},$$
$$\mathscr{B} = \{B \in \mathscr{P}(V) : B \text{ is a basis for } \mathscr{V}\}.$$

We assert that $(V, \mathscr{I})$, $(V, \mathscr{S})$, and $(V, \mathscr{B})$ are all exchange systems. For if J_1 and J_2 are independent sets of equal cardinality k and if $x_1 \in J_1 + (J_1 \cap J_2)$, then $J_1 + \{x_1\}$ is a basis for a $(k-1)$-dimensional subspace of $\mathscr{V}$ which cannot contain J_2, since J_2 spans some k-dimensional subspace. Hence $J_1 + \{x_1, x_2\}$ is independent for some $x_2 \in J_2 + (J_1 \cap J_2)$. Thus $(V, \mathscr{I})$ is an exchange system, and by Exercise A4, so is $(V, \mathscr{B})$. If S_1 and S_2 are spanning k-subsets of $\mathscr{V}$ and $x_1 \in S_1 + (S_1 \cap S_2)$, then $\dim\langle S_1 + \{x_1\}\rangle = \dim(\mathscr{V})$ or $\dim(\mathscr{V}) - 1$. In the first case, choose any $x_2 \in S_2 + (S_1 \cap S_2)$. In the second case, there exists some $x_2 \in S_2 + (S_1 \cap S_2)$ such that $x_2 \notin \langle S_1 + \{x_1\}\rangle$. Hence $S_1 + \{x_1, x_2\}$ spans $\mathscr{V}$.

A8 *Exercise.* Let $\mathscr{A}_1 \in \mathbb{P}(V)$ and let $\mathscr{A}_2 = \{A_2 \in \mathscr{P}(V): A_1 \cap A_2 \neq \varnothing$ for all $A_1 \in \mathscr{A}_1\}$. Show that $(V, \mathscr{A})$ is an exchange system when $\mathscr{A}$ is each of the following:

(a) $\mathscr{A}_2$;
(b) $\mathscr{M}(\mathscr{A}_2)$;
(c) $\{A \in \mathscr{P}(V): A_1 \nsubseteq A$ for all $A_1 \in \mathscr{A}_1\}$.

We conclude this batch of examples of exchange systems with one that will play an important role in §E below. Let $\Gamma = (V, \mathscr{E})$ be a graph and let $\mathscr{F}$ be an independent subset of $\mathscr{E}$ (in the sense of Chapter V). Then the set $\bigcup_{E \in \mathscr{F}} E$ is called a **matched set**. Clearly the largest matched sets of Γ have cardinality $2\alpha_1(\Gamma)$.

A9 Proposition. *Let $\Gamma = (V, \mathscr{E})$ be a graph and let $\mathscr{A}$ be the collection of largest matched sets of Γ. Then $(V, \mathscr{A})$ is an exchange system.*

PROOF. Let A_1 and A_2 be distinct matched sets and let $x_1 \in A_1 + (A_1 \cap A_2)$. Let $\mathscr{F}_i$ be an independent edge set such that

$$A_i = \bigcup_{E \in \mathscr{F}_i} E, \quad i = 1, 2.$$

There exists $x_2 \in A_1 \cap A_2$ such that $\{x_1, x_2\} \in \mathscr{F}_1$. With this beginning, we construct the sequence $x_1, x_2, \ldots, x_k$ of distinct vertices of Γ so that $\{x_i, x_{i+1}\} \in \mathscr{F}_1$ if i is odd and $\{x_i, x_{i+1}\} \in \mathscr{F}_2$ if i is even. Since these vertices are distinct, this process necessarily terminates at some vertex x_k. Let

$$\mathscr{E}' = \{\{x_i, x_{i+1}\}: i = 1, \ldots, k-1\}.$$

If k is even, then $|\mathscr{E}' \cap \mathscr{F}_1| = \frac{1}{2}k$ while $|\mathscr{E}' \cap \mathscr{F}_2| = \frac{1}{2}k - 1$. Hence $|\mathscr{F}_2 + \mathscr{E}'| = |\mathscr{F}_2| + 1 > \alpha_1(\Gamma)$, which is impossible since, as one easily verifies, $\mathscr{F}_2 + \mathscr{E}'$ is independent. Therefore k is odd, and so $x_k \in A_2 + (A_1 \cap A_2)$. In this case $|\mathscr{E}' \cap \mathscr{F}_1| = |\mathscr{E}' \cap \mathscr{F}_2|$, and so $|\mathscr{F}_1 + \mathscr{E}'| = |\mathscr{F}_1|$. Since $\mathscr{F}_1 + \mathscr{E}'$ is a largest independent edge set, we have

$$A_1 + \{x_1, x_k\} = \sum_{E \in \mathscr{F}_1 + \mathscr{E}'} E \in \mathscr{A}.$$

Thus $(V, \mathscr{A})$ is an exchange system (of type 1). $\square$

A maximal independent set of edges of a graph need not be a largest independent set. By adapting the above proof, however, one can show the following:

A10 Exercise. *Every maximal matched set of a graph* Γ *is a largest matched set of* Γ.

A11 Lemma. *Let* $\Lambda = (V, \mathscr{A})$ *where* $\mathscr{A} \subseteq \mathscr{P}_k(V)$. *Then* Λ *is an exchange system if and only if* $(V, \bigcup_{A \in \mathscr{A}} \mathscr{P}(A))$ *is an exchange system.*

Proof. If $(V, \bigcup_{A \in \mathscr{A}} \mathscr{P}(A))$ is an exchange system, then so is Λ, by Exercise A4.

Conversely, suppose that Λ is an exchange system. It suffices to prove that $(V, \bigcup_{A \in \mathscr{A}} \mathscr{P}_{k-1}(A))$ is an exchange system; the rest follows by repeated application of this result and Exercise A4. Let $A_1, A_2 \in \bigcup_{A \in \mathscr{A}} \mathscr{P}_{k-1}(A)$, and let $x_1 \in A_1 + (A_1 \cap A_2)$.

Case 1: there exists $y \in V$ *such that* $A_1 + \{y\}$, $A_2 + \{y\} \in \mathscr{A}$. Then $x_1 \in (A_1 + \{y\}) + [(A_1 + \{y\}) \cap (A_2 + \{y\})]$. Since Λ is an exchange system, there exists $x_2 \in (A_2 + \{y\}) + [(A_1 + \{y\}) \cap (A_2 + \{y\})]$ such that $A_1 + \{y, x_1, x_2\} \in \mathscr{A}$. Since $x_2 \notin A_1 + \{y\}$, we have

$$A_1 + \{x_1, x_2\} \in \mathscr{P}_{k-1}(A_1 + \{y, x_1, x_2\})$$

as required.

Case 2: for all $y \in V$, *either* $A_1 + \{y\} \notin \mathscr{A}$ *or* $A_2 + \{y\} \notin \mathscr{A}$. For $i = 1, 2$, there exists y_i such that $A_i + \{y_i\} \in \mathscr{A}$. If $y_1 \in A_2$, then $A_1 + \{x_1, y_1\}$ is a $(k-1)$-subset of $A_1 + \{y_1\}$ as required. Hence suppose $y_1 \notin A_2$. Since Λ is an exchange system, the triple $(A_1 + \{y_1\}, A_2 + \{y_2\}, y_1)$ yields $x_2 \in (A_2 + \{y_2\}) + [(A_1 + \{y_1\}) \cap (A_2 + \{y_2\})]$ such that $A_1 + \{x_2\} = (A_1 + \{y_1\}) + \{y_1, x_2\} \in \mathscr{A}$. We know that $x_2 \neq y_2$, or else Case 1 would apply. Hence $x_2 \in A_2 + (A_1 \cap A_2)$, while $A_1 + \{x_1, x_2\} \in \mathscr{P}_{k-1}(A_1 + \{x_2\})$. □

Let $w: V \to \{x \in \mathbb{R} : x \geq 0\}$. For each $U \in \mathscr{P}(V)$, define $w(U) = \sum_{u \in U} w(u)$. Thus $w: \mathscr{P}(V) \to \{x \in \mathbb{R} : x \geq 0\}$ is a "weight function." Given a set system $\Lambda = (V, \mathscr{A})$, we seek a procedure for determining a "heaviest block"; that is, a block $A_0 \in \mathscr{A}$ such that $w(A_0) \geq w(A)$ for all $A \in \mathscr{A}$. The procedure we shall describe is very naïve. J. Edmonds [e.2] has named it the **greedy algorithm**. It goes as follows.

Enlarging $\mathscr{A}$ to include all subsets of every set in $\mathscr{A}$ yields an equivalent problem. We therefore make the assumption that

A12 $$\mathscr{A} = \bigcup_{A \in \mathscr{A}} \mathscr{P}(A).$$

Choose $x_0, x_1, \ldots \in V$ so that for each i, $\{x_0, \ldots, x_i\} \in \mathscr{A}$ and

$$w(x_i) = \max\{w(x) : x \in V + \{x_0, \ldots, x_{i-1}\}; \{x_0, \ldots, x_{i-1}, x\} \in \mathscr{A}\}.$$

The procedure terminates when a set $A_0 = \{x_0, \ldots, x_k\}$ in $\mathscr{A}$ has been constructed and $A_0 \cup \{x\} \notin \mathscr{A}$ for all $x \in V + A_0$. Certainly unless some conditions are imposed on Λ, there is no reason to believe that A_0 is a heaviest block. We can only be sure that A_0 is a "local maximum" in the sense of the following exercise.

A13 *Exercise.* Let $A_0 \in \mathscr{A}$ be a set obtained by application of the greedy algorithm. Let $A_1 \in \mathscr{A}$ and suppose that $|A_0 + A_1| \leq 2$. Then $w(A_1) \leq w(A_0)$.

The next result generalizes a theorem of D. Gale [g.1].

A14 Theorem. *Let* $\Lambda = (V, \mathscr{A})$ *be a set system satisfying* A12, *and let* $\mathscr{B}$ *be the collection of largest sets in* $\mathscr{A}$. *The greedy algorithm yields a heaviest set in* Λ *for every weight function if and only if the following two conditions hold:*

(a) *For every* $A \in \mathscr{A}$, *there exists* $B \in \mathscr{B}$ *such that* $A \subseteq B$.

(b) $(V, \mathscr{B})$ *is an exchange system.*

Proof. Suppose that Λ satisfies conditions (a) and (b), and let w be any weight function. Since $U_1 \subseteq U_2 \subseteq V$ implies $w(U_1) \leq w(U_2)$, it is clear by (a) that the greedy algorithm will always terminate with a set in $\mathscr{B}$. Also, some heaviest set must belong to $\mathscr{B}$.

Suppose the greedy algorithm terminates with the set $B_0 = \{x_0, \ldots, x_k\}$, where $w(x_0) \geq \ldots \geq w(x_k)$. Let $B_1 = \{y_0, \ldots, y_k\}$ be any set in $\mathscr{B}$, and assume $w(y_0) \geq \ldots \geq w(y_k)$. We show that $w(x_i) \geq w(y_i)$ for $i = 0, \ldots, k$, and hence $w(B_0) \geq w(B_1)$. For let h be the smallest index if any for which $w(x_h) < w(y_h)$. By the greedy algorithm, $h \geq 1$, and we define nonempty sets $A_1 = \{x_0, \ldots, x_h\}$ and $A_2 = \{y_0, \ldots, y_h\} \in \bigcup_{B \in \mathscr{B}} \mathscr{P}_h(B)$. Note that $x_h \in A_1 + (A_1 \cap A_2)$ since $w(y_{h-1}) \geq w(y_h) > w(x_h)$.

Since $(V, \mathscr{B})$ is an exchange system by (b), Lemma A11 yields that $(V, \bigcup_{B \in \mathscr{B}} \mathscr{P}(B))$ is an exchange system, and so with respect to the triple (A_1, A_2, x_h), we obtain $z \in A_2 + (A_1 \cap A_2)$ such that $A_1 + \{x_h, z\} \subseteq B'$ for some $B' \in \mathscr{B}$. But $z = y_i$ for some $i \leq h$, and so $w(\{x_0, \ldots, x_{h-1}, z\}) = w(A_1) - w(x_h) + w(y_i) > w(A_1)$, contrary to the greedy algorithm.

Conversely, suppose that either (a) or (b) fails. If (a) fails, choose a maximal set $A \in \mathscr{A}$ contained in no set in $\mathscr{B}$ and choose $B \in \mathscr{B}$. If (b) fails, choose $B, B' \in \mathscr{B}$ and $x_1 \in B' + (B \cap B')$ such that the triple (B', B, x_1) yields no appropriate x_2 required of an exchange system, and let $A = B' + \{x_1\}$. Let $1 < t < |B|/|A|$, and let the weight function w be defined by

$$w(x) = \begin{cases} t & \text{if } x \in A; \\ 1 & \text{if } x \in B + (A \cap B); \\ 0 & \text{if } x \in V + (A \cup B). \end{cases}$$

Since $w(x_1) = 0$, A is contained in no $(|A| + 1)$-set in $\mathscr{A}$ of greater weight. Hence the greedy algorithm terminates with some set A' such that $A \subseteq A'$ and $w(A) = w(A')$. However, $w(A) = |A|t < |B|$, although B has weight $w(B) = |B| + (t - 1)|A \cap B| \geq |B|$.

As an application of the greedy algorithm, the reader is referred to the "connector problem" or "minimum tree problem." (See Berge [b.5, p. 470].) In this case one seeks to link together various cities by a single (connected) communications network as cheaply as possible, where to each pair of cities some cost function assigns the cost of joining them. One procedure is to use a weight function which is the reciprocal of the cost function to construct a spanning forest (the cities being the vertices) whose edge set is a heaviest set. Alternatively (though less efficiently if the number of vertices exceeds 4) is to use the given cost function as a weight function to construct a heaviest spanning coforest—then throw it away and use what is left.

XB Matroids

In this section four types of set systems will be presented. It will become evident that they are very closely interrelated. The first three will turn out to be types of exchange systems. The fourth one generally is not, but it is this one that will be given the name of "matroid."

A set system $(V, \mathscr{A})$ is called an **independence system** if it satisfies conditions B1 and B2:

B1 $\mathscr{A} = \bigcup_{A \in \mathscr{A}} \mathscr{P}(A)$;

B2 *If* $A_1, A_2 \in \mathscr{A}$ *and if* $|A_1| < |A_2|$, *then there exists* $x \in A_2 + (A_1 \cap A_2)$ *such that* $A_1 + \{x\} \in \mathscr{A}$.

B3 Proposition. *Every independence system is an exchange system.*

Proof. Let $\Lambda = (V, \mathscr{A})$ be an independence system. Let $A_1, A_2 \in \mathscr{A}$ such that $|A_1| = |A_2|$ and let $x_1 \in A_1 + (A_1 \cap A_2)$. By condition B1, $A_1 + \{x_1\} \in \mathscr{A}$. By condition B2, there exists $x_2 \in A_2 + (A_1 \cap A_2)$ such that $A_1 + \{x_1, x_2\} \in \mathscr{A}$. □

A set system $(V, \mathscr{A})$ is called a **spanning system** if it satisfies conditions B4 and B5:

B4 *If* $A_1 \in \mathscr{A}$ *and* $A_1 \subseteq A_2 \subseteq V$, *then* $A_2 \in \mathscr{A}$;

B5 *If* $A_1, A_2 \in \mathscr{A}$ *and if* $|A_1| < |A_2|$, *then there exists* $x \in A_2 + (A_1 \cap A_2)$ *such that* $A_2 + \{x\} \in \mathscr{A}$.

One proves straightforwardly:

B6 Lemma. Λ *is an independence system if and only if* $\hat{\Lambda}$ *is a spanning system.*

By this lemma, Proposition B3, and Corollary A3, we have

B7 Corollary. *Every spanning system is an exchange system.*

An exchange system of constant blocksize is called a **basis system.**

B8 *Example*. In Example A7, $(V, \mathscr{I})$ is an independence system, $(V, \mathscr{S})$ is a spanning system, and $(V, \mathscr{B})$ is a basis system. Obviously the terminology of vector spaces has inspired the terminology of matroid theory.

B9 *Exercise*. Which of the exchange systems in Exercise A8 are independence systems, spanning systems, or basis systems?

The next result shows how one can always obtain an independence system from a basis system and vice-versa. Its corollary in a similar way relates spanning systems to basis systems.

B10 Proposition. *Let $\mathscr{I}, \mathscr{B} \subseteq \mathscr{P}(V)$. Then the following two statements are equivalent:*

(a) *$(V, \mathscr{I})$ is an independence system and $\mathscr{B}$ is the collection of largest sets in $\mathscr{I}$;*

(b) *$(V, \mathscr{B})$ is a basis system and $\mathscr{I} = \bigcup_{B \in \mathscr{B}} \mathscr{P}(B)$.*

Proof. If we assume (a), then $(V, \mathscr{B})$ is an exchange system by B3 and A4, and hence $(V, \mathscr{B})$ is a basis system. If $A \in \mathscr{I}$, then by repeated application of condition B2, a set $B \in \mathscr{B}$ is obtained such that $A \subseteq B$.

Conversely, let (b) be assumed. It is immediate that $\mathscr{B}$ is the collection of largest sets in $\mathscr{I}$, and hence condition B1 holds for $\mathscr{I}$. To verify B2, let $A_1, A_2 \in \mathscr{I}$ and suppose $|A_1| < |A_2|$. We may pick $B \in \mathscr{B}$ such that $A_1 \subseteq B$, and pick $x \in B + A_1$. If $x \in A_2$, then since $A_1 + \{x\} \in \mathscr{I}$, we are done. If $x \notin A_2$, then let A_2' be a subset of A_2 of cardinality $|A_1| + 1$. By Lemma A11, $(V, \mathscr{I})$ is an exchange system, and so for the triple $(A_1 + \{x\}, A_2', x)$ we obtain $y \in A_2' + (A_1 \cap A_2') \subseteq A_2 + (A_1 \cap A_2)$ such that $A_1 + \{y\} \subseteq A_1 + \{x, y\} \in \mathscr{I}$. □

B11 Corollary. *Let $\mathscr{S}, \mathscr{B} \subseteq \mathscr{P}(V)$. Then the following two statements are equivalent:*

(a) *$(V, \mathscr{S})$ is a spanning system and $\mathscr{B}$ is the collection of smallest sets in $\mathscr{S}$.*

(b) *$(V, \mathscr{B})$ is a basis system and $\mathscr{S} = \{S \in \mathscr{P}(V): S \supseteq B \text{ for some } B \in \mathscr{B}\}$.*

Proof. Use the proposition and Lemma B6. □

B12 Corollary. *The greedy algorithm with weight function identically* 1 *always yields a largest set in an independence system.*

The complete designs (cf. A6) are basis systems. Examples of basis systems coming from graph theory are those given by the spanning forests, the spanning coforests (cf. A5), and the largest matched sets (cf. A9). There are many other examples of basis systems, but these are particularly "natural" ones. Using the above proposition and corollary, the reader should determine from these basis systems the associated independence systems and spanning systems. Note in particular that if $(V, \mathscr{B})$ is the basis system of the largest matched sets of a graph, then by Exercise A10, the blocks of the associated independence system include all the matched sets of the graph.

Let $(V, \mathscr{A})$ be a set system. A triple (A_1, A_2, y) will be called **admissible** for $(V, \mathscr{A})$ if $A_1, A_2 \in \mathscr{A}$, $A_1 \neq A_2$, and $y \in A_1 \cap A_2$. The 4-tuple (A_1, A_2, x, y) will be called **admissible** if (A_1, A_2, y) is an admissible triple and $x \in A_1 + (A_1 \cap A_2)$.

Let us consider the following conditions applicable to a set system $(V, \mathscr{A})$:

B13 *$\mathscr{A}$ is incommensurable.*

B14 *Given the admissible triple (A_1, A_2, y), there exists $A_3 \in \mathscr{A}$ such that $A_3 \neq \varnothing$ and $A_3 \subseteq (A_1 \cup A_2) + \{y\}$.*

B15 *Given the admissible 4-tuple (A_1, A_2, x, y), there exists $A_3 \in \mathscr{A}$ such that $x \in A_3$ and $A_3 \subseteq (A_1 \cup A_2) + \{y\}$.* (See Figure B16.)

B16

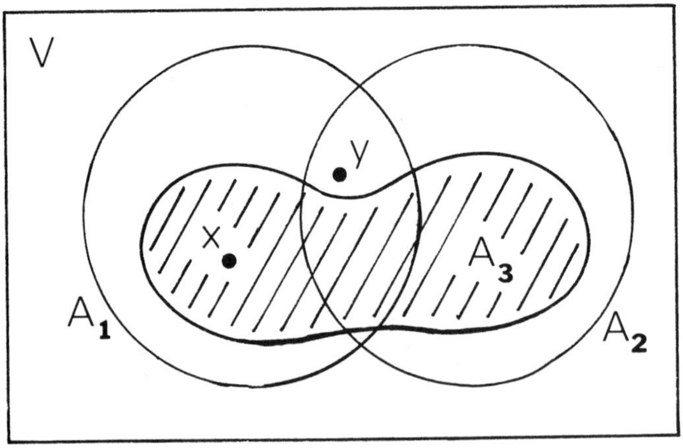

B17 Lemma. *Let $(V, \mathscr{A})$ be a set system.*

(**a**) *If $(V, \mathscr{A})$ satisfies* B14, *then $(V, \mathscr{M}(\mathscr{A}))$ satisfies* B13 *and* B15.

(**b**) *If $(V, \mathscr{A})$ satisfies* B15, *then $(V, \mathscr{M}(\mathscr{A}))$ satisfies* B13 *and* B14, *and every set in $\mathscr{A}$ is a union of sets in $\mathscr{M}(\mathscr{A})$.*

Proof. (a) It is obvious that $\mathscr{M}(\mathscr{A})$ is incommensurable. Supposing B15 to fail for $(V, \mathscr{M}(\mathscr{A}))$, consider the (nonempty) set of admissible 4-tuples (M_1, M_2, x, y) for $(V, \mathscr{M}(\mathscr{A}))$ such that there exists no $M \in \mathscr{M}(\mathscr{A})$ with the properties that both $x \in M$ and $M \subseteq (M_1 \cup M_2) + \{y\}$. Among these, select a 4-tuple (M_1, M_2, x, y) with $|M_1 \cup M_2|$ as small as possible.

By B14, there exists $A \in \mathscr{A}$ such that $\varnothing \neq A \subseteq (M_1 \cup M_2) + \{y\}$. The set A contains some set $M \in \mathscr{M}(\mathscr{A})$, and by our choice of the 4-tuple, $x \notin M$. Since $\mathscr{M}(\mathscr{A})$ is incommensurable, $M \nsubseteq M_1$, and so there exists $z \in M \cap M_2$ with $z \notin M_1$. The 4-tuple (M_2, M, y, z) is admissible. By the minimality of

$|M_1 \cup M_2|$, there exists $M' \in \mathscr{M}(\mathscr{A})$ such that $y \in M'$ and $M' \subseteq (M_2 \cup M) + \{z\}$. Now (M_1, M', x, y) is admissible for $(V, \mathscr{M}(\mathscr{A}))$, and $M_1 \cup M' \subset M_1 \cup M_2$ since $z \notin M_1 \cup M'$. Again by the minimality assumption, there exists $M'' \in \mathscr{M}(\mathscr{A})$ such that $x \in M''$ and $M'' \subseteq (M_1 \cup M') + \{y\} \subseteq (M_1 \cup M_2) + \{y\}$, which provides a contradiction.

(b) It is immediate that $(V, \mathscr{M}(\mathscr{A}))$ satisfies B13 and that $(V, \mathscr{A})$ satisfies B14. Hence $(V, \mathscr{M}(\mathscr{A}))$ satisfies B15 by part (a), whence it also satisfies B14.

If the remainder of the assertion is false, we pick a smallest nonempty set $A \in \mathscr{A}$ which is not a union of sets in $\mathscr{M}(\mathscr{A})$. Thus $\mathscr{M}(\mathscr{A}) \cap \mathscr{P}(A)$ fails to cover A; there exists $x \in A$ such that $x \notin \bigcup_{M \in \mathscr{M}(\mathscr{A}) \cap \mathscr{P}(A)} M$. However, A contains some set $M \in \mathscr{M}(\mathscr{A})$. Letting $y \in M$, we form the admissible 4-tuple (A, M, x, y) for $(V, \mathscr{A})$. There exists a set $A' \in \mathscr{A}$ such that $x \in A'$ and $A' \subseteq A + \{y\}$. Since $|A'| < |A|$, A' is a union of sets in $\mathscr{M}(\mathscr{A})$. One of these sets M' contains x, and $x \in M' \subseteq A' \subseteq A$, giving a contradiction. □

If a set system $(V, \mathscr{A})$ satisfies B13, then $\mathscr{A} = \mathscr{M}(\mathscr{A})$, and the above lemma implies that it satisfies B14 if and only if it satisfies B15. Thus we define a **cycle system** to be a set system $(V, \mathscr{A})$ satisfying B13 and either B14 or B15, with the added condition that $\mathscr{A} \neq \{\varnothing\}$.

Example. The complete design $(V, \mathscr{P}_k(V))$ is a cycle system if $1 \leq k \leq |V|$. Despite this example, cycle systems need not generally be exchange systems, as mentioned earlier. Note, too, that $(V, \varnothing)$ is a cycle system, but by definition, $(V, \{\varnothing\})$ is not a cycle system.

The following example is the prototype for cycle systems just as Example A7 was the prototype for the other three systems studied in this section.

B18 *Example.* For a set U and an *arbitrary* field $\mathbb{F}$, we define the **support function** $\sigma\colon \mathbb{F}^U \to \mathscr{P}(U)$ by

$$\sigma(h) = \{x \in U\colon h(x) \neq 0\}, \quad \text{for } h \in \mathbb{F}^U.$$

(The only difference between this definition and those in §*II*A and §*IV*A is that here $\mathbb{F}$ is not necessarily $\mathbb{K}$ or $\mathbb{Q}$.) Let $\mathscr{L}$ be a subspace of $\mathbb{F}^U$, and let $\mathscr{A} = \{\sigma(h)\colon h \in \mathscr{S}\}$. Let us verify that the set system $(U, \mathscr{A})$ satisfies condition B14. Given distinct supports $A_1, A_2 \in \mathscr{A}$ and $y \in A_1 \cap A_2$, there must exist distinct functions $h_1, h_2 \in \mathscr{S}$ having A_1 and A_2 as their respective supports. Thus $h = h_1(y)h_2 - h_2(y)h_1 \in \mathscr{S}$ and $h(y) = 0$. Hence $A = \sigma(h) \subseteq (A_1 \cup A_2) + \{y\}$. Since $\sigma(h) \supseteq A_1 + A_2$, $A \neq \varnothing$. Thus $(U, \mathscr{M}(\mathscr{A}))$ is a cycle system by Lemma B17a, and $\sigma(h)$ is a union of minimal nonempty supports of functions in $\mathscr{S}$ by B17b. (Cf. *IV*A6(a,b).)

An important special case of the above example arises when $\mathbb{F} = \mathbb{K}$. In this case, the function h is given by $h = h_1 + h_2$. Since by *I*B2, $\mathscr{P}(U)$ and $\mathbb{K}^U$ are identified, each function may be identified with its support. Thus $A = A_1 + A_2$. In fact, *given any subspace $\mathscr{A}$ of $\mathscr{P}(U)$*, it becomes easy to verify directly that $(U, \mathscr{M}(\mathscr{A}))$ *is a cycle system*. Condition B13 holds by definition, and if (M_1, M_2, x, y) is an admissible 4-tuple, then certainly $x \in M_1 + M_2 \subseteq (M_1 \cup M_2) + \{y\}$. Since $\mathscr{A}$ is a subspace, $M_1 + M_2 \in \mathscr{A}$. By *IIC*1, $M_1 + M_2$ is a sum of pairwise-disjoint elements of $\mathscr{M}(\mathscr{A})$. One of these, say M, contains x. Thus $x \in M \subseteq (M_1 \cup M_2) + \{y\}$, and we have verified B14 for $(U, \mathscr{M}(\mathscr{A}))$.

If $\Gamma = (V, f, E)$ is a multigraph, then $\mathscr{Z}(\Gamma)$ and $\mathscr{Z}^{\perp}(\Gamma)$ are two very important subspaces of $\mathscr{P}(E)$. Thus $(E, \mathscr{M}(\mathscr{Z}(\Gamma)))$ and $(E, \mathscr{M}(\mathscr{Z}^{\perp}(\Gamma)))$ are very important cycle systems. Their blocks are respectively the elementary cycles and the elementary cocycles of Γ. Note that here the *blocks* of Γ become the *vertices* of the cycle system.

B19 Proposition. *Let $\mathscr{I}, \mathscr{M} \subseteq \mathscr{P}(V)$. Then the following two statements are equivalent:*

(**a**) *$(V, \mathscr{I})$ is an independence system and $\mathscr{M} = \mathscr{M}(\mathscr{P}(V) + \mathscr{I})$.*

(**b**) *$(V, \mathscr{M})$ is a cycle system and $\mathscr{I} = \{J \in \mathscr{P}(V): J \not\supseteq M$ for all $M \in \mathscr{M}\}$.*

Proof. If we assume (a), then it is immediate both that $\mathscr{I}$ consists of those sets which contain no set in $\mathscr{M}$ and that $\mathscr{M}$ is incommensurable. To complete the proof of (b), it suffices to show that $(V, \mathscr{M})$ satisfies the condition B14. Let (M_1, M_2, y) be an admissible triple.

First suppose that $(M_1 \cup M_2) + \{y\} \in \mathscr{I}$. Since $\mathscr{M}$ is incommensurable, we may select $x \in M_1 + (M_1 \cap M_2)$. Since $M_1 + \{x\} \subset M_1$, $M_1 + \{x\} \in \mathscr{I}$. Since $M_2 \not\subseteq M_1$, we have $|M_1 + \{x\}| < |(M_1 \cup M_2) + \{y\}|$. By B2 applied repeatedly, we obtain eventually that $(M_1 \cup M_2) + \{x\} \in \mathscr{I}$, which is impossible since $M_2 \subseteq (M_1 \cup M_2) + \{x\}$. Hence $(M_1 \cup M_2) + \{y\} \notin \mathscr{I}$ and there exists $M \in \mathscr{M}$ such that $\varnothing \neq M \subseteq (M_1 \cup M_2) + \{y\}$.

Conversely, let (b) be assumed. Clearly $\mathscr{M} = \mathscr{M}(\mathscr{P}(V) + \mathscr{I})$, and $\mathscr{I}$ is closed with respect to subsets; i.e., B1 holds for $\mathscr{I}$. It remains to verify B2, which if false implies the existence of $J_1, J_2 \in \mathscr{I}$ with $|J_1| < |J_2|$ such that $J_1 + \{x\} \notin \mathscr{I}$ for all $x \in J_2 + (J_1 \cap J_2)$. We select such sets J_1 and J_2 with $|J_1 + (J_1 \cap J_2)|$ as small as possible. Certainly $J_1 \not\subseteq J_2$.

Let $y \in J_1 + (J_1 \cap J_2)$. Then $J_1 + \{y\}$ is independent, and by our minimality assumption, $J_1 + \{y, x_1\} \in \mathscr{I}$ for some $x_1 \in J_2 + (J_1 \cap J_2)$. Since

$$|(J_1 + \{y, x_1\}) + [(J_1 + \{y, x_1\}) \cap J_2]| < |J_1 + (J_1 \cap J_2)|$$

and $|J_1 + \{y, x_1\}| < |J_2|$, there exists $x_2 \in J_2 + (J_1 \cap J_2) + \{x_1\}$ such that $J_1 + \{y, x_1, x_2\} \in \mathscr{I}$. Hence $J_1 + \{y, x_2\} \in \mathscr{I}$. By our assumption, since for $i = 1, 2$ we have $J_1 + \{x_i\} \notin \mathscr{I}$, there exists $M_i \in \mathscr{M}$ such that $\{y, x_i\} \subseteq M_i \subseteq J_1 + \{x_i\}$. Thus (M_1, M_2, x_1, y) is admissible. It follows that there exists $M \in \mathscr{M}$ such that $M \subseteq M_1 \cup M_2 + \{y\} \subseteq J_1 + \{y, x_1, x_2\} \in \mathscr{I}$, which is impossible. □

B20 *Example.* Let $\mathscr{M}$ be the collection of minimal dependent sets in Example A7. Then $(V, \mathscr{M})$ is a cycle system.

We have defined and considered four kinds of set systems in this section: independence systems, spanning systems, basis systems, and cycle systems.

By means of Proposition B10, Corollary B11, and Proposition B19, we have seen how any set system of one of these kinds uniquely determines a set system of each of the other three kinds and that in each case, by reversing this determination process, one recovers the original system. In the next section three more equivalent structures will be introduced. Each one of these seven objects has been designated as a "matroid" somewhere in the literature. We will choose to use the term "matroid" to designate a cycle system. While this choice appears from the logical point of view to be arbitrary, it seems to make a difference in the way one thinks of matroids, in particular, in the way that one abstracts the concepts of linear algebra.

To abstract the concepts of linear algebra for independence systems, basis systems, and spanning systems, we identify, as in Example A7, the vertices of these systems with the vectors in a vector space and reuse the terms "independent set," "spanning set," and "basis" in this context. The sets in the corresponding cycle system then become the minimal dependent sets (cf. B20). One may think of them as the minimal sets of vectors satisfying a nontrivial relation. On the other hand, the prototype for cycle systems was Example B18. Here via the support function, each block corresponds to a single vector rather than to a set of vectors. The vertices of this system correspond to the coordinates of this coordinatized $|V|$-dimensional vector space in the following sense. If x is a vertex and M is a block in a cycle system, then to say $x \in M$ means that M has a nonzero x-coordinate while $x \notin M$ means that the x-coordinate of M is 0. Of course, both interpretations lack the full algebraic power of the prototype vector space. Nonetheless, many of the results from linear algebra can still be carried over, and thinking in terms of the appropriate prototype often helps in understanding and proving these results.

Henceforth the terms **matroid** *and* **cycle system** *will be synonymous.* If $\Lambda = (V, \mathscr{M})$ is a matroid, then the elements of $\mathscr{M}$ will be called **cycles** of Λ. Following Proposition B19, we let $\mathscr{I}(\Lambda) = \{J \in \mathscr{P}(V): J \not\supseteq M \text{ for all } M \in \mathscr{M}\}$. The elements of $\mathscr{I}(\Lambda)$ are called the **independent sets** of Λ. Following Proposition B10, we let $\mathscr{B}(\Lambda)$ be the collection of largest independent sets. The elements of $\mathscr{B}(\Lambda)$ are called **bases** of Λ. The **rank** of Λ, denoted by $r(\Lambda)$, is the cardinality of a basis. Following Corollary B11, we let $\mathscr{S}(\Lambda)$ be the collection of supersets of bases. Its elements are called the **spanning sets** of Λ. Of course $(V, \mathscr{I}(\Lambda))$ is an independence system, etc.

B21 *Example.* If $\Lambda = (V, \varnothing)$, then $\mathscr{I}(\Lambda) = \mathscr{P}(V)$ and $\mathscr{B}(\Lambda) = \mathscr{S}(\Lambda) = \{V\}$.

We have already noted that the complete designs of positive blocksize are matroids; they are called **complete matroids**.

B22 Exercise. *If* $\Lambda = (V, \mathscr{P}_k(V))$ *for* $k > 0$,

(a) Show that $\mathscr{B}(\Lambda) = \mathscr{P}_{k-1}(V)$.

(b) Find $\mathscr{I}(\Lambda)$ and $\mathscr{S}(\Lambda)$.

The following example of a matroid will turn out to be very important.

B23 *Example.* Consider the set system $\Lambda = (V, \mathcal{M})$, where $\hat{\Lambda}$ denotes the projective plane of order 2 (cf. *IX*C12). Recall that $|V| = |\mathcal{M}| = 7$. One verifies straightforwardly that Λ is a matroid; it may be helpful to refer to Figure B24, where the six straight line segments and the arc of a circle represent the lines of $\hat{\Lambda}$. Since every 2-subset of V is contained in some line, every

B24

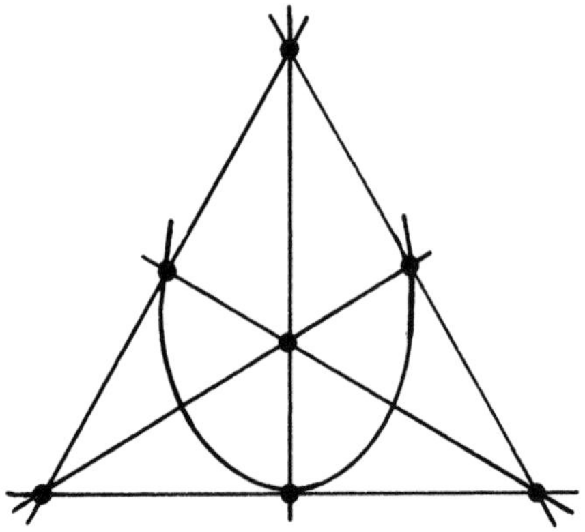

5-subset contains the complement of a line. Every 4-subset either is the complement of a line (i.e., a cycle) or contains a line. The latter 4-subsets together with all k-subsets for $k < 4$ comprise $\mathcal{I}(\Lambda)$. $\mathcal{B}(\Lambda)$ is the set of 4-subsets which contain a line, and $r(\Lambda) = 4$. $\mathcal{S}(\Lambda)$ consists of $\mathcal{B}(\Lambda)$ together with all k-subsets for $k > 4$. This matroid is called the **Fano matroid of rank** 4.

There is a second matroid often associated with the projective plane of order 2, called the **Fano matroid of rank** 3. Here the cycles are all lines and all complements of lines.

B25 *Exercise.* If Λ is the Fano matroid of rank 3, determine $\mathcal{I}(\Lambda)$, $\mathcal{B}(\Lambda)$, and $\mathcal{S}(\Lambda)$.

B26 Exercise. *Let* $\Gamma = (V, f, E)$ *be a multigraph. Let* $\Lambda = (E, \mathcal{M}(\mathcal{Z}(\Gamma)))$. Show that $\mathcal{B}(\Lambda)$ *consists of the edge sets of the spanning forests of* Γ. *Similarly* show that *if* $\Lambda = (E, \mathcal{M}(\mathcal{Z}^{\perp}(\Gamma)))$, *then* $\mathcal{B}(\Lambda)$ *consists of the edge sets of the spanning coforests of* Γ.

If $\Gamma = (V, f, E)$ is a multigraph, then $(E, \mathcal{M}(\mathcal{Z}(\Gamma)))$ is called the **cycle matroid** of Γ and $(E, \mathcal{M}(\mathcal{Z}^{\perp}(\Gamma)))$ is called the **cocycle matroid** of Γ. By Exercise B26, *III*C11, and *III*C18 one derives

B27
$$r(E, \mathcal{M}(\mathcal{Z}(\Gamma))) = \nu_0(\Gamma) - \nu_{-1}(\Gamma);$$
$$r(E, \mathcal{M}(\mathcal{Z}^{\perp}(\Gamma))) = \nu_1(\Gamma) - \nu_0(\Gamma) + \nu_{-1}(\Gamma).$$

The next proposition generalizes parts of the graph-theoretical results *III*C11 and *III*C18.

B28 Proposition. *Let $\Lambda = (V, \mathscr{M})$ be a matroid.*

(a) *If $B \in \mathscr{B}(\Lambda)$, then to each $x \in V + B$ there corresponds a unique cycle $M_x \in \mathscr{M}$ such that $x \in M_x \subseteq B + \{x\}$.*

(b) *If $M \in \mathscr{M}$, then to each $x \in M$ there corresponds a basis $B_x \in \mathscr{B}(\Lambda)$ such that $M \subseteq B_x + \{x\}$.*

(c) *Let $S \in \mathscr{P}(V)$. Then $S \in \mathscr{S}(\Lambda)$ if and only if to each $x \in V + S$ there corresponds a cycle M_x such that $x \in M_x \subseteq S + \{x\}$.*

PROOF. (a) Let $B \in \mathscr{B}(\Lambda)$ and let $x \in V + B$. Since $B + \{x\} \notin \mathscr{I}(\Lambda)$, there exists $M_x \in \mathscr{M}$ such that $M_x \subseteq B + \{x\}$ by Proposition B19. Since $M_x \nsubseteq B$, we must have $x \in M_x$. Now suppose that $x \in M \subseteq B + \{x\}$. If $M \neq M_x$, then (M, M_x, x) is an admissible triple. There exists $M' \in \mathscr{M}$ such that $M' \subseteq (M \cup M_x) + \{x\} \subseteq B$, which is impossible.

(b) Let $M \in \mathscr{M}$ and let $x \in M$. Then $M + \{x\} \in \mathscr{I}(\Lambda)$, and hence $M + \{x\} \subseteq B_x$ for some $B_x \in \mathscr{B}(\Lambda)$. Since $M \nsubseteq B_x$, we have $x \notin B_x$. Hence $M \subseteq B_x + \{x\}$.

(c) Let $S \in \mathscr{S}(\Lambda)$ and $x \in V + S$. By Corollary B11, S contains some basis B. By part (a) above, $x \in M_x \subseteq B + \{x\} \subseteq S + \{x\}$ for some $M_x \in \mathscr{M}$.

Conversely, suppose that $S \in \mathscr{P}(V)$, and that to each $x \in V + S$ there corresponds a cycle M_x such that $x \in M_x \subseteq S + \{x\}$. It suffices to prove that S contains a basis. Let J be a maximal independent subset of S. We wish to show that $J + \{x\} \notin \mathscr{I}(\Lambda)$ for all $x \in V + J$. This is the case if $x \in S$, by definition of J. Hence suppose $x \in V + S$.

By our assumption, there exists $M_x \in \mathscr{M}$ such that $x \in M_x \subseteq S + \{x\}$. Subject to this condition, let M_x be chosen so that $|M_x + (M_x \cap J)|$ is as small as possible. Indeed $|M_x + (M_x \cap J)| > 1$ or else $J + \{x\} \supseteq M_x \notin \mathscr{I}(\Lambda)$. Hence there exists $y \in M_x + (M_x \cap J) + \{x\}$. Since $M_x \subseteq S + \{x\}$, $y \in S$. Therefore $J + \{y\} \notin \mathscr{I}(\Lambda)$ and $y \in M_y \subseteq J + \{y\}$ for some $M_y \in \mathscr{M}$. Since (M_x, M_y, x, y) is admissible, there exists $M \in \mathscr{M}$ such that $x \in M \subseteq (M_x \cup M_y) + \{y\} \subseteq S + \{x\}$. However, $[M + (M \cap J)] \subset [M_x + (M_x \cap J)]$ since the former set does not contain y. This contradicts the minimality of $M_x + (M_x + J)$. □

The following lemma generalizes the principle from linear algebra that if $\mathscr{A}_1$ and $\mathscr{A}_2$ are finite-dimensional subspaces of some vector space, if $\dim(\mathscr{A}_1) \leq \dim(\mathscr{A}_2)$, and if $\mathscr{A}_2 \subseteq \mathscr{A}_1$, then $\mathscr{A}_1 = \mathscr{A}_2$.

B29 Lemma. *If $\Lambda_i = (V, \mathscr{M}_i)$ is a matroid for $i = 1, 2$ with $r(\Lambda_1) \leq r(\Lambda_2)$ and if for each $M \in \mathscr{M}_1$ one can write $M = \bigcup_{N \in \mathscr{N}} N$ for some $\mathscr{N} \subseteq \mathscr{M}_2$, then $\Lambda_1 = \Lambda_2$.*

PROOF. Let $J \in \mathscr{P}(V)$. If $J \notin \mathscr{I}(\Lambda_1)$, then $J \supseteq M$ for some $M \in \mathscr{M}_1$, whence $J \supseteq N$ for some $N \in \mathscr{M}_2$. This implies that $J \notin \mathscr{I}(\Lambda_2)$. We have proved that $\mathscr{I}(\Lambda_2) \subseteq \mathscr{I}(\Lambda_1)$.

If $B \in \mathscr{B}(\Lambda_2)$, then $B \in \mathscr{I}(\Lambda_1)$ by the preceding paragraph, and so $|B| \leq r(\Lambda_1) \leq r(\Lambda_2) = |B|$. Thus $B \in \mathscr{B}(\Lambda_1)$ and so $\mathscr{B}(\Lambda_2) \subseteq \mathscr{B}(\Lambda_1)$.

Let $N \in \mathscr{M}_2$ and let $x \in N$. By Proposition B28, there exists $B \in \mathscr{B}(\Lambda_2)$ such that $N \subseteq B + \{x\}$, and N is the unique cycle of Λ_2 such that $x \in N \subseteq B + \{x\}$. Since $B \in \mathscr{B}(\Lambda_1)$, there exists a cycle $M \in \mathscr{M}_1$ such that $x \in M \subseteq B + \{x\}$. Since, moreover, M is a union of cycles of Λ_2 and since N is the only cycle of Λ_2 contained in $B + \{x\}$, $N = M$. Thus $N \in \mathscr{M}_1$, and so $\mathscr{M}_2 \subseteq \mathscr{M}_1$.

To obtain the reverse inclusion, let $M \in \mathscr{M}_1$. By hypothesis there exists $N \in \mathscr{M}_2$ such that $N \subseteq M$. But $N \in \mathscr{M}_1$, and so $N = M$. Thus $\mathscr{M}_1 = \mathscr{M}_2$. □

XC Rank and Closure

In the previous section we alluded to the other structures equivalent to matroids. Unlike the four structures discussed there, these three new structures are not systems. In fact, one of them will be a lattice while the other two will be ordered pairs consisting of a set and a function. Each of these structures will have its own axioms.

Let V be a set and let r be a selection of $\mathscr{P}(V)$. We say that (V, r) is a **rank structure** if for all $U \in \mathscr{P}(V)$ and all $x_1, x_2 \in V + U$:

C1 $r(\varnothing) = 0$;

C2 $0 \leq r(U + \{x_1\}) - r(U) \leq 1$;

C3 $r(U) = r(U + \{x_1\}) = r(U + \{x_2\})$ *implies* $r(U) = r(U + \{x_1, x_2\})$.

Let us first prove some elementary properties of rank structures.

C4 Lemma. *Let (V, r) be a rank structure and let $U \in \mathscr{P}(V)$. Then:*

(a) *If $T \subseteq U$, then $r(T) \leq r(U) \leq r(T) + |U| - |T|$;*

(b) $0 \leq r(U) \leq |U|$;

(c) *If $W \subseteq V + U$ and $r(U) = r(U + \{w\})$ for all $w \in W$, then $r(U) = r(U + W)$.*

Proof. (a) Let $U + T = \{u_1, \ldots, u_k\}$. By C2 we have

$$0 \leq r(T + \{u_1\}) - r(T) \leq 1,$$
$$0 \leq r(T + \{u_1, \ldots, u_i\}) - r(T + \{u_1, \ldots, u_{i-1}\}) \leq 1, \quad i = 2, \ldots, k.$$

Summing these k inequalities gives

$$0 \leq r(U) - r(T) \leq |U| - |T|$$

as required.

(b) This follows from part (a) above by letting $T = \varnothing$ and invoking C1.

(c) The result is trivial if $|W| = 1$. Suppose $W = \{w_1, \ldots, w_k\}$ for some $k \geq 2$, and proceed by induction. We assume that

$$r(U) = r(U + \{w_1, \ldots, w_{k-1}\}) = r(U + \{w_1, \ldots, w_{k-2}, w_k\}).$$

Then by C3, $r(U) = r(U + \{w_1, \ldots, w_{k-2}\}) = r(U + W)$. □

C5 Exercise. *If* (V, r) *is a rank structure*, prove that

(a) *If* $U_1 \subseteq U_2$, $x \in V + U_2$, *and* $r(U_1 + \{x\}) = r(U_1)$, *then* $r(U_2 + \{x\}) = r(U_2)$.

(b) $r(U_1 \cup U_2) + r(U_1 \cap U_2) \leq r(U_1) + r(U_2)$ *for all* $U_1, U_2 \in \mathscr{P}(V)$.

We are now prepared to relate rank structures to matroids—via independence systems.

C6 Proposition. *Let* $\mathscr{I} \subseteq \mathscr{P}(V)$ *and let r be a selection of* $\mathscr{P}(V)$. *Then the following two statements are equivalent:*

(a) $(V, \mathscr{I})$ *is an independence system and*

$$r(U) = \max\{|J| : J \in \mathscr{I}; J \subseteq U\}, \quad U \in \mathscr{P}(V);$$

(b) (V, r) *is a rank structure and*

$$\mathscr{I} = \{J \in \mathscr{P}(V) : r(J) = |J|\}.$$

Proof. Assuming (a), we have C1 immediately. Let $U \in \mathscr{P}(V)$ and $x \in V + U$ be given. By definition, clearly $r(U) \leq r(U + \{x\})$. Now pick $J \subseteq U + \{x\}$ such that $|J| = r(U + \{x\})$. If $x \notin J$, then $r(U) = |J|$. If $x \in J$, then $J + \{x\} \in \mathscr{I}$ and $r(U) \geq |J + \{x\}|$. In either case, we have $r(U) \geq |J| - 1 = r(U + \{x\}) - 1$. Thus C2 holds.

To verify C3, let J be a largest set in $\mathscr{I}$ such that $J \subseteq U$. If $r(U) < r(U + \{x_1, x_2\})$, then there exists a largest set $J' \in \mathscr{I}$ such that $J' \subseteq U + \{x_1, x_2\}$ and $|J| < |J'|$. Invoking B2, we assert the existence of a vertex $x \in J' + (J \cap J')$ such that $J + \{x\} \in \mathscr{I}$. By the maximality of $|J|$, $x \notin U$. Hence $x = x_i$ for $i = 1$ or 2. Since $J + \{x_i\} \subseteq U + \{x_i\}$, we have $r(U) = |J| < |J + \{x_i\}| \leq r(U + \{x_i\})$.

It is immediate that $\mathscr{I} = \{J \in \mathscr{P}(V) : r(J) = |J|\}$.

Conversely, let (b) be assumed. Let $J \in \mathscr{I}$ and suppose $J' \subseteq J$. By C4a, $|J| = r(J) \leq r(J') + |J| - |J'|$. Hence $|J'| \leq r(J')$ which combined with C4b yields $J' \in \mathscr{I}$. We have verified B1 for $(V, \mathscr{I})$.

To prove that B2 holds, suppose $J_1, J_2 \in \mathscr{I}$ with $|J_1| < |J_2|$ but that $J_1 + \{x\} \notin \mathscr{I}$ for all $x \in J_2 + (J_1 \cap J_2)$. By C2, this means that $r(J_1) = r(J_1 + \{x\})$ for all $x \in J_2 + (J_1 \cap J_2)$, and so by C4(a,c), $|J_1| = r(J_1) = r(J_1 + J_2 + (J_1 \cap J_2)) \geq r(J_2) = |J_2|$, which is a contradiction.

If $U \in \mathscr{P}(V) + \{\varnothing\}$, let J be a largest set in $\mathscr{I}$ such that $J \subseteq U$. If $r(J) < r(U)$, then by C4c, $r(J) < r(J + \{x\})$ for some $x \in U + J$. But $r(J) = |J|$ and so by C2, $r(J + \{x\}) = |J| + 1$. This implies that $J + \{x\} \in \mathscr{I}$, contrary to the maximality of $|J|$. Hence $r(U) = \max\{|J| : J \in \mathscr{I}; J \subseteq U\}$. □

If $\Lambda = (V, \mathscr{M})$ is a matroid, it now makes sense to define the **rank function** $r: \mathscr{P}(V) \to \mathbb{N}$ of Λ by

$$r(U) = \max\{|J| : J \in \mathscr{I}(\Lambda); J \subseteq U\}.$$

Since the bases are the largest sets B such that $r(B) = |B|$, our definition is consistent with the definition of the rank of a matroid as defined in the previous section.

The next result enables us to relate the rank function of a matroid to the cycles of the matroid directly.

C7 Corollary. *Let $\Lambda = (V, \mathcal{M})$ be a matroid with rank function r. Let $U \in \mathcal{P}(V)$ and $x \in V + U$. There exists a cycle $M \in \mathcal{M}$ such that $x \in M \subseteq U + \{x\}$ if and only if $r(U) = r(U + \{x\})$.*

PROOF. Suppose that $x \in M \subseteq U + \{x\}$ for some $M \in \mathcal{M}$. Pick a largest independent set J of Λ such that $J \subseteq U + \{x\}$. Then $|J| = r(U + \{x\})$ by the above proposition. By Proposition B19a, $M + \{x\} \in \mathcal{I}(\Lambda)$. Since $M + \{x\} \subseteq U$, if $|M + \{x\}| = |J|$, then $r(U) \geq |J|$, and by C2, $r(U) = r(U + \{x\})$. On the other hand, if $|M + \{x\}| < |J|$, then by repeated applications of B2, one constructs an independent set J' by adjoining to $M + \{x\}$ elements of J, one by one. Thus $|J'| = |J|$. Furthermore, if $x \in J'$, then $M \subseteq J'$, which is impossible. Thus $J' \subseteq U$ and, as above, $r(U) = r(U + \{x\})$.

Conversely, suppose $r(U) = r(U + \{x\})$. Then by the proposition, $r(U) = |J|$ for some $J \in \mathcal{I}(\Lambda)$ such that $J \subseteq U$. If $J + \{x\} \in \mathcal{I}(\Lambda)$, then $r(U + \{x\}) \geq |J + \{x\}| > r(U)$. Hence $J + \{x\}$ is not independent and so $M \subseteq J + \{x\}$ for some $M \in \mathcal{M}$. By Proposition B19, $M \nsubseteq J$. Hence $x \in M$. □

C8 Exercise. *Let r be the rank function of the matroid $\Lambda = (V, \mathcal{M})$. Let $U \in \mathcal{P}(V)$.* Show that:

(**a**) $r(U) = r(\Lambda)$ *if and only if* $U \in \mathcal{S}(\Lambda)$.

(**b**) $r(U) = |U| - 1 = r(U + \{x\})$ *for all* $x \in U$ *if and only if* $U \in \mathcal{M}$.

Let V be a set and let $c \colon \mathcal{P}(V) \to \mathcal{P}(V)$. We say that (V, c) is a **closure structure** if the following three conditions hold:

C9 $U \subseteq c(U)$, for all $U \in \mathcal{P}(V)$;

C10 If $U_1 \subseteq c(U_2)$, then $c(U_1) \subseteq c(U_2)$, for all U_1, $U_2 \in \mathcal{P}(V)$;

C11 If $U \in \mathcal{P}(V)$ and $x_1, x_2 \in V + U$ and if $x_2 \in c(U + \{x_1\})$ while $x_2 \notin c(U)$, then $x_1 \in c(U + \{x_2\})$.

If (V, c) is a closure structure, a set $U \in \mathcal{P}(V)$ is said to be **closed** if $c(U) = U$.

Let us first consider some elementary properties of closure structures.

C12 Exercise. *Let (V, c) be a closure structure. If U_1, $U_2 \in \mathcal{P}(V)$,* prove:

(**a**) *V is closed.*

(**b**) *$c(U)$ is closed for all $U \in \mathcal{P}(V)$.*

(**c**) *If $U_1 \subseteq U_2$, then $c(U_1) \subseteq c(U_2)$.*

(**d**) $c(U_1 \cap U_2) \subseteq c(U_1) \cap c(U_2)$, *and equality holds if* U_1 *and* U_2 *are closed.*

(**e**) $c(U_1) \cup c(U_2) \subseteq c(U_1 \cup U_2)$.

(**f**) *If* $U \in \mathscr{P}(V)$ *and* $x_1, x_2 \in V + U$, *then* $c(c(U + \{x_1\}) + \{x_2\}) = c(U + \{x_1, x_2\})$.

The closure operator in a topological space satisfies conditions C9 and C10. However, the function c in a closure structure (V, c) need not induce a topology on V. In particular we do not necessarily have that $c(\varnothing) = \varnothing$ nor does equality always hold in C12e, both of which always hold for topological closure operators. The following examples shows how a closure structure may fail to determine a topological space in each of these respects.

C13 *Example.* Let m be an integer such that $0 \leq m \leq |V|$ and define $c: \mathscr{P}(V) \to \mathscr{P}(V)$ by

$$c(U) = \begin{cases} U & \text{if } |U| < m; \\ V & \text{if } |U| \geq m. \end{cases}$$

One verifies easily that (V, c) is a closure structure. If $1 < m < |V|$, it is possible to choose sets $U_1, U_2 \in \mathscr{P}(V)$ such that $|U_1|, |U_2| < m$ but $m \leq |U_1 \cup U_2| < |V|$. Observe that $c(U_1) \cup c(U_2) = U_1 \cup U_2 \subset V = c(U_1 \cup U_2)$. In the case that $m = 0$, we have $c(\varnothing) = V$.

C14 Lemma. *Let* (V, c) *be a closure structure. Let* $U \in \mathscr{P}(V)$ *and* $x_1 \in V + U$. *If for some set* $S \in \mathscr{P}(V)$ *it holds that* $c(S) \subset c(U) \subseteq c(S + \{x_1\})$, *then* $U \nsubseteq c(S)$ *and* $c(U) = c(S + \{x_1\}) = c(S + \{x_2\})$ *for all* $x_2 \in U + (U \cap c(S))$.

Proof. If $U \subseteq c(S)$, then by C10, $c(U) \subseteq c(S) \subset c(U)$, which is absurd. Hence we may arbitrarily choose $x_2 \in U + (U \cap c(S))$.

Since $x_2 \in U \subseteq c(S + \{x_1\})$ and $x_2 \notin c(S)$, we have by C11 that $x_1 \in c(S + \{x_2\})$. Thus $S + \{x_1\} \subseteq c(S + \{x_2\})$. Also $S + \{x_2\} \subseteq c(U)$. Finally by C10, $c(S + \{x_2\}) \subseteq c(U) \subseteq c(S + \{x_1\}) \subseteq c(S + \{x_2\})$, and so equality holds throughout. □

We now relate closure structures to matroids by relating them to rank structures.

C15 Proposition. *Let* V *be a set and let* $c: \mathscr{P}(V) \to \mathscr{P}(V)$ *and* $r: \mathscr{P}(V) \to \mathbb{N}$ *be functions. Then the following two statements are equivalent:*

(**a**) (V, c) *is a closure structure and*

$$r(U) = \min\{|T| : T \in \mathscr{P}(V);\ c(T) = c(U)\}, \quad \textit{for all } U \in \mathscr{P}(V).$$

(**b**) (V, r) *is a rank structure and*

$$c(U) = U + \{x \in V + U : r(U) = r(U + \{x\})\}, \quad \textit{for all } U \in \mathscr{P}(V).$$

PROOF. Assuming (a) we deduce trivially from the definition of r in terms of c that $r(\varnothing) = 0$. Observe that if two sets have the same closure, then they have equal rank. We assert that

C16 $\quad r(U) = \min\{|T| : T \subseteq U; c(T) = c(U)\}$ for all $U \in \mathscr{P}(V)$.

For let T be a subset of V such that $c(T) = c(U)$ and $|T| = r(U)$; subject to these conditions, let $|T + (T \cap U)|$ be as small as possible. If this quantity is positive, let $x_1 \in T + (T \cap U)$ and let $S = T + \{x_1\}$. By C12c, $c(S) \subseteq c(T) = c(U)$, but since $|S| < |T| = r(U)$, we must have $c(S) \subset c(U)$. Hence by C10 and C12b, $U \nsubseteq c(S)$. Let $x_2 \in U + (U + c(S))$. Since $c(U) = c(S + \{x_1\})$, Lemma C14 implies that $c(S + \{x_2\}) = c(U)$. Furthermore, $r(U) = |T| = |S + \{x_2\}|$. However, $|(S + \{x_2\}) + [(S + \{x_2\}) \cap U]| < |T + (T \cap U)|$, contrary to our assumption. This proves C16.

To verify C2, let $U \in \mathscr{P}(V)$ and $x \in V + U$. By C16 we choose $T \subseteq U + \{x\}$ such that $r(U + \{x\}) = |T|$ and $c(U + \{x\}) = c(T)$. If $T \subseteq U$, then $c(U + \{x\}) = c(T) \subseteq c(U)$. Hence $c(U) = c(U + \{x\})$ and $r(U) = r(U + \{x\})$. Otherwise, $x \in T$. Let $S = T + \{x\}$, and so $S \subseteq U$. If $c(S) \subset c(U)$, then by Lemma C14, $c(U) = c(S + \{x\}) = c(T) = c(U + \{x\})$, and so $r(U) = r(U + \{x\})$. If $c(S) = c(U)$, then $r(U) \leq |S| < |T| = r(U + \{x\})$. Hence $0 \leq r(U + \{x\}) - r(U)$.

By C16 we may select $T_0 \subseteq U$ such that $c(T_0) = c(U)$ and $r(U) = |T_0|$. By C9, $T_0 + \{x\} \subseteq U + \{x\} \subseteq c(T_0) + \{x\} \subseteq c(T_0 + \{x\})$. Hence $c(T_0 + x) = c(U + \{x\})$ by C12b,c. Hence $r(U + \{x\}) \leq |T_0 + \{x\}| = |T_0| + 1 = r(U) + 1$, as required.

To verify C3, let $x_1, x_2 \in V + U$. For $i = 1, 2$, suppose that $r(U) = r(U + \{x_i\})$. By C16 there exists $T_i \subseteq U + \{x_i\}$ such that $c(T_i) = c(U + \{x_i\})$ and $r(U + \{x_i\}) = |T_i|$. If $T_i \subseteq U$ for some i, we may take $T_1 = T_2 = T$. In this case $U + \{x_1, x_2\} \subseteq c(T)$, and so by C10, $c(U + \{x_1, x_2\}) \subseteq c(T) = c(U) \subseteq c(U + \{x_1, x_2\})$. Hence $r(U + \{x_1, x_2\}) = |T|$ by C2. Otherwise, let $S_i = T_i + \{x_i\}$. Since $S_i \subseteq U$, $c(S_i) \subseteq c(U)$. If $c(S_i) = c(U)$, then $r(U) \leq |S_i| < |T_i| = r(U + \{x_i\})$, contrary to assumption. Hence $c(S_i) \subset c(U) \subseteq c(U + \{x_i\}) = c(S_i + \{x_i\})$. By Lemma C14, $c(U) = c(U + \{x_i\})$. It follows that $U + \{x_1, x_2\} \subseteq c(U)$, and so $c(U + \{x_1, x_2\}) \subseteq c(U)$. Hence $r(U + \{x_1, x_2\}) = r(U)$. Thus (V, r) is a rank structure.

Finally let $x \in c(U)$. Then $c(U \cup \{x\}) = c(U)$ and so $r(U \cup \{x\}) = r(U)$. Conversely, let $x \in V + U$ and suppose $r(U + \{x\}) = r(U)$. By C16 we may select $T \subseteq U + \{x\}$ such that $c(T) = c(U + \{x\})$ and $|T| = r(U + \{x\})$. As in our verification of C2, we get $c(T) = c(U)$, and so $x \in c(U)$.

Conversely, let (b) be assumed. Then C9 is immediate. Let U_1, $U_2 \in \mathscr{P}(V)$ and suppose that $U_1 \subseteq c(U_2)$. If C10 fails, we may select $x \in c(U_1)$ such that $x \notin c(U_2)$. Since $x \in c(U_1) + U_1$, we have $r(U_1) = r(U_1 + \{x\})$. By Exercises C4a, C5a, then C4c, $r(U_2 + \{x\}) \leq r(c(U_2) + \{x\}) = r(c(U_2)) = r(U_2)$, whence $x \in c(U_2)$, giving a contradiction. Hence C10 holds.

Let $x_1, x_2 \in V + U$. Suppose that $x_2 \in c(U + \{x_1\})$ and $x_2 \notin c(U)$, but that $x_1 \notin c(U + \{x_2\})$. By C2,

C17 $$r(U + \{x_1, x_2\}) - r(U + \{x_1\}) = 0,$$

C18 $$r(U + \{x_2\}) - r(U) = 1,$$

C19 $$r(U + \{x_1, x_2\}) - r(U + \{x_2\}) = 1.$$

By successive substitutions of C18, C19, then C17,

$$\begin{aligned} r(U) &= r(U + \{x_2\}) - 1 \\ &= r(U + \{x_1, x_2\}) - 2 \\ &= r(U + \{x_1\}) - 2, \end{aligned}$$

contrary to C2. Hence (V, c) is a closure structure.

Finally, let us define $s: \mathscr{P}(V) \to \mathbb{N}$ by

$$s(U) = \min\{|T| : T \in \mathscr{P}(U);\, c(T) = c(U)\}.$$

Since (a) implies (b) in this proposition, (V, s) is a rank structure, and it suffices to prove $r = s$. We proceed by induction on $|U|$. By C1, $r(\varnothing) = 0 = s(\varnothing)$. Suppose that $r(U) = s(U)$ for some set $U \in \mathscr{P}(V)$ but that $r(U + \{x\}) \neq s(U + \{x\})$ for some $x \in V + U$. By C2 exactly one of the following two equations holds:

$$r(U + \{x\}) = r(U); \qquad s(U + \{x\}) = s(U).$$

However, if $x \in c(U)$, then both equations hold, while if $x \notin c(U)$ then both equations fail. We arrive at a contradiction either way. Hence $r = s$. □

If $\Lambda = (V, \mathscr{M})$ is a matroid whose rank function is r, it now makes sense to define the **closure operator** $c: \mathscr{P}(V) \to \mathscr{P}(V)$ of Λ by

$$c(U) = U + \{x \in V + U : r(U) = r(U + \{x\})\}, \quad \text{for all } u \in \mathscr{P}(V).$$

We let $\mathscr{C}(\Lambda)$ denote the collection of closed sets of the closure structure (V, c).

The following result directly relates the closure operator of a matroid to the blocks of the matroid.

C20 Corollary. *Let $\Lambda = (V, \mathscr{M})$ be a matroid. Let $c: \mathscr{P}(V) \to \mathscr{P}(V)$ be given by*

$$c(U) = U + \{x \in V + U : x \in M \subseteq U + \{x\} \text{ for some } M \in \mathscr{M}\}.$$

Then c is the closure operator of Λ.

Proof. Let r be the rank function of Λ and let $U \in \mathscr{P}(V)$. By Corollary C7,

$$\begin{aligned} &\{x \in V + U : x \in M \subseteq U + \{x\} \text{ for some } M \in \mathscr{M}\} \\ &\qquad = \{x \in V + U : r(U) = r(U + \{x\}).\end{aligned}$$

The result now follows immediately from the above proposition. □

C21 *Example.* In Example B20, let c be the closure operator of $\Lambda = (V, \mathscr{M})$ and let $U \in \mathscr{P}(V)$. Then $c(U)$ is the intersection of V with the subspace of

$\mathscr{V}$ spanned by U, and $\mathscr{C}(\Lambda)$ is the collection of intersections of V with subspaces of $\mathscr{V}$.

C22 *Example.* Let $\Lambda = (E, \mathscr{M}(\mathscr{Z}(\Gamma)))$ be the cycle matroid of $\Gamma = (V, f, E)$. Then $r(F)$ is the cardinality of a largest subset of $F \in \mathscr{P}(E)$ containing no cycle. Thus $r(F)$ is the number of edges in a spanning forest of Γ_F, and (cf. B27) we have $r(F) = \nu_0(\Gamma_F) - \nu_{-1}(\Gamma_F)$. On the other hand,

$$c(F) = F + \{e \in E + F : e \in Z \subseteq F + \{e\} \text{ for some } Z \in \mathscr{Z}(\Gamma)\}.$$

In other words, if $e \in c(F)$ then the vertices in $f(e)$ are joined by a path all of whose edges are in F. If $W_1, \ldots, W_k$ are the vertex sets of the components of Γ_F, then $c(F)$ is the set of edges of $\bigoplus_{i=1}^k \Gamma_{W_i}$.

C23 *Exercise.* Determine the closed sets of the cocycle matroid of a multigraph.

A matroid $\Lambda = (V, \mathscr{M})$ is called a **geometry** if $\mathscr{P}_0(V) + \mathscr{P}_1(V) \subseteq \mathscr{C}(\Lambda)$. (Crapo and Rota [c.5] use the term "pregeometry" for a matroid.) Both of the Fano matroids, for example, are geometries. Their closed sets are precisely the "geometric objects" such as the points, the lines, the entire plane, and of course $\varnothing$. The same is true of the matroid in Example C21.

The interest in geometries lies in the fact that they correspond in a nice way to a class of lattices, as we shall see at the end of this section. Furthermore, any matroid need be only slightly modified to yield a geometry: first delete all vertices in the closure of $\varnothing$; second observe that the relation $y \sim x$ if $y \in c(\{x\})$ is an equivalence relation. (To prove symmetry, use C11 with $U = \varnothing$.) Then identify all vertices within each equivalence class.

To understand what is going on, let us just for the moment admit loops in our multigraphs. The loops would form $c(\varnothing)$ in the cycle matroid of Γ. Our first step is analogous to deleting all loops. An edge e' is in the closure of the edge e if and only if $f(e) = f(e') \in \mathscr{P}_2(E)$. Our second step is analogous to replacing a multigraph by its underlying graph. Finally, just as the graph underlying a multigraph preserves most of the interesting structure of the multigraph, the geometry underlying a matroid preserves most of the interesting structure of that matroid.

C24 Exercise. Show that *a matroid is a geometry if and only if the cardinality of every block is at least* 3. (In particular, *the cycle matroid of a graph is a geometry.*)

C25 *Exercise.* Show that the rank of a geometry with a nonempty vertex set is at least 2.

A lattice is said to be **semimodular** if whenever both x and y are successors of $x \wedge y$, then $x \vee y$ is a successor of both x and y. If $L = (U, \leq)$ is semimodular, then the sublattice $L_{[x,y]} = (\{u \in U : x \leq u \leq y\}, \leq)$ is clearly semimodular. A **matroid lattice** $(U, \leq)$ is a semimodular lattice with the further condition that every element $x \in U$ can be expressed as $x = \bigvee_{a \in A} a$ where A is a set of atoms. In particular, A may be taken as $\{a \in V: a \text{ is an atom}; a \leq x\}$ (see [b.9]). The lattice shown in Figure C26a is semimodular, but the elements b_1, b_2, and 1 are not joins of a set of atoms. In the lattice shown in Figure C26b, every element is a join of atoms, but the lattice is not semimodular since $a_1 \vee a_2 = 1$ is not a successor of a_1 or a_2.

C26

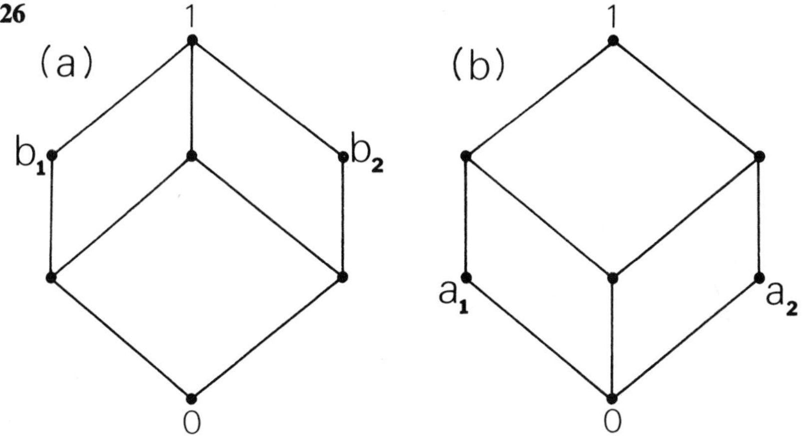

C27 *Exercise.* Verify that every Boolean lattice is a matroid lattice.

C28 Lemma. *Let $L = (U, \leq)$ be a semimodular lattice and let $x, y \in U$.*

(a) *If $x \leq y$, then all maximal chains in $L_{[x,y]}$ have the same length.*

(b) *If x is an atom and $x \nleq y$, then $x \vee y$ is a successor of y.*

PROOF. (a) We proceed by induction on $|U|$. The result holds if $|U| \leq 2$. If it is not the case that $x = 0$ and $y = 1$, then since $L_{[x,y]}$ is semimodular, all maximal chains in $L_{[x,y]}$ have the same length. Now consider maximal chains $0 = x_0, \ldots, x_h = 1$ and $0 = y_0, \ldots, y_k = 1$ in L. If $x_1 \neq y_1$, let $z = x_1 \vee y_1$. Since both x_1 and y_1 are successors of 0, z must be a successor of both x_1 and y_1. If $z = z_2, z_3, \ldots, z_m, 1$ is a maximal chain in $L_{[z,1]}$, then $x_1, z_2, \ldots, z_m, 1$ and $x_1, x_2, \ldots, x_h, 1$ are maximal chains in $L_{[x_1,1]}$. By the induction hypothesis $h = m$. Similarly one shows that $k = m$. If $x_1 = y_1$, the above argument yields $h = k$ directly, whence the result.

(b) Assuming that x is an atom and $x \nleq y$, we proceed by induction on the length of a maximal chain in $L_{[0,y]}$. The assertion is trivial if $y = 0$. If $y > 0$, then y is a successor of some element $w \in U$. Since $x \nleq w$, the induction hypothesis implies that $x \vee w$ is a successor of w. Since L is semimodular, $(x \vee w) \vee y$ is a successor of both $(x \vee w)$ and y. But $(x \vee w) \vee y = x \vee (w \vee y) = x \vee y$ as required. □

If $\mathscr{C}$ is the collection of closed sets of the closure structure (V, c), then $(\mathscr{C}, \subseteq)$ is a partially-ordered subset of $(\mathscr{P}(V), \subseteq)$. We shall see that $(\mathscr{C}, \subseteq)$

is in fact a lattice, although it need not be a sublattice of $(\mathscr{P}(V), \subseteq)$. If U_1, $U_2 \in \mathscr{C}$, their meet in $(\mathscr{P}(U), \subseteq)$ is of course $U_1 \cap U_2$, and by C12d, $U_1 \cap U_2 = c(U_1 \cap U_2) \in \mathscr{C}$. The join of U_1 and U_2 in $\mathscr{P}(V)$ is, of course $U_1 \cup U_2$, but in $(\mathscr{C}, \subseteq)$ it is $c(U_1 \cup U_2)$ which, as we already know, need not equal $U_1 \cup U_2$. Since $(\mathscr{C}, \subseteq)$ characterizes a closure structure, it characterizes the matroid derivable from it; the matroid is unique up to isomorphism.

C29 Lemma. *Let $\mathscr{C}$ be the collection of closed sets of a closure structure (V, c), and let $U_1, U_2 \in \mathscr{C}$. Then U_2 is a successor of U_1 in $(\mathscr{C}, \subseteq)$ if and only if $U_2 = c(U_1 + \{x\})$ for some $x \in U_2 + U_1$.*

Proof. Suppose that U_2 is a successor of U_1 and choose $x \in U_2 + U_1$. Since $U_1 + \{x\} \subseteq U_2$, we have $U_1 \subset c(U_1 + \{x\}) \subseteq c(U_2) = U_2$. Since U_2 is a successor of U_1, we cannot have $c(U_1 + \{x\}) \subset U_2$.

Conversely, suppose that $U_2 = c(U_1 + \{x_1\})$ for some $x_1 \in U_2 + U_1$. We must show that if $U_1 \subset T \subseteq U_2$ for some $T \in \mathscr{C}$, then $T = U_2$. Let $x_2 \in T + U_1$. Thus $x_2 \in c(U_1 + \{x_1\})$, but $x_2 \notin c(U_1)$. By C11, $x_1 \in c(U_1 + \{x_2\})$. Thus $U_1 + \{x_1\} \subseteq c(U_1 + \{x_2\}) \subseteq c(T) = T$, and so $U_2 = c(U_1 + \{x_1\}) \subseteq T$ by C10. ☐

C30 Theorem. *Let L be a lattice. Then the following are equivalent:*

(**a**) *L is isomorphic to the partially-ordered collection of the closed sets of a closure structure derived from a geometry.*

(**b**) *L is a matroid lattice.*

Moreover, if either condition holds, then the geometry in (a) *is unique up to isomorphism.*

In proving that (a) implies (b) it is not necessary to assume that our matroid is in fact a geometry. We prove instead the following slightly stronger result.

C31 Lemma. *If $\mathscr{C}$ is the collection of closed sets of the closure structure (V, c), then $(\mathscr{C}, \subseteq)$ is a matroid lattice with $U_1 \wedge U_2 = U_1 \cap U_2$ and $U_1 \vee U_2 = c(U_1 \cup U_2)$.*

Proof. Let $\mathscr{C}$ be the collection of closed sets of the closure structure (V, c). We have already remarked that $(\mathscr{C}, \subseteq)$ is a partially-ordered subset of the lattice $(\mathscr{P}(V), \subseteq)$. Let $U_1, U_2 \in \mathscr{C}$. Since $U_1 \cap U_1 \in \mathscr{C}$ by C12d, we have proved (cf. *II*B38) that $U_1 \wedge U_2$ is defined in $(\mathscr{C}, \subseteq)$ and equals $U_1 \cap U_2$. We assert that $U_1 \vee U_2$ also exists and that it equals $c(U_1 \cup U_2)$. By C12b, $c(U_1 \cup U_2) \in \mathscr{C}$ and by C9, $U_1 \subseteq c(U_1 \cup U_2)$ and $U_2 \subseteq c(U_1 \cup U_2)$. If $U_1 \subseteq C$ and $U_2 \subseteq C$ for some $C \in \mathscr{C}$, then $U_1 \cup U_2 \subseteq C$ and by C10, $c(U_1 \cup U_2) \subseteq c(C) = C$. Hence $(\mathscr{C}, \subseteq)$ is a lattice.

We observe that for $U_1, \ldots, U_k \in \mathscr{C}$, $\bigwedge_{i=1}^{n} U_i = \bigcap_{i=1}^{n} U_i$ and $\bigvee_{i=1}^{n} U_i = c(\bigcup_{i=1}^{n} U_i)$. In particular, $0 = \bigwedge_{U \in \mathscr{C}} U = c(\varnothing)$. We next show that if $x \in V + c(\varnothing)$, then $c(\{x\})$ is an atom of $(\mathscr{C}, \subseteq)$. For suppose $c(\varnothing) < U \leq$

$c(\{x\})$. For any $y \in U + c(\varnothing)$, we have $c(\varnothing) < c(\{y\}) \leq c(\{x\})$. Thus $y \in c(\varnothing + \{x\})$ and $y \notin c(\varnothing)$. By C11, $x \in c(\{y\})$ and so $c(\{x\}) = c(\{y\})$. If $U \in \mathscr{C}$, then by C10, $c(\{x\}) \subseteq U$ for all $x \in U$. Thus $U = c(\varnothing) + \bigcup_{x \in U + c(\varnothing)} \{x\} = \bigcup_{x \in U + c(\varnothing)} c(\{x\}) = c(\bigcup_{x \in U + c(\varnothing)} c(\{x\})) = \bigvee_{x \in U + c(\varnothing)} c(\{x\})$. Thus each element of $\mathscr{C}$ is the join of a set of atoms.

To prove semimodularity, let $U_1, U_2 \in \mathscr{C}$ be distinct successors of $U = U_1 \cap U_2$. By C29, there exists $x_i \in U_i + U$ such that $U_i = c(U + \{x_i\})$ for $i = 1, 2$. Suppose $x_1 \in U_2$. Then $U \subset U_1 = c(U + \{x_1\}) \subseteq c(U_2) = U_2$, and since U_2 is a successor of U, we have $U_2 = U_1$, contrary to assumption. Thus $x_1 \notin U_2$ and similarly $x_2 \notin U_1$. By C29 again, $c(U_1 + \{x_2\}) = c(U + \{x_1, x_2\}) = c(U_2 + \{x_1\})$ is a successor of both U_1 and U_2. It remains only to show that $U_1 \vee U_2 = c(U + \{x_1, x_2\})$. Clearly since $U_i \subseteq c(U + \{x_1, x_2\})$ for $i = 1, 2$, we have $U_1 \vee U_2 \subseteq c(U + \{x_1, x_2\})$. Since $U_i \subset U_1 \vee U_2 \subseteq c(U + \{x_1, x_2\})$ and $c(U + \{x_1, x_2\})$ is a successor of U_i, the required equality holds. □

Proof of Theorem C30: (b) implies (a). Suppose that $L = (S, \leq)$ is a matroid lattice. Let V denote the set of atoms of L and define a function $c: \mathscr{P}(V) \to \mathscr{P}(V)$ by

$$c(U) = \{x \in V: x \leq \bigvee_{a \in U} a\}, \quad U \in \mathscr{P}(V).$$

We first show that (V, c) is a closure structure. Let $U \in \mathscr{P}(V)$ and let $z = \bigvee_{a \in U} a$. Clearly $x \leq z$ for all $x \in U$, which implies that $U \subseteq c(U)$. To verify C10, let $U_1, U_2 \in \mathscr{P}(V)$, suppose $U_1 \subseteq c(U_2)$, and let $x \in c(U_1)$. If $a \in U_1$, then $a \leq \bigvee_{b \in U_2} b$. Hence $x \leq \bigvee_{a \in U_1} a \leq \bigvee_{b \in U_2} b$ and so $x \in c(U_2)$. To verify C11, let $x_1, x_2 \in V + U$ and suppose $x_2 \in c(U + \{x_1\})$ but $x_2 \notin c(U)$; that is, $x_2 \leq z \vee x_1$ and $x_2 \nleq z$. Hence $z < z \vee x_2 \leq z \vee x_1$. This implies that $x_1 \nleq z$, and so by Lemma C28b, $z \vee x_1$ is a successor of z. Hence $z \vee x_2 = z \vee x_1$, or $x_1 \leq z \vee x_2$, which means that $x_1 \in c(U + \{x_2\})$.

Since the join over an empty set is 0, $c(\varnothing) = \varnothing$. If $x \in V$, then $c(\{x\}) = \{x\}$. The unique matroid $(V, \mathscr{M})$ of which c is the closure operator is thus a geometry.

Let $\mathscr{C}$ be the collection of closed sets of the closure structure (V, c). By Lemma C31, we know that $(\mathscr{C}, \subseteq)$ is a lattice, and we now establish a lattice-isomorphism h from $(\mathscr{C}, \subseteq)$ to $(S, \leq)$. For each set $U \in \mathscr{C}$, let $h(U) = \bigvee_{a \in U} a$. If $s \in S$ and $U = \{a \in V: a \leq s\}$, then $c(U) = U \in \mathscr{C}$ and $h(U) = s$. Hence h is a surjection. Clearly if $U_1, U_2 \in \mathscr{C}$ and $U_1 \neq U_2$, then by the definitions of c and h, $h(U_1) \neq h(U_2)$. Hence h is a bijection. We leave as a straightforward exercise for the reader the following three assertions if $U_1, U_2 \in \mathscr{C}$:

$$U_1 \subseteq U_2 \Rightarrow h(U_1) \leq h(U_2);$$

$$h(U_1 \cap U_2) = h(U_1) \wedge h(U_2);$$

$$h(U_1 \cup U_2) = h(U_1) \vee h(U_2).$$

The proof is then complete. □

A **rank function** r can now be defined on the set of elements of a semi-modular lattice, $L = (S, \leq)$. Specifically $r(x)$ for $x \in S$ is the length of any maximal chain in $L_{[0,x]}$.

C32 *Exercise.* Let r be the rank function of the geometry $\Lambda = (V, \mathscr{M})$ and let h be the lattice-isomorphism in the above proof. If r' is the rank function of the lattice L (same proof), show that $r(U) = r'(h(U))$ for all closed sets U of Λ.

A lattice $L = (S, \leq)$ is said to be **relatively complemented** if $L_{[x,y]}$ is complemented for all $x, y \in S$, $x \leq y$.

C33 *Exercise.* (a) Show that matroid lattices are relatively complemented. (b) Interpret this notion in the context of geometries.

XD Orthogonality and Minors

Let V be an n-set and let $\Lambda = (V, \mathscr{M})$ be a matroid of rank r. We define

$$\mathscr{B}^{\perp}(\Lambda) = \{B \in \mathscr{P}(V): V + B \in \mathscr{B}(\Lambda)\}.$$

By Corollary A3, it follows that $(V, \mathscr{B}^{\perp}(\Lambda))$ is a basis system. If we now define

D1 $$\mathscr{I}^{\perp}(\Lambda) = \bigcup_{B \in \mathscr{B}^{\perp}(\Lambda)} \mathscr{P}(B) \quad \text{and} \quad \mathscr{M}^{\perp} = \mathscr{M}(\mathscr{P}(V) + \mathscr{I}^{\perp}(\Lambda)),$$

then by Propositions B10 and B19, $\Lambda^{\perp} = (V, \mathscr{M}^{\perp})$ is a matroid of rank $n - r$; it is called the **matroid orthogonal to** Λ.

It is immediate that for any matroid $\Lambda = (V, \mathscr{M})$,

D2 (a) $(\Lambda^{\perp})^{\perp} = \Lambda$;
(b) $r(\Lambda) + r(\Lambda^{\perp}) = |V|$;
(c) *Λ is isomorphic to Θ if and only if $\Lambda^{\perp}$ is isomorphic to $\Theta^{\perp}$.*

Example. Recall that edge sets of the spanning forests of a multigraph $\Gamma = (V, f, E)$ are the complements of the edge sets of the spanning coforests of Λ and that these collections are the collections of bases for the cycle matroid of Γ and the cocycle matroid of Γ, respectively. (Cf. Exercise B26.) These two matroids are therefore orthogonal. In symbols,

D3 $$(E, \mathscr{M}(\mathscr{Z}(\Gamma)))^{\perp} = (E, \mathscr{M}(\mathscr{Z}^{\perp}(\Gamma))).$$

In addition to D1, one can easily prove the following:

D4 Exercise. Prove:
(a) *The following are equivalent:*
(i) $U \in \mathscr{I}(\Lambda)$;
(ii) $V + U \in \mathscr{S}(\Lambda^{\perp})$;
(iii) $U \cap B = \varnothing$ *for some* $B \in \mathscr{B}^{\perp}(\Lambda)$.

(b) *$\mathscr{B}(\Lambda^\perp)$ is the collection of minimal sets which meet each cycle of Λ.*
(c) *$\mathscr{M}^\perp$ is the collection of minimal sets which meet every basis of Λ.*
(d) *$\mathscr{I}(\Lambda^\perp) = \{J \in \mathscr{P}(V): J \cap B = \varnothing$ for some $B \in \mathscr{B}(\Lambda)\}$.*
(e) *If $r^\perp$ is the rank function of $\Lambda^\perp$ and $U \in \mathscr{P}(V)$, then $r^\perp(U) = |U| + r(V + U) - r(V)$.*

If $U = V$ in D4e, we obtain D2b.

In the above terminology we can restate Proposition B28b as follows: *If $M \in \mathscr{M}$, then to each $x \in M$ there corresponds a basis B of $\Lambda^\perp$ such that $M \cap B = \{x\}$.* Using this fact, we now characterize $\mathscr{M}^\perp$ directly in terms of $\mathscr{M}$, rather than via $\mathscr{I}(\Lambda)$, $\mathscr{B}(\Lambda)$, $\mathscr{B}^\perp(\Lambda)$, and $\mathscr{I}^\perp(\Lambda)$.

D5 Proposition. *Let $\Lambda = (V, \mathscr{M})$ and let $\mathscr{A} = \{A \in \mathscr{P}(V): |A \cap M| \neq 1$ for all $M \in \mathscr{M}\}$. Then $\mathscr{M}^\perp = \mathscr{M}(\mathscr{A})$, and each set in $\mathscr{A}$ is a union of sets in $\mathscr{M}^\perp$.*

PROOF. Proceeding naively, let $\mathscr{N} = \mathscr{M}(\mathscr{A})$. Our first step is to prove that $(V, \mathscr{A})$ satisfies condition B15. We will actually prove the stronger result: if (A_1, A_2, y) is an admissible triple for $(V, \mathscr{A})$, then there exists $A \in \mathscr{A}$ such that $A_1 + A_2 \subseteq A \subseteq (A_1 \cup A_2) + \{y\}$. Thus if (A_1, A_2, x, y) is an admissible 4-tuple, then $x \in A$. Let (A_1, A_2, y) be an admissible triple and let

$$X = \{x \in (A_1 \cup A_2) + \{y\}: M_x \cap [(A_1 \cup A_2) + \{y\}] = \{x\} \text{ for some } M_x \in \mathscr{M}\}.$$

If $x \in X \cap A_1$ and $x \notin A_2$, then for some $M_x \in \mathscr{M}$, either $M_x \cap A_1 = \{x\}$ or $M_x \cap A_2 = \{y\}$, contrary to the definition of $\mathscr{A}$. The argument is the same if $x \in X \cap A_2$ and $x \notin A_1$. Hence $X \cap (A_1 + A_2) = \varnothing$. We let $A = (A_1 \cup A_2) + X + \{y\}$. Since $A_1 + A_2 \subseteq A$, it remains to show that $A \in \mathscr{A}$.

Suppose $A \notin \mathscr{A}$, and select $M \in \mathscr{M}$ such that $|X \cap M|$ is as small as possible subject to the condition that $|A \cap M| = 1$. Let us say $A \cap M = \{z\}$. If $x \in X \cap M$, then (M, M_x, z, x) is admissible for Λ, and so there exists $M' \in \mathscr{M}$ such that $z \in M' \subseteq (M \cup M_x) + \{x\}$. Now $|A \cap M'| = 1$, but $(X \cap M') \subset (X \cap M)$ since $x \notin X \cap M'$, giving a contradiction. Hence $X \cap M = \varnothing$, which implies that $M \cap [(A_1 \cup A_2) + \{y\}] = \{z\}$, and so $z \in X$ by definition. Since $X \cap A = \varnothing$, this is a contradiction.

Since $(V, \mathscr{A})$ satisfies B15, it follows by Lemma B17b that $\Theta = (V, \mathscr{N})$ is a matroid and that every set in $\mathscr{A}$ is a union of sets in $\mathscr{N}$. It remains to show that $\Theta = \Lambda^\perp$. We shall do this by showing that $\mathscr{B}(\Lambda) = \mathscr{B}(\Theta^\perp)$.

Let $B \in \mathscr{B}(\Lambda)$ and fix $x \in B$. By Proposition B28a, to each $v \in V + B$ there corresponds a unique cycle $M_v \in \mathscr{M}$ such that $(V + B) \cap M_v = \{v\}$. We define $N_x = \{v \in V + B: x \in M_v\} + \{x\}$. Thus $N_x \neq \varnothing$, since certainly $x \in N_x$.

Suppose that $N_x \notin \mathscr{A}$. Then $|N_x \cap M| = 1$ for some $M \in \mathscr{M}$, and we may select M subject to this condition so that $|(V + B) \cap M|$ is as small as possible. If $N_x \cap M = \{x\}$, then $(V + B) \cap M \neq \varnothing$ or else $M \subseteq B$, which is impossible. Hence let $y \in (V + B) \cap M$. Then $y \notin N_x$ and so $x \notin M_y$. Therefore the 4-tuple (M, M_y, x, y) is admissible. On the other hand, if $N_x \cap M = \{y\}$ for some $y \in V + B$, then $x \in M_y$ and $x \notin M$. Hence

(M_y, M, x, y) is admissible. Either way, there exists a cycle $M' \in \mathscr{M}$ such that $x \in M' \subseteq (M \cup M_y) + \{y\}$. Thus $|N_x \cap M'| = 1$. But then $(V + B) \cap M' \subseteq [(V + B) \cap M] + \{y\}$, contrary to our minimality assumption. We have shown that $N_x \in \mathscr{A}$.

If $\varnothing \subset N' \subset N_x$, one easily verifies that $|N' \cap M_v| = 1$ for some $v \in N_x + \{x\}$. Hence N_x is a minimal set in $\mathscr{A}$; i.e., $N_x \in \mathscr{N}$.

Let $N \in \mathscr{N}$ and let $v \in N$. If $V + B \supseteq N$, then $N \cap M_v = \{v\}$, contrary to the definition of $\mathscr{N}$. Hence $V + B$ contains no cycle of Θ. Thus $V + B \in \mathscr{I}(\Theta)$. To prove that $V + B$ is a maximal independent set of Θ, we note that for any $x \in B$, we have $N_x \subseteq (V + B) + \{x\}$, and we have just shown that N_x is a cycle of Θ. Hence $V + B \in \mathscr{B}(\Theta)$, and so $B \in \mathscr{B}(\Theta^\perp)$.

Let

$$\mathscr{A}' = \{A \in \mathscr{P}(V): |A \cap N| \neq 1 \text{ for all } N \in \mathscr{N}\}$$

and consider the set system $\Phi = (V, \mathscr{N}')$ where $\mathscr{N}' = \mathscr{M}(\mathscr{A}')$. By the above argument, Φ is a matroid, and every set in $\mathscr{A}'$ is a union of sets in $\mathscr{N}'$. Just as we have shown that $\mathscr{B}(\Lambda) \subseteq \mathscr{B}(\Theta^\perp)$, it follows that $\mathscr{B}(\Theta) \subseteq \mathscr{B}(\Phi^\perp)$. Hence by D2a, $\mathscr{B}(\Lambda) \subseteq \mathscr{B}(\Theta^\perp) \subseteq \mathscr{B}(\Phi)$, whence

D6 $$r(\Lambda) = r(\Phi).$$

Let $M \in \mathscr{M}$. Since $|A \cap M| \neq 1$ for all $A \in \mathscr{A}$, it holds that $|N \cap M| \neq 1$ for all $N \in \mathscr{N}$. Thus $M \in \mathscr{A}'$. Hence $\mathscr{M} \subseteq \mathscr{A}'$, and so every set in M is also a union of sets in $\mathscr{N}'$. This together with D6 yields $\Lambda = \Phi$, by Lemma B29. Hence $\mathscr{B}(\Lambda) = \mathscr{B}(\Theta^\perp)$. □

D7 *Exercise.* Using only the results of the first three sections of this chapter (including the exercises and examples), give direct proofs of Propositions *III*C11 and *III*C18, resorting as little to "graph theory" as possible.

D8 *Exercise.* (a) Describe all matroids with five vertices. (b) Describe all matroids with n vertices having ranks $0, 1, n - 1$, and n.

Before proceeding further we present a summary of the relationships between the eight set systems associated with a matroid established thus far. The references in parentheses indicate the first explicit appearance of the result.

D9 Theorem. *Let* $\Lambda = (V, \mathscr{M})$ *be a matroid.* (We suppress the argument Λ and write $\mathscr{I}^\perp, \mathscr{B}^\perp, \mathscr{S}^\perp$ for $\mathscr{I}(\Lambda^\perp), \mathscr{B}(\Lambda^\perp), \mathscr{S}(\Lambda^\perp)$, respectively.)

(a) $\mathscr{I} = \{A \in \mathscr{P}(V): A \subseteq B$ *for some* $B \in \mathscr{B}\}$; (B10)
$\mathscr{S} = \{A \in \mathscr{P}(V): A \supseteq B$ *for some* $B \in \mathscr{B}\}$; (B11)
$\mathscr{B}$ *is the collection of largest sets of* $\mathscr{I}$; (B10)
$\mathscr{B}$ *is the collection of smallest sets of* $\mathscr{S}$. (B11)

(b) $\mathscr{B} = \mathscr{I} \cap \mathscr{S}$.

(c) *$\mathcal{M}$ is the collection of minimal nonempty sets of $\mathcal{P}(V) + \mathcal{I}$;*
$\mathcal{I} = \{A \in \mathcal{P}(V): A \not\supseteq M$ for all $M \in \mathcal{M}\}$. (B19)
(d) *$\mathcal{B}^\perp$ is the collection of complements of sets of $\mathcal{B}$;*
$\mathcal{I}^\perp$ is the collection of complements of sets of $\mathcal{S}$; (D4a)
$\mathcal{S}^\perp$ is the collection of complements of sets in $\mathcal{I}$.
(e) *$\mathcal{B}^\perp$ is the collection of minimal sets which meet every set in $\mathcal{M}$;* (D4b)
$\mathcal{S}^\perp$ is the collection of sets which meet every set in $\mathcal{M}$;
$\mathcal{M}^\perp$ is the collection of minimal sets which meet every set in $\mathcal{B}$; (D4c)
$\mathcal{I}^\perp$ is the collection of sets that avoid some set in $\mathcal{B}$. (D4d)
(f) *$\mathcal{M}^\perp$ is the collection of minimal sets which meet no set in $\mathcal{M}$ in a 1-set.* (D5)

In §*II*A we introduced the projection $\pi_U: \mathcal{P}(V) \to \mathcal{P}(U)$ (where $U \subseteq V$) given by $\pi_U(S) = U \cap S$ for all $S \in \mathcal{P}(V)$. If $\Lambda = (V, \mathcal{E})$ is a set system and if $U \subseteq V$, we shall use the symbol $\Lambda_{[U]}$ to denote the set system $(U, \mathcal{M}(\pi_U[\mathcal{E}]))$. We shall concern ourselves with the systems of the form $\Lambda_{[U]}$ and Λ_U, the latter being truly a subsystem of Λ.

D10 Proposition. *If $\Lambda = (V, \mathcal{M})$ is a matroid and if $U \in \mathcal{P}(V)$, then Λ_U and $\Lambda_{[U]}$ are matroids. Furthermore, if $M \in \mathcal{M}$, then $M \cap U$ is a union of cycles of $\Lambda_{[U]}$.*

Proof. Let us write $\Lambda_U = (U, \mathcal{M}_1)$. Recall that $\mathcal{M}_1 = \mathcal{M} \cap \mathcal{P}(U)$. Since $\mathcal{M}$ is incommensurable, so is $\mathcal{M}_1$. To verify B15 for Λ_U, let (M_1, M_2, x, y) be an admissible 4-tuple. Since Λ is a matroid, $x \in M \subseteq (M_1 \cup M_2) + \{y\}$ for some $M \in \mathcal{M}$. Clearly $M \in \mathcal{M}_1$.

To prove that $\Lambda_{[U]}$ is a matroid, let (A_1, A_2, x, y) be an admissible 4-tuple for the system $(U, \pi_U[\mathcal{M}])$. There exist distinct $M_1, M_2 \in \mathcal{M}$ such that $A_1 = M_1 \cap U$ and $A_2 = M_2 \cap U$. Then (M_1, M_2, x, y) is an admissible 4-tuple for Λ, and so $x \in M \subseteq (M_1 \cup M_2) + \{y\}$ for some $M \in \mathcal{M}$. Let $A = M \cap U$. Then $x \in A$, and $A \subseteq (A_1 \cup A_2) + \{y\}$. Hence $(U, \pi_U[\mathcal{M}])$ satisfies B15. The result follows by B17b. □

D11 Exercise. *Let $\Lambda = (V, \mathcal{M})$ be a matroid, let $x \in V$, and let $\Lambda_{[V+\{x\}]} = (V + \{x\}, \mathcal{L})$.* Prove that:

(a) *If $L \in \mathcal{L}$, then either $L \in \mathcal{M}$ or $L + \{x\} \in \mathcal{M}$.*
(b) *If $M \in \mathcal{M}$ and $x \in M$, then $M + \{x\} \in \mathcal{L}$.*
(c) *If $x \notin M$, and $M \in \mathcal{M}$, then either $M \in \mathcal{L}$ or $M = L_1 \cup L_2$ where $L_i + \{x\} \in \mathcal{M}$ for $i = 1, 2$.*

If Λ is one of the other kinds of systems studied in §B, it does not always hold that both Λ_U and $\Lambda_{[U]}$ are the same kind of system as Λ. We glance at this phenomenon in the next exercise, which together with Exercise D21 below, presents further arguments in favor of the selection of cycle systems as matroids.

D12 Exercise. *Let* $\Lambda = (V, \mathscr{M})$ *be a matroid with rank function* r *and let* $U \in \mathscr{P}(V)$. Prove:

(a) $\mathscr{I}(\Lambda_U) = \mathscr{I}(\Lambda) \cap \mathscr{P}(U)$. *Thus if* $\Theta = (V, \mathscr{I})$ *is an independence system, then so is* Θ_U.

(b) $\mathscr{S}(\Lambda_{[U]}) = \pi_U[\mathscr{S}(\Lambda)]$. *Thus if* $\Theta = (V, \mathscr{S})$ *is a spanning system, then so is* $\Theta_{[U]}$.

(c) *The rank function* r_U *of* Λ_U *is given by* $r_U = r_{|\mathscr{P}(U)}$.

(d) *The rank function* $r_{[U]}$ *of* $\Lambda_{[U]}$ *is given by*

$$r_{[U]}(S) = r(V + U + S) - r(V + U), \quad S \in \mathscr{P}(U).$$

(e) $\mathscr{B}(\Lambda_U) = \{B \cap U: B \in \mathscr{B}(\Lambda); |B \cap U| = r_U(\Lambda_U)\}$.

(f) $\mathscr{B}(\Lambda_{[U]}) = \{B \cap U: B \in \mathscr{B}(\Lambda): |B \cap U| = r_{[U]}(\Lambda_{[U]})\}$.

(g) *If* $J_1 \in \mathscr{I}(\Lambda_{[U]})$ *and* $J_2 \in \mathscr{I}(\Lambda_{V+U})$, *then* $J_1 + J_2 \in \mathscr{I}(\Lambda)$.

D13 Proposition. *Let* $\Lambda = (V, \mathscr{M})$ *be a matroid and let* $U \in \mathscr{P}(V)$. *Then*

(a) $(\Lambda_{[U]})^{\perp} = (\Lambda^{\perp})_U$;

(b) $(\Lambda_U)^{\perp} = (\Lambda^{\perp})_{[U]}$.

PROOF. (a) We show that the two matroids in question have the same collection of independent sets by means of the following string of equivalent statements.

$$\begin{aligned} J \in \mathscr{I}((\Lambda_{[U]})^{\perp}) &\Leftrightarrow J \in \mathscr{I}^{\perp}(\Lambda_{[U]}) \\ &\Leftrightarrow U + J \in \mathscr{S}(\Lambda_{[U]}) && \text{by D9d} \\ &\Leftrightarrow U + J = S \cap U \quad \text{for some } S \in \mathscr{S}(\Lambda) && \text{by D12b} \\ &\Leftrightarrow J = (V + S) \cap U \quad \text{for some } S \in \mathscr{S}(\Lambda) \\ &\Leftrightarrow J = J' \cap U \quad \text{for some } J' \in \mathscr{I}^{\perp}(\Lambda) && \text{by D9d} \\ &\Leftrightarrow J \subseteq U \quad \text{and} \quad J \in \mathscr{I}^{\perp}(\Lambda) && \text{by B1} \\ &\Leftrightarrow J \in \mathscr{I}((\Lambda^{\perp})_U). \end{aligned}$$

(b) Let $\Theta = \Lambda^{\perp}$. By D2a and part (a) above, $(\Lambda_U)^{\perp} = ((\Theta^{\perp})_U)^{\perp} = ((\Theta_{[U]}{}^{\perp})^{\perp}) = (\Lambda^{\perp})_{[U]}$. □

D14 Proposition. *Let* $\Lambda = (V, \mathscr{M})$ *and let* $T \subseteq U \subseteq V$. *Then*

(a) $(\Lambda_U)_T = \Lambda_T$;

(b) $(\Lambda_{[U]})_{[T]} = \Lambda_{[T]}$;

(c) $(\Lambda_{[U]})_T = (\Lambda_{V+U+T})_{[T]}$;

(d) $(\Lambda_U)_{[T]} = (\Lambda_{[V+U+T]})_T$.

PROOF. (a) and (b) are immediate consequences of the definitions. To prove (c) we show that the two matroids in question have the same cycles, again by means of a string of equivalent statements.

$$\begin{aligned} M \text{ is a cycle of } (\Lambda_{[U]})_T &\Leftrightarrow M \subseteq T \quad \text{and} \quad M \in \mathscr{M}(\pi_U[\mathscr{M}]) \\ &\Leftrightarrow M \subseteq T \quad \text{and} \quad M \in \mathscr{M}(\{M' \cap U: M' \in \mathscr{M}\}) \\ &\Leftrightarrow M \in \mathscr{M}(\{M' \cap T: M' \in \mathscr{M} \text{ and } M' \subseteq V + U + T\}) \\ &\Leftrightarrow M \in \mathscr{M}(\{M' \cap T: M' \text{ is a cycle of } \Lambda_{V+U+T}\}) \\ &\Leftrightarrow M \text{ is a cycle of } (\Lambda_{V+U+T})_{[T]}. \end{aligned}$$

To prove (d), let $U' = V + U + T$. Then $T \subseteq U'$ and $U = V + U' + T$. By part (c), $(\Lambda_U)_{[T]} = (\Lambda_{V+U'+T})_{[T]} = (\Lambda_{[U']})_T = (\Lambda_{[V+U+T]})_T$. □

By successive applications of the various parts of the above proposition we obtain:

D15 Corollary. *Let $\Lambda^{(i)} = (V_i, \mathcal{M}_i)$ for $i = 0, \ldots, n$ be a sequence of matroids such that for each $i = 1, \ldots, n$, $V_i \subseteq V_{i-1}$ and either $\Lambda^{(i)} = \Lambda^{(i-1)}_{V_i}$ or $\Lambda^{(i)} = \Lambda^{(i-1)}_{[V_i]}$. Then there exist $T, U \in \mathcal{P}(V_0)$ containing V_n such that $\Lambda^{(n)} = (\Lambda^{(0)}_{[T]})_{V_n} = (\Lambda^{(0)}_U)_{[V_n]}$.*

If $\Lambda = (V, \mathcal{M})$ is a matroid, then any matroid of the form $(\Lambda_U)_{[T]}$, or equivalently of the form $(\Lambda_{[U]})_T$, where $T \subseteq U \subseteq V$, is called a **minor** of Λ.

D16 *Exercise.* Let Γ be a graph and let Θ be a subcontraction of Γ.

(a) Show that the cycle matroid (cocycle matroid) of Θ is a minor of the cycle matroid (cocycle matroid) of Γ.

(b) Show that every minor of the cocycle matroid of Γ is the cocycle matroid of a subcontraction of Γ.

(c) Find necessary and sufficient conditions under which a minor of the cycle matroid of Γ is the cycle matroid of a subcontraction of Γ.

D17 *Exercise.* Show that all minors of a complete matroid are complete matroids and give a formula for their rank.

D18 Proposition. *A set system is a matroid if and only if all of its components are matroids.*

Proof. Let $\Lambda = \bigoplus_{i=1}^n \Lambda_i$, where $\Lambda_i = (V_i, \mathcal{M}_i)$ for $i = 1, \ldots, n$, and $\Lambda = (V, \mathcal{M})$. Thus $\Lambda_i = \Lambda_{V_i}$ for $i = 1, \ldots, n$. If Λ is a matroid then so is Λ_{V_i}, by Proposition D10; in particular, so is any component.

Conversely, suppose Λ_i is a matroid for $i = 1, \ldots, n$. Since $\mathcal{M} = \sum_{i=1}^n \mathcal{M}_i$ and $\mathcal{M}_i \cap \mathcal{M}_j = \varnothing$ for $i \neq j$, it is obvious that $\mathcal{M}$ is incommensurable. Let (M_1, M_2, x, y) be an admissible 4-tuple for Λ. Then $M_1 \in \mathcal{M}_i$ and $M_2 \in \mathcal{M}_j$ for some indices i, j. Since $y \in M_1 \cap M_2$, we must have $i = j$. Condition B15 now holds for $\mathcal{M}$ since it is assumed to hold for $\mathcal{M}_i$. □

If $\Lambda = (V, \mathcal{E})$ is a set system, a subsystem of Λ which is a matroid is called a **submatroid** of Λ. For example, if Λ is a matroid and $U \subseteq V$, then (Proposition D10) Λ_U is a submatroid and so is any direct summand of Λ. In general the matroid $\Lambda_{[U]}$ is not a subsystem of Λ with the following notable exception.

D19 Exercise. *Let $\Lambda = (V, \mathcal{M})$ be a matroid and let $U \in \mathcal{P}(V) + \{\varnothing\}$.* Prove:

(a) *Λ_U is a direct summand of Λ if and only if $\Lambda_U = \Lambda_{[U]}$.*

(b) *If $r(\Lambda_U) = r(\Lambda_{[U]})$, then Λ_U is a direct summand of Λ.* [*Hint*: use D10 followed by B29 and part (a).]

(c) *If* $r(\Lambda_U) + r(\Lambda_{V+U}) = r(\Lambda)$, *then* Λ_U *is a direct summand of* Λ. [*Hint*: use D12d.]

D20 Exercise. *If* $\Lambda_1, \ldots, \Lambda_n$ *are matroids whose vertex sets are pairwise-disjoint*, prove that

$$\bigoplus_{i=1}^{n} \Lambda_i^{\perp} = \left(\bigoplus_{i=1}^{n} \Lambda_i\right)^{\perp}.$$

The next exercise shows why one cannot substitute the words "independence system," "basis system," or "spanning system" for "matroid" in D18 or D19.

D21 Exercise. *Let* $\Lambda = (V, \mathscr{M})$ *be a matroid with rank function* r. *Let* $\Lambda = \Lambda_1 \oplus \Lambda_2$, *where* $\Lambda_i = (V_i, \mathscr{M}_i)$ *and* r_i *is the rank function for* Λ_i $(i = 1, 2)$. Show that

(**a**) $\mathscr{I}(\Lambda) = \{J_1 \cup J_2 : J_i \in \mathscr{I}(\Lambda_i), i = 1, 2\}$;
(**b**) $\mathscr{B}(\Lambda) = \{B_1 \cup B_2 : B_i \in \mathscr{B}(\Lambda_i), i = 1, 2\}$;
(**c**) $\mathscr{S}(\Lambda) = \{S_1 \cup S_2 : S_i \in \mathscr{S}(\Lambda_i), i = 1, 2\}$;
(**d**) $r(U) = r_1(U \cap V_1) + r_2(U \cap V_2)$, *for all* $U \in \mathscr{P}(V)$.

It was pointed out above in §B that if $\mathscr{A}$ is a subspace of $\mathscr{P}(V)$, then $(\mathrm{Fnd}(\mathscr{A}), \mathscr{M}(\mathscr{A}))$ is an example of a matroid. A necessary and sufficient condition for $(\mathrm{Fnd}(\mathscr{A}), \mathscr{M}(\mathscr{A}))$ to be connected is given by Proposition *II*C22. The next proposition shows that this same condition characterizes connectedness for arbitrary matroids.

D22 Proposition. *A necessary and sufficient condition for a matroid* $(V, \mathscr{M})$ *to be connected is that for every* $x_1, x_2 \in V$, *there exists* $M \in \mathscr{M}$ *such that* $\{x_1, x_2\} \subseteq M$.

PROOF: *Necessity*. If $(V, \mathscr{M}) = (V_1, \mathscr{M}_1) \oplus (V_2, \mathscr{M}_2)$ and if one can select $x_1 \in V_1$ and $x_2 \in V_2$, then clearly $\{x_1, x_2\} \nsubseteq M$ for all $M \in \mathscr{M}_1 + \mathscr{M}_2 = \mathscr{M}$.

Sufficiency. Suppose that $\Lambda = (V, \mathscr{M})$ is connected, and let $x_1 \in V$. It follows that each vertex of Λ belongs to some cycle. Let $U = \bigcup\{M \in \mathscr{M} : x_1 \in M\}$. We must show $U = V$.

We first show that if $M \in \mathscr{M}$ and $M \cap U \neq \varnothing$, then $M \subseteq U$. For if this were not so, one could select $M_1, M_2 \in \mathscr{M}$ such that $x_1 \in M_1$, $z \in M_2 + (M_2 \cap U)$, and $M_1 \cap M_2 \neq \varnothing$ hold while $|M_1 \cup M_2|$ is as small as possible. Note that $x_1 \notin M_2$. Hence for any $y \in M_1 \cap M_2$, the 4-tuple (M_1, M_2, x_1, y) is admissible. Since Λ is a matroid, there exists $M \in \mathscr{M}$ such that $x_1 \in M \subseteq (M_1 \cup M_2) + \{y\}$. Since $\mathscr{M}$ is incommensurable, $M \nsubseteq M_1$, and so $M \cap M_2 \neq \varnothing$. By our minimality assumption, $|M \cup M_2| \geq |M_1 \cup M_2|$. This implies that $M \supseteq M_1 + (M_1 \cap M_2)$. Since (M_2, M_1, z, y) is also admissible, there exists $M' \in \mathscr{M}$ such that $z \in M' \subseteq (M_1 \cup M_2) + \{y\}$. Since $M' \nsubseteq M_2$, we have $M' \cap (M_1 + (M_1 \cap M_2)) \neq \varnothing$. It follows that $M \cap M' \neq \varnothing$ and that $M \cup M' \subseteq (M_1 \cup M_2) + \{y\}$, contrary to our minimality assumption.

We have shown that $\pi_U[\mathcal{M}] \subseteq \mathcal{M}$ and in fact that $\Lambda_U = \Lambda_{[U]}$. Hence Λ_U is a direct summand of Λ. Since $U \neq \varnothing$ and Λ is connected, $\Lambda_U = \Lambda$. Hence $U = V$ as required. □

D23 *Exercise.* Describe all connected matroids with not more than six vertices.

D24 *Exercise.* (a) Under what conditions are complete matroids connected? (b) Under what conditions is the matroid of matched sets (of a multigraph) connected?

Analogous to Tutte connectivity for multigraphs (cf. §*VI*E), one can define m-connectedness for matroids (see [t.8]). If $\Lambda = (V, \mathcal{M})$ is a matroid with rank function r, we say that Λ is **m-separated** if there exists $U \in \mathcal{P}(V)$ such that

(a) $m < \min\{|U|, |V + U|\}$;

(b) $m = r(U) + r(V + U) - r(V)$.

The **connectivity** of Λ is given by

$$\tau(\Lambda) = \min\{m \in \mathbb{N} : \Lambda \text{ is } m\text{-separated}\}$$

with $\tau(\Lambda) = \infty$ if Λ is m-separated for no $m \in \mathbb{N}$.

D25 Proposition. *For any matroid* Λ, $\tau(\Lambda) = \tau(\Lambda^\perp)$.

Proof. Let $r^\perp$ denote the rank function of $\Lambda^\perp$, and let $U \in \mathcal{P}(V)$. Then by D4e,

$$\begin{aligned} &r^\perp(U) + r^\perp(V + U) - r^\perp(V) \\ &\quad = (|U| + r(V + U) - r(V)) \\ &\qquad + (|V + U| + r(U) - r(V)) - (|V| + r(\varnothing) - r(V)) \\ &\quad = r(U) + r(V + U) - r(V), \end{aligned}$$

and the result follows. □

D26 *Exercise.* Show that the matroid Λ is connected if and only if $\tau(\Lambda) \geq 1$.

D27 *Exercise.* Compute $\tau(\Lambda)$ for (a) the complete matroid of rank r with n vertices, and (b) the two Fano matroids.

XE Transversal Matroids

This section is a synthesis of results extracted from the following foundational papers: L. Mirsky and H. Perfect [m.11], H. Perfect [p.2], and R. Rado [r.1].

Let $\Lambda = (V, f, E)$ be a system. A **transversal** of Λ is a subset of V of the form $\lambda[F]$ where $F \subseteq E$ and $\lambda: F \to \bigcup_{e \in F} f(e)$ is an LDR of Λ_F. The largest transversals of Λ have cardinality $|E|$ if and only if Λ admits an LDR.

Recall that if $\mathscr{B}$ is the collection of largest matched sets of a graph $\Gamma = (V, \mathscr{E})$, then $(V, \mathscr{B})$ is a basis system. The matroid Θ such that $\mathscr{B}(\Theta) = \mathscr{B}$ is called the **matching matroid** of Γ.

E1 Lemma. *Let $\mathscr{T}$ be the set of transversals of a system $\Lambda = (V, f, E)$. Then $(V, \mathscr{T})$ is an independence system.*

PROOF. Let $\Gamma = ([V, E], \mathscr{F})$ be the bipartite graph of the system Λ. Let $\Theta = (V \cup E, \mathscr{M})$ denote the matching matroid of Γ. We shall show that $\mathscr{T} = \mathscr{I}(\Theta_V)$.

A subset $T \subseteq V$ is in $\mathscr{T}$ if and only if for some $E' \subseteq E$ and some matching $\mathscr{F}' \subseteq \mathscr{F}$, we have $T \cup E' = \bigcup_{F \in \mathscr{F}'} F$; that is to say, $T \cup E'$ is a matched set of Γ. By Exercise A10, this is equivalent to saying that $T \cup E'$ is contained in some member of $\mathscr{B}$, or equivalently, $T \cup E' \in \mathscr{I}(\Theta)$. Hence $T \in \mathscr{I}(\Theta) \cap \mathscr{P}(V) = \mathscr{I}(\Theta_V)$, by Exercise D12a. Conversely, if $T \in \mathscr{I}(\Theta_V)$, then $T \in \mathscr{I}(\Theta)$. Hence T is contained in some member of $\mathscr{B}$, and using the definition of $\mathscr{B}$, we return through the above chain of equivalent statements. □

If $\Lambda = (V, f, E)$ is a system, then the matroid Θ such that $\mathscr{I}(\Theta)$ is the set of all transversals of Λ is called the **transversal matroid** of Λ. Clearly $\mathscr{B}(\Theta)$ is the set of largest transversals. Thus $r(\Theta) \leq |E|$, with equality holding if and only if Λ admits an LDR.

E2 Exercise. *If Θ is the transversal matroid of a system $\Lambda = (V, f, E)$, prove that the rank function r of Θ is given by*

$$r(U) = |U| - \max\{|S| - |\bigcup_{x \in S} f^*(x)| : S \subseteq U\}, \quad \text{for all } U \in \mathscr{P}(V).$$

E3 Lemma. *Let Θ be the matching matroid of a bipartite graph $\Gamma = ([V_1, V_2], \mathscr{E})$. Then $\Theta = \Theta_{V_1} \oplus \Theta_{V_2}$, and $r(\Theta) = 2r(\Theta_{V_1}) = 2r(\Theta_{V_2})$.*

PROOF. By definition of the matching matroid, $r(\Theta) = 2\alpha_1(\Gamma)$. It follows by Exercise D12c that $r(\Theta_{V_1}) = r(\Theta_{V_2}) = \alpha_1(\Gamma)$. By Exercise D19c, Θ_{V_1} and Θ_{V_2} are direct summands of Θ, whence the lemma. □

An immediate consequence of this lemma is the following:

E4 Lemma. *Let Θ be the transversal matroid of the system Λ and let Φ be the transversal matroid of Λ^*. Then $\Theta \oplus \Phi$ is the matching matroid of the bipartite graph of Λ, and $r(\Theta) = r(\Phi)$.*

E5 Proposition. *Every transversal matroid of rank b is the transversal matroid of a system with exactly b blocks.*

PROOF. Let Θ be the transversal matroid of $\Lambda = (V, f, E)$ and suppose $r(\Theta) = b$. Let Φ be the transversal matroid of Λ^*, and let $F \subseteq E$ be a basis for Φ. Let $U \subseteq V$ be a basis for Θ. By Lemma E4, $\Theta \oplus \Phi$ is the matching matroid of the bipartite graph of Λ. By Exercise D21b, $U \cup F$ is a basis of $\Theta \oplus \Phi$. That means that $U \cup F$ is the set of vertices of a matching from F onto U. We have shown that U is a transversal of the system $(V, f_{|F}, F)$. Hence every transversal of Λ is also a transversal of $(V, f_{|F}, F)$. Finally, any transversal of $(V, f_{|F}, F)$ is clearly a transversal of Λ. □

Certainly the present discussion of matchings, LDR's, etc., must have suggested to the reader that transversal matroids are related to the Phillip Hall Theorem (VD1). The next proposition gives the connection. A system and a matroid are given which have the same vertex set V. A necessary and sufficient condition is given for the existence of an LDR which has a particular property relative to the matroid. Note that if the matroid is $(V, \varnothing)$, then its rank function is merely the cardinality function, and the result reduces to the Phillip Hall Theorem.

E6 Proposition (R. Rado [r.1]). *Let $\Lambda = (V, f, E)$ be a system and let $\Theta = (V, \mathscr{M})$ be a matroid with rank function $\mathring{r}$. A necessary and sufficient condition for there to exist an LDR λ of Λ such that $\lambda[E] \in \mathscr{I}(\Theta)$ is that*

$$r\Big(\bigcup_{e \in F} f(e)\Big) \geq |F| \quad \textit{for all } F \in \mathscr{P}(E).$$

PROOF: *Necessity.* Let λ be an LDR of Λ and suppose that $\lambda[E] \in \mathscr{I}(\Theta)$. If $F \subseteq E$, then certainly $\lambda[F] \in \mathscr{I}(\Theta)$. Since $\lambda[F] \subseteq \bigcup_{e \in F} f(e)$, we have $|F| = |\lambda[F]| = r(\lambda[F]) \leq r(\bigcup_{e \in F} f(e))$.

Sufficiency. We assume the condition to hold and proceed by induction on the number of blocks in Λ. To begin, suppose that $E = \{e\}$. Since $r(f(e)) > 0$, it is clear that the required LDR exists. Suppose therefore that $|E| > 1$ and, as induction hypothesis, that the condition is sufficient for all matroids with fewer than $|E|$ blocks.

Case 1: $r(\bigcup_{e \in F} f(e)) > |F|$ *whenever* $\varnothing \subset F \subset E$. Let $e_0 \in F$. Since $r(f(e_0)) > 1$, there exists $x_0 \in f(e_0)$ such that $\{x_0\} \in \mathscr{I}(\Theta)$. Let $\Lambda' = (V + \{x_0\}, f', E + \{e_0\})$, where $f'(e) = f(e) \cap (V + \{x_0\})$ for all $e \in E + \{e_0\}$. Let $F \in \mathscr{P}(E + \{e_0\})$, and let $r_{[V + \{x_0\}]}$ denote the rank function of $\Theta_{[V + \{x_0\}]}$. By D12d, we have

$$r_{[V+\{x_0\}]}\Big(\bigcup_{e \in F} f'(e)\Big) = r\Big(V + (V + \{x_0\}) + \bigcup_{e \in F} f'(e)\Big) - r(V + (V + \{x_0\}))$$

$$= r\Big(\{x_0\} + \bigcup_{e \in F} f'(e)\Big) - 1$$

$$= r\Big(\{x_0\} \cup \bigcup_{e \in F} f(e)\Big) - 1$$

$$\geq r\Big(\bigcup_{e \in F} f(e)\Big) - 1 \geq |F|.$$

By the induction hypothesis, Λ' admits an LDR λ' such that $\lambda'[E + \{e_0\}] \in \mathscr{I}(\Theta_{[V+\{x_0\}]})$. Defining $\lambda: E \to V$ by

$$\lambda(e) = \begin{cases} \lambda(e) & \text{if } e \in E + \{e_0\}, \\ x_0 & \text{if } e = e_0, \end{cases}$$

we have that λ is an LDR of Λ. Finally, since $\{x_0\} \in \mathscr{I}(\Theta)$ and $\lambda(E) + \{x_0\} \in \mathscr{I}(\Theta_{[V+\{x_0\}]})$, we have by D12g that $\lambda(E) \in \mathscr{I}(\Theta)$.

Case 2: there exists E_1 such that $r(\bigcup_{e \in E_1} f(e)) = |E_1|$ and $\varnothing \subset E_1 \subset E$. Let $\Lambda_1 = (V_1, f_1, E_1) = \Lambda_{E_1}$. Let $V_2 = V + V_1$, $E_2 = E + E_1$, and $f_2(e) = f(e) \cap V_2$ for all $e \in E_2$. Let $\Lambda_2 = (V_2, f_2, E_2)$. By D12c, we may apply the induction hypothesis to Λ_1; there exists an LDR λ_1 of Λ_1 such that $\lambda_1[E_1] \in \mathscr{I}(\Theta_{V_1})$.

If $F \in \mathscr{P}(E_2)$, then by D12d,

$$\begin{aligned} r_{[V_2]}(\bigcup_{e \in F} f_2(e)) &= r(V + V_2 + \bigcup_{e \in F} f_2(e)) - r(V + V_2) \\ &= r([\bigcup_{e \in E_1} f(e)] + [\bigcup_{e \in F} f_2(e)]) - r(V_1) \\ &= r(\bigcup_{e \in E_1 + F} f(e)) - r(V_1) \\ &\geq |E_1 + F| - |E_1| = |F|. \end{aligned}$$

We apply the induction hypothesis to Λ_2; there exists an LDR λ_2 of Λ_2 such that $\lambda_2[E_2] \in \mathscr{I}(\Theta_{[V_2]})$. One easily verifies that the function λ given by: $\lambda(e) = \lambda_i(e)$ if $e \in E_i$ $(i = 1, 2)$ is an LDR of Λ. It follows from D12g that $\lambda[E] = \lambda[E_1] + \lambda[E_2] \in \mathscr{I}(\Theta)$. □

The above proposition affords us a characterization of a pair of systems admitting an LCR which is "nicer" than Proposition *V*D4. Observe in the next result that the subscripts 1 and 2 are clearly interchangeable.

E7 Corollary. *A necessary and sufficient condition for the pair of systems $\Lambda_i = (V, f_i, E_i)$ for $i = 1, 2$ with $|E_1| = |E_2|$ to admit an LCR is that*

$$r_1(\bigcup_{e \in F} f_2(e)) \geq |F|, \quad \textit{for all } F \in \mathscr{P}(E_2),$$

where r_1 is the rank function of the transversal matroid of Λ_1.

Proof. Let Θ be the transversal matroid of Λ_1. By the above proposition, the condition given in this corollary is equivalent to the existence of an LDR λ_2 of Λ_2 such that $\lambda_2[E_2] \in \mathscr{I}(\Theta)$, i.e., such that $\lambda[E_2]$ is a transversal of Λ_1. This is equivalent by definition to the existence of an LDR λ_1 of Λ_1 such that $\lambda_1[E_1] = \lambda_2[E_2]$. □

E8 *Exercise.* Derive Proposition $VD4$ from the above Corollary. [*Hint*: use Exercise E2.]

E9 *Exercise.* For $i = 1, 2$, let $\Lambda_i = (V, f_i, E_i)$ where $|E_1| = |E_2| = m$, and let r_i denote the rank function of the transversal matroid of Λ_i. Prove that a necessary and sufficient condition for the existence of an LCR for the systems Λ_1 and Λ_2 is that $r_1(S) + r_2(V + S) \geq m$ for all $S \subseteq V$.

E10 *Exercise.* Let $\Lambda_1, \ldots, \Lambda_n$ be systems with the same vertex set V. Let $\mathscr{T}_i$ be the collection of transversals of Λ_i $(i = 1, \ldots, n)$. Under the assumption that $(V, \bigcap_{i=1}^{n-1} \mathscr{T}_i)$ is an independence system, state and prove a necessary and sufficient condition for the existence of an LCR $(\lambda_1, \ldots, \lambda_n)$ for the systems $\Lambda_1, \ldots, \Lambda_n$.

The above exercise should give some insight into the difficulty in formulating a "nice" necessary and sufficient condition for the existence of an LCR for more than two systems. The condition that the intersection of the transversals of all but one of the systems form an independence system is indeed rather strong. Unfortunately it appears indispensible. Let us look at a simple case.

Example. Let $V = \{x_0, x_1, x_2, x_3\}$. For $i = 1, 2, 3$, let $E_i = \{x_0, x_i\}$, let $\mathscr{E}_i = \{E_i, V + E_i\}$, and let Λ_i be the set system $(V, \mathscr{E}_i)$. One easily verifies that the pair of systems Λ_1 and Λ_2 admits exactly two LCR's (λ_1, λ_2) and (μ_1, μ_2), where $\lambda_i(E_i) = x_0$, $\lambda_i(V + E_i) = x_3$ and $\mu_1(E_1) = x_1 = \mu_2(V + E_2)$, $\mu_1(V + E_1) = x_2 = \mu_2(E_2)$. The largest transversals common to Λ_1 and Λ_2 are $T_0 = \{x_0, x_3\}$ and $T_1 = \{x_1, x_2\}$. We note that $(V, \{T_0, T_1\})$ is not an independence system since Condition B2 fails. Furthermore, neither T_0 nor T_1 is a transversal of Λ_3. By the symmetry of this example, it is clear that any two of Λ_1, Λ_2, and Λ_3 admit an LCR, but there exists no LCR for all three systems.

Having considered a class of matroids closely related to the Phillip Hall Theorem, namely transversal matroids, we turn to a class of matroids related to the Menger Theorem.

E11 Proposition. *Let (V, D) be a directed graph, let $Z \subseteq V$, and let*

$$\mathscr{A} = \{A \in \mathscr{P}(V)\text{: there exists a } |A|\text{-family of pairwise-disjoint } AZ\text{-paths}\}.$$

Then $(V, \mathscr{A})$ is an independence system, and the matroid corresponding to it has rank $|Z|$.

Proof. Condition B1 is satisfied by $(V, \mathscr{A})$, since clearly any subset of a set in $\mathscr{A}$ also belongs to $\mathscr{A}$. Clearly if $A \in \mathscr{A}$, then no edge of an AZ-path can be of the form (z, x) where $z \in Z$. Without loss of generality we may assume that (V, D) admits no such edges. It follows that a vertex in Z can only be a terminal vertex of a path in (V, D).

Suppose now that when $\mathscr{A}$ is defined in terms of some set $Z \subseteq V$, then $(V, \mathscr{A})$ fails to satisfy B2. Clearly $|Z| > 1$, for if $|Z| = 1$, then $\mathscr{A} \subseteq \mathscr{P}_1(V)$ and B2 holds vacuously. We may therefore assume that $|Z|$ is minimal with respect to the condition that $(V, \mathscr{A})$ does not satisfy B2. It follows that for some sets $A_1, A_2 \in \mathscr{A}$ with $|A_1| < |A_2|$, we have that $A_1 + \{x\} \notin \mathscr{A}$ for all $x \in A_2 + (A_1 \cap A_2)$.

Suppose that $A_1 \cap Z \neq \varnothing$. Let $Z' = Z + (A_1 \cap Z)$. Then $0 < |Z'| \leq |Z|$. We define

$$\mathscr{A}' = \{A \in \mathscr{P}(V)\text{: there exists a } |A|\text{-family of pairwise-disjoint } AZ'\text{-paths}\}.$$

Letting $A_1' = A_1 + (A_1 \cap Z)$, we have $A_1' \in \mathscr{A}'$. Consider a $|A_2|$-family of pairwise-disjoint A_2Z-paths, and let A_2' be the set of initial vertices of those paths in the family which terminate in Z'. Clearly $A_2' \in \mathscr{A}'$, and $|A_2'| \geq |A_2| - |A_1 \cap Z| > |A_1| - |A_1 \cap Z| = |A_1'|$. By the minimality of $|Z|$, there exists $x \in A_2' + (A_1' \cap A_2')$ such that $A_1' + \{x\} \in \mathscr{A}'$. Since no $(A_1' + \{x\})Z'$-path can meet $Z + Z'$, it is immediate that $A_1 + \{x\} \in \mathscr{A}$, contrary to assumption. Hence $A_1 \cap Z = \varnothing$.

For each $A \in \mathscr{A}$, let us define $d(A)$ to be the minimum, taken over all $|A|$-families of pairwise-disjoint AZ-paths, of the sum of the lengths of the paths in the family. We may assume without loss of generality that if $|A| < |A_2|$ and if $A + \{x\} \notin \mathscr{A}$ for all $x \in A_2 + (A \cap A_2)$, then $d(A) \geq d(A_1)$. Since $A_1 \cap Z = \varnothing$, we have $d(A_1) \geq |A_1|$.

Since $|A_1| < |A_2|$, we note that A_1 does not separate Z from A_2; i.e., there exists an A_2Z-path which avoids A_1. Let w denote the initial vertex of such a path, and note that $A_1 + \{w\} \notin \mathscr{A}$. Let us write $m = |A_1|$. If $w \notin Z$, then $(A_1 + \{w\}) \cap Z = \varnothing$ and we may invoke Dirac's generalization of the Menger Theorem, namely Theorem *VI*B2, to obtain a set $T \subseteq V \cup D$ such that $|T| < m + 1$ and T separates Z from $A_1 + \{w\}$. If, on the other hand, $w \in Z$, then there exists no m-family of pairwise-disjoint $A_1(Z + \{w\})$-paths. Again by *VI*B2, there exists a set $T' \subseteq V \cup D$ such that $|T'| < m$ and T' separates $Z + \{w\}$ from A_1. In this case, let $T = T' + \{w\}$. In either case, $|T| \leq m$ and $T \neq A_1$. We define

$$S = (T \cap V) \cup \{v \in V\text{: } (u, v) \in T \cap D \text{ for some } u \in V\}.$$

Thus $S \subseteq V$, $|S| \leq |T|$, and S separates Z from $A_1 + \{w\}$. Since S also separates Z from A_1 and since $A_1 \in \mathscr{A}$, we have in fact that $|S| = |T| = |A_1| = m$.

Let $\Pi_1, \ldots, \Pi_m$ be an m-family of pairwise-disjoint A_1Z-paths, and let s_i be the first vertex of S encountered on Π_i when one proceeds from A_1 $(i = 1, \ldots, m)$. By considering the portions of these paths between S and Z, we conclude that $S \in \mathscr{A}$. Moreover, since $S \neq A_1$, we have $d(S) < d(A_1)$. By our assumption of the minimality of $d(A_1)$, we infer that $S + \{x\} \in \mathscr{A}$ for some $x \in A_2 + (S \cap A_2)$. Clearly $x \in A_2 + (A_1 \cap A_2)$. There exists an $(m + 1)$-family of pairwise-disjoint $(S + \{x\})Z$-paths. Let us denote them by

$\Sigma_1, \ldots, \Sigma_{m+1}$. Since $|S| = m$, we may suppose that s_i is the initial vertex of Σ_i for $i = 1, \ldots, m$ and that x is the initial vertex of Σ_{m+1}.

Let $\Pi_1', \ldots, \Pi_m'$ be the m-family of A_1S-paths obtained by truncating the paths $\Pi_1, \ldots, \Pi_m$, respectively. We assert that if Π_i' and Σ_j have a common vertex, then $i = j$ and that vertex must be s_i. For if they shared some other common vertex $t \notin S$, then consider the path formed by proceeding from A_1 along Π_i' to t and then following Σ_j from t to Z. This path would be an A_1Z-path which avoids S.

We may now construct an $(m + 1)$-family of pairwise-disjoint $(A_1 + \{x\})Z$-paths as follows. For $i = 1, \ldots, m$, form the path consisting of Π_i' followed by Σ_i. (These are joined at s_i.) To this family add the path Σ_{m+1}. Hence $A_1 + \{x\} \in \mathscr{A}$.

Since $Z \in \mathscr{A}$ while no element of $\mathscr{A}$ has cardinality greater than $|Z|$, it is clear that the matroid Λ with $\mathscr{I}(\Lambda) = \mathscr{A}$ has rank $|Z|$. □

Let (V, D) be a directed graph, and let $Z \subseteq V$. The matroid $\Lambda = (V, \mathscr{M})$ such that $\mathscr{I}(\Lambda) = \{A \in \mathscr{P}(V)$: there exists a $|A|$-family of pairwise-disjoint AZ-paths$\}$ is called a **strict gammoid**, and any submatroid Λ_U for $U \subseteq V$ is called a **gammoid**. Thus $\Theta = (U, \mathscr{N})$ is a gammoid if $\mathscr{I}(\Theta) = \{A \in \mathscr{P}(U)$: there exists an $|A|$-family of pairwise disjoint AZ-paths in $(V, D)\}$; this follows from Exercise D12a. This section concludes with some results relating transversal matroids and gammoids, and so, in a sense, relating the Philip Hall Theorem and the Menger Theorem.

E12 Lemma. *Every transversal matroid is a gammoid.*

Proof. Let Θ be a transversal matroid. By Proposition E5, Θ is the transversal matroid of a system $\Lambda = (V, f, E)$, where $|E| = r(\Theta)$. Consider the directed graph $(V \cup E, D)$, where $D = \{(x, e): x \in f(e)\}$. Let $\mathscr{A} = \{U \in \mathscr{P}(V)$: there exists a $|U|$-family of pairwise-disjoint UE-paths in $(V \cup E, D)\}$. By definition, the matroid $\Phi = (V, \mathscr{M})$ such that $\mathscr{I}(\Phi) = \mathscr{A}$ is a gammoid, and by Proposition E11, $r(\Phi) = |E| = r(\Theta)$. We leave it to the reader to complete the verification of the fact that $\Theta = \Phi$. □

E13 Theorem. *Let Θ be a matroid. Then Θ is a transversal matroid if and only if $\Theta^\perp$ is a strict gammoid.*

Proof. By Proposition E5, we may suppose that Θ is the transversal matroid of a system $\Lambda = (V, f, E)$ and that $|E| = r(\Theta)$. Hence Λ admits an LDR λ, and the image $B = \lambda[E]$ is a basis of Θ. We form the directed graph (V, D), where $D = \{(x, y): x \in B; y \in (V + B) \cap f(\lambda^{-1}(x))\}$. From (V, D) we form the strict gammoid Φ such that $\mathscr{I}(\Phi) = \{A \in \mathscr{P}(V)$: there exists a $|A|$-family of pairwise-disjoint $A(V + B)$-paths$\}$. We will show that $\Theta^\perp = \Phi$ by showing that the two matroids have the same bases.

Suppose $V + A \in \mathscr{B}(\Theta^\perp)$, or equivalently (cf. D9d) that $A \in \mathscr{B}(\Theta)$. Then $A = \kappa[E]$ for some LDR κ of Λ, and we consider the bijection $\beta = \kappa\lambda^{-1}: B \to A$. Since $\beta(x) \in f(\lambda^{-1}(x))$ for all $x \in B$, we have that either $\beta(x) = x$ or $(x, \beta(x)) \in D$. Proposition *VI*B7 implies the existence in (V, D) of a $|V + A|$-family of pairwise-disjoint $(V + A)(V + B)$-paths, and so $V + A \in \mathscr{B}(\Phi)$. On the other hand, if $V + A \in \mathscr{B}(\Phi)$, retracing our argument in the reverse direction yields a bijection $\beta: B \to A$ such that $\beta(x) \in f(\lambda^{-1}(x))$ for all $x \in B$. Let $\kappa = \beta\lambda$. Then $\kappa: E \to A$ is a bijection, and for each $e \in E$, $\kappa(e) = \beta(\lambda(e)) \in f(\lambda^{-1}(\lambda(e))) = f(e)$. Hence κ is an LDR of Λ. Since $|E| = r(\Theta)$, A is a largest transversal of Θ, whence $A \in \mathscr{B}(\Theta)$. Thus $V + A \in \mathscr{B}(\Theta^\perp)$.

Conversely, suppose that the matroid Φ is a strict gammoid formed from a directed graph (V, D) and a set $Z \subseteq V$, as in the definition of a gammoid. Let E be a $|V + Z|$-set disjoint from all other sets considered thus far, and let $\lambda: E \to V + Z$ be a bijection. We form the system $\Lambda = (V, f, E)$ by defining

$$f(e) = \{x \in Z: (\lambda(e), x) \in D\} \cup \{\lambda(e)\}.$$

Let Θ be the transversal matroid of Λ. We will show that $\Theta^\perp = \Phi$.

By our construction, λ is clearly an LDR of Λ, and $V + Z = \lambda[E]$ is a largest transversal of Λ. Let

$$D_1 = \{(x, y) \in D: x \in V + Z; y \in Z \cap f(\lambda^{-1}(x))\}.$$

The directed graph (V, D_1) and the set Z yield a strict gammoid, and by the argument of the first half of this proof, this strict gammoid is none other than $\Theta^\perp$. A close look at our terminology yields that for $x \in V + Z$, $f(\lambda^{-1}(x)) = \{y \in Z: (x, y) \in D\} \cup \{x\}$, and so $D_1 = \{(x, y) \in D: x \in V + Z\}$. We see that no edge of (V, D) coming from $D + D_1$ can occur in any path in a $|Z|$-family of pairwise-disjoint AZ-paths, where $A \in \mathscr{P}_{|Z|}(V)$. It follows that (V, D) and (V, D_1), with the same set Z, yield the same gammoid. □

E14 *Exercise.* Show that
(a) any minor of a gammoid is a gammoid;
(b) any gammoid may be obtained from a transversal matroid by successive operations of taking minors or taking orthogonal complements.

XF Representability

Let $\Lambda = (V, \mathscr{M})$ be a matroid, and let $\mathbb{F}$ be a field. A **Whitney function** for Λ over $\mathbb{F}$ is a function $w: V \to \mathscr{V}$, where $\mathscr{V}$ is a vector space over $\mathbb{F}$, such that for each set $U \in \mathscr{P}(V)$, $w[U]$ is a basis for $\mathscr{V}$ if and only if $U \in \mathscr{B}(\Lambda)$. Equivalently, w is a Whitney function if $w[V]$ spans $\mathscr{V}$ and $w[U]$ is independent in $\mathscr{V}$ if and only if $U \in \mathscr{I}(\Lambda)$.

F1 *Example.* If $\Lambda = (E, \mathscr{M}(\mathscr{Z}(\Gamma)))$ is the cycle matroid of a multigraph Γ, then the function $w: E \to \mathscr{P}(E)/\mathscr{Z}(\Gamma)$, where $w(e)$ is the coset $\{e\} + \mathscr{Z}(\Gamma)$ for each $e \in E$, is a Whitney function. This is shown by the following sequence of equivalent statements. Let $F \subseteq E$.

$$
\begin{aligned}
F \in \mathscr{I}(\Lambda) &\Leftrightarrow \Gamma_F \text{ is a subforest of } \Gamma. \\
&\Leftrightarrow \text{If } G \subseteq F \text{ and } G \in \mathscr{Z}(\Gamma), \text{ then } G = \varnothing. \\
&\Leftrightarrow \text{If } \textstyle\sum_{e \in G} (\{e\} + \mathscr{Z}(\Gamma)) = \mathscr{Z}(\Gamma), \text{ then } G = \varnothing. \\
&\Leftrightarrow \text{If } \textstyle\sum_{e \in F} a_e(\{e\} + \mathscr{Z}(\Gamma)) = \mathscr{Z}(\Gamma), \text{ then } a_e = 0 \text{ for all } e \in F. \\
&\Leftrightarrow \{\{e\} + \mathscr{Z}(\Gamma) \colon e \in F\} \text{ is independent in } \mathscr{P}(E)/\mathscr{Z}(\Gamma). \\
&\Leftrightarrow w[F] \text{ is independent in } \mathscr{P}(E)/\mathscr{Z}(\Gamma).
\end{aligned}
$$

Moreover,

$$
\begin{aligned}
\dim(\mathscr{P}(E)/\mathscr{Z}(\Gamma)) &= \dim(\mathscr{P}(E)) - \dim(\mathscr{Z}(\Gamma)) \\
&= \nu_1(\Gamma) - (\nu_1(\Gamma) - \nu_0(\Gamma) + \nu_{-1}(\Gamma)) = r(\Lambda)
\end{aligned}
$$

by B27. Hence w maps bases only onto bases.

F2 *Exercise.* Given a vector space $\mathscr{V}$, a set V, and any function $w \in \mathscr{V}^V$, show that w is a Whitney function for some matroid $\Lambda = (V, \mathscr{M})$ and that Λ is unique up to isomorphism.

Again let $\Lambda = (V, \mathscr{M})$ be a matroid and let $\mathbb{F}$ be a field. A **Tutte subspace** for Λ over $\mathbb{F}$ is a subspace $\mathscr{T} \subseteq \mathbb{F}^V$ such that $\mathscr{M} = \mathscr{M}(\sigma[\mathscr{T}])$, i.e., $\mathscr{M}$ is the collection of minimal, nonempty subsets of V which are supports of functions in $\mathscr{T}$. This concept can be illustrated as well by Example F1, using the fact (cf. *I*B2) that the support function is an isomorphism from $\mathbb{K}^E$ to $\mathscr{P}(E)$.

F3 Lemma. *Let $\Lambda = (V, \mathscr{M})$ be a matroid. Let $\mathbb{F}$ be a field and let $\mathscr{T}$ be a subspace of $\mathbb{F}^V$. Let the function $w\colon V \to \mathbb{F}^V/\mathscr{T}$ be given by $w(x) = t_x + \mathscr{T}$, where $t_x(y) = 0$ if $y \neq x$ and $t_x(x) = 1$* (cf. *I*B1 and *I*B4). *Then $\mathscr{T}$ is a Tutte subspace for Λ if and only if w is a Whitney function for Λ.*

Proof. Suppose that $\mathscr{T}$ is a Tutte subspace for Λ. Since the image $w[V]$ spans $\mathbb{F}^V/\mathscr{T}$, one proves that w is a Whitney function for Λ by a sequence of equivalent statements identical to that of Example F2 above, except that V replaces E, $\mathbb{F}^V$ replaces $\mathscr{P}(E)$, and $\mathscr{T}$ replaces $\mathscr{Z}(\Gamma)$.

Conversely, suppose that w is a Whitney function for Λ, and let $M \in \mathscr{M}$. Then $M + \{x\} \in \mathscr{I}(\Lambda)$ for all $x \in M$. Hence $w[M + \{x\}]$ is an independent set of vectors in $\mathbb{F}^V/\mathscr{T}$. This means that for any indexed set $\{a_y \colon y \in M + \{x\}\}$ of elements of $\mathbb{F}$, $\sum_{y \in M + \{x\}} a_y t_y \in \mathscr{T}$ if and only if $a_y = 0$ for all $y \in M + \{x\}$. This shows that no proper subset of M is the support of a function in $\mathscr{T}$. On the other hand, since $M \notin \mathscr{I}(\Lambda)$, the definition of w and the previous argument imply that $\sum_{y \in M} a_y t_y \in \mathscr{T}$ for some indexed set $\{a_y \colon y \in M\}$ of *nonzero* elements of $\mathbb{F}$. Thus M is the support of $\sum_{y \in M} a_y t_y$. □

Let $w\colon V \to \mathscr{V}$ be a Whitney function. If $\mathscr{T}$ is the kernel of the extension by linearity of w to $\mathbb{F}^V \to \mathscr{V}$, i.e., $w(f) = \sum_{x \in V} f(x) w(x)$ for $f \in \mathbb{F}^V$, then this lemma may be formulated more succinctly as follows:

F4 Proposition. *A matroid admits a Tutte subspace over* $\mathbb{F}$ *if and only if it admits a Whitney function over* $\mathbb{F}$.

If a matroid admits a Tutte subspace over $\mathbb{F}$ or, equivalently, a Whitney function over $\mathbb{F}$, then it is said to be **representable** over $\mathbb{F}$. If $\mathscr{T}$ is a Tutte subspace of the matroid Λ and if $w\colon V \to \mathbb{F}^V/\mathscr{T}$ is the corresponding Whitney function, then $r(\Lambda) = \dim(\mathbb{F}^V/\mathscr{T}) = |V| - \dim(\mathscr{T})$. Thus

F5 $$r(\Lambda) + \dim(\mathscr{T}) = |V|.$$

If V is a set and $\mathscr{A}$ is a subspace of $\mathbb{F}^V$, then we write $\mathscr{A}^\perp$ to indicate the orthogonal complement of $\mathscr{A}$ under the standard inner product, i.e.,

$$\mathscr{A}^\perp = \{g \in \mathscr{V} : \sum_{x \in V} f(x)g(x) = 0 \text{ for all } f \in \mathscr{A}\}.$$

The inner product of §*II*A is the special case of this notion when $\mathbb{F} = \mathbb{K}$. Equation *II*A6 still applies.

F6 Proposition. *Let* $\Lambda = (V, \mathscr{M})$ *be a matroid which is representable over* $\mathbb{F}$.

(a) *If* $\mathscr{T} \subseteq \mathbb{F}^V$ *is a Tutte subspace for* Λ, *then* $\mathscr{T}^\perp$ *is a Tutte subspace for* $\Lambda^\perp$.

(b) *Every minor of* Λ *is representable over* $\mathbb{F}$.

PROOF. (a) Let $\mathscr{T}$ be a Tutte subspace for Λ and let $\mathscr{N}$ be the set of minimal nonempty supports of functions in $\mathscr{T}^\perp$. Following Example B18, $\Theta = (V, \mathscr{N})$ is a matroid, and $\mathscr{T}^\perp$ is a Tutte subspace for Θ.

Let $M \in \mathscr{M}$ and $N \in \mathscr{N}$. Then $M = \sigma(f)$ and $N = \sigma(g)$ for some $f \in \mathscr{T}$ and $g \in \mathscr{T}^\perp$. If there exisits $y \in M \cap N$, then $0 = \sum_{x \in V + \{y\}} f(x)g(x) + f(y)g(y)$. Thus $f(x)g(x) \neq 0$ for some $x \in V + \{y\}$, which implies that $|M \cap N| > 1$. Thus $\mathscr{N} \subseteq \{A \in \mathscr{P}(V) : |A \cap M| \neq 1 \text{ for all } M \in \mathscr{M}\}$. By Proposition D5, every set in $\mathscr{N}$ is a union of sets in $\mathscr{M}^\perp$. We have by F5 and D2b that $r(\Theta) = |V| - \dim(\mathscr{T}^\perp) = \dim(\mathscr{T}) = |V| - r(\Lambda) = r(\Lambda^\perp)$, whence $\Theta = \Lambda^\perp$ by Lemma B29.

(b) If w is a Whitney function for Λ and $U \subseteq V$, we assert that $w_{|U}$ is a Whitney function for Λ_U. This is clear since by Exercise D12a, a set $J \in \mathscr{I}(\Lambda_U)$ if and only if $J \in \mathscr{I}(\Lambda)$ and $J \subseteq U$, which is equivalent to saying that $w_{|U}[J] = w[J]$ is independent in $\mathbb{F}^U$. Hence Λ_U is representable over $\mathbb{F}$. This result and part (a) above imply that $((\Lambda^\perp)_U)^\perp$ is representable over $\mathbb{F}$. Hence by Proposition D13a, $\Lambda_{[U]}$ is representable over $\mathbb{F}$. It follows that any minor of a matroid representable over $\mathbb{F}$ is representable over $\mathbb{F}$. □

We have shown that if a matroid Λ is representable over $\mathbb{F}$, then so is $\Lambda^\perp$. A matroid is said to be **regular** if it is representable over every field. We therefore have

F7 Corollary. *A matroid* Λ *is regular if and only if* $\Lambda^\perp$ *is regular.*

F8 Exercise. Show that *the two Fano matroids are not regular by showing that they are not representable over* $\mathbb{Q}$.

F9 Proposition. *The cycle and cocycle matroids of any multigraph are regular.*

Proof. Let $\mathbb{F}$ be a field and let $\Gamma = (V, f, E)$ be a multigraph. For each $x \in V$, we choose a function $i_x\colon E \to \mathbb{F}$ satisfying

$$\begin{aligned} i_x(e) &= 0 \quad \text{if } x \notin f(e);\\ i_x(e) &= \pm 1 \quad \text{if } x \in f(e);\\ i_x(e)i_y(e) &= -1 \quad \text{if } f(e) = \{x, y\}. \end{aligned}$$

The function i_x for $x \in V$ in effect orients the edges in the vertex cocycle of x.

Let $\varphi\colon \mathbb{F}^E \to \mathbb{F}^V$ be defined so that for each $h \in \mathbb{F}^E$, its image $\varphi(h)$ is given by $\varphi(h)(x) = \sum_{e \in E} i_x(e)h(e)$. It is immediate that φ is a linear transformation. Let $\mathscr{T}$ denote the kernel of φ. We will show that the minimal nonempty supports of functions in $\mathscr{T}$ are precisely the elementary cycles of Γ; that is, we show that $\mathscr{T}$ is a Tutte subspace over $\mathbb{F}$ for $(E, \mathscr{M}(\mathscr{Z}(\Gamma)))$.

Let Z be an elementary cycle of Γ where Γ_Z is the elementary circuit $x_0e_1x_1e_2\ldots e_nx_n = x_0$. Let us define

$$h(e) = \begin{cases} 1 & \text{if } e = e_1;\\ -i_{x_j}(e_j)h(e_j)i_{x_j}(e_{j+1}) & \text{if } e = e_{j+1}\ (j = 1, \ldots, n-1);\\ 0 & \text{otherwise.} \end{cases}$$

Clearly $\sigma(h) = Z$, and one can verify that h is well-defined and that $\varphi(h)(x) = 0$ for all $x \in V$. Thus $h \in \mathscr{T}$. That Z is a minimal, nonempty support follows since Z is an elementary cycle. If $h \in \mathscr{T}$, then since $\varphi(h) = 0$, we must have $\check{\rho}(\Gamma_{\sigma(h)}) \geq 2$. Thus $\Gamma_{\sigma(h)}$ contains an elementary circuit, by *III*A6a. Hence $\sigma(h)$ contains an elementary cycle. □

It was remarked in the above proof that the functions i_x serve to orient the edges with which x is incident. On this "oriented multigraph" the function φ is in effect a boundary operator (cf. §*IV*B). Its kernel, the Tutte subspace, looks like a flow space. Indeed if $\mathbb{F} = \mathbb{Q}$ and if Γ is a graph, then this is precisely what happens. Let us sketch how the notions of §*IV*A and §*IV*B can be exploited to prove the more limited result, that the cycle and cocycle matroids of a graph $\Theta = (V, \mathscr{E})$ are realizable over $\mathbb{Q}$.

Letting $W = (V \times V) + \{(x, x)\colon x \in V\}$ as in §*IV*B, we orient the edges by defining $i\colon \mathscr{E} \to W$ so that if $E = \{x, y\} \in \mathscr{E}$, then $i(E) \in \{(x, y), (y, x)\}$. The function i induces a natural imbedding of $\mathbb{Q}^{\mathscr{E}}$ in $\mathbb{Q}^W$ when the elements of $\mathbb{Q}^{\mathscr{E}}$ are extended by zero. Abusing notation, let $\mathbb{Q}^{\mathscr{E}}$ also denote this subspace of $\mathbb{Q}^W$, and let $\mathscr{T} = \mathbb{Q}^{\mathscr{E}} \cap \ker(\partial)$, where ∂ is the boundary operator (§*IV*B). By Exercise *IV*A18 and Proposition *IV*B8, $\mathscr{T}$ is a unimodular subspace, and a function $h \in \mathscr{T}$ is elementary if and only if it assumes values of ± 1 on some elementary cycle of Θ and the value 0 elsewhere. Thus $\mathscr{M}(\sigma[\mathscr{T}]) = \mathscr{M}(\mathscr{Z}(\Theta))$. Actually we have shown that $(\mathscr{E}, \mathscr{M}(\mathscr{Z}(\Theta)))$ has a *unimodular* Tutte subspace over $\mathbb{Q}$. By Proposition F6a and *IV*A19, the cocycle matroid of Θ also has a unimodular subspace over $\mathbb{Q}$.

The next exercise shows that when $\mathbb{Q}$ is replaced by an arbitrary field $\mathbb{F}$

in Proposition *IVA*6 and if $\mathscr{S}$ is a subspace of $\mathbb{F}^V$, then any function $h \in \mathscr{S}$ has a decomposition satisfying conditions (a) and (b) of *IVA*6. If $\mathbb{F}$ is an ordered field, then condition (c) also holds, but we shall not require this fact.

F10 Exercise. *Let $\mathscr{S}$ be a subspace of $\mathbb{F}^V$ and let $h \in \mathscr{S}$.* Prove that $h = \sum_{i=1}^m h_i$ *where for $i = 1, \ldots, m$, $h_i \in \mathscr{S}$* and:

(**a**) $\sigma(h_i) \in \mathscr{M}(\{\sigma(g): g \in \mathscr{S}\}) = \mathscr{M}(\sigma[\mathscr{S}])$;
(**b**) $\sigma(h_i) \subseteq \sigma(h)$.

Realizability of a matroid over $\mathbb{Q}$ is *almost* a sufficient condition for regularity, as we now see.

F11 Proposition. *If some Tutte subspace over $\mathbb{Q}$ for the matroid Λ is unimodular, then Λ is regular.*

Proof. Let $\mathscr{T}$ be a unimodular Tutte subspace over $\mathbb{Q}$ for $\Lambda = (V, \mathscr{M})$ and let $\mathbb{F}$ be an arbitrary field. By Proposition *IVA*19, $\mathscr{T}^\perp$ is a unimodular subspace of $\mathbb{Q}^V$, and by F6a, $\mathscr{T}^\perp$ is a Tutte subspace over $\mathbb{Q}$ for $\Lambda^\perp$. We consider a mapping $h \mapsto \bar{h}$ from the set $\{h \in \mathbb{Q}^V: h[V] \subseteq \{-1, 0, 1\}\}$ into $\mathbb{F}^V$ (cf. §*IVA*) whereby $h(x) = \bar{h}(x)$ for all $x \in V$; i.e., the integers $-1, 0, 1$ are reinterpreted as elements of $\mathbb{F}$, with the understanding that $-1 = 1$ if $\mathbb{F}$ has characteristic 2. Clearly $\sigma(h) = \sigma(\bar{h})$. Let $\mathscr{S}_1$ denote the subspace of $\mathbb{F}^V$ spanned by the set

$$\{\bar{h} \in \mathbb{F}^V: h \in \mathscr{T};\ h \text{ is elementary}\},$$

and let $\mathscr{S}_2$ be the subspace spanned by

$$\{\bar{h} \in \mathbb{F}^V: h \in \mathscr{T}^\perp;\ h \text{ is elementary}\}.$$

Following Example B18, the set systems $\Theta_j = (V, \mathscr{M}(\sigma[\mathscr{S}_j]))$ for $j = 1, 2$ are matroids.

If $M \in \mathscr{M}$, then $M = \sigma(h)$ for some $h \in \mathscr{T}$. By the definition we may assume that h is elementary. Hence $\bar{h} \in \mathscr{S}_1$, and by Exercise F10, $M = \sigma(\bar{h})$ is a union of cycles of Θ_1. In particular if a subset of V contains a cycle of Λ, then it contains a cycle of Θ_1. Equivalently, every independent set of Θ_1 is an independent set of Λ, whence

F12 $$r(\Theta_1) \leq r(\Lambda).$$

By a similar argument, by picking an elementary function in $\mathscr{T}^\perp$, we deduce

F13 $$r(\Theta_2) \leq r(\Lambda^\perp).$$

Let $h_1 \in \mathscr{T}$ and $h_2 \in \mathscr{T}^\perp$ be elementary. Then $\sum_{x \in V} h_1(x)h_2(x) = 0$, and so $\sum_{x \in V} \bar{h}_1(x)\bar{h}_2(x) = 0$. By the definitions of $\mathscr{S}_1$ and $\mathscr{S}_2$, it follows that $\mathscr{S}_2 \subseteq \mathscr{S}_1^\perp$. Hence

F14 $$\dim(\mathscr{S}_2) \leq \dim(\mathscr{S}_1^\perp).$$

Combining successively F12, D2b, F13, F5, and F14, we obtain:

$$\begin{aligned} r(\Theta_1) \leq r(\Lambda) = |V| - r(\Lambda^\perp) \leq |V| - r(\Theta_2) = \dim(\mathscr{S}_2) \leq \dim((\mathscr{S}_1)^\perp) \\ = |V| - \dim(\mathscr{S}_1) = r(\Theta_1), \end{aligned}$$

whence $r(\Theta_1) = r(\Lambda)$. By our previous remarks, we may invoke Lemma B29, concluding that $\Lambda = \Theta_1$. Thus Λ is representable over $\mathbb{F}$. □

F15 Exercise. *Let* $\Lambda = (V, \mathscr{M})$ *be a matroid and suppose that* $\Lambda = \bigoplus_{i=1}^{k} \Lambda_i$, *where* $\Lambda_i = (V_i, \mathscr{M}_i)$ *for* $i = 1, \ldots, k$. Prove that

(a) *If* $\mathscr{T}$ *is a Tutte subspace over* $\mathbb{F}$ *for* Λ, *and* $U \subseteq V$, *then* $\{h_{|U} : h \in \mathscr{T}\}$ *is a Tutte subspace over* $\mathbb{F}$ *for* $\Lambda_{[U]}$. *In particular, if* $U = V_i$, *then* $\{h_{|U} : h \in \mathscr{T}\} = \mathscr{T} \cap \mathbb{F}^{V_i}$.

(b) *If* $\mathscr{T}_i$ *is a Tutte subspace over* $\mathbb{F}$ *for* Λ_i $(i = 1, \ldots, k)$, *then* $\bigoplus_{i=1}^{k} \mathscr{T}_i$ *is a Tutte subspace over* $\mathbb{F}$ *for* Λ.

(c) *If* $\mathscr{T}_i$ *is a subspace of* $\mathbb{Q}^{V_i}$ $(i = 1, \ldots, k)$, *then* $\bigoplus_{i=1}^{k} \mathscr{T}_i$ *is unimodular if and only if* $\mathscr{T}_i$ *is unimodular, for* $i = 1, \ldots, k$. (Cf. Exercise *IV*A18.)

Of particular interest is the question of whether a given matroid Λ is representable over the field $\mathbb{K}$. If so, then Λ is called a **binary matroid**. We have seen in Example F1 that the cycle matroid of a multigraph is a binary matroid. By Proposition F6a (even without Proposition F9), the cocycle matroid of a multigraph is therefore binary, too. Let us first consider two easy characterizations of binary matroids.

F16 Proposition. *A matroid* $\Lambda = (V, \mathscr{M})$ *is binary if and only if* $\mathscr{M} = \mathscr{M}(\mathscr{A})$ *for some subspace* $\mathscr{A}$ *of* $\mathscr{P}(V)$.

Proof. This result is an immediate consequence of the definition and the fact (*I*B2 and *I*B3) that the support function $\sigma: \mathbb{K}^V \to \mathscr{P}(V)$ is a vector space isomorphism. □

F17 Corollary. *Let* $\Lambda = (V, \mathscr{M})$ *be a matroid.*

(a) *If* Λ *is binary and* $M_1, \ldots, M_k \in \mathscr{M}$, *then* $\sum_{i=1}^{k} M_i$ *can be expressed as the sum of pairwise-disjoint cycles of* Λ.

(b) *Suppose that* $M_1 + M_2$ *can be expressed as the sum of pairwise-disjoint cycles of* Λ *whenever* $M_1, M_2 \in \mathscr{M}$. *Then* Λ *is binary.*

The proof is a restatement of various properties of the vector space $(\mathscr{P}(V), +)$ over $\mathbb{K}$; the details are left to the reader.

A more interesting characterization of binary matroids will now be given in terms of a particular "forbidden" minor.

F18 *Example.* Let V be a 4-set. Then the complete matroid $(V, \mathscr{P}_3(V))$ is not binary. This is evident by the criterion of Corollary F17a. This particular matroid of rank 2 will be denoted by Θ_2. One easily checks that every matroid with 3 or fewer vertices is binary.

F19 Theorem (W. T. Tutte [t.6]). *A matroid is binary if and only if no minor is isomorphic to* Θ_2.

Proof. By Proposition F6b, every minor of a binary matroid is also binary. Since Θ_2 is not binary, it cannot be a minor of a binary matroid.

Given a matroid which is not binary, some minor $\Lambda = (V, \mathscr{M})$ is also not binary. We choose Λ so that $|V|$ is as small as possible. Thus every proper minor of Λ is a binary matroid. By Corollary F17b, there exist cycles M_1, $M_2 \in \mathscr{M}$ such that $M_1 + M_2$ cannot be expressed as a sum of pairwise-disjoint cycles of Λ. In particular, $M_1 + M_2 \notin \mathscr{M}$. Since M_1 and M_2 are also cycles of $\Theta_{M_1 \cup M_2}$, it follows that $\Theta_{M_1 \cup M_2}$ cannot be a binary matroid. By the minimality of $|V|$, we have $M_1 \cup M_2 = V$. By our assumption, $M_1 \cap M_2 \neq \varnothing$.

We first show that $M_1 + M_2 \in \mathscr{I}(\Lambda)$. For suppose that $M \subseteq M_1 + M_2$ for some $M \in \mathscr{M}$. Then $M \subset M_1 + M_2$ since $M_1 + M_2 \notin \mathscr{M}$. Hence for some $i = 1, 2$, we have $M \cup M_i \neq V$; let us say $M \cup M_2 \subset V$. Since $\Lambda_{M \cup M_2}$ is binary, we have by Corollary F17a that $M + M_2$ can be expressed as the sum of pairwise-disjoint cycles $N_1, \ldots, N_k \in \mathscr{M}$, and $N_i \subseteq M + M_2$ for $i = 1, \ldots, k$. Clearly $M_1 + (M + M_2) = M_1 + N_1 + \ldots + N_k$, and $M_1 \cup N_1 \cup \ldots \cup N_k = M_1 \cup (M + M_2) \subset V$, since M must contain at least one element of $M_2 + (M_1 \cap M_2)$. Applying F17a again, we have that $M_1 + N_1 + \ldots + N_k$ is a sum of pairwise-disjoint cycles $N_1', \ldots, N_{k'}' \in \mathscr{M}$. Hence

$$M_1 + M_2 = N_1' + \ldots + N_{k'}' + M.$$

But $M \cap N_i' = \varnothing$ for $i = 1, \ldots, k'$, contrary to our initial assumption concerning $M_1 + M_2$.

Let $x \in M_1 \cap M_2$. Thus $\Lambda_{[V + \{x\}]} = (V + \{x\}, \mathscr{L})$ is a binary matroid. By Exercise D11b, $M_1 + \{x\}$ and $M_2 + \{x\}$ belong to $\mathscr{L}$, and so $M_1 + M_2 = (M_1 + \{x\}) + (M_2 + \{x\})$ is a sum of pairwise-disjoint cycles $L_1, \ldots, L_m \in \mathscr{L}$. Since $L_i \subseteq M_1 + M_2$, we have $L_i \notin \mathscr{M}$. Hence by D11a, $L_i + \{x\} \in \mathscr{M}$ for $i = 1, \ldots, m$. If $m \geq 2$, then $(L_1 + \{x\}, L_2 + \{x\}, x)$ is an admissible triple, yielding a cycle $M \subseteq L_1 \cup L_2 \subseteq M_1 + M_2$, which is impossible. Thus $M_1 + M_2 + \{x\} = L_1 + \{x\} \in \mathscr{M}$ for all $x \in M_1 \cap M_2$. Hence $M_1 + M_2 \in \mathscr{B}(\Lambda)$. By the incommensurability of $\mathscr{M}$, $M_1 \nsubseteq M_1 + M_2 + \{x\}$, which implies that there exists some vertex $w \in (M_1 \cap M_2) + \{x\}$.

Let $M_3 = M_1 + M_2 + \{x\}$ and $M_4 = M_1 + M_2 + \{w\}$. Since $M_3 + M_4 = \{w, x\} \subset M_1$, $M_3 + M_4 \in \mathscr{I}(\Lambda)$, and so $M_3 + M_4$ is not a sum of pairwise-disjoint cycles. Hence $M_3 \cup M_4 = V$, whence $M_1 \cap M_2 = \{w, x\}$.

Since M_1 and M_3 are a pair of cycles of Λ whose union is V, $M_1 + M_3$ cannot be expressed as a union of pairwise-disjoint cycles. Hence all of the arguments applied in this proof to M_1 and M_2 may also be applied to M_1 and M_3. In particular, $|M_1 \cap M_3| = 2$. One vertex in $M_1 \cap M_3$ is x; let y denote the other vertex. Since $M_1 = M_2 + M_3 + \{x\} \subseteq M_2 \cup M_3$, we have $M_1 = (M_1 \cap M_2) \cup (M_1 \cap M_3) = \{w, x, y\}$. Similarly, working with M_2 and M_3, we deduce $M_2 \cap M_3 = \{x, z\}$ for some $z \in V$ and $M_2 = \{w, x, z\}$. Thus $M_3 = \{x, y, z\}$ and $M_4 = \{w, y, z\}$. We have shown that Λ is Θ_2. □

F20 *Exercise.* Show that the matroid Θ_2 is representable over every field except $\mathbb{K}$.

F21 *Exercise.* Show that every binary matroid with not more than six vertices is regular.

There exist a number of interesting and powerful results on representability of matroids which have not been presented in this section. By combining some further results of W. T. Tutte [t.6] with Proposition F11, we can formulate the following strong characterization of regular matroids.

F22 Theorem. *Let* Λ *be a matroid. Then the following are equivalent:*

(a) Λ *is regular.*

(b) Λ *admits a unimodular Tutte subspace* (*over* $\mathbb{Q}$).

(c) Λ *is binary and no minor of* Λ *is a Fano matroid.*

M. J. Piff and D. J. A. Welsh [p.5] have shown that every gammoid, and hence every transversal matroid, is representable over all sufficiently large fields. We close by presenting in the form of an exercise an example of a matroid which is representable over no field at all. The reader may recognize part (b) as part of the classical theorem of Desargues.

F23 *Exercise.* Let V consist of the 10 points and let $\mathscr{L}$ consist of the 9 lines shown in Figure F24. Let $\mathbb{F}$ be an arbitrary field.

(a) Show that $(V, \mathscr{P}_3(V) + \mathscr{L})$ is a basis system.

(b) Show that if $V \subseteq \mathbb{F}^3$ and if each set $L \in \mathscr{L}$ were contained in a line in $\mathbb{F}^3$, then $\{a'', b'', c''\}$ would be contained in a line in $\mathbb{F}^3$.

(c) Let Λ be the matroid such that $\mathscr{B}(\Lambda) = \mathscr{P}_3(V) + \mathscr{L}$. Show that there exists no Whitney function over $\mathbb{F}$ for Λ.

(d) In the case that $\mathbb{F} = \mathbb{K}$, describe the minors of Λ which are isomorphic to Θ_2.

F24

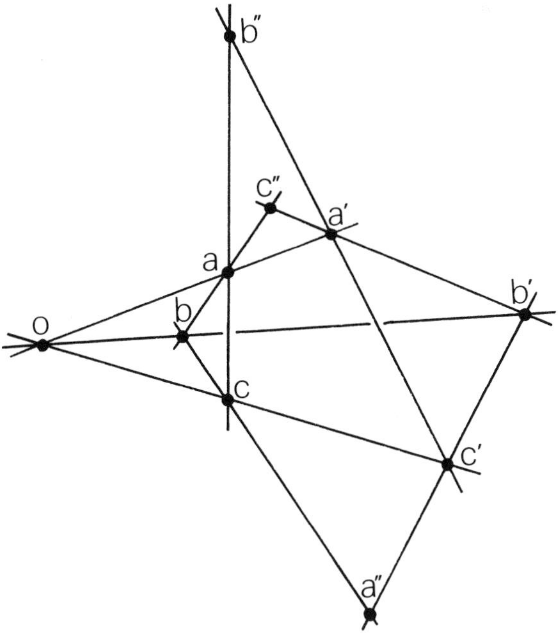

CHAPTER XI

Enumeration Theory

Some basic principles of enumeration were presented in §*I*E. Several well-known methods of enumeration are presented in the four sections of this chapter.

The second section presents generating functions and recurrence techniques. It is necessarily preceded by a section on the theory of formal power series, due originally to E. T. Bell [b.2]. In the third section we present Polya's "Fundamental Theorem of Combinatorial Enumeration," which enables one to count isomorphism classes of objects rather than just the objects themselves. The final section introduces inversion techniques, which generalize the Principle of Inclusion-Exclusion.

XIA Formal Power Series

In this section we shall be concerned with the vector space $\mathbb{C}^{\mathbb{N}}$ of infinite sequences of complex numbers. Bowing to tradition as well as for reasons of convenience, we shall denote a sequence $a\colon \mathbb{N} \to \mathbb{C}$ by the "formal power series"

$$a = a(x) = \sum_{j=0}^{\infty} a_j x^j$$

where a_j denotes $a(j)$ for all $j \in \mathbb{N}$. Viewed abstractly, the symbol x is merely a sort of placeholder. Its presence, however, will help underline many analogies to the theory of analytic functions, which we will derive quite independently for formal power series.

We have in $\mathbb{C}^{\mathbb{N}}$ the basic operations of addition:

A1 $$a(x) + b(x) = \sum_{j=0}^{\infty} a_j x^j + \sum_{j=0}^{\infty} b_j x^j = \sum_{j=0}^{\infty} (a_j + b_j) x^j = (a + b)(x)$$

for all $a, b \in \mathbb{C}^{\mathbb{N}}$, and scalar multiplication:

A2
$$za(x) = z \sum_{j=0}^{\infty} a_j x^j = \sum_{j=0}^{\infty} za_j x^j = (za)(x)$$

for all $z \in \mathbb{C}$. With respect to these two operations, $\mathbb{C}^{\mathbb{N}}$ is a vector space, with additive identity 0 being the sequence whose every value is 0.

If $a, b \in \mathbb{C}^{\mathbb{N}}$, we define a product

A3
$$(ab)(x) = \sum_{k=0}^{\infty} \left(\sum_{j=0}^{k} a_j b_{k-j} \right) x^k$$

which is clearly commutative. To verify that it is associative, let $a, b, c \in \mathbb{C}^{\mathbb{N}}$ and observe that by A3 the term $((ab)c)_n$ is

$$\begin{aligned} \sum_{k=0}^{n} (ab)_k c_{n-k} &= \sum_{k=0}^{n} \left(\sum_{j=0}^{k} a_j b_{k-j} \right) c_{n-k} \\ &= \sum_{j=0}^{n} \sum_{k=j}^{n} a_j b_{k-j} c_{n-k} \\ &= \sum_{j=0}^{n} a_j \sum_{k=0}^{n-j} b_k c_{n-(j+k)} \\ &= \sum_{j=0}^{n} a_j (bc)_{n-j} \end{aligned}$$

which is precisely $(a(bc))_n$. The multiplicative identity 1 is the sequence that assigns 1 to 0 and assigns 0 to everything else. One can straightforwardly verify that multiplication distributes over addition and that $(z(ab))(x) = ((za)b)(x)$ for all $z \in \mathbb{C}$ and $a, b \in \mathbb{C}^{\mathbb{N}}$. We thus obtain the standard algebraic result:

A4 Proposition. *Under the operations* A1, A2, *and* A3, $\mathbb{C}^{\mathbb{N}}$ *is a commutative algebra over* $\mathbb{C}$.

If $n \in \mathbb{N}$, a formal power series $a \in \mathbb{C}^{\mathbb{N}}$ such that $a_n \neq 0$ but $a_j = 0$ for all $j > n$ is called a **polynomial of degree** n, while 0 is by definition the polynomial of degree -1. We let $\mathbb{M}_n$ denote the set of polynomials of degree less than n. If $a \in \mathbb{M}_m$ and $b \in \mathbb{M}_n$, then $ab \in \mathbb{M}_{m+n-1}$. Clearly $\bigcup_{n=0}^{\infty} \mathbb{M}_n$ (whose elements are called **polynomials**) is an infinite-dimensional subalgebra of $\mathbb{C}^{\mathbb{N}}$, while $\mathbb{M}_n$ is an n-dimensional subspace of $\mathbb{C}^{\mathbb{N}}$. Finally $a \mapsto a_0$ defines an algebra-isomorphism from $\mathbb{M}_1$ onto $\mathbb{C}$; in this case we identify a with a_0 and regard $\mathbb{C}$ as a subalgebra of $\mathbb{C}^{\mathbb{N}}$. Thus A2 becomes a special case of A3.

An element of an algebra is called a **unit** if it has a multiplicative inverse. Surely all the elements of $\mathbb{C} + \{0\}$ are units. Since the following is a standard algebraic result, we omit the proof.

A5 Proposition. *If $a \in \mathbb{C}^{\mathbb{N}}$, then a is a unit if and only if $a_0 \neq 0$.*

If a is a unit of $\mathbb{C}^{\mathbb{N}}$, its inverse will be denoted by $1/a$. If a and b are polynomials and $a_0 \neq 0$, then $b/a = b(1/a)$ is a **rational function.**

A6 Exercise. Show that *if a is a polynomial of degree m with $a_0 \neq 0$, if b is a polynomial of degree n, and if $m \leq n$, then $b/a = d + c/a$ where d is a polynomial of degree $n - m$ and $c \in \mathbb{M}_m$.*

If $z \in \mathbb{C}$ and $n > 0$, then $1 - zx^n$ is a unit by Proposition A5. It is particularly useful to know the form of its inverse.

A7 Lemma. $1/(1 - zx^n) = \sum_{j=0}^{\infty} z^j x^{nj}$ *for all $z \in \mathbb{C}$ and $n > 0$.*

PROOF. By A3,

$$(1 - zx^n) \sum_{j=0}^{\infty} z^j x^{nj} = \sum_{j=0}^{\infty} z^j x^{nj} - \sum_{j=0}^{\infty} z^{j+1} x^{n(j+1)} = 1. \qquad \square$$

The reader surely must have noticed the formal similarity between the formal power series $\sum_{j=0}^{\infty} z^j x^{nj}$ and the Taylor series expansion of $1/(1 - zx^n)$ about zero for $|x| < (1/|z|)^{1/n}$. In particular, we have

A8 *Examples.*

(a) $1/(1 - x) = \sum_{j=0}^{\infty} x^j$;
(b) $1/(1 + x) = \sum_{j=0}^{\infty} (-1)^j x^j$;
(c) $1/(1 - x^2) = 1/(1 - x)(1 + x) = \sum_{j=0}^{\infty} x^{2j}$.

A9 Exercise. Prove:

(a) $[a/(1 - x)]_n = \sum_{j=0}^{n} a_j$ *for any $a \in \mathbb{C}^{\mathbb{N}}$*;
(b) $1/(1 - x)^n = \sum_{j=0}^{\infty} \binom{n+j-1}{j} x^j$;
(c) $x^n/(1 - x)^{n+1} = \sum_{j=n}^{\infty} \binom{j}{n} x^j$.

We define the function $D: \mathbb{C}^{\mathbb{N}} \to \mathbb{C}^{\mathbb{N}}$ by $D(a)(x) = \sum_{j=0}^{\infty} (j + 1)a_{j+1}x^j$. It is easy to verify that $D(a + zb) = D(a) + zD(b)$ for all $a, b \in \mathbb{C}^{\mathbb{N}}$ and all $z \in \mathbb{C}$, and so D is linear. The analogies between the function D and the differential operator go much further. We list some of these in the following exercise; observe that the proofs rely in no way upon any results from the calculus.

A10 Exercise. Prove *for all $a, b \in \mathbb{C}^{\mathbb{N}}$ and $k \in \mathbb{N}$,*

(a) $D^k(a)(x) = \sum_{j=0}^{\infty} [(j + k)!/j!]a_{j+k}x^j$;
(b) $a(x) = \sum_{j=0}^{\infty} \{[D^j(a)]_0/j!\}x^j$;
(c) $D(ab) = D(a)b + aD(b)$;
(d) $D(1/b) = -D(b)/b^2$; [*Hint*: use (c) to compute $D(b(1/b))$.]
(e) $D(a/b) = (D(a)b - aD(b))/b^2$.
(f) $D(a^n) = na^{n-1}D(a)$ *for all $n \in \mathbb{Z}$.*

A11 Exercise. *Let* $a, b \in \mathbb{C}^{\mathbb{N}}$. Prove that *if a is a polynomial or if* $b_0 = 0$, *then* $\sum_{j=0}^{\infty} a_j(b(x))^j \in \mathbb{C}^{\mathbb{N}}$.

When $\sum_{j=0}^{\infty} a_j(b(x))^j$ is a formal power series, it is denoted by $a \circ b$ and is called the **composition** of a by b. As in the calculus, composition is associative and a "chain rule" applies.

A12 Exercise. Prove *for all* $a, b, c \in \mathbb{C}^{\mathbb{N}}$ *with* $b_0 = c_0 = 0$,

(a) $(a \circ b) \circ c = a \circ (b \circ c)$;

(b) $D(a \circ b) = (D(a) \circ b)D(b)$.

The identity for composition is the polynomial x. The condition given in Exercise A11 is sufficient but not necessary for the existence of $a \circ b$. For example, if $a_n = b_n = 1/n!$ then $a \circ b$ has the form of the MacLauren series for $\exp(\exp(x))$, yet a and b are not polynomials and $a_0 = b_0 = 1$.

The concluding result of this section gives necessary and sufficient conditions for the existence of an nth root of a formal power series.

A13 Proposition. *Let* $a \in \mathbb{C}^{\mathbb{N}}$ *and let* $n \geq 2$ *be an integer. There exists* $b \in \mathbb{C}^{\mathbb{N}}$ *such that* $a = b^n$ *if and only if the least* j *for which* $a_j \neq 0$ *is an integral multiple of* n.

In lieu of a proof, we give a demonstration for $n = 2$, leaving the generalization to the reader as an *Exercise*.

Let $a = b^2$ and let k be the least integer for which $b_k \neq 0$. If j is the least integer such that $a_j \neq 0$, then clearly $j = 2k$.

Conversely, suppose $2k$ is the least integer such that $a_{2k} \neq 0$. Then $a(x) = x^{2k} \sum_{j=0}^{\infty} c_j x^j$, where $c_j = a_{j-2k}$ and $c_0 \neq 0$. It suffices to show that $c(x) = \sum_{j=0}^{\infty} c_j x^j = [f(x)]^2$ for some $f \in \mathbb{C}^{\mathbb{N}}$. We let $f_0 = \pm\sqrt{c_0}$ and define

$$f_j = \frac{1}{2f_0}\left(c_j - \sum_{m=1}^{j-1} f_m f_{j-m}\right).$$

By solving this equation for c_j, one sees that $c_j = ([f(x)]^2)_j$.

A14 Exercise. Prove *for all* $a \in \mathbb{C}^{\mathbb{N}}$ *and all* $q \in \mathbb{Q}$ *such that* a^q *is defined, we have* $D(a^q) = qa^{q-1}D(a)$. [*Hint*: first consider the case where $q = 1/n$ for $n \in \mathbb{N} + \{0\}$.]

A15 *Exercise.* Define $A: \mathbb{C}^{\mathbb{N}} \to \mathbb{C}^{\mathbb{N}}$ by $A(a) = \sum_{k=1}^{\infty} (a_{k-1}/k)x^k$. Show that

(a) A is a linear injection but is not surjective.

(b) $D(A(a)) = a$ for all $a \in \mathbb{C}^{\mathbb{N}}$.

(c) $A(D(a)) \neq a$ for some $a \in \mathbb{C}^{\mathbb{N}}$, and characterize the formal power series a having this property.

XIB Generating Functions

If $a, b \in \mathbb{C}^{\mathbb{N}}$, we say that $b(x) = \sum_{j=0}^{\infty} b_j x^j$ is a **generating function** for the sequence a if there exists a vector space isomorphism $\varphi: \mathbb{C}^{\mathbb{N}} \to \mathbb{C}^{\mathbb{N}}$ such that $\varphi(a) = b$. (Thus φ respects the two operations defined by A1 and A2.) Obviously a given sequence may have many different generating functions. In this section we shall confine our attention to the two most commonly used generating functions. If φ is the identity function, then $\varphi(a)$ is called the **ordinary generating function** for a. If φ is given by $\varphi(a) = \sum_{j=0}^{\infty} (a_j/j!)x^j$, then $\varphi(a)$ is called the **exponential generating function** for a. In the latter case,

$$\varphi(a(x))\varphi(b(x)) = \sum_{k=0}^{\infty} \left(\sum_{j=0}^{k} \binom{k}{j} a_j b_{k-j} \right) x^k/k!$$

while

$$\varphi(a(x)b(x)) = \sum_{k=0}^{\infty} \left(\sum_{j=0}^{k} a_j b_{k-j} \right) x^k/k!,$$

and so φ is not an algebra-isomorphism.

B1 *Examples.*

(a) The ordinary generating function for the sequence of binomial coefficients $\binom{n}{j}$ for $j = 0, 1, \ldots, n$ is $(1 + x)^n$.

(b) By Exercise A9b, $1/(1 - x)^n$ is the ordinary generating function for the sequence $|\mathbb{S}_j(X)|$ for $j \in \mathbb{N}$, where X is an n-set.

(c) If $z \in \mathbb{C}$ and $a_j = z^j$ for all $j \in \mathbb{N}$, then the ordinary generating function for a is $1/(1 - zx)$. The exponential generating function for a is $\sum_{j=0}^{n} (zx)^j/j!$, which we recognize as the form of the Taylor series about 0 for the function $\exp(zx)$. As the following exercise will show, it is not unreasonable to designate this exponential generating function by the symbol $\exp(zx)$.

B2 *Exercise.* Prove by the rules of formal power series:

(a) $\exp(z_1 x)\exp(z_2 x) = \exp((z_1 + z_2)x)$ for all $z_1, z_2 \in \mathbb{C}$;

(b) $[\exp(zx)]^q = \exp(qzx)$ for all $z \in \mathbb{C}$ and $q \in \mathbb{Q}$;

(c) $D^n(\exp(zx)) = z^n \exp(zx)$ for all $z \in \mathbb{C}$ and $n \in \mathbb{N}$.

Analogous to Exercise A10b, one easily shows

B3 Lemma. *If b is the exponential generating function for a, then* $b(x) = \sum_{j=0}^{\infty} [D^j(a)]_0 x^j$.

B4 *Example.* Recall (*I*E17) the nth derangement number $D_n = n! \sum_{j=0}^{n} (-1)^j/j!$. The ordinary generating function for the sequence of derangement numbers does not have a particularly elegant form, but let us consider its exponential generating function:

$$\sum_{n=0}^{\infty} \frac{D_n}{n!} x^n = \sum_{n=0}^{\infty} \left(\sum_{j=0}^{n} \frac{(-1)^j}{j!} \right) x^n$$

$$= \left(\sum_{j=0}^{\infty} \frac{(-x)^j}{j!} \right) \left(\sum_{j=0}^{\infty} x^j \right) \qquad \text{by A3}$$

$$= \exp(-x)/(1 - x).$$

B5 Lemma. $\sum_{j=k}^{n} \binom{n}{j}\binom{j}{k} = \binom{n}{k} 2^{n-k}$ *where* $0 \le k \le n$.

If U is an n-set, then Lemma B5 merely equates two different enumerations of the set

$$\{(S, T): S \subseteq T \subseteq U; |S| = k\}.$$

It may be instructive, however, to give a proof using generating functions.

PROOF OF LEMMA B5. By Example B1a, we have

$$(1 + x)^n = \sum_{j=0}^{n} \binom{n}{j} x^j,$$

to which we apply D^k (cf. A10a) and divide by $k!$:

$$\frac{n!}{k!\,(n-k)!}(1 + x)^{n-k} = \sum_{j=k}^{n} \binom{n}{j} \frac{j!}{k!\,(j-k)!} x^{j-k}.$$

By Exercise A11, we may compose both sides of this equation with the constant polynomial 1, which yields the lemma. □

B6 *Exercise.* Prove $\sum_{j=k}^{n} (-1)^j \binom{n}{j}\binom{j}{k} = 0$, where $0 \le k \le n$.

Let $k \in \mathbb{N}$. A **recurrence** may be defined as a function $f\colon \mathbb{N} \times \mathbb{C}^{\mathbb{N}} \to \mathbb{C}$. If $a \in \mathbb{C}^{\mathbb{N}}$, we say that a satisfies the recurrence f if

$$f(n, a_0, a_1, \ldots, a_n, 0, 0, \ldots) = 0 \quad \text{for all } n \ge k,$$

and a is said to be a "solution" of the recurrence. In general, if a recurrence has a solution, the solution need not be unique. In the examples which follow, however, once the terms $a_0, a_1, \ldots, a_{k-1}$ are specified, the solution will be unique. The uniqueness problem in its full generality is not within the scope of this text; the interested reader is advised to consult [r.5]. Our intention for the remainder of this section is to show how recurrences can be exploited to obtain enumeration formulas for various combinatorial objects. Most of our examples are objects already encountered in the early chapters of this book.

The easiest type of recurrence is the linear homeogeneous kind. Specifically, if the sequence a satisfies such a recurrence, then there exist $c_0, \ldots, c_k \in \mathbb{C}$ such that

B7 $$\sum_{j=0}^{k} c_j a_{n-j} = 0$$

for all $n \geq k$. As an example of this kind, we consider the sequence whose first two terms are 1 and every subsequent term is the sum of its two immediate predecessors. Thus $k = 2$ and B7 assumes the form

B8 $$a_n - a_{n-1} - a_{n-2} = 0 \quad \text{for } n \geq 2.$$

From B8 we obtain the power series equation

$$\sum_{j=2}^{\infty} a_j x^j - \sum_{j=2}^{\infty} a_{j-1} x^j - \sum_{j=2}^{\infty} a_{j-2} x^j = 0$$

whence

$$\sum_{j=2}^{\infty} a_j x^j - x \sum_{j=1}^{\infty} a_j x^j - x^2 \sum_{j=0}^{\infty} a_j x^j = 0.$$

Since $a(x) = \sum_{j=0}^{\infty} a_j x^j$ by definition, we obtain with $a_0 = a_1 = 1$,

$$a(x) - (1 + x) - x(a(x) - 1) - x^2 a(x) = 0$$

and solving yields $a(x) = 1/(1 - x - x^2)$. The roots of $1 - x - x^2$ are

$$r_1 = \frac{-1 + \sqrt{5}}{2} \quad \text{and} \quad r_2 = \frac{-1 - \sqrt{5}}{2}.$$

Hence

$$\begin{aligned} a(x) &= \frac{-1}{(r_1 - x)(r_2 - x)} \\ &= \frac{1}{r_1 - r_2}\left(\frac{1}{r_1}\cdot\frac{1}{1 - (x/r_1)} - \frac{1}{r_2}\cdot\frac{1}{1 - (x/r_2)}\right) \\ &= \frac{1}{\sqrt{5}}\left[\frac{1}{r_1}\sum_{j=0}^{\infty}\left(\frac{x}{r_1}\right)^j - \frac{1}{r_2}\sum_{j=0}^{\infty}\left(\frac{x}{r_2}\right)^j\right] \\ &= \frac{1}{\sqrt{5}}\sum_{j=0}^{\infty}\frac{r_2^{j+1} - r_1^{j+1}}{(r_1 r_2)^{j+1}} x^j \\ &= \sum_{j=0}^{\infty}\frac{(-1)^j}{\sqrt{5}}(r_1^{j+1} - r_2^{j+1}) x^j. \end{aligned}$$

The number $a_j = (-1)^j(r_1^{j+1} - r_2^{j+1})/\sqrt{5}$ is called the jth **Fibonacci number**, after the medieval Italian mathematician who first observed some of these numbers' arithmetic properties. The recurrence B8 with $a_0 = a_1 = 1$ obviously generates a sequence of positive integers, which makes the above formula for a_j rather surprising.

A linear but nonhomogeneous recurrence arises from the problem of determining the ordinary generating function for the sequence a where $a_n = \sum_{j=0}^{n} j^2$ for $n \in \mathbb{N}$. We have the recurrence $a_n - a_{n-1} - n^2 = 0$ with $a_0 = 0$, which yields in the manner of the previous example

$$\sum_{n=1}^{\infty} a_n x^n = \sum_{n=1}^{\infty} a_{n-1} x^n + \sum_{n=1}^{\infty} n^2 x^n.$$

Thus $a(x) = xa(x) + \sum_{n=1}^{\infty} n^2 x^n$, whence

B9
$$a(x) = \sum_{n=1}^{\infty} n^2 x^n/(1 - x)$$

and the problem reduces to expressing $\sum_{n=1}^{\infty} n^2 x^n$ as a rational function. This may be done in two ways and we illustrate them both.

Method 1. Beginning with $1/(1 - x) = \sum_{n=0}^{\infty} x^n$, we apply D and then "multiply by x," obtaining

$$\frac{x}{(1 - x)^2} = \sum_{n=1}^{\infty} n x^n.$$

Repeating this pair of operations yields

$$\frac{x + x^2}{(1 - x)^3} = \sum_{n=1}^{\infty} n^2 x^n.$$

Method 2. From Exercise A9b (with $n = 3$) and A9c (with $n = 2$) we obtain:

$$\begin{aligned}
\frac{x + x^2}{(1 - x)^3} &= x \sum_{j=0}^{\infty} \binom{2 + j}{j} x^j + \sum_{j=2}^{\infty} \binom{j}{2} x^j \\
&= \sum_{j=1}^{\infty} \binom{j + 1}{j - 1} x^j + \sum_{j=2}^{\infty} \binom{j}{2} x^j \\
&= x + \sum_{j=2}^{\infty} \left[\binom{j + 1}{2} + \binom{j}{2}\right] x^j \\
&= \sum_{j=1}^{\infty} j^2 x^j.
\end{aligned}$$

Either way, B9 becomes $a(x) = (x + x^2)/(1 - x)^4$. By pursuing these calculations, one can obtain a nice formula for $\sum_{j=0}^{n} j^2$.

$$\begin{aligned}
a(x) &= \frac{x + x^2}{(1 - x)^4} = \frac{1}{(1 - x)^2} - \frac{3}{(1 - x)^3} + \frac{2}{(1 - x)^4} \\
&= \sum_{j=0}^{\infty} \left[\binom{j + 1}{j} - 3\binom{j + 2}{j} + 2\binom{j + 3}{j}\right] x^j \qquad \text{by A9b} \\
&= \sum_{j=0}^{\infty} \frac{j(j + 1)(2j + 1)}{6} x^j.
\end{aligned}$$

Thus

B10
$$\sum_{j=0}^{n} j^2 = \frac{n(n+1)(2n+1)}{6},$$

as one can also verify by induction.

B11 *Exercise.* Obtain ordinary generating functions for $a_n = \sum_{j=0}^{n} j$ and $a_n = \sum_{j=0}^{n} j^3$ as well as formulas for these sums similar to B10.

For our next example, we consider the recurrence

B12
$$a_n - na_{n-1} - (-1)^n = 0,$$

with the condition $a_0 = 1$. If b is the exponential generating function for a, then

$$b(x) - 1 = \sum_{n=1}^{\infty} \frac{a_n}{n!} x^n = \sum_{n=1}^{\infty} \frac{a_{n-1}}{(n-1)!} x^n + \sum_{n=1}^{\infty} \frac{(-1)^n}{n!} x^n.$$

Thus

$$b(x) - 1 = xb(x) + \exp(-x) - 1$$

whence

$$b(x) = \exp(-x)/(1 - x).$$

Comparing this equation with example B4 implies that $a_n = D_n$ and that B12 holds for the derangement numbers. Another recurrence for the derangement numbers is *I*E19a.

B13 *Exercise.* Show that if $a \in \mathbb{C}^{\mathbb{N}}$ and if

(a) $a_n - na_{n-1} - 1 = 0$ with $a_0 = 1$, then $a_n = n! \sum_{j=0}^{n} 1/j!$

(b) $a_n - 2a_{n-1} - 3 = 0$ with $a_0 = 1$, then $a_n = 2^{n+2} - 3$.

Let $*$ denote a binary operation on some set X, let $x_1, \ldots, x_n \in X$, and let a_n denote the number of ways that $n - 1$ (pairs of) parentheses can be inserted in the expression $x_1 * x_2 * \ldots * x_n$ so that just two terms are combined at a time. Thus $a_2 = 1$, and $a_3 = 2$ since we can have $(x_1 * x_2) * x_3$ and $x_1 * (x_2 * x_3)$. The reader can easily check that $a_4 = 5$. It is reasonable to adopt the convention that $a_1 = 1$. Under this assumption we derive the nonlinear recurrence $a_n = \sum_{j=1}^{n-1} a_j a_{n-j}$. With the further convention that $a_0 = 0$, we obtain

$$a_n = \sum_{j=0}^{n} a_j a_{n-j}, \quad n \geq 2,$$

whence

$$\begin{aligned} a(x) = \sum_{n=0}^{\infty} a_n x^n &= a_1 x + \sum_{n=2}^{\infty} \left(\sum_{j=0}^{n} a_j a_{n-j} \right) x^n \\ &= x + \sum_{n=0}^{\infty} \left(\sum_{j=0}^{n} a_j a_{n-j} \right) x^n \\ &= x + [a(x)]^2. \end{aligned}$$

Thus $[a(x)]^2 - a(x) + 1/4 = 1/4 - x$, and by Proposition A13, this quadratic equation has the two solutions $a(x) = 1/2 \pm \sqrt{1 - 4x}/2$. The boundary condition $a_0 = 0$ determines that

$$a(x) = \frac{1}{2} - \frac{\sqrt{1 - 4x}}{2}.$$

This is the ordinary generating function for the sequence $a_0, a_1, \ldots$, and it remains only to use Exercise A10b and A14 to get a general formula for a_n.

From $D(a) = (1 - 4x)^{-1/2}$ we derive inductively $D^n(a) = 2^{n-1} \cdot 1 \cdot 3 \cdot 5 \cdot \ldots \cdot (2n - 3)(1 - 4x)^{-n+1/2}$ for $n \geq 2$, whence

$$a_n = \frac{[D^n(a)]_0}{n!} = \frac{2^{n-1} \cdot 1 \cdot 3 \cdot \ldots \cdot (2n - 3)}{n!}$$

$$= \frac{2^{n-1}}{n!} \cdot \frac{(2n - 2)!}{2^{n-1}(n - 1)!} = \frac{1}{n}\binom{2n - 2}{n - 1}.$$

For our final application of the notions of this section, we return to the doubly-indexed sequence $p_k(n)$ $(k, n \in \mathbb{N})$, the number of k-partitions of the integer n. We first obtain the ordinary generating function for $p_k(0), p_k(1), \ldots$.

B14 Proposition. *For each integer* $k \geq 1$,

$$\sum_{j=0}^{\infty} p_k(j)x^j = x^k \Big/ \prod_{j=1}^{k} (1 - x^j).$$

Proof. We proceed by induction on k, first noting that $p_k(j) = 0$ when $j < k$ and $p_1(j) = 1$ whenever $j \geq 1$. Thus for $k = 1$,

$$\sum_{j=0}^{\infty} p_1(j)x^j = \sum_{j=1}^{\infty} x^j = x \sum_{j=0}^{\infty} x^j = x/(1 - x).$$

Now let $k \geq 2$, and as induction hypothesis, assume that

B15 $$\sum_{j=0}^{\infty} p_{k-1}(j)x^j = x^{k-1} \Big/ \prod_{j=1}^{k-1} (1 - x^j).$$

For each integer $j \geq k$, the number of k-partitions s of j with $s(1) \geq 1$ is $p_{k-1}(j - 1)$. The k-partitions s of j with $s(1) = 0$ are in one-to-one correspondence $s \leftrightarrow s'$ with the k-partitions s' of $j - k$ where $s'(m) = s(m + 1)$ for all m, since $j = \sum_{m=1}^{\infty} (m + 1)s(m + 1) = \sum_{m=1}^{\infty} (m + 1)s'(m) = \sum_{m=1}^{\infty} ms'(m) + k$. Thus we have the recurrence

$$p_k(j) = p_{k-1}(j - 1) + p_k(j - k),$$

whence

$$\sum_{j=k}^{\infty} p_k(j)x^j = \sum_{j=k}^{\infty} p_{k-1}(j-1)x^j + \sum_{j=k}^{\infty} p_k(j-k)x^j,$$

or

$$\sum_{j=0}^{\infty} p_k(j)x^j = x\sum_{j=0}^{\infty} p_{k-1}(j)x^j + x^k\sum_{j=0}^{\infty} p_k(j)x^j.$$

By B15, we have

$$(1-x^k)\sum_{j=0}^{\infty} p_k(j)x^j = x\sum_{j=0}^{\infty} p_{k-1}(j)x^j = x^k\Big/\prod_{j=1}^{k-1}(1-x^j),$$

whence the result. □

The following corollary enables us to enumerate the partitions of an integer j into not more than a given number n of parts.

B16 Corollary. *For each integer $n \geq 1$,*

$$\sum_{j=0}^{\infty}\left(\sum_{k=0}^{n} p_k(j)\right)x^j = 1\Big/\prod_{j=1}^{n}(1-x^j).$$

PROOF. The proof is by induction on n. By convention, $p_0(0) = 1$ and $p_0(j) = 0$ for $j \geq 1$. Thus for $n = 1$, $\sum_{k=0}^{1} p_k(j) = 1$ for all j. Hence

$$\sum_{j=0}^{\infty}\left(\sum_{k=0}^{1} p_k(j)\right)x^j = \sum_{j=0}^{\infty} x^j = 1/(1-x).$$

If the identity holds for $n-1$, then by the proposition,

$$\begin{aligned}\sum_{j=0}^{\infty}\left(\sum_{k=0}^{n} p_k(j)\right)x^j &= \sum_{k=0}^{n}\left(\sum_{j=0}^{\infty} p_k(j)x^j\right)\\ &= \sum_{k=0}^{n-1}\left(\sum_{j=0}^{\infty} p_k(j)x^j\right) + \sum_{j=0}^{\infty} p_n(j)x^j\\ &= \frac{1}{\prod_{j=1}^{n-1}(1-x^j)} + \frac{x^n}{\prod_{j=1}^{n}(1-x^j)}\\ &= \frac{1}{\prod_{j=1}^{n}(1-x^j)}.\end{aligned}$$

□

In Chapter I we indicated how difficult it is to enumerate even the 3-partitions of an integer n. Let us now use B14 to determine $p_3(n)$. (Cf. *I*C37.)

B17 Corollary. *For any integer $n \geq 3$,*

$$p_3(n) = \frac{1}{12}\left\{n^2 + 3\left(\left[\frac{n+1}{2}\right] - \left[\frac{n}{2}\right]\right) - 4\left(\left[\frac{n+2}{3}\right] - \left[\frac{n}{3}\right]\right)\right\}.$$

PROOF. By Proposition B14,

$$\sum_{n=0}^{\infty} p_3(n)x_n = \frac{x^3}{(1-x)(1-x^2)(1-x^3)}$$
$$= \frac{1}{72}\left\{\frac{-1}{1-x} + \frac{-18}{(1-x)^2} + \frac{12}{(1-x)^3} + \frac{-9}{1+x} + \frac{8(2+x)}{1+x+x^2}\right\}.$$

Now

$$\frac{2+x}{1+x+x^2} = \frac{2-x-x^2}{1-x^3} = (2-x-x^2)\sum_{n=0}^{\infty} x^{3n} \qquad \text{(by Lemma A7)}$$
$$= (2-x-x^2)\sum_{n=0}^{\infty}\left(\left[\frac{n+3}{3}\right] - \left[\frac{n+2}{3}\right]\right)x^n$$
$$= \sum_{n=0}^{\infty}\left\{2\left[\frac{n+3}{3}\right] - 3\left[\frac{n+2}{3}\right] + \left[\frac{n}{3}\right]\right\}x^n$$
$$= -3\sum_{n=0}^{\infty}\left\{\left[\frac{n+2}{3}\right] - \left[\frac{n}{3}\right]\right\}x^n + 2\sum_{n=0}^{\infty} x^n.$$

Hence by A8 and A9b,

$$\sum_{n=0}^{\infty} p_3(n)x^n = \frac{1}{72}\left\{-\sum_{n=0}^{\infty} x^n - 18\sum_{n=0}^{\infty}(n+1)x^n + 12\sum_{n=0}^{\infty}\frac{n^2+3n+2}{2}x^n\right.$$
$$\left. -9\sum_{n=0}^{\infty}(-1)^n x^n - 24\sum_{n=0}^{\infty}\left(\left[\frac{n+2}{3}\right] - \left[\frac{n}{3}\right]\right)x^n + 16\sum_{n=0}^{\infty} x^n\right\}$$
$$= \frac{1}{72}\left\{9\sum_{n=0}^{\infty}[1-(-1)^n]x^n + 6\sum_{n=0}^{\infty} n^2x^n\right.$$
$$\left. - 24\sum_{n=0}^{\infty}\left(\left[\frac{n+2}{3}\right] - \left[\frac{n}{3}\right]\right)x^n\right\}$$
$$= \frac{1}{12}\sum_{n=0}^{\infty}\left\{3\left(\left[\frac{n+1}{2}\right] - \left[\frac{n}{2}\right]\right) + n^2 - 4\left(\left[\frac{n+2}{2}\right] - \left[\frac{n}{3}\right]\right)\right\}x^n. \qquad \square$$

The following exercise is relatively difficult.

B18 *Exercise.* Show that $p_k(n) = f(n) + g(n)$, where f is a polynomial of degree $k-1$ and g depends only upon the congruence class of n modulo k.

B19 Lemma. *The number of partitions of the integer n having largest summand k is $p_k(n)$.*

PROOF. To each k-partition s of n, there corresponds a unique nondecreasing sequence $n_1, \ldots, n_k$ in which exactly $s(m)$ terms are equal to m. To this sequence corresponds a unique $k \times n_k$ $\{0, 1\}$-matrix $M = [m_{ij}]$ such that

$$m_{ij} = \begin{cases} 1 & \text{if } j \le n_i; \\ 0 & \text{if } j > n_i. \end{cases}$$

(The number of 1's in M is then $\sum_{i=1}^{k} n_i = n$.) Its transpose represents an n_k-partition of n whose largest summand is k. □

Letting $q_k(n)$ denote the number of partitions of n in which no summand exceeds k, we have by Lemma B19 that $q_k(n) = \sum_{j=0}^{k} p_j(n)$. If $k \geq n$, then $q_k(n) = p(n)$ where, as in Chapter I, $p(n)$ denotes the total number of partitions of n. From Corollary B16 it follows immediately that

B20 Proposition. $1/\prod_{j=1}^{k} (1 - x^j) = \sum_{n=0}^{\infty} q_k(n)x^n$.

B21 *Exercise.* Prove $\sum_{n=0}^{\infty} p(n)x^n = 1/\prod_{n=1}^{\infty} (1 - x^n)$.

B22 *Exercise.* Compute $q_3(n)$ for $n \in \mathbb{N}$.

Further References
C. Berge [b.6] and J. Riordan [r.6].

XIC Pólya Theory

As motivation for the study of Pólya's theory of counting, we first present a problem which is not solved by this method, that of counting "labeled trees." It is perhaps one of the oldest graph-enumeration problems. It was first resolved by A. Cayley [c.1] in 1889, who was concerned with the relation between multigraphs and schematic diagrams of chemical structures. Thus we let $V = \{x_1, \ldots, x_n\}$ and ask how many trees have V as vertex set? The problem can be reformulated as follows.

C1 Theorem (Cayley). *The complete graph* K_n *admits* n^{n-2} *distinct subgraphs which are spanning trees.*

Indication of proof. The method sketched here is due to H. Prüfer [p.7]. Let $V = \{1, \ldots, n\}$ denote the vertex set of K_n. We describe a function from the set of spanning subtrees of K_n into the cartesian product V^{n-2} and leave to the reader the proof that this function is a bijection. Let $T_1 = (V, \mathcal{F})$ be a spanning subtree of K_n. Some vertex of T_1 has valence 1 in T_1. Letting V be ordered in the natural way, we fill the first position of the $(n - 2)$-tuple assigned to T_1 with the name x_1 of the unique vertex joined by an edge to the least vertex y_1 of valence 1. Let T_2 denote the subtree of T_1 spanned by $V + \{y_1\}$, and fill the second position of our $(n - 2)$-tuple with the vertex of T_2 joined by an edge to the least vertex of T_2 having valence 1 (in T_2). This process is repeated until an $(n - 2)$-tuple is completed. □

More often than not one is less interested in the labeling of spanning trees than in their graphical structure. For example, $n!/2$ of the trees counted above are paths of length $n - 1$, but they in effect all look alike. We are not interested in counting trees but in counting isomorphism classes of trees. Thus the $n!/2$ paths count as a single class. Clearly the n^{n-2} spanning subtrees of K_n are not divided equally into isomorphism classes. (For example, another class comprises only the n trees consisting of a single vertex of valence $n - 1$ which is joined by an edge to each of the others.) Thus our problem is considerably more difficult than the labeled problem. We find that the groups of automorphisms of systems (§*II*E) play a significant role.

Let G be any group, let e denote its identity, and let X be a set. We say that G **acts** on X if there exists a function $G \times X \to X$ given by $(g, x) \mapsto gx$ such that for all $x \in X$ and all $g_1, g_2 \in G$,

(a) $ex = x$

(b) $(g_2g_1)x = g_2(g_1x)$.

Of particular interest to us will be the case where G is a subgroup of $\Pi(X)$, the group of all permutations on X. In this case, gx becomes $g(x)$ for all $(g, x) \in G \times X$.

Suppose that G acts on X and let $S \subseteq X$. Generalizing a notion from §*I*A, we say that g **fixes** S if $gx \in S$ for all $x \in S$. If every element of G fixes S, then S is a **fixed set** of G. Certainly $\varnothing$ and X are always fixed sets. If $\mathscr{S}$ denotes the collection of fixed sets, then the cells of the fine partition of $\mathscr{S}$ are called the **orbits** of G. If $S \neq X$, we say that G acts **transitively** on S if given $x, y \in S$, there exists $g \in G$ such that $gx = y$. (Cf. §*II*E.) If $x \in X$, we define the **stabilizer** of x to be $G_x = \{g \in G: gx = x\}$. Clearly, G_x is a subgroup of G.

The following is a pair of standard results concerning groups acting on sets.

C2 Exercise. *Let G act on X. Suppose that $S \subseteq X$, and let $x \in S$.* Show that:

(a) *S is an orbit if and only if S is a maximal set on which G acts transitively.*

(b) *If S is an orbit, then $|G_x||S| = |G|$.*

As in §*I*A, an element $x \in X$ is called a fixed-point of g if $gx = x$. The set of fixed-points of g will be denoted by $F(g)$. For example, $x \in \bigcap_{g \in G} F(g)$ if and only if $\{x\}$ is an orbit of G.

C3 Lemma (W. Burnside [b.20], 1911). *If G acts on X, then G has exactly $|G|^{-1} \sum_{g \in G} |F(g)|$ orbits.*

Proof. Let $X_1, \ldots, X_q$ be the orbits of G. Then

$$\begin{aligned}\sum_{g \in G} |F(g)| &= |\{(g, x): g \in G; x \in X; gx = x\}| \\ &= \sum_{x \in X} |G_x| = \sum_{i=1}^{q} \sum_{x \in X_i} |G_x| \\ &= \sum_{i=1}^{q} |X_i|(|G|/|X_i|) \qquad \text{by C2b} \\ &= q|G|,\end{aligned}$$

whence the result. □

If $g \in G$ and G acts on X, then we write $\langle g \rangle$ to denote the (cyclic) subgroup of G generated by $\{g\}$. Clearly $\langle g \rangle$ acts on X; we let $o_j(g)$ denote the number of orbits of $\langle g \rangle$ of cardinality j and let $O(g) = \sum_{j=1}^{\infty} o_j(g)$.

C4 *Example.* Let $X = \{x_0, x_1, \ldots, x_{m-1}\}$ and let $p \in \Pi(X)$ be given by $p(x_i) = x_{i+1}$ for $i = 0, \ldots, m-1$, where subscripts are read modulo m. Then $F(p^i) = \varnothing$ for $i = 1, \ldots, m-1$ while $F(1_V) = X$. Not surprisingly, C3 tells us that $O(p) = 1$. More generally, if $i \not\equiv 0$ (modulo m), then $O(p^i) = \gcd(i, m)$, the greatest common divisor of i and m, and $o_j(p^i) = O(p^i)$ if $j = m/O(p^i)$. Let $r \in \Pi(X)$ be given by $r(x_i) = x_{m-i}$. Then $F(r) = \{x_0\}$ if m is odd, and $F(r) = \{x_0, x_{m/2}\}$ if m is even. Also $r^2 = 1_V$, $o_1(r) = (3 + (-1)^m)/2$, $o_2(r) = [(m-1)/2]$, and $o_j(r) = 0$ for $j > 2$.

Suppose now that G is a subgroup of $\Pi(X)$ and let Y be any set. We consider the relation $\sim$ on Y^X given by: $f_1 \sim f_2$ if and only if there exists a function-isomorphism from f_1 to f_2 of the form $(g, 1_Y)$, where $g \in G$. Clearly $\sim$ is an equivalence relation. The cells of the partition of Y^X induced by $\sim$ are called **patterns** (with respect to G). This partition clearly refines the collection of isomorphism classes of Y^X. Implicit in the equation $f_1 = f_2 g$ is a definition of an action of G on Y^X, where $g(f) = fg$ for all $f \in Y^X$ and $g \in G$. The orbits of G in Y^X are precisely the patterns we have just defined. Clearly the number of such patterns is determined only by $|Y|$ and the action of G. Thus when G is a subgroup of $\Pi(X)$, the number of these patterns may be denoted by $\pi_{|Y|}(G)$.

C5 Lemma. *If G is a subgroup of $\Pi(X)$, then*

$$\pi_{|Y|}(G) = |G|^{-1} \sum_{g \in G}^{\infty} |Y|^{O(g)}.$$

Proof. The group G may be regarded as acting on the set Y^X, taking (g, f) into the composition fg for all $f \in Y^X$. With respect to this action, suppose $f \in F(g)$, and let x_1, x_2 be two elements of some orbit S_i of $\langle g \rangle$. Then $f(g(x)) = f(x)$ for all $x \in X$ and $g^j(x_1) = x_2$ for some j. In particular, $f(x_1) = f(g(x_1)) = f(g(g(x_1))) = \ldots = f(g^j(x_1)) = f(x_2)$. Thus f maps all the elements of S_i onto a common image in Y. We see that $F(g)$ is in one-to-one correspondence with the set $Y^{\{1,\ldots,O(g)\}}$. The lemma now follows from Lemma C3. □

If $Y = \{y_1, \ldots, y_n\}$, it is convenient to regard the elements $y_1, \ldots, y_n$ as indeterminants and consider the commutative ring of multinomials $\mathbb{F}[y_1, \ldots, y_n]$, where $\mathbb{F}$ is any field containing $\mathbb{Q}$. We define a **weight function** $w: Y^X \to \mathbb{F}[y_1, \ldots, y_n]$ given by

$$w(f) = \prod_{x \in X} f(x) \quad \text{for } f \in Y^X,$$

and $w(f)$ is called the **weight** of f. One easily verifies that $w(f) = \prod_{i=1}^{n} y_i^{|f^{-1}[y_i]|}$, and that

C6 $$w(f) = w(fg) \quad \text{for all } g \in \Pi(X), f \in Y^X.$$

It follows that if G is a subgroup of $\Pi(X)$ and if $Q \subseteq Y^X$ is a pattern with respect to G, then *any two functions in Q have the same weight.* One may therefore speak of the "weight of the pattern Q," and let $w(Q) = w(f)$ where $f \in Q$.

We are now ready to prove the main result of this section.

C7 Theorem (G. Pólya [p.6]). *Let G be a subgroup of $\Pi(X)$, and let $\mathscr{Q} \in \mathbb{P}(Y^X)$ be the set of patterns with respect to G. Then*

$$\sum_{Q \in \mathscr{Q}} w(Q) = |G|^{-1} \sum_{g \in G} \prod_{j=1}^{|X|} \left(\sum_{i=1}^{n} y_i^{j} \right)^{o_j(g)}.$$

Proof. Let $\bar{y} \in \mathbb{F}[y_1, \ldots, y_n]$ be of the form $\prod_{i=1}^{n} y_i^{e_i}$. Then $w^{-1}[\bar{y}]$ is a union of patterns with respect to G. The exact number of such patterns is obtained by regarding G as a subgroup of $\Pi(Y^X)$ (see just before C5), restricting G to $w^{-1}[\bar{y}]$, and applying Burnside's lemma (C5). For $g \in G$, the set of "fixed-points of g" becomes

$$F(g) = \{f \in Y^X : f = fg\},$$

and so the number of orbits of G contained in $w^{-1}[\bar{y}]$, that is, the number of patterns contained in $w^{-1}[\bar{y}]$, becomes

$$|G|^{-1} \sum_{g \in G} |F(g) \cap w^{-1}[\bar{y}]|.$$

Hence

$$\sum_{Q \in \mathscr{Q}} w(Q) = |G|^{-1} \sum_{g \in G} \sum_{\bar{y}} |F(g) \cap w^{-1}[\bar{y}]| \bar{y}.$$

It remains to show

C8 $$\sum_{\bar{y}} |F(g) \cap w^{-1}[\bar{y}]| \bar{y} = \prod_{j=1}^{|X|} \left(\sum_{i=1}^{n} y_i^{j} \right)^{o_j(g)} \quad \text{for all } g \in G.$$

Let $g \in G$, and let $X_1, \ldots, X_q$ be the orbits of $\langle g \rangle$ in X. As in the proof of Lemma C5, $f = fg$ if and only if f is constant on X_h for each $h = 1, \ldots, n$. Thus for each $\bar{y}$, $F(g) \cap w^{-1}[\bar{y}] = \{f \in Y^X : f \text{ is constant on } X_h\ (h = 1, \ldots, q);\ w(f) = \bar{y}\}$. If $f \in F(g) \cap w^{-1}[\bar{y}]$, then $\bar{y} = \prod_{i=1}^{n} y_i^{|f^{-1}[y_i]|}$, and the list $f^{-1}[y_1], \ldots, f^{-1}[y_n]$ is an ordered partition of X, except that some of the cells may be empty. Moreover, each cell is a union of orbits of $\langle g \rangle$. Conversely, if $\bar{y}$ is of this form, then $\bar{y} = w(f)$ for some $f \in F(g)$. We therefore can easily verify that the left-hand member of C8 has the form

$$(y_1^{|X_1|} + y_2^{|X_1|} + \ldots + y_n^{|X_1|})(y_1^{|X_2|} + y_2^{|X_2|} + \ldots + y_n^{|X_2|}) \ldots$$
$$\ldots (y_1^{|X_q|} + y_2^{|X_q|} + \ldots + y_n^{|X_q|}),$$

and so

C9 $$\sum_{\bar{y}} |F(g) \cap w^{-1}[\bar{y}]| \bar{y} = \prod_{h=1}^{q} \left(\sum_{i=1}^{n} y_i^{|X_h|} \right).$$

By grouping together those orbits X_h of equal length j, we see that the right-hand member of C9 is just

$$\prod_{j=1}^{\infty} \left(\sum_{i=1}^{n} y_i^j \right)^{o_j(g)},$$

as required. □

Observe that Pólya's theorem reduces to Lemma C5 if we set $y_i = 1$ for all $i = 1, \ldots, n$.

We interrupt the theoretical development at this point in order to demonstrate an application to the easily stated problem, how many necklaces is it possible to string with m beads of n different colors? An equivalent problem in terms of more standard combinatorial objects is, what is the largest number of functions from the set $X = \{x_0, x_1, \ldots, x_{m-1}\}$ of vertices of the circuit Δ_m into an n-set $Y = \{y_1, \ldots, y_n\}$ such that if h_1 and h_2 are any two such functions, then $h_2 \neq h_1 g$ for all $g \in G = G_0(\Delta_m)$? (See §*II*E.)

In this case, G is the dihedral group D_m, of order $2m$, generated by p and r as defined in Example C4. Its elements are

$$r^j p^i, \quad i = 0, 1, \ldots, m-1; j = 0, 1.$$

Since $\langle p \rangle$ is transitive, G has a single orbit in X, and one can easily verify Burnside's lemma by noting that $|F(e)| = m$ and $|F(p^i)| = 0$ for $i = 1, \ldots, m-1$. If m is odd, then $|F(rp^i)| = 1$, while if m is even, then $|F(rp^i)| = 0$ or 2, depending upon the parity of i. Thus $\sum_{g \in D_m} F(g)/|D_m| = (m + m \cdot 1)/2m = 1$.

Let us next use Lemma C5 to count the number of patterns induced by G on Y^X. Referring to Example C4 we have

$$O(e) = m$$

$$O(p^i) = \gcd(i, m) \quad \text{for } i = 1, \ldots, m-1;$$

$$O(rp^i) = \begin{cases} (m+1)/2 & \text{if } m \text{ is odd;} \\ (m+2)/2 & \text{if } m \text{ is even and } i \text{ is even;} \\ m/2 & \text{if } m \text{ is even and } i \text{ is odd.} \end{cases}$$

Hence the number of patterns is

C10 $$\left(n^m + mn^{(m+1)/2} + \sum_{i=1}^{m-1} n^{\gcd(i,m)} \right) \Big/ 2m$$

if m is odd, and

C11 $$\left(n^m + mn^{(m+2)/2}/2 + mn^{m/2}/2 + \sum_{i=1}^{m-1} n^{\gcd(i,m)} \right) \Big/ 2m$$

if m is even.

For example, if $m = 3$, then C10 yields $\pi_n(G) = (n^3 + 3n^2 + 2n)/6$. Thus $\pi_2(G) = 4$. One may consider Y as a set of colors; Figure C12 shows one representative of each of the four patterns we have just counted.

C12

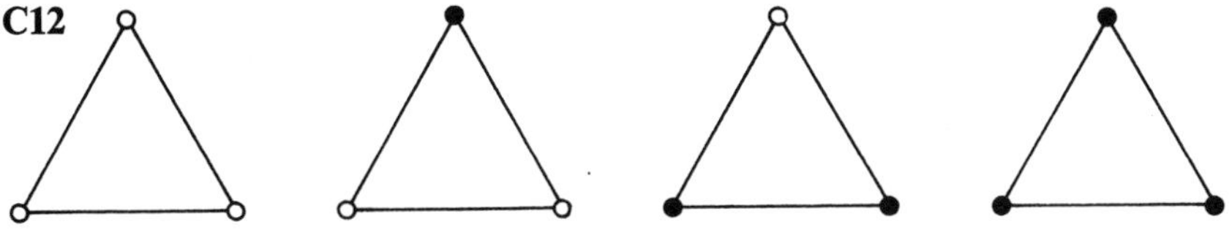

If $m = 4$, then $\pi_n(G)$ is determined by C11:

$$\pi_n(G) = (n^4 + 2n^3 + 3n^2 + 2n)/8.$$

Thus $\pi_2(G) = 6$, and representatives are shown in Figure C13. The reader should list representatives of the 21 patterns when $n = 3$.

C13

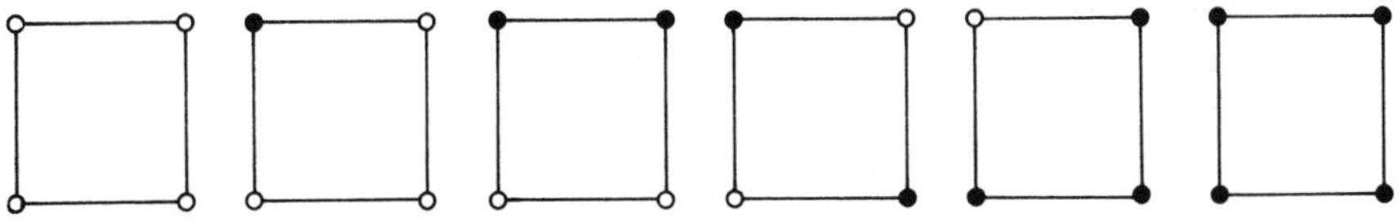

C14 *Exercise.* If m is an odd prime, show that the number of patterns induced by $G = G_0(\Delta_m)$ when Y is an n-set is

$$\pi_n(G) = (n^{m-1} + mn^{(m-1)/2} + m - 1)n/2m.$$

Let us refine the necklace question as follows. Given $k_1, k_2, \ldots \in \mathbb{N}$, how many necklaces can one string using exactly k_i beads of color i, where it is of course understood that $\sum_{i=1}^{\infty} k_i$ is finite? Pólya's theorem addresses itself to this question while the lemmas of this section are inadequate. What we are in fact asking is, how many patterns Q have been given weight $w(Q) = \prod_{i=1}^{\infty} y_i^{k_i}$? The answer is to be found by determining the coefficient of $\prod_{i=1}^{\infty} y_i^{k_i}$ in the right-hand member of the formula in Theorem C7.

For example, how many necklaces can be made with two orange beads, two blue beads and one white bead? With $m = 5$, $k_1 = k_2 = 2$, $k_3 = 1$, and $|G| = 10$, we shudder at the prospect of having to use Póyla's formula, but since in this case $o_3(g) = o_4(g) = 0$ for all $g \in G$, our problem reduces to finding the coefficient of $y_1^2 y_2^2 y_3$ in

$$\{\sum_{g \in G} (y_1 + y_2 + y_3)^{o_1(g)}(y_1^2 + y_2^2 + y_3^2)^{o_2(g)}(y_1^5 + y_2^5 + y_3^5)^{o_5(g)}\}/10$$

$$= \{(y_1 + y_2 + y_3)^5 + 5(y_1 + y_2 + y_3)(y_1^2 + y_2^2 + y_3^2)^2 + 4(y_1^5 + y_2^5 + y_3^5)\}/10.$$

The answer is $(30 + 5 \cdot 2)/10 = 4$. Representatives of these patterns are shown in Figure C15. Note that only one of these corresponds to a vertex 3-coloring of Δ_5, showing that such a vertex coloring is essentially unique.

C15

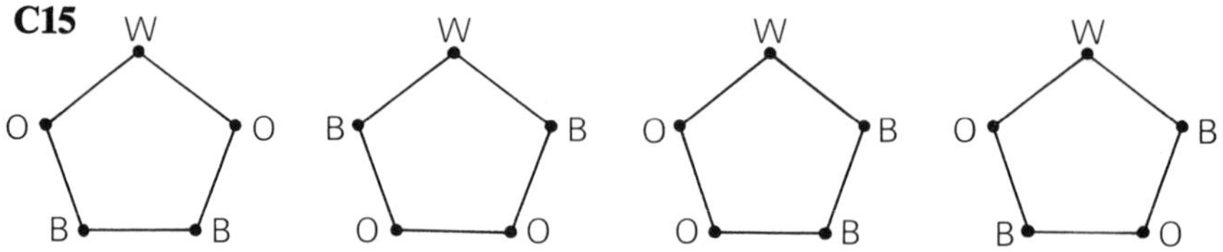

C16 *Exercise.* Show that the number of necklaces that can be made with k_1 orange beads and k_2 blue beads, where $k_1 + k_2$ is an odd prime is

$$\frac{1}{2(k_1 + k_2)}\binom{k_1 + k_2}{k_1} + \frac{1}{2}\binom{(k_1 + k_2 - 1)/2}{[k_1/2]}.$$

No doubt the reader has already surmised that as a practical counting tool, the Pólya Theorem is hardly the epitome of simplicity and efficiency. Even so simple a situation as the necklace problem can lead to a considerable quantity of tedious calculations. In the following exercise we suggest some other patterns to enumerate which are relatively simple. The reader is encouraged to attempt variations on these.

C17 *Exercise.* (a) Count the patterns of functions into an n-set from the set of vertices (respectively, edges, faces) of the regular tetrahedron under all symmetries of the tetrahedron. [Remarks: A decision must be made whether "symmetries" include only physically possible rigid motions in Euclidean 3-space, or whether reflections are also permitted. If reflections are allowed, then G is the symmetric group on a 4-set, and $|G| = 24$. Without reflections, G is the alternating group of order 12. Do the problem both ways. The reader will quickly recognize that the problem is identical when faces instead of vertices are considered. When edges are considered, then $G \cong G_1(K_4)$—if reflections are acceptable. Since K_4 is a connected graph, $G_1(K_4) \cong G_0(K_4)$ by Exercise *III*D2, but notice that the isomorphism refers only to the abstract group structure and not to the action of groups of permutations. For example, the stabilizer of a vertex is of order 6 (the symmetric group on a 3-set), while the stabilizer of an edge is of order 4 (the Klein 4-group).]

(b) Repeat part (a) for the cube. [Arguing via the orthogonal graph, we see that group of symmetries is again the same abstract group for all three cases—vertices, edges, and faces—but is a different permutation group in all three cases. The problem for the regular octahedron is resolved in the process.]

Further Reference

N. G. de Bruijn [b.19].

XID Möbius Functions

A basic notion in §*I*E was the injective function $\mathbb{N}^{\mathscr{P}(U)} \to \mathbb{N}^{\mathscr{P}(U)}$, where U was a finite set, which assigned to each selection s another selection $\bar{s}$. In this section we shall generalize this linear function as well as many of its properties.

On the whole, this section follows the work of G.-C. Rota [r.8], although we shall forgo Rota's fullest generality in some instances.

The partially-ordered set $(\mathscr{P}(U), \subseteq)$ of §*I*E is a Boolean lattice. This notion will be superseded in this section by that of a partially-ordered set $(U, \leq)$, where *U is a finite or infinite set throughout this section, and $(U, \leq)$ is* **locally finite**, i.e., for all $x, y \in U$, the segment

$$U_{[x,y]} = \{z \in U : x \leq z \leq y\}$$

is finite. The collection of finite subsets of U will be denoted by $\mathscr{P}_f(U)$. Thus $(\mathscr{P}_f(U), \subseteq)$ is an example of a locally finite partially-ordered set, but $(\mathscr{P}(U), \subseteq)$ is not locally finite if U is infinite. The next exercise gives another useful example.

D1 Exercise. *For $i = 1, \ldots, n$, let $(U_i, \leq_i)$ be a locally finite partially-ordered set. Let $U = U_1 \times \ldots \times U_n$ and define $(U, \leq)$ by $(x_1, \ldots, x_n) \leq (y_1, \ldots, y_n)$ if $x_i \leq_i y_i$ for $i = 1, \ldots, n$.* Show that *$(U, \leq)$ is a locally finite partially-ordered set.*

Let us extend the "bracket function" $U \times U \to \mathbb{N}$ defined in §*I*E; if $x, y \in U$, then

$$[x, y] = \begin{cases} 1 & \text{if } x \leq y; \\ 0 & \text{if } x \nleq y. \end{cases}$$

If $(U, \leq)$ has a maximum element, then *I*E4 may be generalized; for any selection $s \in \mathbb{S}(U) = \mathbb{N}^U$, we define

$$\bar{s}(x) = \sum_{u \in U} [x, u]s(u) \quad \text{for all } x \in U.$$

Because U is locally finite and has a maximum element, $[x, u] \neq 0$ for only finitely many $u \in U$, and so $\bar{s}(x)$ is always well-defined. Dually, if $(U, \leq)$ has a minimum element, then for each $s \in \mathbb{S}(U)$ we define

$$\tilde{s}(x) = \sum_{u \in U} [u, x]s(u), \quad \text{for all } x \in U.$$

(The selection $\tilde{s}$ generalizes the selection defined in Exercise *I*E23.) Often $(U, \leq)$ is a finite lattice, in which case both the functions $s \mapsto \bar{s}$ and $s \mapsto \tilde{s}$ are always well-defined.

If an ordered field $\mathbb{F}$ replaces $\mathbb{N}$ in the discussion thus far, then the functions $s \mapsto \bar{s}$ and $s \mapsto \tilde{s}$ when well-defined become functions from the vector space $\mathbb{F}^U$ into itself. (We leave it to the reader to show that they are in fact injective linear transformations. Cf. Exercise *I*E7.) Actually the functions $s \mapsto \bar{s}$ and $s \mapsto \tilde{s}$ are injective in any case. Let U have a maximum element and a minimum element in $\mathbb{F}$, and let us define

$$\bar{\mathbb{S}}(U) = \{\bar{s} : s \in \mathbb{S}(U)\}; \qquad \tilde{\mathbb{S}}(U) = \{\tilde{s} : s \in \mathbb{S}(U)\}.$$

The injectivity will be demonstrated constructively; the inverse functions $\bar{s} \mapsto s$ from $\bar{\mathbb{S}}(U)$ onto $\mathbb{S}(U)$ and $\tilde{s} \mapsto s$ from $\tilde{\mathbb{S}}(U)$ onto $\mathbb{S}(U)$ will presently be given explicitly. We will thereby generalize Proposition *I*E6.

The **Möbius function** of a partially-ordered set $(U, \leq)$ is the function $\mu: U \times U \to \mathbb{Z}$ defined inductively as follows:

D2 $$\mu(x, x) = 1 \quad \text{for all } x \in U;$$

D3 $$\mu(x, y) = -\sum_{x \leq u < y} \mu(x, u) \quad \text{for all } x, y \in U.$$

Since the sum over an empty set is 0, it follows from D3 that

D4 $$\mu(x, y) = 0 \quad \text{if } x \nleq y.$$

Let $\delta_{x,y}$ denote the Kronecker delta. We also have

D5 $$\sum_{u \in U} \mu(x, u)[u, y] = \sum_{x \leq u \leq y} \mu(x, u) = \delta_{x,y} \quad \text{for all } x, y \in U.$$

Example. Let X be a finite set, and let us determine the Möbius function for $(\mathscr{P}(X), \subseteq)$. Let $S, T \in \mathscr{P}(X)$. It should quickly become apparent that if $S \subseteq T$, then $\mu(S, T)$ is dependent only upon $|S + T|$, the length of a chain from S to T in the Boolean lattice $(\mathscr{P}(X), \subseteq)$. We prove inductively that

D7 $$\mu(S, T) = (-1)^{|S+T|}[S, T] \quad \text{for all } S, T \in \mathscr{P}(X).$$

By D2 and D3, $\mu(S, S + \{x\}) = -1$ for all $x \in U + S$. Now suppose $\mu(S, R) = (-1)^{|S+R|}[S, R]$ for all sets R such that $S \subseteq R \subset T$ and such that $|R| \leq k$ for some $k < |T|$. If $Q \in \mathscr{P}_{k+1}(T)$ and $S \subseteq Q$, then

$$\mu(S, Q) = -\sum_{S \subseteq R \subset Q} \mu(S, R) = -\sum_{i=0}^{k-|S|} \binom{k+1-|S|}{i-|S|}(-1)^i = (-1)^{k+1-|S|}$$

by Corollary *I*C22, as required.

In the light of the foregoing example, the reader should recognize the following lemma as a generalization of Lemma *I*E5.

D8 Lemma. *If μ is the Möbius function for $(U, \leq)$, then for $x, y \in U$,*

$$\sum_{u \in U} [x, u]\mu(u, y) = \delta_{x,y}.$$

Proof. We define $\tilde{\mu}: U \times U \to \mathbb{Z}$ inductively by

$$\tilde{\mu}(x, x) = 1, \quad \text{for all } x \in U;$$
$$\tilde{\mu}(x, y) = -\sum_{x < u \leq y} \tilde{\mu}(u, y) \quad \text{for all } x, y \in U.$$

Dual to D5 we have

D9 $$\sum_{u \in U} [x, u]\tilde{\mu}(u, y) = \delta_{x,y} \quad \text{for all } x, y \in U.$$

We show that $\mu = \tilde{\mu}$. By D9 and D5, for all $x, y \in U$,

$$\mu(x, y) = \sum_{w \in U} \mu(x, w)\delta_{w,y} = \sum_{w \in U} \mu(x, w) \sum_{z \in U} [w, z]\tilde{\mu}(z, y)$$
$$= \sum_{z \in U} \tilde{\mu}(z, y) \sum_{w \in U} \mu(x, w)[w, z] = \sum_{z \in U} \tilde{\mu}(z, y)\delta_{x,z} = \tilde{\mu}(x, y). \quad \square$$

The following is the main result of this section.

D10 Theorem (Inversion Formulas). *Let μ be the Möbius function of $(U, \leq)$. For any selection $s \in \mathbb{S}(U)$ and for all $x \in U$,*
(a) *$s(x) = \sum_{u \in U} \mu(x, u)\bar{s}(u)$ if $(U, \leq)$ has a maximum element;*
(b) *$s(x) = \sum_{u \in U} \mu(u, x)\tilde{s}(u)$ if $(U, \leq)$ has a minimum element.*

PROOF. (a) Using D5 and the definition of $\bar{s}$, we have

$$\sum_{u \in U} \mu(x, u)\bar{s}(u) = \sum_{u \in U} \mu(x, u) \sum_{y \in U} [u, y]s(y)$$
$$= \sum_{y \in U} s(y) \sum_{u \in U} \mu(x, u)[u, y]$$
$$= \sum_{y \in U} s(y)\delta_{x,y} = s(x).$$

(b) Using Lemma D9 and the definition of $\tilde{s}$, one proceeds in a manner similar to (a) above. The details are left to the reader. $\square$

The above theorem makes evident that the question of determining any two of s, $\bar{s}$, and $\tilde{s}$ when the third selection is given reduces to determining the Möbius function of the appropriate partially-ordered set. For the rest of this section, we shall be concerned mainly with just that, the computation of certain Möbius functions.

D11 Exercise. *Let $(U, \leq)$ be a lattice with Möbius function μ.* Show that *for any $x, y \in U$ with $x \leq y$, the Möbius function of $(U_{[x,y]}, \leq)$ is the restriction of μ to $U_{[x,y]} \times U_{[x,y]}$.* Show that *$U_{[x,y]}$ may be replaced by a set of the form* $\{u \in U: x \leq u\}$ or $\{u \in U: u \leq x\}$.

D12 *Exercise.* Determine the Möbius function for $(\mathscr{P}_f(U), \leq)$.

D13 Proposition. *For $i = 1, \ldots, k$, let $(U_i, \leq_i)$ be a locally finite partially-ordered set with Möbius function μ_i. Let μ be the Möbius function for $(U, \leq)$, where $U = U_1 \times \ldots \times U_k$, and $(x_1, \ldots, x_k) \leq (y_1, \ldots, y_k)$ means that $x_i \leq_i y_i$ for $i = 1, \ldots, k$. Then μ is given by $\mu((x_1, \ldots, x_n), (y_1, \ldots, y_n)) = \prod_{i=1}^{k} \mu_i(x_i, y_i)$.*

PROOF. The proof will be given for $k = 2$; the general case will then follow straightforwardly by induction. By Exercise D1, everything is well-defined.

Clearly $\mu((x_1, x_2), (x_1, x_2)) = 1 = \mu_1(x_1, x_1)\mu_2(x_2, x_2)$. Now suppose $(x_1, x_2) < (y_1, y_2)$ and that $\mu((x_1, x_2), (u_1, u_2)) = \mu_1(x_1, u_1)\mu_2(x_2, u_2)$ whenever

$$(x_1, x_2) \leq (u_1, u_2) < (y_1, y_2).$$

Summing over all (u_1, u_2) satisfying the above inequality, we have by D3,

$$\mu((x_1, x_2), (y_1, y_2)) = -\sum \mu_1(x_1, u_1)\mu_2(x_2, u_2).$$

This in turn equals

$$\begin{aligned} &-\sum_{x_1 \leq u_1 < y_1} \sum_{x_2 \leq u_2 < y_2} \mu_1(x_1, u_1)\mu_2(x_2, u_2) \\ &\qquad -\mu_1(x_1, y_1) \sum_{x_2 \leq u_2 < y_2} \mu_2(x_2, u_2) - \mu_2(x_2, y_2) \sum_{x_1 \leq u_1 < y_1} \mu_1(x_1, u_1) \\ &\qquad = -\mu_1(x_1, y_1)\mu_2(x_2, y_2) + 2\mu_1(x_1, y_1)\mu_2(x_2, y_2) \\ &\qquad = \mu_1(x_1, y_1)\mu_2(x_2, y_2). \end{aligned}$$

□

If $(U_1, \leq_1) = (U_i, \leq_i)$ for $i = 1, \ldots, k$, then $\leq$ will be understood to indicate the partial order given above for $(U^k, \leq)$.

D14 Corollary. *The Möbius function μ for $(\mathbb{N}^k, \leq)$ and for $(\mathbb{Z}^k, \leq)$ is given by*

$$\mu((m_1, \ldots, m_k), (n, \ldots, n_k)) = \begin{cases} (-1)^{\Sigma_i(n_i - m_i)} & \text{if } m_i \leq n_i \leq m_i + 1 \\ & \quad \text{for } i = 1, \ldots, k; \\ 0 & \text{otherwise.} \end{cases}$$

Proof. The Möbius function μ_0 for $(\mathbb{Z}, \leq)$ and $(\mathbb{N}, \leq)$ is clearly given by

$$\mu_0(m, n) = \begin{cases} 1 & \text{if } m = n; \\ -1 & \text{if } n = m + 1; \\ 0 & \text{otherwise.} \end{cases}$$

Now apply the theorem. □

The Möbius function as defined in this chapter is a generalization of a classical function used in number theory. It is the Möbius function for the partially ordered set $(\mathbb{N} + \{0\}, |)$. (Cf. Example *II*B27.)

D15 Proposition. *The Möbius function μ for $(\mathbb{N} + \{0\}, |)$ is given by*

$$\mu(m, n) = \begin{cases} (-1)^j & \text{if } n/m \text{ is a product of } j > 0 \text{ distinct primes;} \\ 1 & \text{if } m = n; \\ 0 & \text{otherwise.} \end{cases}$$

Proof. Let $m, n \in \mathbb{N} + \{0\}$. In the light of D2 and D4, we may assume that $n/m = p_1^{e_1} \ldots p_k^{e_k}$, where $p_1, \ldots, p_k$ are distinct primes and $e_1, \ldots, e_k \in \mathbb{N} + \{0\}$. One easily verifies that the segment $\mathbb{N}_{[1,n]}$ of $(\mathbb{N} + \{0\}, |)$ is isomorphic to the segment $\mathbb{N}^k_{[(0,\ldots,0),(e_1,\ldots,e_k)]}$ of $(\mathbb{N}^k, \leq)$ under the isomorphism $m \mapsto (d_1, \ldots, d_k)$, where $m = p_1^{d_1} \ldots p_k^{d_k}$. It follows from this and Exercise D11 that if μ' is the Möbius function for $(\mathbb{N}^k, \leq)$, then

$$\mu(m, n) = \mu'((d_1, \ldots, d_k), (e_1, \ldots, e_k)).$$

However, μ' can be evaluated easily by means of Corollary D14, and the proposition follows. □

D16 Corollary. *If* $m, n \in \mathbb{N} + \{0\}$ *and if* $m|n$, *then* $\mu(1, m) = \mu(n/m, n)$.

D17 Exercise. Show that $(\mathbb{N} + \{0\}, |)$ *is a locally finite lattice with minimum element* 1.

As an application of the Möbius function for $(\mathbb{N} + \{0\}, |)$, we shall give another proof of Theorem *I*E21, which is a formula for the Euler φ-function.

The function φ is a selection; $\varphi \in \mathbb{S}(\mathbb{N} + \{0\})$. Since $(\mathbb{N} + \{0\}, |)$ has the properties described in Exercise D17, we may compute for any $m \in \mathbb{N} + \{0\}$,

$$\tilde{\varphi}(m) = \sum_{k \in \mathbb{N}+\{0\}} [k, m]\varphi(k) = \sum_{k|m} \varphi(k) = \sum_{k|m} \varphi(m/k).$$

For each divisor k of m, let $S_k = \{j \in \mathbb{N}_{[1,m]}: \gcd(j, m) = k\}$. Then $|S_k| = \varphi(m/k)$ and $\{S_k: k|m\} \in \mathbb{P}(\{1, \ldots, m\})$. It follows that $\tilde{\varphi}(m) = \sum_{k|m} \varphi(m/k) = m$ for all $m \in \mathbb{N} + \{0\}$. For any $n \in \mathbb{N} + \{0\}$, if D denotes the collection of prime divisors of n, then

$$\begin{aligned}
\varphi(n) &= \sum_{m \in \mathbb{N}+\{0\}} \mu(m, n)\tilde{\varphi}(m) && \text{(Theorem D10b)} \\
&= \sum_{m|n} \mu(m, n)m && \text{(Proposition D15)} \\
&= \sum_{m|n} \mu(n/m, n)n/m && \\
&= n \sum_{m|n} \mu(1, m)/m && \text{(Corollary D16)} \\
&= n \sum_{C \in \mathscr{P}(D)} (-1)^{|C|} \Big/ \prod_{p \in C} p && \\
&= n \prod_{p \in D} (1 - 1/p), &&
\end{aligned}$$

the last step requiring the same algebraic manipulation which we also evaded in the proof of *I*E21.

In §*V*F the permanent of a matrix was introduced. Its value was shown in Proposition *V*F1 to be equal to two other enumerations, neither of which was given by an explicit formula. As our final application of Möbius functions, we shall derive a formula for the permanent. The partially-ordered set for which the Möbius function will be computed is the lattice of partitions of a finite set ordered by refinement.

Let $(\mathbb{P}(X), \leq)$be the lattice of partitions of the finite set X, ordered by refinement (cf. Exercise *II*B30). If $\mathscr{Q}, \mathscr{R} \in \mathbb{P}(X)$ and $\mathscr{Q} \leq \mathscr{R}$, then for any $R \in \mathscr{R}$, we have $\{Q \in \mathscr{Q}: Q \subseteq R\} \in \mathbb{P}(R)$. This partition is called the **restriction of $\mathscr{Q}$ to R** and is denoted by $\mathscr{Q}(R)$.

D18 Proposition. *TheMöbius function μ for the lattice* $(\mathbb{P}(X), \leq)$ *is*

$$\mu(\mathscr{Q}, \mathscr{R}) = \prod_{R \in \mathscr{R}} (-1)^{|\mathscr{Q}(R)|-1}(|\mathscr{Q}(R)| - 1)!.$$

PROOF. Let $\mathscr{R} = \{R_1, \ldots, R_k\} \in \mathbb{P}(X)$, and suppose $\mathscr{Q} \leq \mathscr{R}$. For each $i = 1, \ldots, k$, let $\mathscr{Q}_i = \mathscr{Q}(R_i)$. One easily sees that the segment $\mathbb{P}(X)_{[\mathscr{Q},\mathscr{R}]}$ of $(\mathbb{P}(X), \leq)$ is isomorphic to the direct product

$$\mathbb{P}(R_1)_{[\mathscr{Q}_1,\{R_1\}]} \times \ldots \times \mathbb{P}(R_k)_{[\mathscr{Q}_k,\{R_k\}]}$$

of segments, which is in turn a segment of $(\mathbb{P}(R_1), \leq) \times \ldots \times (\mathbb{P}(R_k), \leq)$. Since each segment $\mathbb{P}(R_i)_{[\mathscr{Q}_i,\{R_i\}]}$ is isomorphic to the segment $\mathbb{P}(\mathscr{Q}_i)_{[\mathscr{P}_1(\mathscr{Q}_i),\{\mathscr{Q}_i\}]}$, it follows that $\mathbb{P}(X)_{[\mathscr{Q},\mathscr{R}]}$ is isomorphic to

$$\mathbb{P}(\mathscr{Q}_1)_{[\mathscr{P}_1(\mathscr{Q}_1),\{\mathscr{Q}_1\}]} \times \ldots \times \mathbb{P}(\mathscr{Q}_k)_{[\mathscr{P}_1(\mathscr{Q}_k),\{\mathscr{Q}_k\}]}.$$

Hence $\mu(\mathscr{Q}, \mathscr{R}) = \prod_{i=1}^{k} \mu(\mathscr{P}_1(\mathscr{Q}_i), \{\mathscr{Q}_i\})$ by D13, where μ_i is the Möbius function for $(\mathbb{P}(\mathscr{Q}_i), \leq)$, for $i = 1, \ldots, k$. The proof therefore reduces to proving that for any finite set S, the Möbius function μ_0 of $(\mathbb{P}(S), \leq)$ satisfies

D19 $$\mu_0(\mathscr{P}_1(S), \{S\}) = (-1)^{|S|-1}(|S| - 1)!.$$

We proceed by induction on $|S|$. When $|S| = 1$, both sides of D19 reduce to 1. Let S be given with $|S| > 1$, and as induction hypothesis assume that D19 holds for sets T where $|T| < |S|$.

If $\mathscr{P}_1(S) \leq \mathscr{Q} < \{S\}$, then by the same arguments as before, $\mathbb{P}(S)_{[\mathscr{P}_1(S),\mathscr{Q}]}$ is isomorphic to a cartesian product of segments of the form $\mathbb{P}(T)_{[\mathscr{P}_1(T),\{T\}]}$ where T is a set such that $|T| < |S|$. Hence by the induction hypothesis,

$$\mu_0(\mathscr{P}_1(S), \mathscr{Q}) = \prod_{Q \in \mathscr{Q}} (-1)^{|Q|-1}(|Q| - 1)!.$$

Combining this with D3, we see that we must prove

D20 $$\sum_{\mathscr{Q} \in \mathbb{P}(S)} \prod_{Q \in \mathscr{Q}} (-1)^{|Q|-1}(|Q| - 1)! = 0 \quad \text{for } |S| > 1.$$

Fix $x \in S$. Every partition of S may be regarded as the union of the trivial partition of a set containing x with a partition of the complement of that set. In this way, D20 is seen to be equivalent to

D21 $$\sum_{x \in Q \in \mathscr{P}(S)} \left[(-1)^{|Q|-1}(|Q| - 1)! \sum_{\mathscr{R} \in \mathbb{P}(S+Q)} \prod_{R \in \mathscr{R}} (-1)^{|R|-1}(|R| - 1)!\right] = 0.$$

If $|S + Q| > 1$, then the induction hypothesis applied to D20 implies that the second summation in D21 is 0. That summation clearly reduces to $|S| - 1$ if $|S + Q| = 1$. Hence the left-hand member of D21 reduces to

$$(-1)^{|S|-1}(|S| - 1)! + (-1)^{|S|-2}(|S| - 2)!\,(|S| - 1),$$

where the first term comes from the case $Q = S$. But this expression is identically 0. □

Armed with the above proposition, we proceed to compute perm(A), where $A = [a_{ij}]$ is an $n \times m$ matrix over $\mathbb{C}$. Let $N = \{1, \ldots, n\}$. For each $\mathcal{Q} \in \mathbb{P}(N)$, let

$$H(\mathcal{Q}) = \{h \in N^N : h(i) = h(j) \text{ if and only if } i, j \in Q \text{ for some } Q \in \mathcal{Q}\}.$$

and define

$$t(\mathcal{Q}) = \sum_{h \in H(\mathcal{Q})} \prod_{i=1}^{n} a_{i,h(i)}.$$

We observe that $H(\mathscr{P}_1(N)) = \Pi(N)$, and so $t(\mathscr{P}_1(N)) = \text{perm}(A)$ as defined in §VF.

Let $s\colon \mathbb{P}(N) \to \mathbb{C}$ be given by

D22
$$s(\mathcal{Q}) = \prod_{Q \in \mathcal{Q}} \sum_{j=1}^{m} \prod_{i \in Q} a_{ij}.$$

(If $|Q| = 2$, say $Q = \{i_1, i_2\}$, then $\sum_{j=1}^{m} \prod_{i \in Q} a_{ij}$ reduces to the inner product of the i_1-th and i_2-th rows.) The functions s and t are generalizations of the notion of a selection, since their range is in $\mathbb{C}$ rather than $\mathbb{N}$. Nonetheless, the inversion formula D10a can still be applied, as the reader is now asked to show.

D23 Exercise. *Continuing the above notation,* show that *for all* $\mathcal{Q} \in \mathbb{P}(N)$,

(a) $s(\mathcal{Q}) = \sum_{\mathcal{Q} \le \mathscr{R}} t(\mathscr{R})$ (i.e., $s = \bar{t}$);

(b) $t(\mathcal{Q}) = \sum_{\mathscr{R} \in \mathbb{P}(N)} \mu(\mathcal{Q}, \mathscr{R}) s(\mathscr{R})$.

D24 Proposition. *Let* $A = [a_{ij}]$ *be an* $n \times m$ *matrix over* $\mathbb{C}$ *and let* $N = \{1, \ldots, n\}$. *Then*

$$\text{perm}(A) = \sum_{\mathscr{R} \in \mathbb{P}(N)} \prod_{R \in \mathscr{R}} (-1)^{(|R|-1)}(|R| - 1)! \sum_{j=1}^{m} \prod_{i \in R} a_{ij}.$$

PROOF. From Exercise D23b with $\mathcal{Q} = \mathscr{P}_1(N)$ we have

$$\text{perm}(A) = t(\mathscr{P}_1(N)) = \sum_{R \in \mathbb{P}(N)} \mu(\mathscr{P}_1(N), \mathscr{R}) s(\mathscr{R}),$$

into which we substitute $\mu(\mathscr{P}_1(N), \mathscr{R}) = \prod_{R \in \mathscr{R}} (-1)^{|R|-1}(|R| - 1)!$ from Proposition D18 and substitute $s(\mathscr{R})$ from D22. □

Bibliography

The numbers of the pages where each item is cited are listed in parentheses at the right.

a.0 Alltop, W. O. 1972. An infinite class of 5-designs. *J. Combinatorial Theory* **(A) 12**, 390–395. (241)

a.1 Appel, K., Haken, W., 1976. Every planar map is four colorable. *Bull. Amer. Math. Soc.* **82**, 711–712. (188)

a.2 Auslander, L., Brown, T. A., and Youngs, J. W. T. 1963. The imbedding of graphs in manifolds. *J. Math. Mech.* **12**, 629–634. (203)

b.1 Battle, J., Harary, F., Kodama, Y., and Youngs, J. W. T. 1962. Additivity of the genus of a graph. *Bull. Amer. Math. Soc.* **68**, 565–568. (208)

b.2 Bell, E. T. 1923. Euler algebra. *Trans. Amer. Math. Soc.* **25**, 135–154. (310)

b.3 Berge, C. 1958. Sur le couplage maximum d'un graphe. *C. R. Acad. Sci. Paris* **247**, 258–259. (137)

b.4 Berge, C. 1961. Färbung von Graphen deren sämtliche bzw. ungerade Kreise starr sind (Zusammenfassung), *Wiss. Z. Martin Luther Univ. Halle-Wittenberg. Math.-Nat. Reihe* **10**, 114. (224)

b.5 Berge, C. 1970. *Graphes et hypergraphes*. Paris: Dunod. (18, 138, 270)

b.6 Berge, C. 1971. *Principles of Combinatorics*. New York: Academic Press. (322)

b.7 Bhattacharya, K. N. 1944. A new balanced incomplete block design. *Science and Culture* **9**, 508. (246)

b.8 Biggs, N. L., Lloyd, E. K., and Wilson, R. J. 1976. *Graph Theory, 1736–1936*. Ely House, London: Oxford University Press. (60)

b.9 Birkhoff, G. 1967. *Lattice Theory*. Colloquium Publications 25. Providence, R. I.: Amer. Math. Soc. (285)

b.10 Bose, R. C. 1939. On the construction of balanced incomplete block designs. *Ann. Eugenics* **9**, 353–399. (252)

b.11 Bose, R. C. 1963. Strongly regular graphs, partial geometries, and partially balanced designs. *Pacific J. Math.* **13**, 389–419. (262, 264)

b.12 Bose, R. C. and Clatworthy, W. H. 1955. Some classes of partially balanced designs. *Ann. Math. Stat.* **26**, 217–232. (257, 258, 264)

b.13 Bose, R. C. and Mesner, D. M. 1959. On linear associative algebras corresponding to association schemes of partially balanced designs. *Ann. Math. Stat.* **30**, 21–38. (257, 263)

b.14 Bose, R. C. and Nair, K. R. 1939. Partially balanced incomplete block designs. *Sankhyā* **4**, 337–372. (255, 259)

b.15 Bose, R. C., Parker, E. T., and Shrikhande, S. S. 1960. Further results on the construction of mutually orthogonal Latin squares and the falsity of Euler's conjecture. *Canad. J. Math.* **12**, 189–203. (252)

b.16 Bose, R. C. and Shimamoto, **B**. 1952. Classification and analysis of partially balanced incomplete block designs with associate classes. *J. Amer. Statist. Assoc.* **47**, 151–184. (255)

b.17 Bose, R. C. and Shrikhande, S. S. 1960. On the construction of sets of mutually orthogonal Latin squares and the falsity of a conjecture by Euler. *Trans. Amer. Math. Soc.* **95**, 191–209. (252)

b.18 Brooks, R. L. 1941. On colouring the nodes of a network. *Proc. Cambridge Philos. Soc.* **37**, 194–197. (185)

Brown, T. A.: See a.2.

b.19 de Bruijn, N. G. 1964. Pólya's theory of counting. In: *Applied Combinatorial Theory*, E. F. Beckenbach (ed.), pp. 144–184. New York: Wiley. (328)

b.20 Burnside, W. 1911. *Theory of Groups of Finite Order*. Cambridge, England: Cambridge University Press. (323)

c.1 Cayley, A. 1889. A theorem on trees. *Quart. J. Pure Appl. Math.* **23**, 376–378. (322)

c.2 Chowla, S. and Ryser, H. J. 1950. Combinatorial problems. *Canad. J. Math.* **2**, 93–99. (245)

c.3 Chvátal, V. 1973. New directions in hamiltonian graph theory. In: *New Directions in the Theory of Graphs*, F. Harary (ed.), pp. 65–95. New York: Academic Press. (165)

Clatworthy, W. H.: See b.12.

c.4 Connor, W. S. and Hall, M., Jr. 1953. An imbedding theorem for balanced incomplete block designs. *Canad. J. Math.* **6**, 35–41. (246)

c.5 Crapo, H. H. and Rota, G.-C. 1971. *On the Foundations of Combinatorial Theory: Combinatorial Geometries*. Cambridge, Mass.: M. I. T. Press. (265, 284)

d.1 Descartes, B. 1948. Network colourings. *Math. Gaz.* **32**, 67–69. (187)

d.2 Dilworth, R. P. 1950. A decomposition theorem for partially ordered sets. *Ann. of Math.* (2) **51**, 161–166. (123)

d.3 Dirac, G. A. 1952. Map-colour theorems. *Canad. J. Math.* **4**, 480–490. (179, 206)

d.4 Dirac, G. A. 1957. A theorem of R. L. Brooks and a conjecture of Hadwiger. *Proc. London Math. Soc.* **7**, 161–195. (181, 185) d.5 Dirac, G. A. 1957. Short proof of a map-colour theorem. *Canad. J. Math.* **9**, 225–226. (206)

d.6 Dirac, G. A. 1960. Généralisations du théorème de Menger. *C. R. Acad. Sci. Paris* **250**, 4252–4253. (157)

d.7 Dirac, G. A. 1960. In abstrakten Graphen vorhandene vollständige 4-Graphen und ihre Unterteilungen. *Math. Nachr.* **22**, 61–85. (163)

d.8 Doyle, J. K. and Leska, C. J. *On the existence of one-sided designs.* To appear. (237)

e.1 Edmonds, J. 1960. A combinatorial representation for polyhedral surfaces. *Notices Amer. Math. Soc.* **7**, 646. (198, 208)

e.2 Edmonds, J. 1965. Paths, trees and flowers. *Canad. J. Math.* **17**, 449–467. (136, 268)

e.3 Egarváry, E. *On combinatorial properties of matrices.* Translated by H. W. Kuhn 1953. Office Naval Res. Logist. Project Rept. Dept. of Math.: Princeton University. (147)

e.4 Erdős, P. and Gallai, T. 1959. On maximal paths and circuits of graphs. *Acta Math. Acad. Sci. Hungar.* **10**, 337–356. (138)

e.5 Erdős, P., Rényi, A., and Sós, V. 1966. On a problem in graph theory. *Studia Sci. Math. Hungar.* **1**, 215–235. (260)

e.6 Erdős, P. and Szekeres, G. 1935. A combinatorial problem in geometry. *Compositio Math.* **2**, 463–470. (216)

e.7 Errera, A. 1922. Du coloriage des cartes. *Mathesis* **36**, 56–60. (138)

e.8 Euler, L. 1736. Solutio problematis ad geometriam situs pertinentis. *Commentarii Academiae Petropolitanae* **8**, 128–140. (61)

f.1 Fáry, I. 1948. On straight line representations of planar graphs. *Acta Sci. Math. (Szeged)* **11**, 229–233. (190)

f.2 Folkman, J. 1974. Notes on the Ramsey number N(3, 3, 3, 3). *J. Combinatorial Theory* **(A)16**, 371–379. (223)

f.3 Ford, L. R., Jr. and Fulkerson, D. R. 1962. *Flows in Networks.* Princeton, New Jersey: Princeton University. (117)

Fulkerson, D. R.: See f.3.

g.1 Gale, D. 1968. Optimal assignments in an ordered set. *J. Combinatorial Theory* **4**, 176–180. (269)

g.2 Gallai, T. 1959. Über extreme Punkt- und Kantenmengen. *Ann. Univ. Sci. Budapest Eőtvős Sect. Math.* **2**, 133–135. (140)

g.3 Gallai, T. 1964. Maximale Systeme unabhängiger Kanten. *Math Kut. Int. Kőzl.* **9**, 353–395. (136)

Gallai, T.: See also e.4.

Gleason, A. M.: See g.6.

g.4 Graver, J. E. and Jurkat, W. 1973. The module structure of integral designs. *J. Combinatorial Theory* **(A)15**, 75–90. (241)

g.5 Graver, J. E. and Yackel, J. 1968. Some graph theoretical results associated with Ramsey's Theorem. *J. Combinatorial Theory* **4**, 125–175. (222)

g.6 Greenwood, R. E. and Gleason, A. M. 1955. Combinatorial relations and chromatic graphs. *Canad. J. Math.* **7**, 1–7. (222, 223) g.7 Grözsch, H. 1958. Ein Dreifarbensatz für dreikreisfreie Netze auf der Kugel. *Wiss. Z. Martin Luther Univ. Halle-Wittenberg. Math.-Nat. Reihe* **8**, 109–119. (190)

Haken, W.: See a.1.

h.1 Hall, M., Jr. 1948. Distinct representatives of subsets. *Bull. Amer. Math. Soc.* **54**, 922–926. (149)

h.2 Hall, M., Jr. 1964. Block designs. In: *Applied Combinatorial Mathematics*, F. Beckenbach (ed.), pp. 369–405. New York: Wiley. (246)

h.3 Hall, M., Jr. 1967. *Combinatorial Theory*. Waltham, Mass.: Blaisdell. (150, 251)

Hall, M., Jr.: See also c.4.

h.4 Hall, Phillip. 1935. On representatives of subsets. *J. London Math. Soc.* **10**, 26–30. (142)

h.5 Hanani, H. 1961. The existence and construction of balanced incomplete block designs. *Ann. Math. Stat.* **32**, 361–368. (243)

h.6 Harary, F. and Palmer, E. M. 1968. On the point-group and line-group of a graph. *Acta Math. Acad. Sci. Hungar.* **19**, 263–269. (71)

Harary, F.: See also b.1.

h.7 Heawood, P. J. 1890. Map-colour theorem. *Quart. J. Math.* **24**, 332–338. (201, 204)

h.8 Hoffman, A. J. and Singleton, R. R. 1960. On Moore graphs with diameters 2 and 3. *IBM J. Res. Develop.* **4**, 497–504. (261)

j.1 Jung, H. A. 1970. Eine Verallgemeinerung des *n*-fachen Zusammenhangs für Graphen. *Math. Ann.* **187**, 95–103. (163, 171)

j.2 Jung, H. A. 1973. Über den Zusammenhang von Graphen mit Anwendungen auf symmetrische Graphen. *Math. Ann.* **202**, 307–320. (171)

j.3 Jung, H. A. and Watkins, M. E. 1977. On the connectivities of finite and infinite graphs. *Monatsh. Math.* **83**, 121–131. (171)

Jurkat, W.: See g.4.

k.1 Kalbfleisch, J. G. 1966. *Chromatic Graphs and Ramsey's Theorem*. Doctoral dissertation, University of Waterloo. (222)

k.2 Kemeny, J. G., Snell, J. L., and Thompson, G. L. 1966. *Introduction to Finite Mathematics*. Englewood Cliffs, New Jersey: Prentice-Hall. (16)

k.3 Kempe, A. B. 1879. On the geographical problem of four colours. *Amer. J. Math.* **2**, 193–200. (194)

k.4 Kirkman, T. A. 1847. On a problem in combinatorics. *Cambridge and Dublin Math. J.* **2**, 191–204. (243)

Kodama, Y.: See b.1.

k.5 König, D. 1936. *Theorie der endlichen und unendlichen Graphen*. Leipzig: Akad. Verl. M. B. H. (140)

k.6 Kronk, H. V. and Mitchem, J. 1972. On Dirac's generalization of Brooks' theorem. *Canad. J. Math.* **24**, 805–807. (181)

k.7 Kuratowski, K. 1930. Sur le problème des courbes gauches en topologie. *Fund. Math.* **15**, 271–283. (92)

ℓ.1 Larman, D. G. and Mani, P. 1970. On the existence of certain configurations within graphs and the 1-skeletons of polytopes. *Proc. London Phil. Soc.* (**3**) **20**, 144–160. (163)

Leska, C. J.: See d.8.

ℓ.2 Liu, C. L. 1968. *Introduction to Combinatorial Mathematics*. New York: McGraw Hill. (16)

Lloyd, E. K.: See b.8.

ℓ.3 Lovász, L. 1972. Normal hypergraphs and the perfect graph conjecture. *Discrete Math.* **2**, 253–267. (225)

ℓ.4 Lovász, L. 1972. A characterization of perfect graphs. *J. Combinatorial Theory* **13**, 95–98. (225)

ℓ.5 Lovász, L. 1973. Factors of a graph. In: *Proceedings of the Fourth Southeastern Conference on Combinatorics and Graph Theory*, F. Hoffman, R. B. Levow, R. S. D. Thomas (ed.), pp. 13–22. Winnipeg: Utilitas Math. (138)

m.1 MacLane, S. 1937. A structural characterization of planar combinatorial graphs. *Duke Math. J.* **3**, 340–472. (75)

m.2 Mader, W. 1970. Über den Zusammenhang symmetrischer Graphen. *Arch. Math.* **21**, 331–336. (171)

m.3 Mader, W. 1971. Eine Eigenschaft der Atome endlicher Graphen. *Arch. Math. (Basel)* **22**, 333–336. (171)

m.4 Mader, W. 1973. 1-Faktoren von Graphen. *Math. Ann.* **201**, 269–282. (136)

Mani, P.: See ℓ.1.

m.5 Mann, H. B. 1942. The construction of orthogonal Latin squares. *Ann. Math. Stat.* **13**, 418–423. (252)

m.6 Mann, H. B. and Ryser, H. J. 1953. Systems of distinct representatives. *Amer. Math. Monthly* **60**, 397–401. (149)

m.7 Massey, W. S. 1967. *Algebraic Topology: an Introduction*. New York: Harcourt, Brace and World. (201)

m.8 Mel'nikov, L. S. and Vizing, V. G. 1969. New proof of Brooks' theorem. *J. Combinatorial Theory* **7**, 289–290. (185)

m.9 Menger, K. 1927. Zur allgemeinen Kurventheorie. *Fund. Math.* **10**, 96–115. (154)

m.10 Mesner, D. M. and Watkins, M. E. 1966. Some theorems about n-vertex connected graphs. *J. Math. Mech.* **16**, 321–326. (166)

Mesner, D. M.: See also b.13 and w.6.

m.11 Mirsky, L. and Perfect, H. 1967. Applications of the notion of independence to problems of combinatorial analysis. *J. Combinatorial Theory* **2**, 327–357. (295)

m.12 Mirsky, L. 1975. The combinatorics of arbitrary partitions. *Bull. Inst. Math. Appl.* **11**, 6–9. (217)

Mitchem, J.: See k.6.

Nair, K. R.: See b.14.

n.1 Nordhaus, E. A. 1972. On the girth and genus of a graph. In: *Graph Theory And Applications*, Proc. Conf. Western Michigan University. Y. Alavi, D. R. Lick, A. T. White (ed.), pp. 207–214, Lecture Notes in Math., vol. 303. Berlin: Springer-Verlag. (212)

o.1 Ore, O. 1962. *Theory of Graphs*. Colloq. Pub. 38, Providence, R. I.: Amer. Math. Soc. (66, 165)

o.2 Ore, O. 1967. *The Four-Color Problem*. New York: Academic Press. (187,190)

Palmer, E. M.: See h.6.

p.1 Parker, E. T. 1959. Constructions of some sets of mutually orthogonal Latin squares. *Proc. Amer. Soc.* **10**, 946–949. (252)

Parker, E. T.: See also b.15.

p.2 Perfect, H. 1968. Applications of Menger's graph theorem. *J. Math. Anal. Appl.* **22**, 96–110. (295)

Perfect, H.: See also m.11.

p.3 Petersen, J. 1898. Sur le théorème de Tait. *L'Intermédiaire des mathématiciens* **5**, 225–227. (186)

p.4 Piff, M. J. 1972. *Some Problems in Combinatorial Theory*. Doctoral dissertation. University of Oxford. (160)

p.5 Piff, M. J. and Welsh, D. J. A. 1970. On vector representation of matroids. *J. London Math. Soc.* **(2) 2**, 284–288. (309)

p.6 Pólya, G. 1937. Kombinatorische Anzahlbestimmungen für Gruppen, Graphen, und chemische Verbindungen. *Acta Math.* **68**, 145–254. (325)

p.7 Prüfer, H. 1918. Neuer Beweis eines Satzes über Permutationen. *Arch. Math. Phys.* **(3) 27**, 142–144. (322)

r.1 Rado, R. 1966. Abstract linear dependence. *Colloq. Math.* **14**, 257–264. (295, 297)

r.2 Ramsey, F. P. 1930. On a problem of formal logic. *Proc. London Math. Soc.* **(2) 30**, 264–286. (215)

Rényi, A.: See e.5.

r.3 Ringel, G. 1959. *Färbungsprobleme auf Flächen und Graphen*. Berlin: VEB Deutscher Verlag der Wissenschaften. (204)

r.4 Ringel, G. 1974. *Map Color Theorem*. New York: Springer-Verlag. (204)

r.5 Riordan, J. 1966. *An Introduction to Combinatorial Analysis*. New York: Wiley. (315)

r.6 Riordan, J. 1968. *Combinatorial Identities*. New York: Wiley. (322)

r.7 Rota, G.-C. 1964. On the foundations of combinatorial theory I. Theory of Möbius functions. *Z. Wahrscheinlichkeitstheorie und Verw. Gebiete* **2**, 340–368. (329)

Rota, G.-C.: See also c.5.

r.8. Ryser, H. J. 1950. A note on a combinatorial problem. *Proc. Amer. Math. Soc.* **1**, 422–424. (244, 245)

r.9 Ryser, H. J. 1963. *Combinatorial Mathematics*. Carus Math. Monograph No. 14. New York: Wiley. (144, 147, 233, 234, 242, 252)

r.10 Ryser, H. J. 1968. An extension of a theorem of deBruijn and Erdős on combinatorial designs. *J. Algebra* **10**, 246–261. (234)

Ryser, H. J.: See also c.2 and m.6.

s.1 Saaty, T. L. 1972. Colorful variations on Guthrie's four-color conjecture. *Amer. Math. Monthly* **79**, 2–43. (193)

s.2 Sabidussi, G. 1954. Loewy-groupoids related to linear graphs. *Amer. J. Math.* **76**, 477–487. (71)

s.3 Sabidussi, G. 1958. On a class of fixed-point free graphs. *Proc. Amer. Math. Soc.* **9**, 800–804. (55)

s.4 Schur, I. 1916. Über die Kongruenz $x^m + y^m = z^m$ (mod p). *Jber Deutsch. Math.-Verein* **25**, 114–117. (217)

Shimamoto, T.: See b.16.

s.5 Shrikhande, S. S. 1950. *Construction of Partially Balanced Designs and Related Problems*. Doctoral dissertation. University of North Carolina. (245, 258)

Shrikhande, S. S.: See also b.15 and b.17.

s.6 Sider, N. 1971. *Partial Colorings and Limiting Chromatic Numbers*. Doctoral dissertation. Syracuse University. (207)

Singleton, R. R.: See h.8.

Snell, J. L.: See k.2.

Sós, V.: See e.5.

s.7 Sperner, E. 1928. Ein Satz über Untermengen einer endlichen Menge. *Math. Z.* **27**, 544–548. (37)

Szekeres, G.: See e.6.

t.1 Tait, P. G. 1880. Remarks on the colouring of maps. *Proc. Roy. Soc. Edinburgh* **10**, 501–503. (186, 193)

t.2 Tarry, G. 1900–1901. Le problème des 36 officiers. *C. R. Assoc. franc. Avance. Sci. Nat.* **1**, 122–123; **2**, 170–203. (251, 252)

Thompson, G. L.: See k.2.

t.3 Thompson, W. A., Jr. 1958. A note on PBIB-design matrices. *Ann. Math. Stat.* **29**, 919–922. (257)

t.4 Tutte, W. T. 1947. The factorization of linear graphs. *J. London Math. Soc.* **22**, 107–111. (137)

t.5 Tutte, W. T. 1958. A homotopy theory for matroids, I, II. *Trans. Amer. Math. Soc.* **88**, 144–174. (309)

t.6 Tutte, W. T. 1965. *Lectures on matroids.* J. Res. Nat. Bur. Standards Sect. B 69B, 1–47. (70, 265, 307)

t.7 Tutte, W. T. 1966. *Connectivity in Graphs*. Toronto: University of Toronto Press. (171)

t.8 Tutte, W. T. 1966. Connectivity in matroids. *Canad. J. Math.* **18**, 1301–1324. (295)

t.9 Tutte, W. T. 1971. *Introduction to the Theory of Matroids.* Modern Analytical and Computational Methods in Science and Mathematics 37. New York: Elsevier. (265)

v.1 Vizing, V. G. 1964. On an estimate of the chromatic class of a p-graph. *Diskret. Analiz.* **3**, 25–30. [Russian]. (186)

Vizing, V. G.: See also m.8.

w.1 Watkins, M. E. 1964. *A Characterization of the Planar Geodetic Graph and Some Geodetic Properties of Non-Planar Graphs*. Doctoral dissertation. Yale University. (66)

w.2 Watkins, M. E. 1967. A lower bound for the number of vertices in a graph. *Amer. Math. Monthly* **74**, 297. (167)

w.3 Watkins, M. E. 1968. On the existence of certain disjoint arcs in graphs. *Duke Math. J.* **35**, 231–246. (163)

w.4 Watkins, M. E. 1970. Connectivity in transitive graphs. *J. Combinatorial Theory* **8**, 23–29. (171)

w.5 Watkins, M. E. 1973. Sur les groupes d'automorphismes de systèmes combinatoires. *C. R. Acad. Sci. Paris* **277**, Ser. A, 831–833. (53, 71)

w.6 Watkins, M. E. and Mesner, D. M. 1967. Cycles and connectivity in graphs. *Canad. J. Math.* **19**, 1319–1328. (165)

Watkins, M. E.: See also j.3 and m.10.

Welsh, D. J. A.: See p.5.

w.7 White, A. T. 1973. *Graphs, Groups and Surfaces.* Mathematical Studies 8. Amsterdam: North-Holland. (212)

w.8 Whitney, H. 1932. Congruent graphs and the connectivity of graphs. *Amer. J. Math.* **54**, 150–168. (162)

w.9 Whitney, H. 1932. Non-separable and planar graphs. *Trans. Amer. Math. Soc.* **34**, 339–362. (75, 78, 176)

w.10 Whitney, H. 1933. A set of topological invariants for graphs. *Amer. J. Math.* **55**, 231-235. (71, 74)

w.11 Whitney, H. 1935. On the abstract properties of linear dependence. *Amer. J. Math.* **57**, 509–533. (265)

Wilson, R. J.: See b.8.

w.12 Woodall, D. R. 1970. Square λ-linked designs. *Proc. London Math. Soc.* (3) **20**, 669–687. (234)

w.13 Woodall, D. R. 1977. *Property B and the four-color problem.* (to appear). (187)

Yackel, J.: See g.5.

y.1 Youngs, J. W. T. 1963. Minimal imbeddings and the genus of a graph. *J. Math. Mech.* **12**, 303–316. (202)

Youngs, J. W. T.: See also a.2 and b.1.

Subject Index

Index of Symbols

Special juxtapositions of letters and miscellaneous symbols. (All *letters* in the following are variables.)